AF837765

Handbook of Design and Industry

The twenty-first century has been beset by a global pandemic, war, and increasingly concerning environmental disasters. Designers and industries have been forced to imagine a world in which the only way to move forward is to look back. The design and industry sectors need to understand the role they can play in removing obstacles to social progress and work together to create healthier human societies that can interact with the world in a sustainable way.

This book presents contributions from leading experts that reveal that a better and more prosperous world is achievable through good work and system design. This book consists of chapters that bring together researchers, academics, policy makers, and designers from technology companies and business associations with the objective of developing a focused vision that enhances innovation through design and industry for a better future. Through a transdisciplinary scientific exchange, it lists responses to the challenges of climate change and environmental degradation that will contribute to a more modern, resource-efficient, competitive economy, with smart, sustainable, and inclusive growth, promoting knowledge, inter-sector collaboration, health, education, and a digital society for all. By putting the human at the heart of what can be accomplished, this book investigates better design in the disciplines of work, healthcare, product, system, manufacturing, and industry. The reader will gather an interdisciplinary perspective on what good design can achieve and why it is needed to challenge the climate crisis.

The *Handbook of Design and Industry: Scenarios for Sustainable Futures* is essential reading for researchers and academics in the fields and disciplines of ergonomics/human factors, occupational health and safety, industrial design, product design, industrial engineering, materials engineering, process engineering, computer engineering, communication design, electronics, and telecommunications engineering.

Handbook of Design and Industry

Scenarios for Sustainable Futures

Edited by
Maria João Félix, Rita Assoreira Almendra,
Fernando Moreira da Silva, Fátima Pombo,
and Paulo Jorge de Sousa Cruz

CRC Press
Taylor & Francis Group
Boca Raton London New York

CRC Press is an imprint of the
Taylor & Francis Group, an **informa** business

Designed cover image: Márcio Ribeiro

First edition published 2026
by CRC Press
2385 NW Executive Center Drive, Suite 320, Boca Raton FL 33431

and by CRC Press
4 Park Square, Milton Park, Abingdon, Oxon, OX14 4RN

CRC Press is an imprint of Taylor & Francis Group, LLC

© 2026 selection and editorial matter, Maria João Félix, Rita Assoreira Almendra, Fernando Moreira da Silva, Fátima Pombo, and Paulo Jorge de Sousa Cruz; individual chapters, the contributors

Reasonable efforts have been made to publish reliable data and information, but the author and publisher cannot assume responsibility for the validity of all materials or the consequences of their use. The authors and publishers have attempted to trace the copyright holders of all material reproduced in this publication and apologize to copyright holders if permission to publish in this form has not been obtained. If any copyright material has not been acknowledged please write and let us know so we may rectify in any future reprint.

Except as permitted under U.S. Copyright Law, no part of this book may be reprinted, reproduced, transmitted, or utilized in any form by any electronic, mechanical, or other means, now known or hereafter invented, including photocopying, microfilming, and recording, or in any information storage or retrieval system, without written permission from the publishers.

For permission to photocopy or use material electronically from this work, access www.copyright.com or contact the Copyright Clearance Center, Inc. (CCC), 222 Rosewood Drive, Danvers, MA 01923, 978-750-8400. For works that are not available on CCC please contact mpkbookspermissions@tandf.co.uk

Trademark notice: Product or corporate names may be trademarks or registered trademarks and are used only for identification and explanation without intent to infringe.

Library of Congress Cataloging-in-Publication Data
Names: Félix, Maria João editor | Almendra, Rita Assoreira editor | Silva,
Fernando Moreira da editor | Pombo, Fátima (Professor of design) editor | Cruz, Paulo J. S. editor
Title: Handbook of design and industry : scenarios for sustainable futures /
edited by Maria João Félix, Rita Assoreira Almendra, Fernando Moreira
da Silva, and Fátima Pombo, Paulo Jorge de Sousa Cruz.
Description: First edition. | Boca Raton, FL : CRC Press, 2025. |
Includes bibliographical references and index.
Identifiers: LCCN 2025014624 (print) | LCCN 2025014625 (ebook) |
ISBN 9781032701011 hbk | ISBN 9781032719269 pbk | ISBN 9781032719290 ebk
Subjects: LCSH: Industrial design | Sustainable design | Sustainable engineering
Classification: LCC TS171 .H337 2025 (print) | LCC TS171 (ebook)
LC record available at https://lccn.loc.gov/2025014624
LC ebook record available at https://lccn.loc.gov/2025014625

ISBN: 978-1-032-70101-1 (hbk)
ISBN: 978-1-032-71926-9 (pbk)
ISBN: 978-1-032-71929-0 (ebk)

DOI: 10.1201/9781032719290

Typeset in Times
by codeMantra

Contents

PART 4 Thematic – SPECULATIVE: Subthemes: Futures – Critical Thinking – Design Research

PART 5 Thematic – CONVERT: Subthemes: Production – Industry – Customization

Acknowledgments

The realization of this book, *Handbook of Design and Industry: Scenarios for Sustainable Futures*, was made possible through the generous support and commitment of our sponsors, collaborators, and supporting institutions. In this moment of gratitude, we would like to express our recognition to everyone who contributed to the success of this work.

To our sponsors, whose financial and material support strengthened the structure and quality of this project, we offer our heartfelt thanks. Your vision and understanding of the value of innovation in design and industry are fundamental to our collective progress.

We are immensely grateful to the editorial and scientific team, who shared their knowledge and passion, enriching the content of this book with valuable ideas and perspectives. Your dedication to excellence and professional development has not gone unnoticed.

A special thank you goes to the partner and supporting institutions, including the Lisbon School of Architecture (FA ULisboa), the School of Design of the Polytechnic Institute of Cávado and Ave (IPCA), the Department of Communication and Art of the University of Aveiro, the OIKOS research group of the Research Institute for Design, Media, and Culture (ID+), the School of Applied Arts (ESART) of the Polytechnic Institute of Castelo Branco (IPCB), the Rethink – Research Group on Design for the Territory (CIAUD), the School of Architecture, Art, and Design of the University of Minho (EAAD), and the Design and Technology Group (DeTech) of the Laboratory of Landscape, Heritage, and Territory (Lab2PT). The conference operated under the scope of the Research & Education in Design (REDES) research group from the Centre for Research in Architecture Urbanism and Design (CIAUD). The collaboration of these entities was essential to expand the reach and depth of this publication, allowing the ideas discussed here to resonate beyond the pages of this book.

We also acknowledge the assistance of various entities and volunteers, whose help behind the scenes was vital to the organization and realization of this project. Your hard work and attention to detail ensured a smooth and memorable experience for all involved.

Last but not least, we thank the academic and research institutions that, by supporting their representatives and researchers to participate in Design Commit, underscore the importance of an ongoing commitment to learning and innovation.

Together, we have built a work that reflects the intersection between creativity, technology, and business, paving the way for the future of design and industry. Our sincere thanks to all of you for being part of this journey.

Preface

The Design Commit 2024 conference brought together professionals, researchers, academics, and innovators with a common goal: to explore the role of design and industry in building sustainable futures. This book, titled *Handbook of Design and Industry: Scenarios for Sustainable Futures*, is the result of the dedication and collective effort of the participants and authors, who sought to expand and deepen their ideas presented at the conference.

In this collection, each chapter is an expanded version of the articles presented at Design Commit 2024, developed with the aim of providing new perspectives, additional case studies, deeper discussions, and updated research findings. The result is a work that reflects not only the diversity of approaches but also the commitment to developing sustainable scenarios in the field of design and industry.

Today, the theme of sustainability is more relevant than ever. The practices and technologies discussed throughout this book represent the latest advancements in terms of innovation and adaptation in the industry, offering solutions that have the potential to transform our relationship with resources and the environment.

We would like to express our gratitude to the authors who contributed to this work, to the scientific editorial team that guided the process with dedication, and to all the participants of the Design Commit 2024 conference for making the creation of this book possible. We hope this publication inspires new reflections and concrete actions for a more sustainable future and becomes a valuable source of knowledge and reference for professionals, students, researchers and all those interested in design and sustainability.

Contributors

Annamaria Alabiso
Department of Biology
University of Rome "Tor Vergata"
Rome, Italy

Isabel Duarte de Almeida
CIAUD – Research Centre for Architecture, Urbanism, and Design
Lisbon School of Architecture
University of Lisbon
Lisbon, Portugal

Rita Assoreira Almendra
CIAUD – Research Centre for Architecture, Urbanism, and Design
Lisbon School of Architecture
University of Lisbon
Lisbon, Portugal

Francesca Antinarelli Freitas
Design Department
Politecnico di Milano
Milan, Italy

Jeremy Aston
IDEA – Research in Design and Art
ESAD College of Art and Design
Porto, Portugal

Luciana Barbosa
IDEA – Research in Design and Art
ESAD College of Art and Design,
Porto, Portugal

Susana Barreto
ID+, Research Institute for Design, Media, and Culture
Faculty of Fine Arts
University of Porto
Porto, Portugal

Arianna Bionda
Design Department
Politecnico di Milano
Milan, Italy

Juliana Cardoso Braga
Faculty of Architecture, Urbanism, and Design, Federal University of Uberlândia (FAUeD-UFU)
Uberlândia, Brazil

Ana Leocadia de Souza Brum
Brazil Design Center
Curitiba, Brazil

Giorgio Buratti
Department of Design
Politecnico di Milano
Milan, Italy

Rosanna Caldarella
Design Department
Politecnico di Milano
Milan, Italy

Tatiana Campos
Lab2PT – Landscapes, Heritage, and Territory Laboratory
School of Architecture
University of Minho
Portugal, Portugal

Alba Cappellieri
Design Department
Politecnico di Milano
Milan, Italy

Vítor Carvalho
School of Technology
Polytechnic Institute of Cávado and Ave (IPCA)
Barcelos, Portugal

André Castro
Department of Project Design
CIAUD – Research Centre for Architecture, Urbanism, and Design
Lisbon School of Architecture
University of Lisbon
Lisbon, Portugal

Mariana Ciancia
Design Department, Politecnico di Milano
Milan, Italy

Giovanni Maria Conti
Design Department, Politecnico di Milano
Milan, Italy

Llara Fuente Corripio
Department of Art History and Musicology
University of Oviedo
Oviedo, Spain

Emília Costa
Faculty of Fine Arts
University of Porto
Porto, Portugal

Fiammetta Costa
Department of Design
Politecnico di Milano
Milan, Italy

Paulo Cruz
Lab2PT – Landscapes, Heritage, and Territory Laboratory
School of Architecture
University of Minho
Portugal, Portugal

J. Paulo Davim
TEMA – Centre for Mechanical Technology and Automation
Department of Mechanical Engineering
University of Aveiro
Aveiro, Portugal

Maria João Delgado
CIAUD – Research Centre for Architecture, Urbanism, and Design
Lisbon School of Architecture
University of Lisbon
Lisbon, Portugal

Ana Duque
IDEA – Research in Design and Art
ESAD College of Art and Design
Porto, Portugal

Aguinaldo dos Santos
Department of Design; Design & Sustainability Research Group
Federal University of Paraná
Curitiba, Brazil

Elena Elgani
Department of Design
Politecnico di Milano
Milan, Italy

Giulia Ettori
Department of Design
Politecnico di Milano
Milan, Italy

Maria João Félix
CIAUD – Research Centre for Architecture, Urbanism, and Design
Lisbon School of Architecture
University of Lisbon
Lisbon, Portugal

Catarina Ferreira
CIAUD – Research Centre for Architecture, Urbanism, and Design
Lisbon School of Architecture
University of Lisbon
Lisbon, Portugal

Gláucia de Salles Ferro
Brazil Design Center
Curitiba, Brazil

Bruno Figueiredo
Lab2PT – Landscapes, Heritage, and Territory Laboratory
School of Architecture
University of Minho
Portugal, Portugal

Zhiyong Fu
Academy of Arts & Design
Tsinghua University
Beijing, China

Domenico Di Fuccia
Department of Design
Università degli Studi della Campania "Luigi Vanvitelli"
Caserta, Italy

Francesco Galli
IULM University
Milan, Italy

Katia Goldoni
Department of Design
Politecnico di Milano
Milan, Italy

Tiantian Li
Academy of Arts & Design
Tsinghua University
Beijing, China

José Machado
Department of Mechanical Engineering
University of Minho
Guimarães, Portugal

José Rui Marcelino
Department of Project Design
CIAUD – Research Centre for Architecture, Urbanism, and Design
Lisbon School of Architecture
University of Lisbon
Lisbon, Portugal

Nikole Melo Mendonça
CIAUD – Research Centre for Architecture, Urbanism, and Design
Lisbon School of Architecture
University of Lisbon
Lisbon, Portugal

Matteo Meraviglia
Department of Design
Politecnico di Milano
Milan, Italy

Luciana Migliore
University of Rome Tor Vergata
Rome, Italy

Umberto Monchiero
Design Department
Politecnico di Milano
Milan, Italy

Gianni Montagna
CIAUD – Research Centre for Architecture, Urbanism, and Design
Lisbon School of Architecture
University of Lisbon
Lisbon, Portugal

Fernando Moreira da Silva
CIAUD – Research Centre for Architecture, Urbanism, and Design
Lisbon School of Architecture
University of Lisbon
Lisbon, Portugal

Martina Motta
Design Department
Politecnico di Milano
Milan, Italy

Attilio Nebuloni
Politecnico di Milano
Milan, Italy

Yannara Negre
European University, IADE, UNIDCOM/IADE, Research Unit in Design and Communication, ICNOVA
Lisbon, Portugal

Marco Neves
CIAUD – Research Centre for Architecture, Urbanism, and Design
Lisbon School of Architecture
University of Lisbon
Lisbon, Portugal

David Palma
CIAUD – Research Centre for Architecture, Urbanism, and Design
Lisbon School of Architecture
University of Lisbon
Lisbon, Portugal

Hugo Palmares
ID+, Research Institute for Design, Media, and Culture
School of Design, Polytechnic Institute of Cávado and Ave (IPCA)
Barcelos, Portugal

Luis Carlos Paschoarelli
São Paulo State University
São Paulo, Brazil

Rui Patrício
European University, IADE, UNIDCOM/IADE, Research Unit in Design and Communication, ICNOVA
Lisbon, Portugal

Filipe Pereira
Department of Textile Engineering
University of Minho
Guimarães, Portugal

Juliano Aparecido Pereira
Federal University of Uberlândia
Uberlândia, Brazil

Francesca Piredda
Design Department
Politecnico di Milano
Milan, Italy

Juliana Pires
Faculty of Fine Arts
University of Porto
Porto, Portugal

Laura Pirrone
Design Department
Politecnico di Milano
Milan, Italy

Carolina Polónio
ID+, Research Institute for Design, Media, and Culture
Department of Communication and Art
University of Aveiro
Aveiro, Portugal

Fátima Pombo
ID+, Research Institute for Design, Media, and Culture
Department of Communication and Art
University of Aveiro
Aveiro, Portugal

Andrea Ratti
Design Department
Politecnico di Milano
Milan, Italy

Élmano Ricarte
European University, IADE, UNIDCOM/IADE, Research Unit in Design and Communication, ICNOVA
Lisbon, Portugal

Beatrice Rossato
Design Department
Politecnico di Milano
Milan, Italy

Elisa Schembri
Design Department
Politecnico di Milano
Milan, Italy

Sónia Seixas
CIAUD – Research Centre for Architecture, Urbanism, and Design
Lisbon School of Architecture
University of Lisbon
Lisbon, Portugal

Valentina Sorvillo
University of Campania Luigi Vanvitelli
Naples, Italy

Filomena Soares
Department of Industrial Electronics
University of Minho
Guimarães, Portugal

Irina Suteu
NABA – Nuova Accademia di Belle Arti
Milan, Italy

Livia Tenuta
Design Department
Politecnico di Milano
Milan, Italy

Miguel Terroso
ID+, Research Institute for Design, Media, and Culture
School of Design, Polytechnic Institute of Cávado and Ave (IPCA)
Barcelos, Portugal

Susanna Testa
Design Department
Politecnico di Milano
Milan, Italy

Rosa Vasconcelos
Department of Textile Engineering
University of Minho
Guimarães, Portugal

Beatrice Zagatto
Design Department
Politecnico di Milano
Milan, Italy

About the Editors

Maria João Félix is an academic and designer whose career is marked by a deep commitment to the fields of industrial design, education, industry, circular economy, and sustainability. She has completed a postdoctoral research in Design, titled "Bioneurodesign: from noetics to supermodernity," at the Lisbon School of Architecture, reflecting her dedication to pushing the boundaries of design and practice. In a pivotal role, Maria João Félix serves as the founder and General Chair of the international conference "Design Commit 2024," a platform that bridges the worlds of design and industry, fostering collaboration and innovation. Her leadership in this capacity underscores her passion for advancing design and industrial synergy. Beyond her conference leadership, Maria João Félix is an author, with her book, *Strategic Reflection for Design in Portugal: Systematization of internationally produced thinking on the contours of research in the field of design*, emphasizing her commitment to elevating Portuguese design both nationally and internationally. Maria João Félix's academic journey includes earning her Ph.D. in Design from the Lisbon School of Architecture in 2013. In 2006, she completed her Master's in Industrial Design through a collaboration between the School of Arts and Design of Matosinhos | ESAD and the Faculty of Engineering of the University of Porto | FEUP. Her academic foundation was laid in 1998 when she obtained her degree in Design, specializing in Equipment, from the School of Arts and Design of Matosinhos | ESAD. Her professional journey has been characterized by her dedication to teaching in higher education and her work as an Industrial Designer. Maria João Félix has continuously strived to bridge the gap between academia and industry, emphasizing the importance of adapting educational programs to meet the evolving challenges and market demands within the design industry. From 2012 to 2015, she held the position of a Member of the Council for Evaluation and Quality at the Polytechnic Institute of Cávado and Ave. During this time, she actively coordinated evaluation and quality efforts, ensuring the highest standards at the School of Technology | EST and later taking on the role of Coordinator of the School of Design | ESD. Her responsibilities included overseeing self-assessment processes for undergraduate and master's degree programs, actively contributing to quality committees, and preparing comprehensive reports. Maria João Félix has held significant roles in academia, including Director of the Department of Industrial and Product Design, Director of the Undergraduate Program in Industrial Design, Director of the Master's Program in Design and Product Development, and head of the Scientific Area of Industrial Design. Her influence is evident in the restructuring of curricular programs, aligning them with the practical needs of the design industry. Her commitment to pedagogical excellence led to a public commendation (no. 1446/2011) for achieving the highest overall average among lecturers at the School of Technology of the Polytechnic Institute of Cávado and Ave, where she taught seven courses over 3 years. Currently, Maria João Félix holds the position of Director of the Master's Program in Design and Product Development and is actively involved in the Technical-Scientific and Pedagogical Councils of ESD_IPCA. She is also an author and reviewer for scientific publications in accredited international journals and has played a pivotal role in organizing scientific events in the field of Industrial Design. Maria João Félix embarked on her career as a designer in 2006, creating a diverse portfolio of works that have been showcased at various fairs and events. She was honored with the Mobis 2001 Prize in recognition of her significant contributions to the design industry.

Rita Assoreira Almendra is currently a Full Professor at Lisbon University, Lisbon School of Architecture (FA ULisbon). She has a PhD in Design concluded in 2010 in FAUL; her master's in Design Management was concluded in 2004 in Universidade Católica Portuguesa, Faculdade de Ciências Económicas e Empresariais, and she also has an MBA with marketing specialization, finished in 2002 from the same school. Her graduation in Design from Technical University of Lisbon dates from 1998. She started as a Lecturer at FA ULisbon in 2000 and became an Assistant Professor in 2010 and an Associated Professor in 2016. She is the Head of the Project Design Department; member of the Scientific Council (she was the Vice-president of the Scientific Council until 2021); Coordinator of the 3rd Cycle; Coordinator of the Design Doctoral program; and Coordinator of the Nucleus of Design and a Researcher at CIAUD – Research center of FA ULisbon and Professor at the Doctoral Program of Sciences of Sustainability from University of Lisbon. She has published 20 articles in journals and 30 book chapters, she has also two co-authored books. She is the editor of four book and has made more than 75 papers in conference's proceedings. She supervised 20 PhD theses and has 20 more ongoing (with other supervisors) and supervised more than 70 MSc dissertations. She has been a member of more than 100 Master jury and of 45 PhD defense jury. She organized several conferences. She was a member of various scientific committees of journals and conferences. She is a reviewer for multiple conferences and a few publications. She is a member of FCT's (Fundação para a Ciência e a Tecnologia's) doctoral studentship in the Design area. She was also a member of the jury of Lisbon University's doctoral studentship in design. She was a member of 25 jury from eight different schools to select candidates to the position of assistant professors. She was granted with an FCT doctoral scholarship and has received four awards and/or honors. Her scientific and research interest areas are as follows: design education, design research, social design, design processes, sustainable design, design management, competitiveness and innovation, collaboration, participatory design, and service design.

Fernando Moreira da Silva is a Full Professor Emeritus with Habilitation in the Scientific Area of Design. He holds a doctorate in Built Environment (University of Salford), a PhD in Architecture – Visual Communication (Lisbon Technical University), and an MPhil in Color in Interiors (University of Salford). He also has a post-doctorate in Inclusive Design and Visual Communication Design (University of Salford). Fernando is a coordinator from Lisbon School of Architecture, University of Lisbon (FA/ULisboa) of the International Doctoral Course in Design, together with the Universities of La Campania – "Luigi Vanvitelli" in Naples, Polimi, Florence, and Malaga. He is involved as an integrated researcher at CIAUD – Research Center for Architecture, Urbanism, and Design, being its President between 2009 and 2020. Now, he is the coordinator of the Design Main Group at CIAUD and an Honorary Researcher at SURFACE – Inclusive Design Research Centre, University of Salford (UK). He has coordinated or participated in several research projects in the general field of Design. He has been a supervisor of several master dissertations (more than 150), PhD theses (more than 80) and post-docs (25). He is a member of numerous Master's and Doctorate Jury Panels mainly in Portugal, Brazil, Spain, Italy, and UK and involved in evaluation panels to the ALBAN Program, PROTEC Program, CNPq from Brazil, and the Italian Agency ANVUR, being the Panel Coordinator for the PhD and post-doc scholarships at FCT, between 2007 and 2020. He is a Lecturer in Design (Product, Communication, and Fashion), Human Factors, Ergonomics, Inclusive Design, Fashion and Color, and coordinator of several congresses, seminars, international conferences, and exhibitions. Fernando is an author of three books, many book chapters, articles, papers, and communications in Design, Color, Inclusivity, Fashion, Human Factors, Ergonomics, Research Methodologies, and Accessibility. He is a member of several journal and books scientific committees and an editor or coeditor of five books.

Fátima Pombo is Full Professor of Design at the University of Aveiro (UA), Portugal, and Director of the Research Institute for Design, Media, and Culture [ID+]. Pombo was director of MEDP-Master in Product Design and Engineering (2022–2024) and vice-director of MEDP (2020–2022), founder, coordinator (2018–2023), and senior researcher of the research group Oikos-Design for Ecosystemic Spaces at UA. She was a Visiting Professor at the Department of Architecture at the University of Leuven (KUL), Belgium, for almost a decade until 2016. She also lectured in teaching missions from Erasmus Programme, among others, at Politecnico di Milano, University of Arts, Crafts, and Design in Stockholm, University of Southern Denmark. She developed research in Design at University of Barcelona, in cultural studies at the University of Munich as a postdoctoral student with a fellowship from Alexander von Humboldt Foundation and in Phenomenology and Aesthetics as a PhD student at the University of Heidelberg.

Her current research interests focus on phenomenology of design, interior design, ecosystemic spaces, aesthetics of sustainability, and cultural heritage. Pombo is a sought-after expert, participating in academic activities and serving, among others, as a design jury member for the iF Design Award, Germany, the Henry Van de Velde Label/Design Vlanderen; scholarships award from EU/Brussels; and other juries and examination committees as the Evaluation Panel of FCT-Foundation for Science and Technology, Evaluation Commission of A3ES – Accreditation of Study Programmes. Engaging in international research projects as a principal investigator or senior researcher and in international conferences, Pombo has been invited as a member of scientific committees, organizing committees, co-convenor, reviewer, and chair, namely, among others, from European Academy of Design; International Committee for Design and History Studies; Centre for Research in Architecture, Urbanism, and Design from FA ULisboa; Interiors Forum World; Nordic Design Research Conferences; and Design Commit. She publishes regularly academic works as a single author or in co-authorship with colleagues from Europe, USA, and Latin America. She published novels, children, and youth stories as well as chapters in art books, art catalogs, and numerous dissemination texts about design, music, and visual arts. From 2007 to 2010, she collaborated in the artistic project of commenting classical music concerts with Casa da Música/Orquestra Nacional do Porto; from 2005 to 2006, she collaborated with RDP-Antenna 2 in the artistic project 22 Chronicles-Birds Fantasy; in 2018 and in 2001, she collaborated in the content of two programmes broadcasted in the Portuguese TV about the Portuguese cellist Guilhermina Suggia. Beyond academia, she plays cello, perceiving music as the heartbeat of the world.

Paulo Jorge de Sousa Cruz has been a Professor at the University of Minho since 1989, initially in the Department of Civil Engineering, where he was Head of Department in 2003 and 2004, and, later, at the School of Architecture, Art, and Design (EAAD). He has also been a Full Professor at the School of Architecture, Art, and Design, since 2008; Dean of EAAD (2021–2024 and 2004–2011); Pro-Rector of the University of Minho (2017–2021); and Director of Lab2PT – Laboratory of Landscapes, Heritage, and Territory (2015–2017). The R&D and teaching activities aim to drive the convergence of Architecture, Construction, and Technology through pioneering research and education related to recent breakthroughs in arts, practice, and theory within these multidisciplinary domains. Since the early beginning of his involvement with the School of Architecture, Art, and Design, he devoted efforts to create and promote a set of laboratorial facilities with a vast potential for research, development, and innovation. The Architecture, Construction, and Technology Hub aggregates those facilities and envisages the interdisciplinary collaboration to address emerging challenges (https://actech.uminho.pt/). Integrated Researcher at IN2PAST – Associate Laboratory for Research and Innovation in Heritage, Arts, Sustainability, and Territory (since 2021). He is an integrated researcher at Lab2PT

(Since 2015). He is an author of approximately four hundred scientific publications and supervisor or co-supervisor of 21 PhD students (16 concluded), where he has an active role in the leadership and management of projects.

He has been an Associate Editor of the journal *Structure and Infrastructure Engineering*, Taylor & Francis, since 2005. He is the Founder and President of the International Association of Structures and Architecture (2016–2022) and chaired five international congresses on this topic (ICSA2010, ICSA2013, ICSA2016, ICSA2019, and ICSA2022). He is a member of the Organizing Committee of DESIGN COMMIT 2024 – 1st International Conference on Design & Industry, Braga, Portugal. He was a member of the Organizing Committee of ISIC 2022 – Trends on Construction in the Post-Digital Era, the annual conference of the International Society for Intelligent Construction, held in Guimarães, Portugal. He was a member of the High-Level Independent Panel for selection and monitoring of the Financing Program of the "National Program for Accommodation in Higher Education." Order No. 886/2022 from the Minister of Science, Technology, and Higher Education, published in the Diário da República, 2nd series, Part C, of January 21, 2022. He is also a member of the Advisory Board of the Center for Technological Innovation in Building and Civil Engineering of the University of Coruña, Spain, since 2022.

He was also a part of the Research Assessment Committee 2010–2015 of the Faculty of Architecture and Built Environment of TU Delft, 2016, and member of the Scientific Council of Social Sciences and Humanities of FCT (2013–2016).

He coordinated the "Design, Architecture, and Urbanism" evaluation panel of the FCT Competition for Studentships for Doctoral Research – 2022 and 2021 – and was a co-coordinator in 2020. Several patents illustrate the ability to explore innovative technological, constructional, and structural solutions in a wide range of materials (ceramics, steel-glass, and timber-glass) and technologies and to establish effective collaborations with the construction industry. In 2023, he received in Prague the Best Paper Award in the 12th International Conference on Smart Cities and Green ICT Systems. In 2008, he received the "Special Service Award." In 2005, he received the "Thomas Fitch Rowland" award, given by the ASCE – American Society of Civil Engineers.

Part 1

Thematic – SMART

Subthemes: Transportation – Digital Innovation – Industry Collaboration

1 FLY-PT Project
Collaborative Approaches in Passenger Cabin Design for Air Mobility

José Rui Marcelino, André Castro, and Catarina Ferreira

1.1 INTRODUCTION

In response to increasing urbanization and growing road congestion, UAM is emerging as a promising near-future solution, garnering support from industries worldwide. This momentum is fueled by advancements in sustainable propulsion and energy storage systems, connectivity, automation, and lightweight materials, laying the groundwork for transformative projects like FLY-PT. The FLY.PT project was developed with 17 Portuguese companies during a period of three years. The project pioneers an E-VTOL intermodal transport system, seamlessly combining an autonomous aerial vehicle (drone) for air mobility, an autonomous automotive vehicle (skate) for road mobility, and a modular cabin attachable to either vehicle. This chapter aims to describe the strategy to collaboratively develop the "Passenger Cabin Design", using different design methodologies with multidisciplinary teams from the various partners. One of the main challenges in this project, from the designer's perspective, was how to incorporate inputs from various partners to develop the "Passenger Cabin Design".

The main objective during the initial concept design phase was to develop several passenger cabin concepts, in accordance with the technical specification as well as functional and constructive aspects, based on the knowledge acquired in the previous phases. In this phase, the design team's objective was to propose the general geometry of the cabin, both interior and exterior, as well as the main functional features, such as entrance/doors, windows, seating for passengers, luggage space, lighting, and infotainment. The goal was to develop a unique cabin design to address the complexity of a system that consists of the combination of three "products": drone, cabin, and skateboard.

Initially, a set of differentiated alternatives for the basic geometry and features were generated using a specific matrix method, which allowed for free subsequent development in sketching sessions. From the visuals generated, the most promising concepts were chosen in consortium meetings with all partners, according to well-defined criteria and with a voting methodology. Emphasis was placed on

DOI: 10.1201/9781032719290-2

collaborating with other partners in defining the cabin concepts and also in integrating the Passenger Cabin Design within the drone + skate design, both aesthetically and functionally. This approach was fundamental to bringing consistency to the overall system/product developed within the scope of the project. The different steps of the process will be described in this chapter.

The objective of the project was to develop a "Passenger Cabin Design". The main challenge was to create a methodology that was inclusive of all the partners – different companies with specific technical capabilities – which would allow for a collaborative process. To achieve this goal, first, a morphological table was generated, which would identify the most important cabin elements for the design – i.e., structure, exterior shape, battery storage, lighting and heating, ventilation and air conditioning systems, landing gear, interior seat layout, number of passengers, and luggage space. The morphological table allowed the partners to map all the elements and combine possible solutions at the level of each component. By mapping the different solutions in the table, it was possible to create different "paths" for each set of solutions, structural, aesthetic, seat layout, etc. The most promising paths were chosen during project meetings and served as the basis for the preliminary sketching sessions.

Following the morphological table methodology, designers developed sketching sessions to visualize several concepts. These designs underwent several iterations with input from partners for both the exterior and interior design. The inputs were diverse, from functional issues, such as the location of the cabin's attachment points with the drone and the skateboard, to aerodynamic issues, related to the exterior shape of the cabin and its efficiency in flight and on the road.

After the sketching phase, a Concept Selection Matrix was developed to enable the analysis and voting of the most promising concepts. These concepts were eventually refined and improved in subsequent sketching iterations until they were optimized for the development phase.

There are many challenges in mobility/vertical take-off systems, of which we highlight the following regarding cabin design: psychological barriers, safety, legislation, technology, and operation/infrastructure, among others. In the case of FLY-PT, the interior design of a multimodal vehicle, capable of traveling on the road and in the air, is a multidisciplinary challenge as the two areas have different functional requirements, different safety regulations, and psychological factors associated with the behavior of different passengers. "Design thinkers observe how people behave, how the context of their experience affects their reaction to products and services. They take into account the emotional meaning of things as well as their functional performance" (Brown, 2011:10). Presently, there is no vehicle on the market with these characteristics, multimodal and with vertical take-off capabilities, the most similar vehicles being helicopters for urban operation. There are several psychological factors related to flying, and the biggest challenge in a UAM cabin design is to integrate safety systems that can meet the demanding criteria of both road and aeronautical safety. The interior design must also functionally communicate the safety of the system to the passenger, through the design of the seats, covering materials, lighting, glass area, rigor of construction, and finishes to increase confidence in use. On the other hand, the design must provide functionality for different types of passengers, with notable anthropometric differences, and allow all these populations to use the interior in a safe and comfortable way to provide an excellent travel experience. The shared use by different users also brings new challenges, where easy maintenance, the need for permanent cleaning, and the robustness of the interior are also fundamental. "Inclusion Is Designing the Future- Can we understand the exclusion created by our solutions before we release them into the world, and design something better?" (Holmes & Maeda, 2018). With the advent of automation, interiors have the potential to become a space for rest, work, and leisure, where all travel time is spent on tasks other than driving. This fact is directly related not only to the interior layout (i.e., seating position, luggage space, etc.), but also to the integration of technology in the interior parts, with the possibility of interacting with interior panels, windows, tactile surfaces, etc., promising as a customization interface for each passenger. "Several studies indicate that increasing leg room, knee space, and personal space have a positive effect on the comfort experience. So, leg room and personal space have a priority in the design and also expectations and preflight experiences" (Vink & Brauer, 2011). There are also discomfort factors that must be addressed in this type of project, such as noise, vibrations, flight direction, external visibility, etc., and which are fundamental in defining the initial specification. If in the case of a road vehicle, only accelerations and changes of direction in the horizontal plane are a problem that increases discomfort, in a flight module, we have all dimensions interacting, which makes it essential to study ways to reduce passenger discomfort. The bridge between the aerodynamic factors of the cabin and flight mode control is also fundamental, as well as the rotational speed of the engines and vibration control. Helicopters currently on the market have high levels of noise and vibration that cause high levels of discomfort. This issue is addressed through the use of noise-canceling headphones; however, this solution is not satisfactory for a passenger who wants a quick and comfortable urban journey, just as using a road transport service. The project had to address and integrate these concerns by reducing the propulsion noise level, but also using materials with better acoustic properties, which allow for more effective cabin insulation. The following methods were developed to address the design problems.

1.2 DESIGN METHODS

Theory and practice are continuously looking for tools to increase the effectiveness of projects, the right channels for communication between stakeholders, and the knowledge application and dissemination process. Design balances the perspectives of users, technology, and business; it is, by its nature, integrative.

> Design methods are an opportunity to structure conversations that can help us better understand and empathize with people, and as a result build more meaningful products.
>
> *(Martin & Hanington, 2012)*

The selection of design methods might also condition the way people interact inside the design team. Some processes are more rational, framed, and organized, and others are more emotional. Concept studies focus on aspects such as form, function, aesthetics, poetics, color, materials, and processes. It is based on needs and existing technical and production resources, and in the cultural and socio-economic context in which they operate. This activity generates the visualization of scenarios for future products and services through drawings, illustrations, and physical and virtual models.

1.2.1 Morphological Table

In order to develop the design concepts, the teams used a morphological table approach. Morphological tables are a systematic and structured methodology used in design processes to explore and generate possible solutions to a given problem. The method was developed by Fritz Zwicky (1966) and is widely used in diverse fields such as engineering, architecture, industrial design, and general problem solving, as it involves a process of breaking down a complex design problem. "Decomposition is a natural strategy for attacking large problems. Cognitive limitations force human decision-makers to decompose problems into subproblems" (Herrmann, 2004). The method breaks down a complex design problem into subcomponents, components, or parameters by creating a matrix or table to organize the information, which then can be systematically explored for different combinations or configurations of these components to generate design solutions. These solutions can then be "visualized in a way that allows your thinking to be seen and understood" (Dziersk, 2007). This method also allows different stakeholders to participate and collaborate by using the table as a collector of inputs from every partner. This allows non-designers to participate using a "visual" tool, which helps communicate and generate discussion. The consortium decided to use the methodology with the following steps.

1.2.1.1 Problem Definition

The main problem of this project, from the perspective of the designers, was how to incorporate inputs from various partners to develop the Passenger Cabin Design. The problem was addressed through the development of a specific collaborative methodology which enabled the co-creation of the solutions through a three-phase process in which all partners could suggested ideas, proposed improvements, and validated concepts: "Morphological table" to support the generative phase, "Sketching" to visualize concepts, and "Concept Selection Matrix" to choose and refine concepts collaboratively. Due to the nature of the project – a cabin design for a transport system – different parameters were identified in the first stages of the project, such as functional, physical, and aesthetic. These preliminary parameters, such as number of rotors, drone powertrain, skateboard wheelbase and powertrain, cabin structure, energy storage in the cabin, lighting, seating arrangement, etc., were used to develop the first part of the methodology, the morphological table. Next, the most promising ideas were visualized and combined in sketches. Finally, using a Concept Selection Matrix, the concepts were selected for further development. The process is described as follows.

1.2.1.2 Creation of a Morphological Table

A table/matrix was developed with all the relevant parameters identified as a column (see Figure 1.1). A standard Spreadsheet software was used as a way of organizing the information. Each parameter represents an option or design attribute, which was illustrated with small "icons" that every partner could add, allowing the teams to visually communicate and understand the proposed solutions. Each one collaborated by "filling in" several possibilities for the components, according to their own specialized knowledge (i.e., structural engineers defined types of structures, fixing points; designers defined the number of seats, seat layout, etc.).

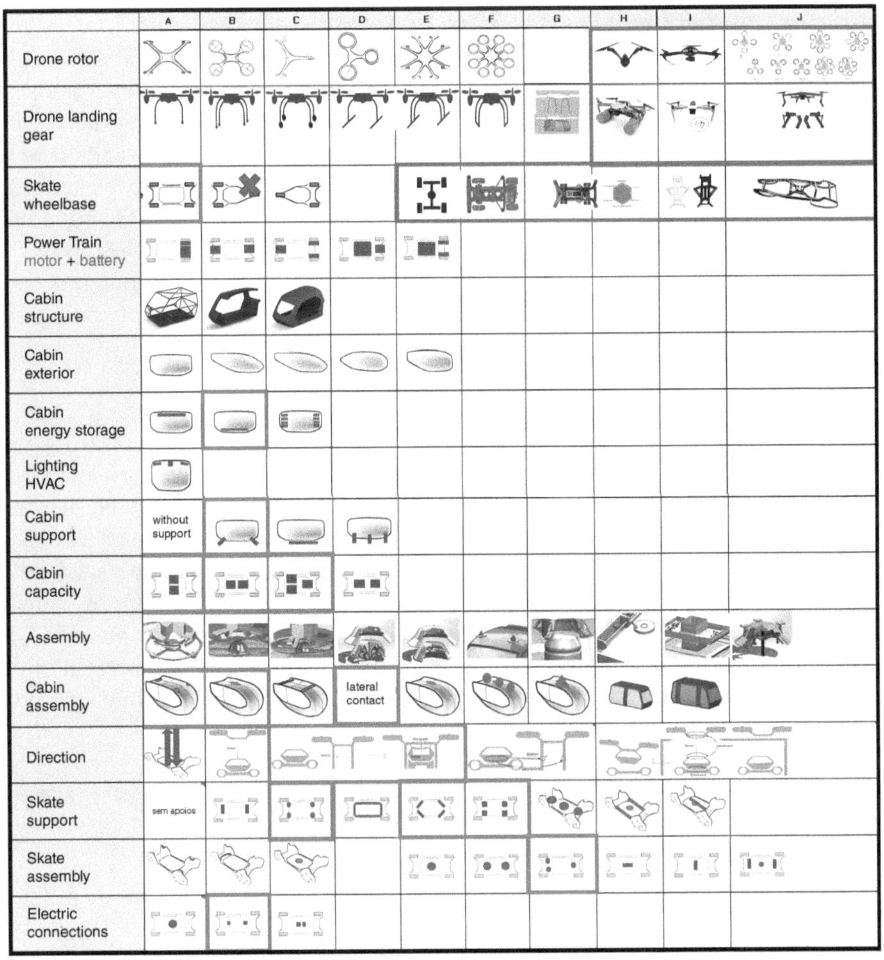

FIGURE 1.1 Morphological table with integration of module configuration options for different PPS. (FLY-PT Project.)

1.2.1.3 Generation of Design Options

For each parameter, different alternatives were created (e.g., number of passengers, number of engines, etc.) and placed in each line as option A, B, C, D, etc. These options were mutually exclusive and covered a wide range of possibilities (see Figure 1.1).

1.2.1.4 Completing the Table

With the help of all partners, the morphological table was filled in to obtain different options for each component. During the consortium meetings with all partners, the different solutions were systematically combined and discussed for each parameter.

1.2.1.5 Exploring Combinations

Based on the different parameters, different combinations were explored considering the viability for the "drone" + "cabin" + "skateboard" design and integration. In Figure 1.1, we see an example of the work developed based on the morphological table methodology, with different components of the product (columns) and different design proposals in each line. Combinations between the different lines were tested to find the most promising concepts. This process is found to be useful to create not only several "solution paths" but also to generate stimuli in others: "Most individuals can think of a new idea when presented with a new stimulus (…) Upon reflection on someone else's ideas, most people are able to generate new ideas" (Eppinger & Ulrich, 2015, p. 109).

1.2.2 Sketching

The "Sketching" methodology in design is a fundamental process for visualizing, exploring, and communicating design ideas and concepts through quick sketches or drawings. "Sketching" is a form of visual representation that allows designers to express their ideas quickly and effectively and to communicate new ideas and/or combinations of ideas to project teams. It is therefore an essential tool for creative development and problem solving in design, playing a crucial role in all phases of the design process, transforming abstract concepts into tangible visual representations, thereby facilitating understanding, communication, and decision-making. Decisions throughout the design process:

> Sketching designs has been shown to be a useful way of planning and considering alternative solutions. The use of lo-fidelity prototyping, especially paper-based sketching, can save time, money and converge to better solutions more quickly.
>
> *(Roberts et al., 2015)*

Based on the morphological table, the designers used a sketching methodology to "visualize" the different alternatives developed. The methodology followed a process that can be described as follows.

1.2.2.1 Generation of Ideas/Alternatives

Based on the information generated in the morphological table, designers freely explored different concepts, solutions, and variations, without focusing on details. To achieve these, it was important to create an "open" environment receptive to experimentation and creative thinking without going into technical/constructive details. "The unmet need acts as the starting point for innovation. The next step is to design a solution around needs and desires of the user" (Best 2006, p. 100).

1.2.2.2 Sketching Support

The designers used various means to make the sketches, such as traditional (pencils, pens, markers, paper) and digital tools (tablets, drawing tables). The choice of support depended on the designer's personal preference and the project context; in the case of FLY-PT, both processes were used.

1.2.2.3 Quick Sketches

The main characteristic of "sketching" is speed. Sketches are usually done quickly and loosely, capturing only the essential elements of the idea. This allows for agile exploration of various options without getting into details. This is what was developed for the project, with dozens of sketches being developed for the cabin and for the integration between cabin, drone, and skateboard.

1.2.2.4 Composition and Layout

Designers organized sketch elements to clearly convey the desired information by creating visual hierarchies, grouping related elements, and defining focal points to direct attention. The sketches were refined and presented in internal meetings with the design team. The most promising solutions underwent several iterations, resulting in ever more refined solutions. These were frequently presented in project meetings to the project partners for feedback.

1.2.2.5 Iteration

The "sketching" process is iterative. As such, the designers created multiple sketches, refining and evolving their ideas over time, receiving feedback from other partners and iterating. Each iteration allowed us to explore different approaches, experiment with alternative solutions, and improve the overall concept. Figure 1.2 presents examples of the sketches developed during the process for the FLY-PT cabin and skateboard.

This process resulted in several sketches/drawings as ideas were explored and refined through "sketching". The designers moved to more advanced stages of development, creating ever more detailed drawings that were presented and selected by partners, and, in later stages, developed into digital models. The process of integrating the cabin with the skateboard and the drone involved defining configurations for two, three, four, or multiple rotors, as shown in Figure 1.3.

Sketches were also specifically developed for the interior of the cabin, considering the three already defined concept directions (based on the number of rotors). The definition of the number of passengers and the location of luggage was also studied

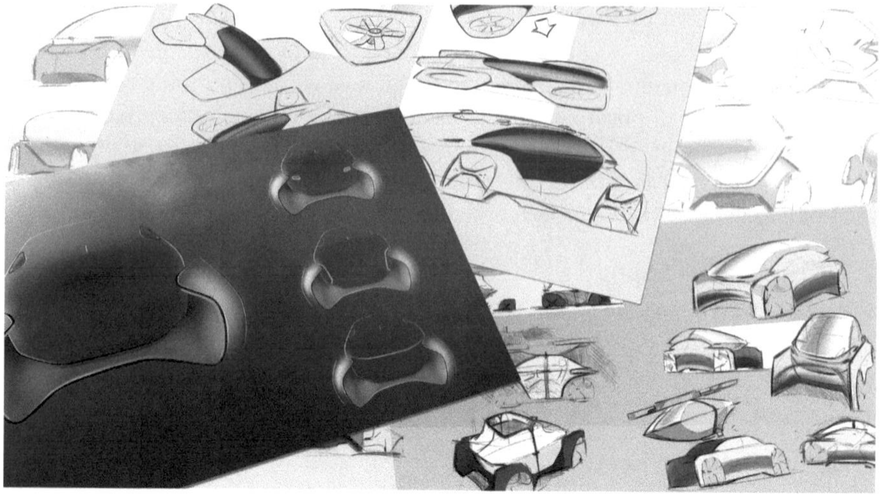

FIGURE 1.2 Cabin concept sketches and skateboard integration. (FLY-PT Project.)

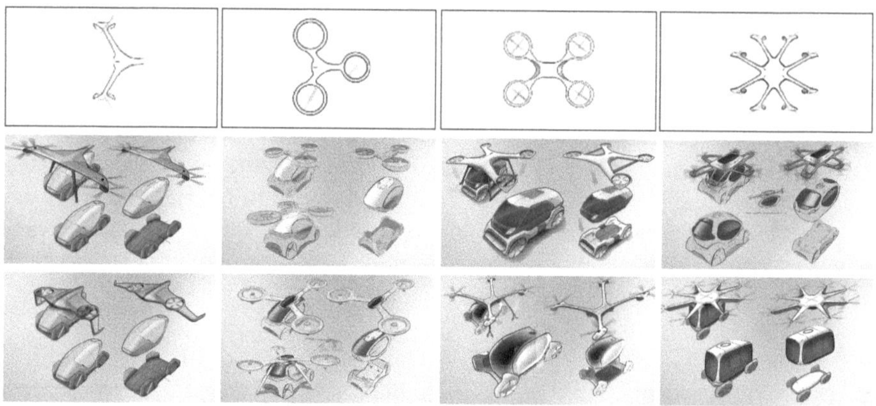

FIGURE 1.3 Drone, cockpit, and skateboard integration sketches based on the number of rotors. (FLY-PT Project.)

at this stage, trying to find different options as defined in the morphological table. The following figures show the refinement of the exterior cabin concept drawings and interior proposals, including seats, dashboard, luggage area, etc.

For the interior, multiple configurations were studied in the distribution of passengers with the possibility of autonomous driving (with two passengers and luggage), autonomous driving with three passengers, pilot and two passengers, among others. The team also explored different forms of construction and aesthetic and formal proposals. Figures 1.4–1.6 show some of the sketches developed with different approaches regarding the number of passengers, morphology, aesthetics, definition of materials and finishes, and construction processes (e.g., monocoque seats integrated into the floor, individual seats, double seats).

FIGURE 1.4 Refinement of exterior cabin and skateboard integration sketches. (FLY-PT Project.)

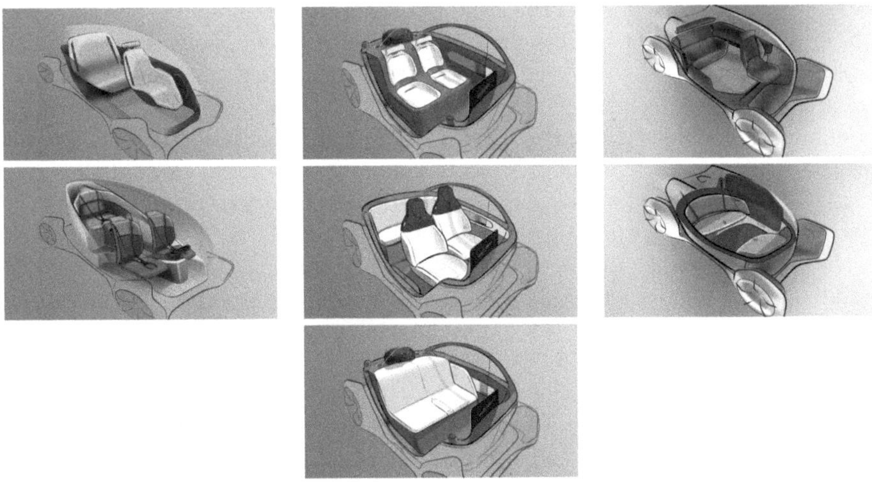

FIGURE 1.5 Interior configuration sketches. (FLY-PT Project.)

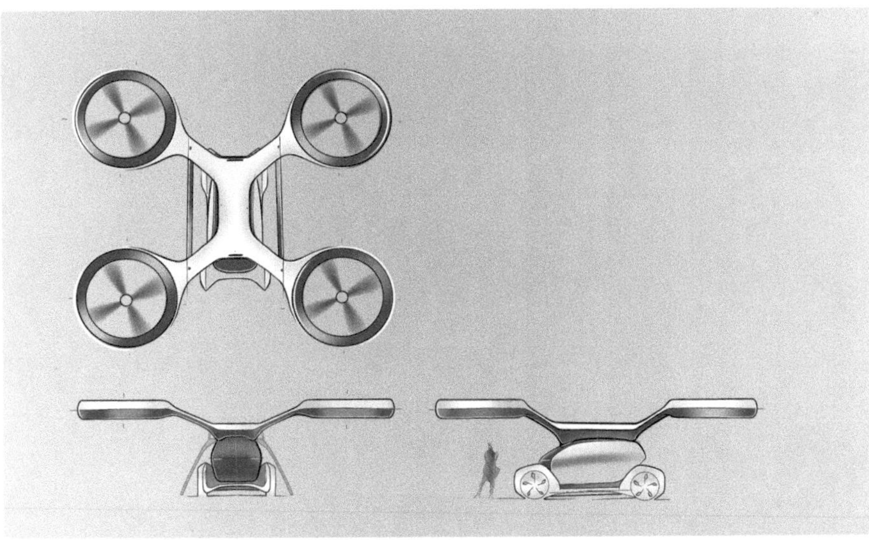

FIGURE 1.6 Exterior cabin and skateboard integration sketches. (FLY-PT Project.)

After several iterations and discussions between partners, the designers developed synthesis designs with the three proposals, including the integration of the drone, cabin, and skateboard, considering the number of rotors (2, 3, or 4). The sketches developed served as a basis for the remaining engineering partners to work (within the scope of other project tasks) and develop preliminary calculations to validate the options.

Overall, the sketches served to propose new concepts, develop new proposals, and allow a quick and intuitive understanding of ideas, facilitating discussion and collaboration. "Good concept generation leaves the team with confidence that the full space of alternatives has been explored. Thorough exploration of alternatives early in the development process greatly reduces the likelihood that the team will stumble upon a superior concept late in the development process" (Eppinger& Ulrich, 2015, p. 99).

Following this process, the designers moved on to a phase of creating (preliminary) 3D digital models. This process was developed with specialized computer-aided design (CAD) software defining the generic geometric shapes, surfaces, and dimensions to represent the physical object digitally, helping to establish proportions, scale, and the general arrangement of components. These models were iteratively refined by adding more details and correcting dimensions and proportions with different iterations of adjustment, team feedback, and evaluation. In Figures 1.7 and 1.8, we can see the synthesis sketches for the exterior cabin.

The refinement of concepts through synthesis drawings and preliminary three-dimensional modeling provided important data for the consortium to choose the best configuration. To this end, a "Concept Selection Matrix" method was developed, as explained below.

FIGURE 1.7 Exterior cabin final sketch. (FLY-PT Project.)

FIGURE 1.8 Exterior cabin final sketch. (FLY-PT Project.)

1.2.3 Concept Selection Matrix

> Concept selection is the process of evaluating concepts with respect to customer needs and other criteria, comparing the relative strengths and weaknesses of the concepts, and selecting one or more concepts for further investigation, testing, or development.
>
> *(Eppinger & Ulrich, 2015, p. 124)*

Following the sketching phase, it was necessary to find a way to decide which concepts to develop. To this end, a Design Selection Matrix method was developed, which is essentially a systematic approach used to evaluate and compare different alternatives or design options based on a set of criteria or factors. The Selection Matrix was the solution found to help make more informed decisions, providing an organized structure to evaluate the strengths and weaknesses of each design option. Considering the different consortium partners, this method was considered appropriate, enabling side-by-side comparison of alternatives and facilitating the selection of the most appropriate design solution: "The technique (Concept Matrix) is applied when needs of stakeholders and requirements have been collected. The idea behind Concept Selection is that a good solution is achieved by analyzing a few alternating designs and comparing them by a set of criteria that fits the problem" (SESG, 2009). The method was developed as follows.

1.2.3.1 Identification of Evaluation Criteria

The criteria considered important to evaluate the design options were defined. The criteria included three types of parameters: Design, Structure, and Manufacturing. For each of these typologies, the following evaluation parameters were defined: Design: Aesthetics, Comfort/Usability, Perceived Safety, and Passenger Interface; Structure: Weight, Stiffness/Structural Cohesion, and Energy Absorption (Impact); and Manufacturing: Cost, Tooling/Mold Complexity, and Assembly.

1.2.3.2 Parameters Definition

Weights were assigned to each criterion/parameter based on their relative importance. The attribution of weights reflected the priority of each criterion in the evaluation process. This step ensured that certain criteria could have a greater impact on the final decision.

1.2.3.3 Rating Scales

A scoring system was defined for each parameter, and each of the concepts was scored by all product–service system (PPS) partners. The defined scale was numerical, from 0 to 3.

1.2.3.4 Matrix Creation

A matrix was constructed with the design options listed as columns and the evaluation criteria as rows. The matrix included the weights assigned to each parameter and the rating scales for evaluation.

1.2.3.5 Evaluation

Each design option was evaluated against each parameter and assigned scores based on the performance or suitability of the option. The total score resulted from the sum of all classifications and the weighted average. Each partner voted separately on the concepts, and the results were compiled, resulting in a final classification weighted with everyone's contribution. A total of eight partners voted for the matrix. The process provided a means of quantifying the performance of each design option against agreed parameters. The strengths and weaknesses of each option, based on the evaluation criteria and based on comments made by the partners, allowed a direct comparison, facilitating communication between partners and a clear communication of the advantages and disadvantages of alternative concepts. Based on this analysis and consortium discussion, a decision was made on the most suitable design option (considering the scores, relative importance of parameters, and feedback from partners). Figure 1.9 depicts the Concept Selection Matrix which was distributed, completed, and voted by partners.

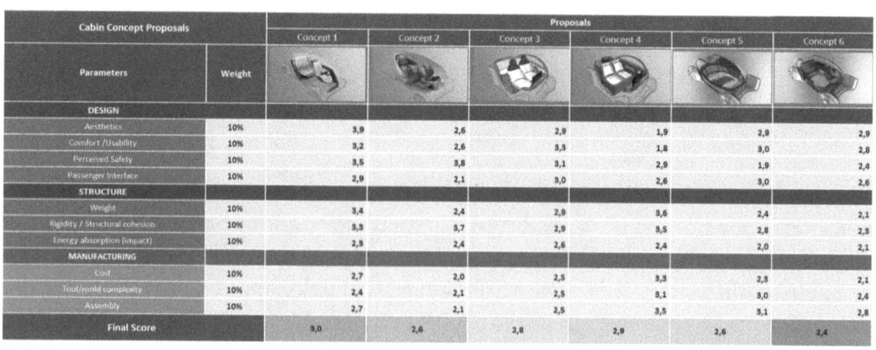

Cabin Concept Proposals		Proposals					
		Concept 1	Concept 2	Concept 3	Concept 4	Concept 5	Concept 6
Parameters	Weight						
DESIGN							
Aesthetics	10%	3,9	2,6	2,9	1,9	2,9	2,9
Comfort /Usability	10%	3,2	2,6	3,3	1,8	3,0	2,8
Perceived Safety	10%	3,5	3,8	3,1	2,9	1,9	2,4
Passenger Interface	10%	2,9	2,1	3,0	2,6	3,0	2,6
STRUCTURE							
Weight	10%	3,4	2,4	2,9	3,6	2,4	2,1
Rigidity / Structural cohesion	10%	3,3	3,7	2,9	3,5	2,8	2,3
Energy absorption (impact)	10%	2,3	2,4	2,6	2,4	2,0	2,1
MANUFACTURING							
Cost	10%	2,7	2,0	2,5	3,3	2,3	2,1
Tool/mold complexity	10%	2,4	2,1	2,5	3,1	3,0	2,4
Assembly	10%	2,7	2,1	2,5	3,5	3,1	2,8
Final Score		3,0	2,6	2,8	2,9	2,6	2,4

FIGURE 1.9 Concept Selection Matrix. (FLY-PT project.)

Based on the methodologies described previously, an interior concept featuring two passengers and cargo was selected for further development, based on results of the previously mentioned criteria (considering the possibility of having an autonomous vehicle, weight reduction, simplification of the number of components, and reduction of cabin volume, among others). In the next phase of the project, the different components of the interior were detailed and reiterated using a sketching method and preliminary 3D CAD models.

> Being at the forefront of any kind of field means that you must have a sense of what is going on, and where new ideas, product and research emerge. For designers this is particularly important... we will consider interdisciplinary Design Research as a process and tool. At least in some parts of the world, science has not only moved from being multidisciplinary to interdisciplinary but is also on its way to become a new discipline in its own right.
>
> *(Laurel 2003, p. 193)*

1.3 CONCLUSIONS

1.3.1 Discussion

The integration of all project partners as active participants is a fundamental part of the design process. The main advantage of this method of co-creation is the ability to involve all stakeholders in the design process to achieve better results. By using "Morphological tables", "Sketching sessions", and a "Concept Selection Matrix", the several project partners were able to collaborate to develop the most promising solution. Figure 1.10 visualizes the process developed during the project, considering the chronological phases and the number of partners involved.

- The process of morphological tables served to integrate different stakeholders in the concept phases (so as not to have the "dictatorship" of the designer or engineers).
- The table allows the visualization of solutions and promotes the combinations of different alternatives.
- The sketching sessions allow us to visually "integrate" the solutions chosen in the previous processes.
- By iterating on the concepts, sketches allow us to progressively incorporate different views and refine ideas into viable concepts.
- The Concept Selection Matrix allows us to integrate a voting-based decision tool, based on an agreed set of criteria. This allows for the inclusion of all partners' technical views.

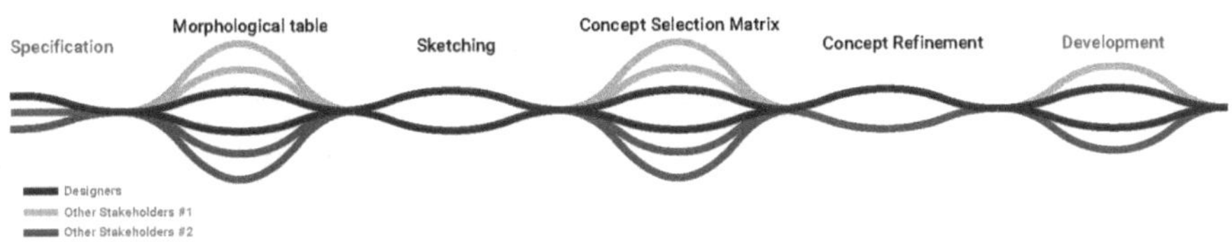

FIGURE 1.10 Collaborative design method overview. (FLY-PT Project.)

- Feedback from the Concept Selection Matrix additionally feeds new sketching and refinement sessions, until the solution is ready for a detailed development agreed upon.
- Although these processes try to integrate different views and opinions, they can also lead to "design by committee" solutions.
- These solutions, typically agreed upon by most partners, can be less innovative, as they tend to be a compromise between different parties' interests and ideas.
- By integrating many opinions and inputs from a variety of experts, the design solution can be diluted and lack a cohesive direction.
- To mitigate these problems, we suggest integrating an innovation management team that should oversee the concept development process that can listen to and integrate opinions but decide on a clear, cohesive approach.

The success of FLY-PT relied heavily on effective communication, shared goals, and a commitment to problem solving in a constructive manner. "Successful innovations require designers and design teams to deal with uncertainty, and this is where the design process supported by method and a sequence of actions plays an important role" (Daalhuizen & Badke-Schaub, 2009; Herrmann, 2004).

The development phase of the project consisted of activities such as 3D CAD modeling, parts engineering, and mockup development and construction. These tasks were addressed in this study, but final project results, including a 1:1 scale mockup of the cabin and a 1:4 scale model of the whole system, are shown in Figures 1.11–1.13.

FIGURE 1.11 Final Mockup. (FLY-PT Project.)

FIGURE 1.12 Final Mockup interior. (FLY-PT Project.)

FIGURE 1.13 Final Mockup interior. (FLY-PT Project.)

As an initial prototype, the FLY-PT project embarks on a path of gradual evolution, prioritizing the refinement of design and engineering to minimize component weight while upholding stringent standards of comfort and safety. The physical mockup of FLY-PT represents the project's inaugural prototype, employing diverse technologies such as metallic structures and composites for external parts, CNC machining, rapid prototyping, and additive manufacturing processes. The design is also poised for development with an eye on serial production, necessitating adaptable processes for varying production scales. Anticipating future needs in the UAM market, such as different service types, diverse markets, and evolving legislation, the design is expected to evolve to streamline operations and reduce aircraft maintenance time.

Acknowledging that flying in an electric vertical take-off and landing aircraft is a novel and unexpected experience for users, the design team tried to develop an intuitive, reassuring environment. The cabin design revolves around the user, emphasizing ergonomics tailored for two passengers and their hand luggage. Innovations like autonomous operation, a "wrap-around" dashboard with a central interface screen, a double seating system, ambient lighting, and expansive glazed surfaces – including doors with lower and upper glass for greater visibility – redefine the vision for autonomous flying vehicles. Safety and informational features such as grab handles, individual lighting, and interface screen displaying route information were incorporated to provide a reassuring environment. Hand luggage finds a secure spot behind the seats, anchored by elastic bands, while the 3-point seat belts introduce a vibrant touch of color with the use of technical woven fabrics and 3D printed materials. The electric architecture and aircraft systems are expected to deliver a highly efficient zero-emission, low-noise flight experience (e.g., when compared to the helicopter).

1.3.2 Conclusions

The objective of this study was to present the design methodologies in generating an E-VTOL Passenger Cabin Design for a collaborative project. The design methods included the creation of a "morphological table" to support the generative phase, "Sketching Sessions" to visualize ideas and concepts, and a "Concept Selection Matrix" to choose and refine the concepts collaboratively. At this stage, the selected concepts were refined and developed into detailed sketches and preliminary 3D models. The process served to integrate different stakeholders in the concept phases, and the validation tables allowed for an inclusive voting procedure based on different partners' opinions and comments. Although we consider these methods to be successful in the project, one can argue that this approach – specifically in the concept phases of the design project – could lead to "design by committee" solutions. Nevertheless, we considered that the collaborative approach ensured a more comprehensive and innovative development process, laying the foundation for the design of better products in a collaborative way. "We study the design process to know what we do and how we do it. To understand it and improve it. To become better designers" (Dubberly, 2004).

1.3.3 Recommendations for Future Work

As future recommendations, it is considered that further study into collaborative design projects should (i) focus on digital collaboration tools which can allow a fast iterative concept development and (ii) focus on presential interactive workshops which could improve the way partners work together (these were not possible during the project, due to the pandemic situation). These meetings can foster creative problem solving, ensuring that all voices are heard and fostering a sense of ownership among all stakeholders. Applying this design method to other projects could potentially bring more detailed findings about the development of collaborative design processes and the connection between collaboration tools and the final design products.

ACKNOWLEDGMENTS

Fly-PT is a collaborative effort involving 17 entities, with support from the European Regional Development Fund.

Cofinanciado por:

COMPETE 2020 PORTUGAL 2020 UNIÃO EUROPEIA Fundo Europeu de Desenvolvimento Regional

REFERENCES

Best, K. (2006). *Design Management: Managing Design Strategy, Process and Implementation.* Lausanne: AVA Publishing.

Brown, T. (2011). *Change by Design: How Design Thinking Transforms Organizations and Inspires Innovation.* New York: HarperCollins.

Daalhuizen, J., & Badke-Schaub, P. (2009). *Design Driven Innovation: How Comunity Based Methodology Can Facilitate the Designer and Foster Innovation* (pp. 4263–4266). Delft: Delft University of Technology, Faculty of Industrial Design Engineering, Section of Design Theory and Methodology.

Dubberly, H. (2004). *How Do You Design?* San Francisco: Dubberly Design Office.

Dziersk, M. (2007). Visual thinking: A leadership strategy. *Design Management Review, 18*(4), pp. 42–49.

Eppinger, S., & Ulrich, K. (2015). *Product Design and Development* (6th ed.). New York: McGraw-Hill Higher Education

Herrmann, J. W. (2004). *Decomposition in Product Development. Technical Report 2004–6.* Maryland: Institute for Systems Research, University of Maryland.

Holmes, K., & Maeda, J. (2018). Mismatch: how inclusion shapes design. In *Mismatch: How Inclusion Shapes Design* (pp. 1–14). Cambridge, MA: MIT Press.

Laurel, B. (2003). *Design Research Methods and Perspectives.* Cambridge, MA: The MIT Press.

Martin, B., & Hanington, B. (2012). *Universal Methods of Design: 100 Ways to Research Complex Problems, Develop Innovative Ideas, and Design Effective Solutions.* Beverly, MA: Rockport Publishers.

Roberts, J., Headleand, C., & Ritsos, P. (2015). Sketching designs using the five design-sheet methodology. In*IEEE Transactions on Visualization and Computer Graphics* (p. 1). https://doi.org/10.1109/TVCG.2015.2467271

SESG (2009). *Concept Selection; Theory and Practice White Paper of SESG Meeting November 6*. Gerrit Muller Buskerud University College.

Vink, P., & Brauer, K. (2011). Aircraft interior comfort and design. In *Aircraft Interior Comfort and Design*. https://doi.org/10.1080/00140139.2012.682446

2 A Research Framework for Designing Waterborne Passenger Mobility Ecosystems That Enable Smart Circular Strategies

Pirrone Laura, Bionda Arianna, and Ratti Andrea

This chapter is an extended version of the article presented in the Design Commit '24 conference proceedings.

2.1 INTRODUCTION

Climate change is a focal issue worldwide, and governments, institutions, organizations, and companies are acting to adopt more sustainable practices, decrease carbon emissions, foster the use of renewable energy, and implement new measures to adapt to the resulting environmental changes.

The transport system is one of the most polluting sectors, accounting for around one quarter of the EU's total greenhouse gas(es) emissions. To tackle the issue, the European Commission has introduced the Sustainable and Smart Mobility Strategy to achieve the green and digital transformation of the transport sector (European Commission, 2019).

Within the mobility sector, waterborne transport can constitute a valuable resource for promoting more sustainable and liveable cities. Waterborne public transport services have become an integral component of urban landscapes in numerous cities worldwide. Using waterways instead of roads can be a sustainable alternative in terms of energy consumption, noise, and pollution emissions, especially in dense urban areas (Campillo et al., 2019; Johansson, 2020). It reduces traffic congestion on roads and can move a higher number of passengers, reducing CO_2 emissions and improving the transport service overall.

Waterborne mobility can be an opportunity for all these cities that are situated around a body of water (i.e., most of the major cities around the world) and are a fundamental transport mode for moving people in urban territories geographically divided by rivers, channels, or lakes. With its natural "water-roads", it required lower investments in terms of infrastructure and routine maintenance compared to

DOI: 10.1201/9781032719290-3

land-based transportation, tied instead to the construction of streets and bridges. Despite that, there are still challenges to overcome to fully realize its sustainable potential.

The maritime sector, particularly in WPM, has been increasingly emphasizing sustainability (Özispa & Arabelen, 2018; Vejvar et al., 2020), with a focus on alternative propulsion systems, fuels, hull design, and operational measures (Grosso et al., 2021). However, existing literature and case studies on the sustainable transition of this specific mobility service are scarce and explore the theme adopting a vertical approach focused on a product development perspective or the supply chain of alternative fuels, exploring eco-design strategies and lifecycle assessment (Ansaloni et al., 2022; Sullivan et al., 2022).

In recent years, a limited but growing body of research has begun to explore the integration of circular economy principles within the maritime sector. These studies primarily focus on two areas: the recycling and repurposing of materials from decommissioned vessels, and the application of Industrial Ecology concepts within the context of ports and cargo shipping (Karimpour et al., 2019; Agarwala, 2023; Faut et al. 2023). While these efforts represent important initial steps, they highlight the need for a more comprehensive approach to embedding circularity across the broader maritime ecosystem.

Digitalization can play a crucial role in promoting sustainable waterways transport solutions. Digital technologies can enable circular strategies (Alcayaga et al., 2019); moreover, they can move the discussion to a required systemic level, encouraging a holistic approach that can expand circularity and support long-term and comprehensive solutions (Li et al., 2023).

The expression "waterborne passenger mobility" (WPM) refers to the transport of people through waterways, and this research will focus on hinterland or intercoastal transportation in urban and suburban areas (Pirrone et al., 2023a). This topic will be addressed through a systemic perspective, which considers the complex interactions among all the actors of the WPM ecosystem, including passengers, ferry and infrastructure operators, regulators, and service providers.

Circular economy is an economic system that aims to change the current linear economy, reshaping the way resources are managed, products are made and used, and materials are treated afterwards. It is based on three principles: eliminate waste and pollution, circulate products and materials (at their highest value), and regenerate nature (Ellen MacArthur Foundation (EMF), 2023).

Circular economy has a systemic nature, and acknowledging the dynamics of the entire ecosystem, circular practices can be more comprehensively supported and facilitated (Konietzko et al., 2020).

The concept of ecosystem is inspired by the network-based organization and complex interdependencies of natural systems. Having different nuances of meanings according to different disciplines or even within the same disciplinary field, from a design and management perspective, an ecosystem can be defined as a self-contained and self-adjusting system of resources based on a network of actors that share a common vision and mutually interact to materialize the same value proposition (Adner, 2017; Moore, 1996; Vink et al., 2021).

Service ecosystem design shifts the discussion towards a more holistic perspective that goes beyond organizations' boundaries and focuses on actors' service exchanges and value co-creation (Adner, 2017).

Digital technologies, such as the Internet of Things, Big Data Analytics, Additive Manufacturing, Blockchain, and Augmented/Virtual Reality, can increase systems integration by blurring the boundaries and fostering ecosystem dynamics. At the same time, they can enable a better management of resources, encouraging circular practices. The novel paradigm of Smart Circular Economy identifies a system that uses digital technologies during the lifecycle phases to implement circular strategies and practices aimed at value creation (Alcayaga et al., 2019; Bressanelli et al., 2022).

The described background highlights two main relevant gaps. First, the knowledge concerning the sustainability of the WPM is still limited and fragmented; moreover, it lacks a systemic perspective that can move the product-focused approach towards a holistic view. Second, the opportunities offered by digital technologies in achieving circularity have never been explored in the design of WPM ecosystems.

This study will aim to develop design strategies for the development of Smart Circular WPM ecosystems, or rather WPM ecosystems that perform circular economy through digital technologies.

By exploring the themes of circular economy, ecosystem, and digitalization within the waterborne mobility sector, the research will answer to the following question: How to design smart circular ecosystems in the WPM service?

The main research question will be addressed by answering three sub-questions: (RQ1) How is the waterborne passenger urban mobility ecosystem constituted? (RQ2) How are digital technologies enabling a circular economy in the WPM ecosystem? What are the related criticalities and barriers? (RQ3) Which design strategies can support the actors of the WPM ecosystem to implement a circular economy through digital technologies?

To assess these research objectives, a mixed-method approach was selected. The first two questions will be answered through literature review and case study research, whereas the third question will be solved with participatory research, aiming to involve the actors of the ecosystem in defining future scenarios, roadmaps, and design strategies for the implementation of a smart and circular WPM ecosystem.

2.2 METHODS

A methodological framework has been designed to answer the research questions. The approach followed to conduct this scientific inquiry is research through design. This type of research aims to create knowledge through an action-reflection approach, which results in the creation of design knowledge instead of project solutions. This action-reflection activity will be applied to enable a mutual interaction between the knowledge collected and developed, and the stakeholders who will be engaged in the study. The output of this process will be design knowledge, methods, frameworks, or guidelines.

Based on a pragmatist epistemology, the inquiry will alternate post-positivism and constructivism, moving back and forth between deduction and induction. Accepting philosophically the continuity between the objectivity of the truth and the subjectivity of multiple realities, pragmatism claims that theories must be linked with experience or practice (Feilzer, 2010). This flexible activity enables one to bridge the gap between general theories and particular design challenges (Dalsgaard, 2014).

Characterized by an evolutionary perspective, this epistemology supports a systemic approach which involves an iterative process of moving between the whole and the parts (Dalsgaard, 2014) and that will be particularly useful for a more comprehensive understanding of the ecosystem dynamics.

Within this epistemology, a mixed-method approach was chosen to alternate quantitative and qualitative methods and support the interdisciplinary nature of this research, which meets both design disciplines and management studies, bringing together multiple perspectives and ways of understanding knowledge (Johansson, 2020) (Figure 2.1).

The methods applied include literature review, case study research, and participatory research, and they were chosen to address different objectives of the study resulting from the research questions.

Inspired by the Double Diamond model of Design Thinking developed by the British Design Council in 2005, the research is structured with two main purposes,

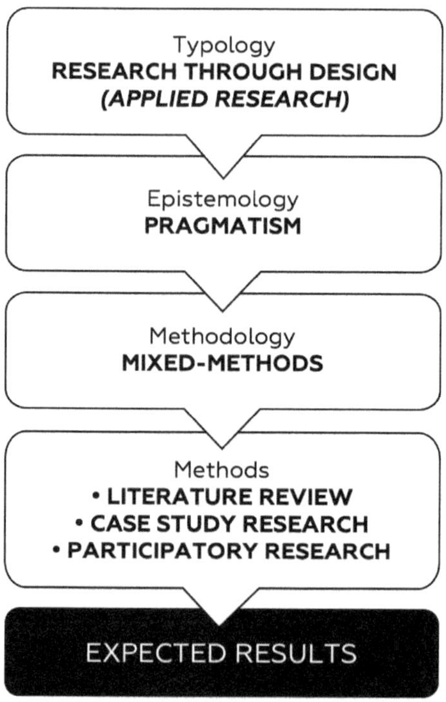

FIGURE 2.1 Research methodology framework. (Designed by the authors.)

which embed activities to undertake for answering the research questions: "explore and define" and "describe and design".

The first research purpose is to explore the scope of the research, understanding, and defining concepts, their interrelations, and the state of the art.

It refers to the first research question (RQ1) and part of the second research question (RQ2).

The second research purpose intends to apply the knowledge gathered to describe hindrances and opportunities to co-design future scenarios and strategies.

This purpose refers to part of the RQ2, and the third research question (RQ3) (Figure 2.2).

Each research question addresses a general aim structured in research objectives.

RQ1: How is the WPM ecosystem constituted? It aims to address the first research gap moving the perspective to a systemic level and understanding the meaning of "ecosystem" in the WPM context.

- Objective #1: Defining and framing the WPM ecosystem.

RQ2: How are digital technologies enabling a circular economy in the WPM ecosystem? What are the related criticalities and barriers? It aims to address the second research gap by defining and describing a framework of the Smart Circular WPM Ecosystem.

- Objective #2: Framing a circular economy and its principles, framing digital technologies, and defining the concept of Smart Circular Economy.
- Objective #3: Defining the "Smart Circular WPM Ecosystem" and its conceptual framework.
- Objective #4: Exploring how digital technologies are enabling circular strategies in the WPM ecosystem.
- Objective #5: Identifying criticalities, barriers, and further opportunities.

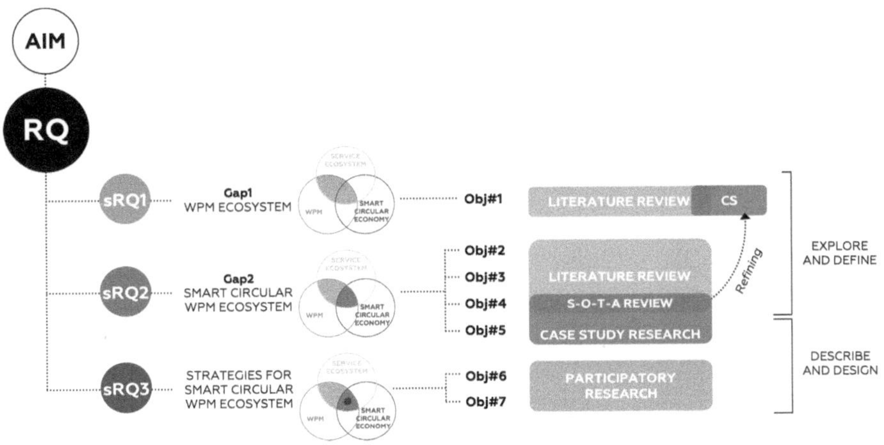

FIGURE 2.2 Research structure diagram. (Designed by the authors.)

RQ3: Which design strategies can support the actors of the WPM ecosystem to implement a circular economy through digital technologies? It aims to develop strategies and actions within the defined framework of the Smart Circular WPM Ecosystem.

- Objective #6: Understanding how criticalities and barriers can be addressed and the respective roles of the ecosystem's actors.
- Objective #7: Developing design strategies that can promote a circular economy through digital technologies in the WPM ecosystem.

The research unfolds in three main phases. The first addresses the definition, mapping, and description of the WPM ecosystem through literature review and case studies. The second focuses on establishing a conceptual framework for the Smart Circular WPM Ecosystem via an initial literature review, validated through subsequent case study research. This iterative process enhances understanding by combining insights from literature and real cases, identifying best practices, barriers, and opportunities. The third phase involves participatory research, including workshops and scenario-building activities with stakeholders from WPM ecosystem cases. This collaborative approach aims to map future scenarios, outline roadmaps, and design strategies supporting smart and circular goals in the WPM ecosystem.

2.3 LITERATURE REVIEW AND CASE STUDY ANALYSIS

The study started with a literature review structured in two main phases. The first phase, more explorative, aimed to understand the wider social and research context, the study's evidence, key concepts, and their relations. The second phase, characterized by a more systematized approach, led to the definition of the disciplinary field, establishing the research positioning within the academic community, and helped to identify the research gaps (Figure 2.3).

After a deep understanding of the identified gaps, the research title, research questions, and associated objectives were formulated.

The literature review was carried out on the Scopus database, combining selected clusters of keywords and using Boolean logic to design search queries. Clusters were defined according to the main topics: WPM, ecosystem perspective, circular economy, and digitalization. The queries adopted for each cluster were:

- WPM: "ferry" OR "inland" OR "urban" OR "public" AND "water* transport*" AND "passenger"
- Servitization: "ecosystem design" OR "service ecosystem" OR "business ecosystem" OR "service system"
- Circular economy: "circular economy" OR "circular strategies" OR "circularity" OR "sustainability"
- Digital technologies: "smart technolog*" OR "digital techonolog*" OR "Industry 4.0"

Nine combinations have been performed among the clusters, obtaining a total of 368 papers, excluding duplicates.

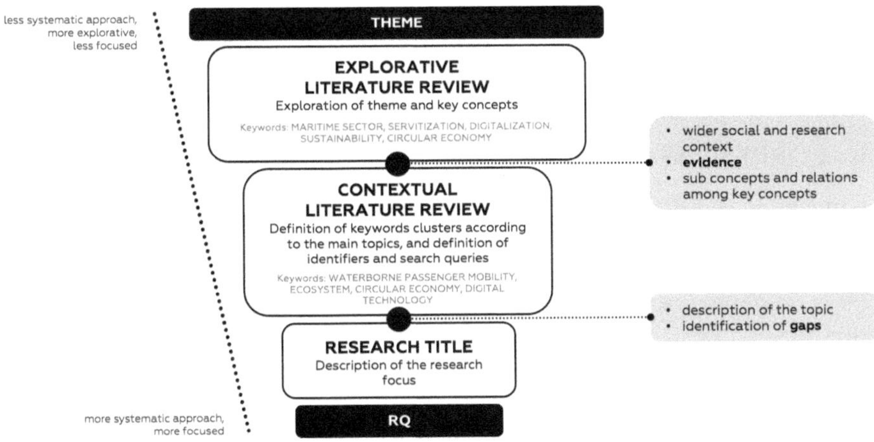

FIGURE 2.3 Literature review process. (Designed by the authors.)

The first screening of results was performed by reading titles and abstracts. The selection was driven by the main research objectives: defining and describing the WPM ecosystem and exploring the Smart Circular Economy, therefore the opportunities that it can offer to the sustainability of the WPM sector. A total of 51 documents were selected after a full-text reading and further analysed.

An additional search for grey literature targeting reports, organizations' websites, and European projects was made using Google's search engine, as complementary to the scientific knowledge and functional to deepen the knowledge on the WPM context.

Results were collected and categorized by themes and mutual relations.

The findings allowed us to define a conceptual framework of the Smart Circular WPM Ecosystem.

Currently, the PhD research is advancing through a primary phase of case study analysis by desk research.

A first analysis has been carried out to map the European region, with a specific focus on understanding the advancements in sustainability and digitalization within the WPM.

An initial selection of five case studies has been performed, focusing on the Northern European area, where the research on the sustainability of the waterborne transport sector is gaining significant importance due to the widespread use of waterways as a daily means of public transport.

These first cases have been selected from public waterway transport systems, whether they were already operational, launched, or in the project development phase. All of them have been developed within the last five years. The selection has been determined based on parameters derived from the outcomes of the literature review phase and is instrumental in delineating the current state of the art and addressing the research question. The selected cases consist of waterborne transport systems exclusively for passengers, operating in urban or suburban areas, that perform circular economy practices, and at least one of those is enabled by digital technologies.

The five cases have been analysed by applying the Smart Circular WPM Ecosystem framework, previously developed from the literature review. The framework was mutually updated in some details based on insights coming from cases.

2.4 CURRENT RESULTS

The literature review phase performed in the first part of the PhD study was fundamental to define the research borders and identify its focus. Knowledge was collected and analysed to position the research within the scientific knowledge and define its disciplinary field. The research has a multidisciplinary nature, therefore mainly crossing Design Studies and Management, moreover, involving Maritime and Transportation, Sustainability, and Information and Communication Technologies Studies (Figure 2.4).

Within the defined disciplinary context, certain gaps were identified. The knowledge on sustainability in the waterborne transport sector is limited and fragmented and lacks a comprehensive systemic perspective, particularly in the design field. The sector principally adopts a vertical approach that focuses on product performance without considering the entire WPM ecosystem, its actors, and the interaction among them. This phenomenon is directly linked to an additional shortage of studies or cases that relate to circular strategies applied to the sector. Indeed, the systemic nature of the circular economy inherently calls for an ecosystem perspective (Konietzko et al., 2020).

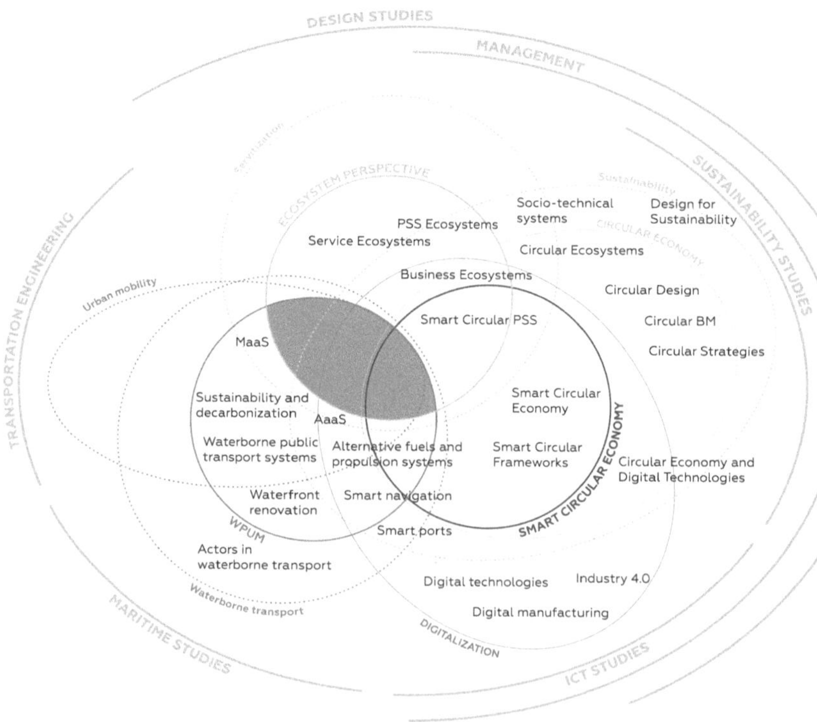

FIGURE 2.4 Positioning map. (Designed by the authors.)

Circular economy deals with complex and interconnected systems, viewing the entire production, consumption, and waste cycle; thereby, it is essential to adopt a broad, holistic approach when trying to implement it.

Moreover, the literature review highlighted significant advancements in the digitalization of the maritime sector. The industry is increasingly adopting digital technologies to enhance traffic monitoring, optimize navigation, reduce emissions, leverage big data analytics, and strengthen cybersecurity (Restrepo-Arias et al., 2022). Nevertheless, it also highlighted a significant gap in the exploration of how these technologies can facilitate the application of circular principles, and associated opportunities and challenges.

The findings allowed us to define the Smart Circular WPM Ecosystem and its conceptual framework. The Smart Circular WPM Ecosystem is a transportation system for passengers on waterways, which performs circular strategies and practices enabled by digital technologies, through its network of actors, to enhance sustainability along its lifecycle. The framework of the Smart Circular WPM Ecosystem interrelates the WPM ecosystem with smart circular strategies to explore possibilities of the use of digital technologies in enabling circular practices (Pirrone et al., 2023b) (Figure 2.5).

The Smart Circular WPM Ecosystem framework shows the complexity of applying Smart Circular Economy to a systemic dimension, where the network can be considered at different levels and from different perspectives. Moreover, the framework establishes a method to analyse emerging practices, uncovering related opportunities and challenges. The framework has been applied to a first selection of five case studies located within Northern Europe to explore the state of the art and identify circular principles inherent in the sustainable innovative action undertaken by the waterborne passenger transport sector. Data gathering involved conducting desk research, which

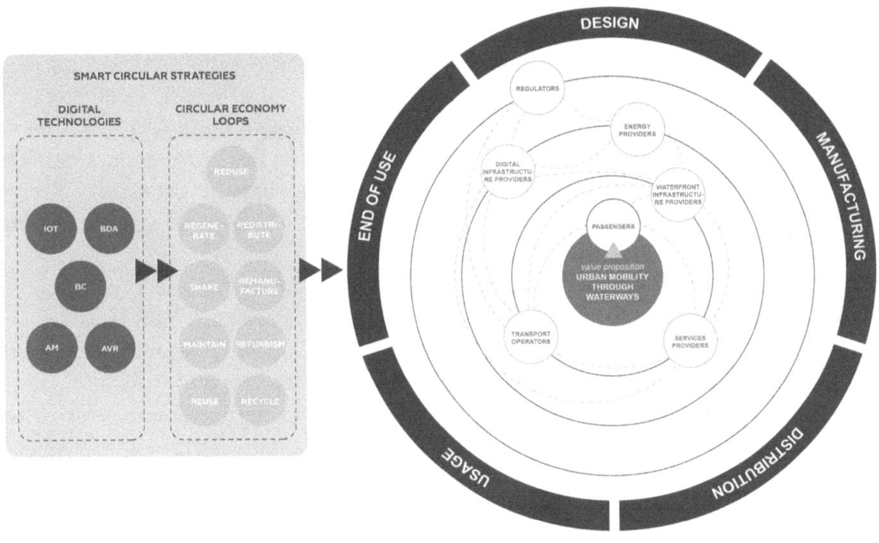

FIGURE 2.5 Smart Circular WPM Ecosystem framework. (Pirrone et al., 2023b.)

included reviewing academic papers, reports, and documents from organizations' websites. Additionally, information was gathered through participation in sector conferences. The application of the framework to the cases helped to examine the circular practices performed and the associated circular principles fulfilled. It identified digital technologies employed, elucidating how these digitalized solutions facilitate circular practices and during which lifecycle phases they operate. Finally, the main categories of actors involved were pointed out, focusing on those who have directly introduced a change within the system or an innovation, to contribute to the activation of the circular practice. Through these first findings, a list of smart circular practices has been defined, encompassing autonomous navigation and charging, intermodal transport platform, virtual testing, share-on-demand service, fleet status monitoring, and modularization and standardization (Pirrone et al., 2023b).

2.5 FUTURE DEVELOPMENTS AND CONTRIBUTIONS

Further studies on case study research will be undertaken to define the WPM ecosystem through field observation and semi-structured interviews with key informants to validate and integrate data collected from the previous phases. It will help to outline further attributes and details on actors, interactions, and touchpoints. Moreover, it will lead to identifying criticalities and barriers linked with the implementation of smart circular practices. The qualitative data gathered from this phase will be coded according to ecosystem features, smart circular characteristics, actors involved, challenges faced, and opportunities to implement other circular principles through digitalization. At the same time, a deeper analysis and mapping of cases performing emerging interesting practices will be carried out to further explore the role of digitalization in fostering a circular economy in the WPM ecosystem.

Afterwards, to investigate the impact of smart circular strategies on actors and their interconnected relationships, participatory research with stakeholders will be conducted. This exploration aims to understand how these actors can influence and be influenced by such strategies, ultimately leading to reimaging, redesigning, and reshaping the ecosystem to promote a sustainable mobility system. Future scenarios and roadmaps will be mapped, thereby outlining design strategies that support those stakeholders in pursuing smart and circular goals in the WPM ecosystem. The intended outcome is to provide a guideline for identifying the necessary innovations in products, services, and ecosystem configurations to support a smart circular transition.

This research establishes theoretical foundations for emerging research avenues in sustainability, mobility systems, and the Smart Circular Economy. It emphasizes and undertakes a multidisciplinary approach encompassing different knowledge coming from Design, Systemic Thinking, Business and Management, and Information and Communication Technologies, and is directly related to the involvement of actors from different industries and fields having certain roles within the waterborne passenger ecosystem. This multidisciplinary and multilevel view is inevitably required when a holistic and ecosystem perspective is considered. Through a literature review and case study analysis, the study aims to outline the concept of the WPM ecosystem, detailing its key characteristics and features across multiple dimensions.

It adopts a zoom-in/zoom-out approach, examining the ecosystem from both micro and macro perspectives to provide a comprehensive understanding.

Furthermore, the research significantly contributes to systematizing knowledge on sustainability in waterborne urban mobility. It ventures into the exploration of the innovative concept of Smart Circular Economy within this sector, a dimension hitherto unexplored.

The study introduces a methodological framework that approaches the design of WPM with a systemic lens, considering the entire ecosystem with its actors, interrelations, and touchpoints. It will identify challenges and barriers in achieving a circular economy while also exploring new opportunities by investigating how digital technologies can trigger circular practices in the WPM system of actors.

On a practical level, the study formulates design strategies to assist actors in the WPM ecosystem in realizing a circular economy through the effective use of digital technologies. The resulting knowledge will contribute substantially to the development of methodologies, guidelines, and tools that can facilitate the design of Smart Circular WPM ecosystems, therefore guiding the sustainable transition of the sector.

REFERENCES

Adner, R. (2017). Ecosystem as structure: an actionable construct for strategy. *Journal of Management*, 43(1), 39–58. https://doi.org/10.1177/0149206316678451

Agarwala, N. (2023). Promoting circular economy in the shipping industry. *Journal of International Maritime Safety, Environmental Affairs, and Shipping*, 7(4). https://doi.org/10.1080/25725084.2023.2276984

Alcayaga, A., Wiener, M., & Hansen, E. G. (2019). Towards a framework of smart-circular systems: an integrative literature review. *Journal of Cleaner Production*, 221, 622–634. https://doi.org/10.1016/j.jclepro.2019.02.085

Ansaloni, G. M. M., Bionda, A., & Rossi, M. (2022). EcoDesign strategies for zero-emission hydrogen fuel vessels scenarios. In *2022 7th International Conference on Smart and Sustainable Technologies (SpliTech)* (pp. 1–6). https://doi.org/10.23919/SpliTech55088.2022.9854279

Bressanelli, G., Adrodegari, F., Pigosso, D. C. A., & Parida, V. (2022). Towards the smart circular economy paradigm: a definition, conceptualization, and research agenda. *Sustainability*, 14(9), 4960. https://doi.org/10.3390/su14094960

Campillo, J., Dominguez-Jimenez, J. A., & Cabrera, J. (2019). Sustainable boat transportation throughout electrification of propulsion systems: challenges and opportunities. In *2019 2nd Latin American Conference on Intelligent Transportation Systems (ITS LATAM)* (pp. 1–6). https://doi.org/10.1109/ITSLATAM.2019.8721330

Dalsgaard, P. (2014). Pragmatism and design thinking. *International Journal of Design*, 8(1), 143–155.

Ellen MacArthur Foundation (EMF). (2023). The circular economy glossary. https://ellenmacarthurfoundation.org/topics/circular-economy-introduction/glossary

European Commission. (2019). Mobility and transport. https://transport.ec.europa.eu/index_en

Faut, L., Soyeur, F., Haezendonck, E., Dooms, M., & De Langen, P. W. (2023). Ensuring circular strategy implementation: the development of circular economy indicators for ports. *Maritime Transport Research*, 4, 100087. https://doi.org/10.1016/j.martra.2023.100087

Feilzer, M. Y. (2010). Doing mixed methods research pragmatically: implications for the rediscovery of pragmatism as a research paradigm. *Journal of Mixed Methods Research*, 4(1), 6–16. https://doi.org/10.1177/1558689809349691

Grosso, M., Marques dos Santos, F. L., Gkoumas, K., Stępniak, M., & Pekár, F. (2021). The role of research and innovation in Europe for the decarbonisation of waterborne transport. *Sustainability*, 13(18), 10447. https://doi.org/10.3390/su131810447

Johansson, M. (2020). Waterborne mobility services in sustainable regional planning – a holistic study approach for new mobility services in North Western Skåne, Sweden. https://share-north.eu/wp-content/uploads/2021/09/Waterbased-mobility-M-Johansson-20201201.pdf

Karimpour, R., Ballini, F., & Ölcer, A. I. (2019). Circular economy approach to facilitate the transition of the port cities into self-sustainable energy ports – a case study in Copenhagen-Malmö Port (CMP). *WMU Journal of Maritime Affairs*, 18(2), 225–247. https://doi.org/10.1007/s13437-019-00170-2

Konietzko, J., Bocken, N., & Hultink, E. J. (2020). Circular ecosystem innovation: an initial set of principles. *Journal of Cleaner Production*, 253, 119942. https://doi.org/10.1016/j.jclepro.2019.119942

Li, A. Q., Lahy, A., Found, P., Kumar, M., & Claes, B. (2023). Developing PSS business ecosystems in the digital era. *Industrial Marketing Management*, 109, 121–134. https://doi.org/10.1016/j.indmarman.2022.12.017

Moore, J. F. (1996). *The Death of Competition: Leadership and Strategy in the Age of Business Ecosystems*. New York: HarperBusiness.

Özispa, N., & Arabelen, G. (2018). Sustainability issues in ports: content analysis and review of the literature (1987–2017). In *SHS Web of Conferences* (vol. 58, p. 01022). https://doi.org/10.1051/shsconf/20185801022

Pirrone, L., Bionda, A., & Ratti, A. (2023a). How digital technologies can support sustainability of the waterborne passenger mobility ecosystem: a case study analysis of smart circular practices in northern Europe. *Sustainability*, 16(1), 353. https://doi.org/10.3390/su16010353

Pirrone, L., Bionda, A., & Ratti, A. (2023b, October 9). Exploring a framework in designing smart circular ecosystems in the waterborne passenger mobility. In *IASDR 2023: Life-Changing Design. IASDR 2023: Life-Changing Design, Politecnico di Milano, Italy*. https://doi.org/10.21606/iasdr.2023.462

Restrepo-Arias, J. F., Branch-Bedoya, J. W., Zapata-Cortes, J. A., Paipa-Sanabria, E. G. & Garnica-López, M. A. (2022). Industry 4.0 technologies applied to inland waterway transport: systematic literature review. *Sensors*, 22, 3708. https://doi.org/10.3390/s22103708

Sullivan, B. P., Ansaloni, G. M. M., Bionda, A., & Rossi, M. (2022). A life cycle perspective to sustainable hydrogen powered maritime systems – functional and technical requirements. *International Journal of Product Lifecycle Management*, 14(2/3), 282. https://doi.org/10.1504/IJPLM.2022.125822

Vejvar, M., Lai, K.-H., & Lo, C. K. Y. (2020). A citation network analysis of sustainability development in liner shipping management: a review of the literature and policy implications. *Maritime Policy & Management*, 47(1), 1–26. https://doi.org/10.1080/03088839.2019.1657971

Vink, J., Koskela-Huotari, K., Tronvoll, B., Edvardsson, B., & Wetter-Edman, K. (2021). Service ecosystem design: propositions, process model, and future research agenda. *Journal of Service Research*, 24(2), 168–186. https://doi.org/10.1177/1094670520952537

3 Mass Immunization Project Developed Collaboratively Using the VAI Automated Artifact

Ana Leocadia de Souza Brum,
Gláucia de Salles Ferro, and
Aguinaldo dos Santos

3.1 INTRODUCTION

This chapter is an extended version of the paper presented in the proceedings of the Design Commit '24 conference. In March 2020, news provided by the World Health Organization (WHO) indicated that the world was impacted by the COVID-19 pandemic. According to WHO, a pandemic is a term used to describe a disease that spreads rapidly across different regions of a country, continent or the world, through sustained transmission from person to person. The severity of the disease is not the determining factor, but rather its capacity for contagion and geographic spread. COVID-19 is the disease caused by the coronavirus called SARS-CoV-2. This information has significantly changed the way people relate to each other in both their personal and professional lives. Furthermore, it exposed and exacerbated existing social and health inequalities, disproportionately affecting vulnerable groups such as the elderly and ethnic minorities. At the same time, in many countries, the pandemic has led to an unprecedented strain on healthcare systems, with overcrowded hospitals and insufficient medical resources. The uneven distribution of vaccines has become a significant challenge, with many low-income countries struggling to access enough doses to vaccinate their populations. Fear became part of the daily routine, encouraging professionals and industries to seek solutions to mitigate the impact of the disease on human life and develop effective vaccines for its treatment. The race to achieve these goals took place on a global scale. By December 2020, some countries had already developed vaccines to combat COVID-19 (such as Great Britain, Argentina, Chile, China, and Brazil). However, the application of vaccines did not always occur quickly enough to achieve mass immunization. Thousands of people die due to this lack of agility. It was in this scenario that the VAI project – VAI in Brazil – emerged, capable of vaccinating a person every 3 minutes through a robotic and automated system. This project differs from other vaccination systems for the following reasons: reduction in the number of health professionals to serve

DOI: 10.1201/9781032719290-4

the population; safety regarding the type, quantity, validity, and conservation of the vaccine; speed of application; immediate validation of information about the user in the country's health system; modular, transportable, safe, and sustainable infrastructure. Although initially designed to handle COVID-19 vaccination, this project also demonstrates its relevance to meeting all global vaccination needs.

With its implementation, thousands of lives can be saved by making vaccines available for hundreds of vaccine-preventable diseases. The VAI project, when implemented, can accelerate the restoration of social normality in the world. Furthermore, the project was developed in an interdisciplinary way, using service design to ensure focus on the user/patient and strategic design as a practice to ensure project implementation. The research presented here is defined based on the following problem: How to develop an automated and innovative vaccination device to meet the urgent need for mass immunization? Therefore, the research objective is defined as follows: developing an automated and innovative vaccination artifact to meet the urgent need for mass immunization. Given this objective, the research scope is established based on key points for the development of the project. They are the use of automation and robotics as a premise for the construction of the artifact, interdisciplinary collaboration, and strategic design for the deepening, differentiation and implementation of the project, the design of services to identify the points of interaction with the patient and the mass vaccination as a starting point and final validation of the project.

This article was structured around the following topics: theoretical review, method, project presentation, results and discussion.

3.2 THEORETICAL REVIEW

This theoretical review highlights the fundamentals of the VAI project. Therefore, its objective is not to exhaustively cover the topics presented, but rather to use them to provide a more appropriate context for the project.

3.2.1 Mass Vaccination

This article seeks to support the understanding of mass vaccination, highlighting it as a condition for rapid immunization of the population, which is the premise for the project in question. From a historical perspective, the practice of vaccination as a preventive measure against diseases originated in India and China around 1000 BC. At that time, the vaccine was introduced to prevent smallpox (Heymann & Aylward, 2006). As the positive results from the application of vaccines advanced, research continued to progress. Around 1796, Edward Jenner, through his studies, managed to make vaccines transportable, since they were initially perishable. With this innovation, some European countries, such as Great Britain, Sweden, Denmark, and Germany, at the beginning of the 19th century, made vaccines mandatory for the entire population. This practice has spread throughout the world.

In Brazil, initiatives promoted by the Brazilian State since the end of the 19th century demonstrate the existence of a culture of immunization, associated with the introduction of vaccines and mass vaccination campaigns (Soárez et al., 2012). Mass vaccination programs are based on the concept of collective immunity. When most of the population is vaccinated, diseases can be eradicated. This immunization occurs as follows: the transmission of diseases from person to person is interrupted by the presence of infected people in contact with vaccinated individuals (Destefano et al., 2018). According to the WHO, mass vaccination programs have been highly successful in reducing the 57 million annual deaths caused by infectious diseases worldwide. Regardless of the efforts of governments around the world, there is resistance from families who believe that vaccines should be avoided. Given this fact, in Brazil, society is already seeing the accelerated return of diseases that previously had low rates of occurrence, such as Measles and Whooping Cough. In this regard, vaccination coverage of at least 95% of the target population for these vaccines becomes urgent (Peres et al., 2022). The VAI project could be a solution for this vaccination coverage if families were convinced to protect children from infectious diseases. Examples like this one in Brazil are spreading around the world. It is up to public bodies and health professionals to make coordinated efforts to convince people of the need for the vaccine.

3.2.2 Scientific Research That Supports the Relevance of the VAI Project

At the time of the initial development of the project, there was no more in-depth research demonstrating aspects necessary for the theoretical basis of the project. Aspects such as: the need to vaccinate the largest number of people at the same time; the need for communication carried out by health agencies to raise awareness among the population about the importance of the vaccine; the practice of mass vaccination for various types of diseases, aiming at avoiding mass contamination. In addition to the aspects indicated, other factors of the same relevance can contribute to the perception of the need for the continuous development of projects such as VAI. Thus, some relevant qualitative data were chosen for this theoretical review. Thus, Montero et al. (2021) report that only when 67% of the population is vaccinated will it be protected. This corresponds to approximately 5 billion people in the world (United Nations Department of Economic and Social Affairs, Population Division, 2024). Consequently, the acceleration vaccination for most of the population for other diseases such as influenza can mitigate rapid contamination and, in this sense, flexible vaccination technologies are essential (Newland et al., 2021). Or that is, these authors indicate that not only COVID 19 should be evaluated for mass vaccination, but other types of diseases, often more common such as influenza. In this sense, Wickmar et al. (2021) indicate that another type of action is fundamental, not only linked to immunization, but to communication that must be clear and objective on the part of public health teams directed to the entire population about the need for rapid mass vaccination. Another factor that occurred as a side effect during the pandemic was the interruption of childhood immunization practices. Such practices were not resumed after the pandemic crisis. This indicates that there will be serious consequences related to suboptimal control of vaccine-preventable diseases among this population segment (Dinleyici et al., 2020)

Although during the pandemic we did not have a clear idea of the full extent of the disease, COVID-19 has prompted the largest global vaccination effort in history. 33 vaccines have already been developed, and more than 10 billion doses have been administered, but new variants of the disease have been developing. Thus, efforts are ongoing to combat the disease (Machado et al., 2022). Despite the good results achieved, the efforts of countries and institutions need to continue so that the global population can be immunized, and public and private leaders must invest in effective vaccine delivery, allocation, distribution and verification strategies (Weintraub et al., 2021).

3.2.3 Interdisciplinary Collaboration

The topic of collaboration covers a wide range of perspectives that can be analyzed. In this article, the focus is on the interdisciplinary collaboration evidenced in the VAI project. In this case, service designers; mechanical and automation engineers; health and safety professionals; public bodies and private companies in the health sector and communication agents form the scope of the collaborative group. Therefore, aiming to better understand the term collaboration, this literature review begins with the perception of Hansen (2009). The author proposes that collaboration should not be conceived as an end, but rather as a means to achieve a specific result. In the case of the VAI project, collaboration is consolidated as a means for building the artifact characterized by an automatic vaccination system. From a design perspective, collaboration can culminate in achieving more efficient solutions (Heemann et al., 2010). This perception occurs considering the depth and polysemy that permeate collaboration, not limited to just the area of activity of designers or other professionals who work in groups. The effectiveness of collaboration is evident when a result is achieved by an interdisciplinary group. In this context, Furtado (2007) presents his perspective on interdisciplinarity:

> Interdisciplinarity symbolizes the most advanced level of interaction between different areas. In this situation, relationships between disciplines would be less hierarchical, providing space for a shared work platform. This involves operating with common concepts and striving to translate specific jargon to facilitate understanding among newly integrated colleagues.
>
> *(Furtado, 2007, p. 242)*

Although it initially appears that collaboration is merely an activity to be carried out together, the 3C model of collaboration (Ellis et al., 1991; Pimentel & Fuks, 2011, p. 25) emerges as the approach most aligned with the desired activity in an interdisciplinary context. This model highlights three fundamental dimensions that precede effective collaboration. First, communication plays a crucial role in the initial negotiation phase between participants, during the acceptance of responsibilities in a project. In the VAI project, this communication was fundamental throughout the process, as the challenges encountered during development were only resolved thanks to the fluid communication that occurred both through digital media and in-person interactions. Then, coordination based on the 3C model was presented through decisions made together with the coordinator. In practical terms, the overall coordination of the VAI project was led by the mechanical engineer, responsible for coordinating all stages of development. This approach is comparable to the role of a conductor who knows the individual skills of each instrumentalist, or in this case, each specialist involved. Finally, cooperation manifests itself when different professionals operate in a vaccination coverage of at least 95% of the target population for these vaccines becomes urgent (Peres et al., 2022). The VAI project could be a solution for this vaccination coverage if families were convinced to protect children from infectious diseases. Examples like this one in Brazil are spreading around the world. It is up to public bodies and health professionals to make coordinated efforts to convince people of the need for the vaccine, jointly following the guidelines proposed by the coordination. In the VAI project, cooperation occurred naturally, as professionals were enthusiastic. Although the 3C model offers effective guidance for collaboration, there is another approach that can further contribute to the success of an interdisciplinary collaborative project.

This is the concept of disciplined collaboration, in which leaders play a fundamental role. From the beginning, it is imperative that they make decisions about when to collaborate and when not to collaborate (Hansen, 2009, p. 17). From this perspective, leaders must evaluate collaboration opportunities, identify possible barriers, and devise solutions to overcome these challenges. In the VAI project, disciplined collaboration was organized based on the project stage. In this sense, the first phase of project idealization included the effective participation of engineers and a designer. Meanwhile, other stakeholders were busy with other individual tasks. In the subsequent stage, suppliers and engineers naturally worked together. In the final phase, everyone collaborated in a coordinated way, respecting the appropriate moments of interaction. As you can see, in this preliminary stage of the project presented, collaborative concepts proved to be appropriate. For the subsequent stage, which includes testing with people and vaccines in addition to implementation for society, collaborative solutions are expected to continue. Such solutions may occur in different ways due to the peculiarities of the stakeholders. In any case, collaboration is an essential pillar for the success of a project of this complexity.

3.2.4 Service Design

Service design in an interdisciplinary collaborative project plays a relevant role, as its mission is to research and understand how users can be well served in implementing the proposed solution. To do so, it is necessary to understand its concept as a theory and professional practice. In this sense, Stickdorn and Schneider (2010, p. 31) propose the following reflection: service design is an interdisciplinary practice that combines different methods and tools from diverse disciplines. It is a constantly evolving approach. Other authors (Moritz, 2005; Brown, 2008; Lockwood, 2010; Saco & Gonçalves, 2008; Kalbach, 2016; Gonçalves & Santos, 2021) expand this perception:

> Service Design helps to innovate (create new) or improve (existing) services to make them more useful, usable, desirable for clients and efficient as well as effective for organizations. It´s a new, holistic, multidisciplinary, integrative field.
>
> *(Moritz, 2005, p. 4)*

Although Moritz (2005) presents a more comprehensive view, Brown (2008) proposes a human-centered service design process, consisting of three development phases: inspiration, ideation, and implementation. In contrast, Saco and Gonçalves (2008) argue that the service designer must be prepared to deal with conflicting objectives, such as the visual aesthetics of an artifact, sustainability, customer satisfaction, the feeling of accomplishment and problem solving. Kalbach (2016) reminds us that we live in a service-based economy and that is why it is a growing field in a process of continuous improvement. This author further states that "The goal of service design is to take deliberate actions that create, deliver and sustain service experiences positive over time in a consistent and repeated manner" (Kalbach, 2016, p. 227). At the same time, the service needs to be seen as a whole, combining different parts into a system in an integrated, interconnected, and harmonious way (Gonçalves & Santos, 2021). The sum of the perspectives of the highlighted authors emphasizes the relevant role of the service designer as a facilitator of ideas, capable of leading to an ideal solution for a project, always subject to user validation. In this sense, the role of the service designer in the VAI project is fundamental due to the following factors: to ensure a good user experience throughout the process that begins with communication and ends with the vaccine being administered in the cabin by a robot; through ergonomic and adaptability analysis targeting different user profiles; for the continuous evaluation of the vaccination process after its implementation. In addition to these factors, the development of a blueprint with the analysis of stakeholders' points of contact at each stage of project implementation shows the effectiveness of the service designer's contribution to a project of this nature.

3.2.5 Strategic Design

Strategic design plays a fundamental role in complex projects, that is, in projects involving different levels of action, different types of stakeholders and interests. According to Mozota et al. (2011), strategic design must take into account issues inherent to the project itself and issues relating to the external environment. The same authors present the understanding that the designer responsible for activities related to strategic design must integrate the needs of each phase of implementation. In this sense, according to Zurlo (2010), the following should be outlined: priorities; the different facets of design such as graphics, products, services, and interiors; the budget; communication and its interfaces and, also, the intersection with other activities that complement the project. This understanding resonates with the perception of Celaschi and Deserti (2007) who summarize the function of strategic design in its relevance in the process that can lead to innovation, new conceptual solutions, and the construction of desirable scenarios. In operational terms, Zurlo (2010) suggests that the designer acts with his ability to see, predict and do. Thus, seeing is related to the in-depth understanding of reality that occurs through different research analyses. Predicting is related to monitoring, experimentation, and interpretation of acquired data. Doing is related to the moment of carrying out and implementing solutions.

In the VAI project, strategic design can ensure that the project continuity steps occur effectively. The steps are communication and acceptance for participation by institutional stakeholders, real testing with vaccines and users, collaboration between multidisciplinary teams, improvement of the artifact, implementation test with the final public, general communication for the application of mass vaccination and, finally, collecting the results.

3.3 METHOD

The authors chose action research and design science methods to develop this project. The reason for these choices is that, in action research, researchers participated actively and directly with the object of study. This is a phenomenological investigation of an exploratory nature. This method was also chosen because it adopts a collaborative, critical, and strategic approach, a necessary condition for the development of a complex project such as VAI. On the other hand, design science was chosen because it is a method focused on the premises of design that is primarily characterized by the development and evaluation of the effectiveness of an artifact that seeks to solve a problem. In the project presented in this article, the artifact is represented by the kiosk containing automatic vaccination booths. Therefore, the data collection protocol for this research followed the following guidelines: secondary data research on the need for mass vaccination, interviews with some health professionals and brief bibliographic review on the topics: mass vaccination, scientific research that supports the relevance of the VAI Project, interdisciplinary collaboration, service design and strategic design. These themes were chosen to support and guarantee the need for development, the quality, and effectiveness of the project and its implementation. After the initial data collection, the roles of each stakeholder and their responsibilities in the project were defined: definition of the research problem; action planning; sketch of the artifact using specific software; validation of the artifact through simulation in the virtual environment; construction of the artifact prototype; partial prototype testing and definition of subsequent actions and stakeholders necessary to implement the project.

3.4 THE VAI PROJECT

Aiming to better understand how the VAI project was developed, we present topics to 4.4 with their respective names:

3.4.1 Team

The project was developed from June 2020 to December 2021 by the following team:

- 1 Financial manager – Administrator – Function of planning and controlling financial aspects directly linked to the project, supporting demands arising from investors and development partners;
- 1 accounting office responsible for accounting, fiscal and tax demands in a transparent manner and support to managers for the good development of the project, aiming at partnerships and investments;
- 1 project leader – Mechanical engineer – Function of project creation, coordination, development of suppliers and technical partnerships, support to stakeholders for technical demands, carrying out tests on prototypes and final implementation;

- 1 technical project development team – 1 Electrical Engineer, 1 Mechanical Engineer and 1 Industrial Automation Engineer;
- 1 service designer – coordination, articulation of stakeholders, maintaining focus on the user in all stages of development and future implementation, mapping the interaction of the artifact with the user, planning;
- 1 project consultant – Administrator – role in opening investment channels and international expansion.

3.4.2 General Project Detail

The general details of the project are shown in Table 3.1.

Work coordination was conducted under the guidance of the lead mechanical engineer. At the same time, the planning was prepared based on the results of regular meetings with the entire team. During the decision-making process, documentation was digitally archived and shared among team members. The project was designed with an emphasis on sustainability in social, environmental, and economic aspects. When necessary, it may be adapted to the cultural and institutional aspects also foreseen in the broader conception of the term sustainability.

3.4.3 Technical Details of the Project

From a technical point of view, the VAI project was designed to meet demands in different situations. The most favorable situation would be to install only vaccination booths instead of a complete kiosk in a closed environment protected from adverse weather conditions. The least favorable situation would be to install kiosks with vaccination booths in open places such as parks, parking lots, and sports fields. Under these conditions, the kiosks were designed to be dismantled and transportable by truck, as long as they were produced in the same region. In this sense, the aim is to develop local suppliers to supply parts, robots, and equipment and assemble the kiosks only with the assistance of the team that created and holds the patents. This way, the kiosks would have a lower cost and could be installed more quickly. Table 3.2 presents the general technical details of the project.

TABLE 3.1
General Information About the Project

Item	Description
Prototype runtime	2,000 hours
Professionals directly involved	6
Partner company	robot manufacturing company,
Partner company	syringe and needle handling machine design
Patents filed in Brazil	2
Approximate cost of the project in 2021	US$110,000
Visual identity of the project	Registered trademark in Brazil

Source: The authors (2021).

TABLE 3.2
Technical Aspects of the Project

Item	Description
Artifact tape	Vaccination booths installed in a demountable kiosk Metal
Material kiosk	Structures and reforestation wood panels with anti-bacterial finish, easy to clean and flame-retardant treatment. All parts are modular
Coverage	Sandwich-type metal tiles with acoustic and sound protection
Floor	Reforestation wood with vinyl finish, easy to clean and non-slippery
Energy sustainable	Use of solar panels to capture energy during the day and supply this energy for the entire period of use of the vaccination booths in the kiosk
Sustainability environmental	The kiosk is assembled, disassembled and transported whenever necessary, allowing it to be reused in different locations
Sustainability social	Modular vaccination booths can be adjusted to meet the demands of children, PWDs and the elderly population. Collaborative robot (Cobot), sensors, vision system, control
Equipment internal	Software, air conditioning; cooler kept at the ideal temperature for supplying vaccines; dispenser for vaccines, syringes, needles, cotton and disposal materials
Equipment external	A panel (tablet) for patient identification, a comfortable chair, a wall with a hole where the vaccine will be administered, an entertainment panel while the patient waits for the robot procedure
Colors	Predominantly green with stickers on the outside showing the VAI brands and sponsors' brands

Source: Axis Robotics (2024).

The vaccination system offered through kiosks or booths installed indoors works as follows: the patient is received by a nursing technician who directs him to one of the vaccination booths. Each kiosk is modular and can have from two to eight booths installed, depending on the space and local demands. On the outside of the cabin, there is an identification panel. In this panel, the patient enters their documentary data and, based on the database connected to the local health system, the patient is notified about all the information related to the vaccine that will be administered to them. From there, the patient sits in the chair, leaning their arm against the wall with a hole where the vaccine will be administered (on the right or left side) and waits motionless for the robot to carry out the application procedure. The sensor system will identify whether the patient made any movement when applying the vaccine. If this occurs, the robot deactivates the process, waits again for the patient to remain still before making the second attempt. Figure 3.1 shows the process in summary.

3.4.4 Visual Presentation-of-the-Project

Vaccination booths, when installed indoors, will comply with the guidelines given by the site managers, such as hospitals, clinics, clubs, bus stations, train stations, and airports. However, the booths installed in external kiosks will have the visual configuration as shown in Figure 3.2.

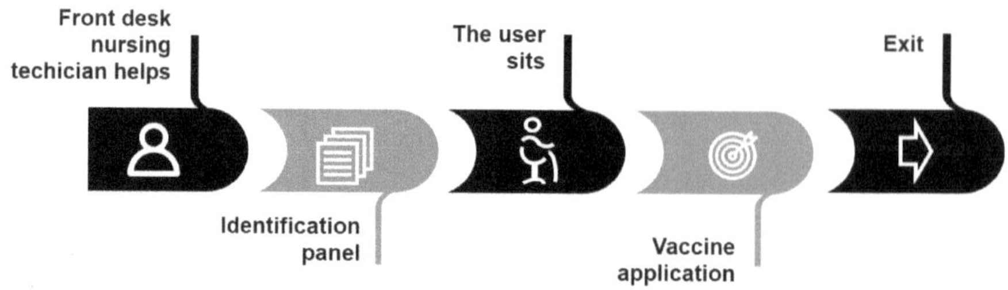

FIGURE 3.1 Schematic flow of vaccination. (The authors, 2024.)

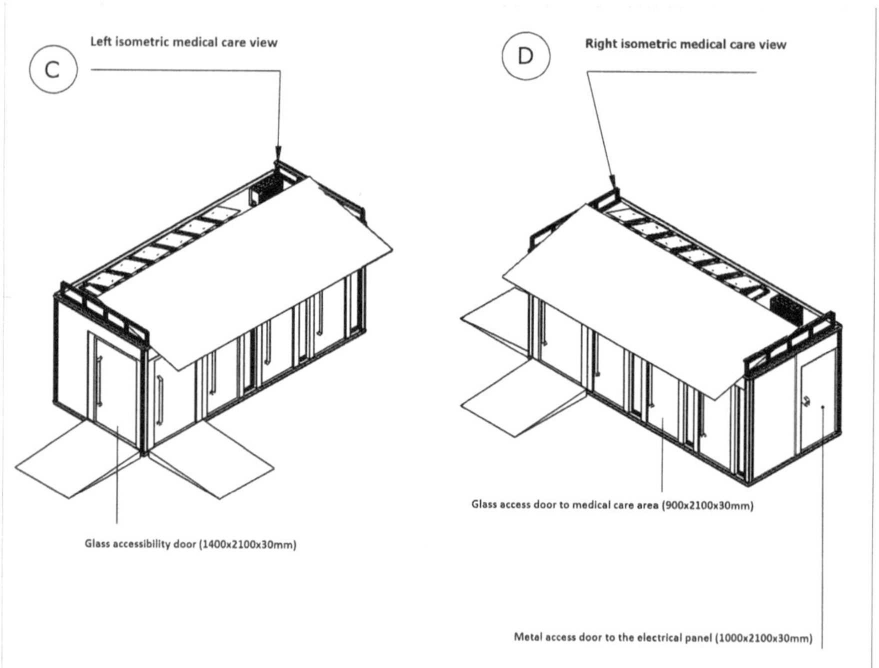

FIGURE 3.2 VAI vaccination kiosk. (The authors, 2024.)

Figure 3.2 shows a sketch of the kiosk with two vaccination stations for patients without special needs and a specific station for wheelchair users. Additionally, the vaccination stations are fully adjustable to meet other patient profiles with special needs, such as individuals of short stature, people with obesity, children, and amputees.

In addition to the adaptation features mentioned, there is the ability to customize the vaccination method by modifying the automation system. In this way, vaccination stations are able to administer vaccines via drops, inhalation, or any other method that may be developed in the future.

3.4.5 Blueprint of User Information during the Vaccination Process

Wide publicity to the population about the possibility of automated vaccination available in public and private places, identifying the path on site to access the kiosks and identifying the kiosks, among other factors, form the user's points of contact with the mass vaccination system. Such factors are part of the scope of collaboration between the service designer and the strategic designer on the project. Such interactions are presented in Table 3.3 through a blueprint.

Table 3.3 summarizes the phases of implementing mass vaccination for society; however, the actions that precede this delivery must be meticulously planned using the tools and concepts of strategic design with different audiences and in accordance with their respective demands.

3.4.6 Main Stakeholders Necessary for the Success of Project Implementation

Although the VAI project is in the development phase, it will only become viable if public and private institutions work collaboratively to promote mass vaccination in an automated way. Therefore, Table 3.4 presents the main stakeholders and the fundamental role of each one in this implementation.

Table 3.4 indicates the possibility of action by the stakeholders, but each country, state, or city will be able to define which partnerships are necessary to serve the population. Probably in countries with fewer public resources, partnerships can be made with other types of institutions such as NGOs, religious organizations and foundations focused on the social and health aspects of the population.

3.4.7 Continuity of the VAI Project

The VAI project was developed up to the pre-testing phase. The final phase of project validation and adjustments, including testing with the population, did not occur due to lack of investment. Because of this, the development team decided to create a business plan to present to potential investors. This plan's main issues are initial partnership with companies to continue the project or investment of 200 thousand dollars; second, the project could be sold or manufactured in partnership with a metallurgical industry and robotics. In terms of commercialization, the kiosk can be rented to companies to meet specific demands or sold to public and private companies that have constant demand for vaccination. Furthermore, once definitions of investments and future partnerships are defined, strategic design will be used to ensure successful implementation. The demands of each new stakeholder who becomes part of the project will be taken into account.

3.5 RESULTS

The results achieved by the VAI project are evidenced by the high level of interdisciplinary collaboration between the professionals involved and between partner companies producing equipment. At the same time, compliance with the individual

TABLE 3.3
Blueprint of User Interaction with the Automated Vaccination System

User Contact Points	1st Stage Responsible	2nd Stage Responsible	3rd Stage Responsible	4th Stage Responsible	5th Stage Responsible	6th Stage Responsible
Access and search for information at media – vaccine app, posters, TV, radio website, social media, influencers	In collaboration with the stakeholders involved					
Online scheduling in the system connecting to the health system or in health centers, hospitals, pharmacies, fire stations, and other public places. Also, at the vaccine application kiosk		In collaboration with the stakeholders involved				
Creation of on-site wayfinding to easily find vaccination kiosks, clearly visible with adhesive			Collaborative project team in partnership with local managers			
1st contact with the kiosk – service provided by a nursing technician. The aim is to provide security and reliability.				Health professional linked to and trained by public health bodies		

(Continued)

TABLE 3.3 (*Continued*)
Blueprint of User Interaction with the Automated Vaccination System

User Contact Points	1st Stage Responsible	2nd Stage Responsible	3rd Stage Responsible	4th Stage Responsible	5th Stage Responsible	6th Stage Responsible
The nursing technician assists the user who enters documentary data into the system through a service totem. The patient is informed by the system about which vaccine will be administered. If the user agrees, they confirm and go to the service station.					Nursing technician trained and users	
At the service station, the patient sits, positions themselves with the arm where the vaccine will be given and remains motionless. In the meanwhile, the user can watch an informative video about the importance of immunization or another entertaining video. At the same time, the robot prepares the vaccine and then the patient receives it.						The nursing technician, the user and the entire team are responsive, for the smooth functioning of the project

Source: The authors (2024).

TABLE 3.4
Main Stakeholders for the Project Implementation

Stakeholder	Necessary Action
Government agencies health (local and federal)	Validate the project with support from specialized professionals during the testing phase. Enable the connection of user data collected at the vaccination kiosk with the public health system. Publicize the availability of kiosks. Provide nursing technicians to work in the kiosks. Be responsible for the correct disposal of infected materials after vaccination. Ensure the safety of the location where the kiosks are installed.
Public vaccination sites – bus and metro stations, airports, health centers and hospitals	Guarantee of access and adequate space for assembling the kiosk or providing internal space on the premises of these locations for administering the vaccine.
Industries	These industries employ hundreds of professionals and therefore can provide an internal or external space to offer vaccination to their employees. They may also be responsible for hiring a nursing assistant to provide services.
Private hospitals and healthcare service companies	These companies already operate in the health sector and, therefore, are committed to maintaining health through vaccination. They may provide space for mass vaccination and health professionals for monitoring.
Public and private communication companies	As this is a topic of public interest, communication companies must commit to publicizing and educating the population about the need for vaccination and the possibility of accessing the vaccine quickly and conveniently through VAI.

Source: The authors (2024).

roles of each professional and leader contributed to speeding up the development of the artifact. Unfortunately, the project cannot count on the support of a health professional during the development period, because, in the same period, these professionals were meeting the demands of the pandemic.

However, the partial implementation of the artifact included a pre-test that was designed only to test the robot's effectiveness in accessing the correct location in the cabin in which the patient's arm was positioned. At the same time, the pre-test successfully simulated the entire operation of capturing the syringe, the needle, discarding the needle cover, and sucking out a liquid that simulated being the vaccine at the right dose. For safety reasons, no vaccine was administered to any user in this pre-test, as the ideal simulation depended on the support of a doctor. In this pre-test,

20 volunteer users simulated the initial identification process. They then sat in a chair facing a monitor with an entertainment function and waited for the vaccine, positioning their arm against the panel that contains an opening through which the vaccine is administered by the robot. This vaccination process took place without problems or delays, demonstrating the minimal desired efficiency of the automated structure. It took each user 3 minutes from identification to vaccination. However, from the point of view of human behavior, four people expressed concerns about administering the vaccine by a robot, while another 16 considered the approach safer. Additional tests were conducted with 10 users who moved purposefully during the vaccine application. The aim was to test the security of the automated process. The result confirmed that the robot only started the procedure when the patient remained still. The total period of this automated structure safety test, measured from user identification to vaccine reapplication, took 5 minutes.

Within the scope of institutional partnerships, attempts were made with mixed companies (public–private), private companies and an innovation center. In all cases, potential partners demonstrated interest in the project, making the partnership's implementation conditional on approval by the Brazilian Ministry of Health. Regarding investors, there was dissatisfaction, as they did not show specific interest in the project, but rather in the acquisition of the development company, damaging the negotiations and making it impossible to continue negotiations for the VAI project.

3.6 DISCUSSION

The VAI project is characterized by a complex artifact that involves robotics and automation. It was initially designed to meet the emergency need arising from the global health scenario that occurred in the period between 2020 and 2022. The pandemic caused by the COVID-19 virus and its variants introduced the world to a health crisis of a magnitude rarely seen. The distinction between the pandemics of past centuries and the one caused by COVID-19 lies in the fact that the world has experienced an era of rapid technological advances. In the past, responses to pandemics were slow. In the 21st century, it was expected that the vaccine response to the pandemic would be faster. And that really happened, 33 vaccines were developed in the world. Technological developments and research in the health sector have contributed to the development of vaccines. However, other problems have arisen, including the fact that some countries do not have enough health professionals to administer vaccines to the largest number of people at the same time, that is, to promote mass vaccination. According to WHO, more than 15 million people have died from COVID-19. On the other hand, 3 million deaths can be prevented through vaccines. And not just to combat COVID-19. In this sense, the VAI project, developed by a collaborative interdisciplinary team, is aligned with the technological vanguard in the health area that aims to partially solve the demand for mass vaccination. In the artifact called kiosk, each person is identified and vaccinated in approximately 3 minutes. This facilitates better tracking of vaccination coverage, identification of gaps and monitoring of adverse effects. Each kiosk can contain modules of up to four vaccination booths and has only one technical-level healthcare professional to manage the process. This way, four people can be vaccinated at the same time in 3 minutes. The more kiosks that are installed, the more people can be vaccinated. This is especially important in densely populated areas or during large-scale immunization campaigns. Another relevant issue is that these automated vaccination units can be deployed to remote or underserved areas. This enhances access to vaccinations for people who might otherwise have difficulty reaching a healthcare facility.

At the same time, the VAI project not only optimizes the mass vaccination process, but also helps other demands in the healthcare sector, as doctors, nurses and technicians stopped caring for patients with other diseases during the pandemic period to vaccinate the population against COVID-19. This absence of health professionals in hospitals, clinics, and health posts generated health complications in patients with acute and chronic illnesses. There were also countless deaths that could have been prevented. In fact, the child population has stopped being vaccinated against common childhood diseases, such as whooping cough, mumps, and chickenpox. This fact causes a serious public health problem for children around the world to this day. The scenario presented involving the pandemic caused by COVID-19 already denotes the potential contribution of the VAI project to the mass immunization of the world population. At the same time, it's possible to reduce labor costs and the expenses associated with training and maintaining a large team of healthcare workers, especially in resource-limited settings.

Another advantage of adopting an automated vaccination system is the fact that it is possible to ensure social distancing, reducing the need for close human interaction during vaccination.

However, this potential increases when the VAI project starts to be seen as a permanent solution for all types of immunization. In this sense, places such as ports, bus stations, airports, train stations, hospitals, medical clinics and public places in general can have a VAI kiosk permanently installed. Therefore, if there is an adequate quantity of vaccines, both the most developed and the least favored nations have the capacity to offer immunization against any vaccine-preventable disease to their population in a practical, efficient, safe manner and in accordance with the regulations from health authorities. At the same time, there is the possibility of achieving a positive economic and social impact through the widespread implementation of VAI kiosks or booths. It would certainly mitigate most of the health challenges involving the application of vaccines currently found in foreign countries with little infrastructure or in war situations. Potential restrictions on the use of the VAI project are subject to public health policies, locations with extreme negative temperatures, lack of investment in the health sector and other external factors. The complexity of the VAI project indicates limitations to be overcome and opportunities to be achieved. The authors of this article believe that through strategic design, planned implementation can help overcome barriers and achieve innovative, viable, and satisfactory solutions for governments, institutions, and society. We are living in the decade of pandemics; if we can implement innovative projects like VAI, we can mitigate the impact of diseases in the world.

The VAI project has two patents in Brazil and awaits potential public and/or private companies for the final improvement of the project and its consequent implementation.

ACKNOWLEDGMENTS

We express our gratitude to Axis Robotics for granting us the opportunity to prepare this article for an international scientific event. We also want to extend our thanks to VM Projects, which played a fundamental role as a collaborating company in the development of a significant part of the VAI project. Furthermore, we would like to thank UR Robots for providing its robot to carry out the tests, thus contributing to the success of our research.

REFERENCES

Axis Robotics. (2024). Inovação Tecnológica Industrial. Available at: https://www.axisrobotics.com.br (Accessed: 10 May 2024).

Brown, T. (2008). Design thinking. Harvard Business Review. https://hbr.org/2008/06/design-thinking

Celaschi, F., & Deserti, A. (2007). *Design e innovazione: Strumenti e pratiche per la ricerca applicata*. Rome: Carocci Editore.

Destefano, F., Offit, P. A., & Fisher, A. (2018). Vaccine safety. *Plotkin's Vaccines*, *7*, 1584–1600. https://doi.org/10.1016/B978-0-323-35761-6.00082-1

Dinleyici, E. C., Ray, B., Marco, A. P. S., Pierre, V. D., & Flor, M. M. (2020). Vaccines and routine immunization strategies during the COVID-19 pandemic. *Human Vaccines & Immunotherapeutics*, *17*(2), 400–407. https://doi.org/10.1080/21645515.2020.1804776

Ellis, C. A., Gibbs, S. J., & Rein, G. L. (1991). Groupware: Some issues and experiences. *Communications of the ACM*, *34*(1), 39–58. https://doi.org/10.1145/99977.99987

Furtado, J. P. (2007). Reference teams: An institutional arrangement for leveraging collaboration between disciplines and professions. *Interface*, *11*(22), 239–255. https://www.scielo.br/j/icse/a/NMxT747jtM8xfpFsxWshvyt/?lang=pt#

Gonçalves, M. S., & Santos, A. (2021). Service design for social innovation: Reflections on sustainability. *Projética*, *12*(2), 108–133. https://doi.org/10.5433/2236-2207.2021v12n2p108

Hansen, T. M. (2009). *Collaboration: How Leaders avoid the Traps, Build Common Ground, and Reap Big Results*. Boston, MA: Harvard Business Review Press.

Heemann, A., Lima, P. J. V., & Corrêa, J. S. (2010). Bases for applying collaboration in design. *Estudos em Design*, *18*(2), 1338–1349. https://eed.emnuvens.com.br/design/article/view/57

Heymann, D. L., & Aylward, R. B. (2006). Mass vaccination: When and why. In *Vaccination: GlobalAspects—ProgressandObstacles. CurrentTopicsinMicrobiologyandImmunology*, vol. 304, pp. 1–16. https://link.springer.com/chapter/10.1007/3-540-36583-4_1

Kalbach, J. (2016). *Mapping Experiences: A Complete Guide to Creating Value through Journeys, Blueprints, and Diagrams*. Sebastopol, CA: O'Reilly Media.

Lockwood, T. (Ed.). (2010). *Design Thinking: Integrating Innovation, Customer Experience and Brand Value*. New York: Allworth Press.

Machado, B. A. S., Katharine, V. S. H., Larissa, M.S. F., Vinícius, C. P., LuIs, A. B. M., Leone, P. C. S. A., Marcelo, A. M., Roberto, B. (2022). The importance of vaccination in the context of the COVID-19 pandemic: A brief update regarding the use of vaccines. *Vaccines*, *10*(4), 591. https://doi.org/10.3390/vaccines10040591

Montero, G. C., Martinez, F. O., Bravo, C., Carranza, T. D., Trujillo, S. L., Lahoz, G. M. A., Guijarro, G. L., Honduvilla, G. N., Asúnsolo, A., Bujan, J. et al. (2021). An updated review of SARS-CoV-2 vaccines and the importance of effective vaccination programs in pandemic times. *Vaccines*, *9*(5), 433. https://doi.org/10.3390/vaccines9050433

Moritz, S. (2005). *Service Design: Practical Access to an Evolving Field.* Cologne: Köln International School of Design (KISD). https://issuu.com/st_moritz/docs/pa2servicedesign

Mozota, B. B., Klöpsch, C., & Costa, F. C. X. (2011). *Gestão do design: Usando o design para construir valor de marca e inovação corporativa.* Porto Alegre: Bookman.

Newland, M., Durham, D., Asher, J., Treanor, J. J., Seals, J., Donis, R. O. & Johnson, R. A. (2021). Improving pandemic preparedness through better, faster influenza vaccines. *Expert Review of Vaccines.* Published online 1 Mar 2021. https://doi.org/10.1080/14760584.2021.1886931

Peres, G. P., Ferraz, J. G., Matos, A. F. M., & Zöllner, M. S. A. (2022). Cobertura vacinal e o retorno do sarampo no Brasil: Uma análise comparativa. *The Brazilian Journal of Infectious Diseases*, *26*(2). https://doi.org/10.1016/j.bjid.2022.102504

Pimentel, M., & Fuks, H. (Eds.). (2011). *Sistemas colaborativos.* Rio de Janeiro: Campus Elsevier.

Saco, R. M., & Gonçalves, A. P. (2008). Service design: An appraisal. *Design Management Review*, *19*(1), 10–19. https://doi.org/10.1111/j.1948-7169.2008.tb00101.x

Soárez, P. C., Sartori, A. M. C., Santos, A., Itria, A., Novaes, H. M. D., & Martelli, C. M. T. (2012). Contributions from the systematic review of economic evaluations: The case of childhood hepatitis A vaccination in Brazil. *Cadernos de Saúde Pública*, *28*(2), 211–228. https://doi.org/10.1590/S0102-311X2012000200002

Stickdorn, M., & Schneider, J. (2010). *This is Service Design Thinking: Basics, Tools, Cases.* Amsterdam: Bis Publishers.

United Nations Department of Economic and Social Affairs, Population Division. (2024). *World Population Prospects 2024: Summary of Results.* New York: https://population.un.org/wpp/Publications/Files/WPP2024_Summary_of_Results.pdf

Weintraub, R. L., Subramanian, L., Karlage, A., Ahmad, I. & Rosenberg, J. (2021). COVID-19 vaccine to vaccination: Why leaders must invest in delivery strategies now. *Health Affairs*, *40*(1), 33–41. https://doi.org/10.1377/hlthaff.2020.01523

Wickmar, L., Ghio, D. S., Tang, Y. M., Keyworth, C., Stanescu, S., Westbrook, J., Jenkinson, E., Kassianos, P. A., Scanlan, D., Garnett, N. et al. (2021). A rapid systematic review of public responses to health messages encouraging vaccination against infectious diseases in a pandemic or epidemic. *Vaccines*, 9(2), 72. https://doi.org/10.3390/vaccines9020072

Zurlo, F. (2010). Design strategico. In *XXI Secolo, vol. IV, Gli spazi e le arti* (Vol. IV, pp. 503–513). Rome: Enciclopedia Treccani.

4 Integrating Artificial Intelligence in Jewellery Design

Opportunities and Challenges

Livia Tenuta, Susanna Testa, Francesca Antinarelli Freitas, Beatrice Rossato, and Alba Cappellieri

4.1 INTRODUCTION

Significant development in science, industry, urbanisation, and secularisation characterised human history from the mid-18th century to the contemporary. Alongside all the benefits this progress has brought, unsustainable ways of living have emerged (Walker, 2020, pp. 268–294). We have entered an age of polycrisis (United Nations, 2023, p. 3), and awareness of the changes is crucial to offering concrete responses. Among the significant developments, the advent of the digital age caused noteworthy social and cultural changes in short periods (Figoli et al., 2022, p. 27). Nowadays, AI is increasingly being used to solve global problems (Raper et al., 2022, p. 1).

One of the most critical challenges AI is helping to address is sustainability, highlighting its potential to monitor and optimise various practices, such as reducing greenhouse carbon emissions by up to 10% (Das & Jayaraman, 2022, p. 2) or processing large datasets with increased accuracy and efficiency (Zia et al., 2022). Described as "the process of creating intelligent machines that can imitate or exceed human abilities in specific tasks" (Elfar & Dawood, 2023, p. 106), AI is transforming industries such as healthcare, aviation, online services, agriculture, education, gaming, marketing, finance, and design, and it has emerged as a possible and fascinating solution to tackle the sustainable challenge. For example, in the aviation sector, AI is being utilised to minimise climate impact by reducing the formation of contrails. By collecting data, companies are developing predictive maps to identify alternative flight routes for pilots. Additionally, AI enables predictive maintenance, allowing for repairs on aircraft components before they fail, optimising downtime when planes are less active. In the realm of recycling, AI is transforming operations under Extended Producer Responsibility laws. These technologies automate material sorting in recycling facilities, significantly increasing recycling rates. In an online

DOI: 10.1201/9781032719290-5

business, issues like insufficient resources, long waiting times, and the inefficiencies of manual customer service can lead to frequent errors in processing. AI-driven customer service solutions address these challenges by providing uninterrupted support, reducing response times, lowering error rates, and enhancing overall customer satisfaction and purchase intentions. Furthermore, AI supports evidence-based decision-making processes. It enables the creation of generalisable predictive systems that can help mitigate pandemic risks at the national level. The application of AI in these sectors demonstrates that an increased utilisation of this technology leads to the digitalisation and optimisation of tasks and resources.

The fashion system is not exempt from the fascination of AI, which is increasingly adopted as a helpful tool to enhance sustainability efforts in supply chain management, creative design, sales and promotion, waste control, and data analysis (Ramos et al., 2023). This digital transformation, widely demonstrated in the previously mentioned sectors, is facilitated through AI-driven processes, which have been effectively integrated into fashion design practices as well (Pournader et al., 2021).

Indeed, the design process in the fashion sector leads to sustainable challenges that could be solved by integrating AI-driven tools. Among the main challenges, the most significant consist of enriching the ideation and development of creative concepts in line with the target and customer preferences (Harreis et al., 2023), which would not result in unsold products, the overall reduction of the production encouraging high-quality projects than fast designs, the use of sustainable materials which can be recycled and used for future projects, the identification of product's defects before the production stage, and the reduction of the vast amount of energy and materials used for completing tasks.

This chapter is divided into five main sections: the literature review, the methodology, the analysis, the discussion, and the conclusions, and it aims to highlight opportunities and limitations of using AI in the jewellery design process and to explore strategies that prioritise a post-growth economy with modern, resource-efficient, smart, and inclusive principles. It focuses on enhancing creativity, sustainability, and education, promoting a new digital scenario that presents a new role for the designer. Before delving into the topic, we would like to acknowledge that this chapter is an extended version of the article presented in the proceedings of the Design Commit '24 conference.

4.2 LITERATURE REVIEW

AI plays a crucial role in advancing sustainable development within the fashion industry (Ramos et al., 2023), offering innovative solutions to address sustainability challenges throughout the design process. According to Ramos et al. (2023), AI is increasingly being adopted by jewellery designers to improve various aspects of the design workflow, such as customisation, prototyping, and material optimisation. These AI-driven tools enable designers to create highly personalised pieces tailored to individual preferences, contributing to more efficient and sustainable design practices.

At the early design stages, the quantity and diversity of conceptual design ideas are essential for designers to explore new design opportunities (Zhu & Luo, 2022). A specific area of the design field, represented by the generative design method, is helping designers generate ideas faster and more efficiently. Generative design is a computer-aided design technique and software category that uses AI to optimise the design process (Chen et al., 2023, p. 1). Therefore, the symbiotic human–computer relationship behind generative design can become an important tool to solve complex sustainable problems, e.g., related to the amount of energy used or unsold productions. Consequently, considering the significant innovative contributions and applications in the design process, AI Generated Content (AGC) has received tremendous attention within the past few years (Ramesh et al., 2021). Thanks to the emergence and advancement of large foundation models, which enable the rapid development of domain-specific models, AGCs are having great success in the creative industry and can produce high-quality content quickly. They are commonly employed to produce various types of content, including images, texts, audio, and video, that can be efficiently integrated into the design process. These models are trained to generate digital images from text descriptions using a dataset of text–images and to create more complex and realistic images to support the creative design process (Chen et al., 2023, p. 1).

In the context of AGC, the role of the prompt designer emerges as a possible key player in this iterative process. These creatives have the role of formulating textual prompts and inputting images that guide AI toward results that are as specific and timely as possible, incorporating style, tone, and creativity (Rossato et al., 2024).

In particular, in the jewellery world, designers have started to study how these technologies can extend their capabilities in their design process, enhancing the creative process through a more sustainable approach. AGC and AI-driven parametric tools allow jewellery designers to generate innovative designs faster, propose customised solutions to clients that fit better with the target, and make the design process more efficient and digitised. Therefore, this reshaping of the industry by integrating these strategies in the design process allows jewellery designers to explore new possibilities, create innovative designs, and push the boundaries of creativity more sustainably. Despite its popularity, AI and the advent of Artificial Intelligence-Generated Content (AIGC) have raised many concerns regarding the environmental impacts caused by more efficient and, consequently, faster production. Also, from a social perspective, issues that arise would be privacy, bias, toxicity, misinformation, intellectual property (IP), and potential misuse of technology (Chen et al., 2023, p. 2).

Within the realm of AI implementation for a sustainable development in the jewellery industry, a team of researchers at Microsoft is working to revolutionise the jewellery design process by reshaping conventional approaches within the industry. They have employed machine learning to develop an AI system designed to automate the design workflow. By instructing the AI to generate a diverse array of designs, they aim to enhance the creative possibilities available to jewellery designers, aligning with the broader trend of integrating advanced technologies to foster innovation and sustainability in design (Bubeck et al., 2023, p. 13).

4.3 METHODOLOGY

Given the complexity of the design process involving AI implementation, we decided to apply a methodology comprising desk research focused on three key stages: research, development, and communication. This approach aims to identify the opportunities and challenges associated with using AI in the jewellery design process. Information was gathered from websites, jewellery studies, scientific papers, international journals, and books.

We have decided to start from the initial stage of the design practice because AI's impact on the process outputs increases if applied in the early phases, considering that the most polarising design decisions are made here (Figoli et al., 2022a, p. 9). The research phase in the jewellery design process consists of three different main actions:

- the collection of data on the target audience (primarily relied on demographic data such as age group, gender, and location)
- the collection of data on customer preferences (using historical data, artistic influence, social media, and economic indicators)
- the collection of data on global trends

The research phase transitions into the ideation stage, where we explored how a mood board featuring relevant images and keywords can be created using AI-driven programs. We identified the case study "Human vs. Midjourney: What Happened When We Used AI to Create Mood Boards" by the Big Human team, which utilises Midjourney, a generative AI program, to enhance this process.

For the development phase of the jewellery design process, the following steps were considered: sketching and prototyping the jewellery pieces, which can be created either digitally or physically. The potential applications of AI in sketching and prototyping were examined, including the use of text-to-image synthetic models, which can convert text prompts into images. Additionally, we reviewed some AI jewellery generators commonly used in this sector, such as StarryAI, utilised by companies and designers. For the prototyping stage, we considered the tangible representation of the jewellery design, including 3D printing and other AI-driven prototyping technologies, as well as their sustainability impacts. We also included case studies of companies implementing AI strategies in the prototyping of their products, highlighting some of the benefits of these implementations.

Lastly, we analysed the communication phase of the jewellery pieces, focusing on how designers' strategies and AI-driven digital techniques can more effectively promote and communicate their products to end clients.

4.4 ANALYSIS

Starting from the research stage, which can be considered the earliest phase of the research process, companies and designers gather and store consumer information that helps to analyse, model, personalise, predict, and influence their potential customers' behaviour in order to gain a competitive business advantage (Gkikas & Theodoridis, 2022, p. 148). In the jewellery industry, AI is especially used for target audience analysis, customer preferences analysis (Tenuta et al., 2023, p. 2), and trend analysis. By researching customer preferences or future trends at the beginning of the process, AI algorithms enable designers to collect data more efficiently and tailor their design concepts to specific target audiences. Moreover, AI can identify real-time impact and can process vast amounts of data from various sources, such as historical data, artistic influences, social media, economic indicators, scientific research, and articles. Predicting new global trends is crucial for both companies and designers.

The importance of the research phase and the collection of data before the ideation of specific concepts in the design process has been demonstrated in a study conducted by Salesforce that surveyed 14,300 consumers and business buyers globally; personalised experiences are considered a critical factor in the purchasing decisions of 84% of shoppers. The research found that 80% of customers now think the experience a company provides to be as important as its products and services (Salesforce, 2023).

By utilising a fusion of Design AI and Computer-Aided Design (CAD) software, the jewellery industry seeks to meet consumers' expectations, offering personalised experiences. Companies and designers use AI technology to empower customer data, allowing customers to actively participate in the design development and eventually co-create their jewellery pieces. By 2024, the total amount of data created, captured, copied, and consumed worldwide is forecast to increase rapidly (Gkikas & Theodoridis, 2022, p. 149).

Today, companies are approaching AI data-driven methods for the research process. For instance, Gucci uses AI to examine consumer trends and identify potential product prospects. Their AI-integrated mobile application furnishes sales personnel with insights into product specifics and customer buying patterns.

The research stage is followed by the development stage, which includes the sketching and prototyping of the jewel. The ideation step begins with creating a mood board that combines evocative images and keywords. Designers collect ideas, images, materials, inspirations, and additional creative design elements to create an inspirational board. Typically, creating a mood board is still largely conventional and can be defined as time-consuming since producing colour and style variations requires a long time. Generative design and AI can quickly generate a wide range of design concepts, allowing designers to select and proceed with the AGC they consider more feasible, as in the case of "Human vs Midjourney: What Happened When We Used AI to Create Mood Boards" by Big Human team (Big Human, 2023). Their work involves designing products to help technology and humanity evolve responsibly.

To understand if AI can positively affect the creation of creative concepts in their design process, they experimented with Midjourney, an AI image generator. The process consisted of four steps: creating mood boards following traditional visual research methodology – the keywords chosen were Efficient, Techno Magical, Infinite Boundless, and Augmented; using the research process to define a new perspective and creating three distinct territories in line with the brand values – the territories were Ethereal Science, Prism Glitch and Executable; replicating the visual directions entering prompts on Midjourney and Dall-E related to their fundamental brand values; and compiling the various images they received back from the software into the same three visual territories (Big Human, 2023).

From this investigation, the Big Human team realised that, when using Midjourney to create mood boards, they spent more time finessing the prompt than looking at the visuals. The prompt engineering took their design features out of the creative process, giving a more mechanical appearance. Also, they realised that originality is not part of AI's nature. Midjourney can only show things that already exist. Instead, humans can develop and generate new visions. In conclusion, they felt limited and uninspired. They chose to turn to Are.na to develop their mood boards, an online social networking community and creative research platform. It led them to unexplored territories and new scenarios that they had never experienced before. Finding connections through other people's creative thinking brings more innovative designs (Big Human, 2023).

As part of the development phase, the digital or physical representation of a jewel may involve the use of an AI image generator platform. For instance, StarryAI supports designers in jewellery creation with generative AI and presents a five-step guide from the concept to the creation phase. The guide starts with suggesting to designers to choose the correct design software between generative design software, AI-powered design tools such as Adobe Illustrator or AI art generators, to set the design parameters, to let AI work and refine the range of design concepts generated and create a prototype (Starryai, 2023). This process allows the designer to refine concepts, visualise the designs better, and make any modifications before the final production. From a design perspective, it is interesting to notice that here, the designer's need consists in setting the parameters, which opens the discussion about the AI still requiring relevant human input.

For the sustainability aspect of the prototype stage, we have examined the use of recycled metals in the AI-driven process of 3D printing for physical prototypes. For instance, companies from the jewellery sector, such as Tiffany & Co. and Boucheron, are using AI to design personalised jewellery through 3D metal printing, creating distinctive designs and prototypes. In particular, Nick Koss, the founder of Volund Jewelry, stated that computers have a unique ability to recreate natural patterns and that humans play a critical role in how machines generate those patterns. Here again, the need for the designer to support the technology is reiterated. Another example is the company Imaginarium, a leader in 3D printing and advanced manufacturing technology, which houses the most extensive setup of 3D printers (Imaginarium, 2023). It offers over 140 materials, including recycled metals for 3D printing, and 1600 designs brought to life daily. The company works in the jewellery and healthcare industries, creating a variety of jewellery pieces.

Some benefits of using AI in the development phase of the jewellery design process are cost-effectiveness, material optimisation, and quality enhancement, but they also come with challenges. AI can predict the cost of a specific design based on the materials involved and the type of manufacturing required for the production. It helps designers create more economically efficient designs without compromising quality. Thus, AI can also help designers optimise the use of materials, especially precious materials, including precious metals such as gold and platinum, and precious gemstones such as diamonds, emeralds, sapphires, and rubies. Therefore, through the analysis of specific designs, AI tools can propose methods to reduce the usage of gold or diamonds without influencing the aesthetics or quality of the piece of jewellery.

Unfortunately, AI lacks when it comes to gemstone selection. AI algorithms cannot select gemstones depending on their quality, colour tone, or dimensions. AI would show the designer the available material list of stones, considering any stone in any size to fulfil the design requests. However, there could be no availability, and they could not exist in nature. This means that designers or gemmologists still need to source available stones in the colour and shape needed. Then, AI offers a superior alternative in terms of quality enhancement for the jewellery pieces, analysing and detecting potential structural weaknesses, and suggesting intelligent improvements.

On the other hand, designing jewels requires precision and technical knowledge, and the supervision of a master designer or goldsmith is necessary during the process. AI can serve as a support, but cannot technically replace human contribution. For instance, AI cannot recognise a jewel's practical wearability. It cannot predict the correct dimension or thickness of a jewel. Lastly, AI cannot select the appropriate materials and techniques to produce the jewel. For instance, while AI algorithms can generate sophisticated and intricate pattern designs that were previously unimaginable, those designs may not effectively translate into jewellery pieces due to the limitations posed by specific manufacturing methods.

The final phase of the design process, the communication phase, incorporates various case studies. The first case study we have analysed is the Italian jewellery brand Pannalù by Susanna Testa. The designer uses AI for the brand's advertising campaigns to enhance creativity and inclusivity, as shown in Figure 4.1. In the discussion on the role of AI in creative fields, the author expresses a belief in the significant potential of AI as a creative tool. AI is seen as an opportunity to experiment and push the boundaries of creativity. One aspect the author finds particularly fascinating about AI is its unpredictability and the element of surprise that emerges in interaction. This unpredictability, she suggests, spontaneously stimulates creativity, sometimes in entirely novel ways. "AI enables the construction of worlds and the shaping of imaginative visions in previously unimaginable ways, all through language and words" (Tancorre, 2023). However, the designer also emphasises the importance of using AI ethically and responsibly. The challenge lies in integrating this technology into the creative process without losing the human essence at the core of art and creativity. She refers to the experiences at Pannalù Paprika Farm, where AI was used not to replace human creativity but to amplify it, to experiment and explore new possibilities.

FIGURE 4.1 AI-driven advertisement campaign by Pannalù.

Second, we have analysed the case study that involves the American jewellery company Brilliant Earth, which allows customers to try on jewellery pieces virtually, create personalised engagement rings, and find matching wedding rings online. In fact, they can find their favourite ring directly from their phone and select the option "Virtual Try On." After uploading a picture of their hand and customising every ring element, they can save and share the design.

Another case study is represented by Maria Tash, who is known for revolutionising the fine jewellery and luxury piercing industries. After more than five years of experimentation, she launched the Tash Studio software. This online platform allows consumers to try different piercing patterns from home. They can virtually simulate their ears with different skin tones, try different compositions, combine different outfits, and be creatively inspired. With the technological advancements of AI, its use expands not only to the sales and retail experience but also to the creative design process (Tenuta et al., 2023, p. 2).

The previous case studies show how integrating AI into the communication phase can benefit designers and companies. AI technology can help designers save time, improve content quality, and boost campaign performances based on data and user preferences. AI can help identify good strategies to improve the user experience and increase customer satisfaction, resulting in a more engaging relationship between the designer and the client.

4.5 DISCUSSION

From the data analysis of the jewellery design process's research, development, and communication phases, it has emerged that employing AI and AI-driven tools leads to opportunities and several challenges. For this study, we have organised and elaborated the results following three main approaches: sustainable, creative, and educative.

From a social and environmentally sustainable perspective, the design process presents some opportunities and limitations. From an environmental point of view, AI integration into the jewellery design process leads to a reduction of waste but also to an increase in the amount of energy required for specific processes, such as generating complex designs. The prototyping phase especially shows the positive impact of reducing waste through the digitalisation of tasks. AI-driven tools can efficiently minimise the materials required for a specific prototype and select suitable materials. For the same reason, in the communication phase, AI-driven tools can help to reduce waste by analysing and establishing the best selection of materials that will be involved in the process of the advertisement campaign. However, using AI in the design process can also have negative impacts. Generally, it involves significant computing power and substantial energy required, often sourced from fossil fuels, resulting in a detrimental environmental impact. Some scholars assessing the energy needed to "train" AI, meaning to operate advanced models of AI that are beneficial for natural language understanding and processing, have observed that it can lead to the emission of 284 metric tons of carbon dioxide equivalent. These emissions are nearly five times the average lifetime emissions of an American car, including its production (CIVS, ONG 2.0, AIx IA, & Università degli Studi di Bari

Aldo Moro, 2021, p. 56). In summary, the discussion between the potential of AI in reducing waste and the consequential increase in energy used is still open. At a global scale, we must consider that, even if AI has demonstrated to be able to reduce waste for each design product process, improving the efficiency, it is more critical to try to reduce the overall number of pieces produced instead of accelerating the process making the production faster and more efficient which is not in line with the concept of sustainability. This touches on a crucial debate in AI, sustainability, and production ethics. It raises a fundamental question about the direction in which technology, particularly AI, should be steering industrial and economic practices. AI's role in increasing production efficiency is undeniable. It can optimise every stage of the manufacturing process, from design to logistics, significantly reducing waste and resource consumption per product unit. These improvements are often celebrated for their potential to make industries more environmentally friendly by lowering their carbon footprint and reducing the waste generated during production. However, this technological efficiency may paradoxically contribute to a cycle of increased consumption and production. This phenomenon is sometimes referred to as the "rebound effect" or "Jevons Paradox," where increases in efficiency lower the per-unit cost of production and can lead to an overall increase in resource consumption. In this case, the efficiency gained through AI could produce and consume more products, which might negate the sustainability benefits through increased total production and consumption. Focusing on responsible production emphasises the need for a shift towards sustainability that is not solely efficiency-oriented but also considers the volume of production and consumption. It suggests a model emphasising producing fewer, higher-quality items that meet needs without encouraging overconsumption. This approach aligns with the circular economy principles, emphasising minimising waste and maximising resources. Reflecting on this, it becomes clear that technology alone cannot solve the complex problems of environmental degradation and resource depletion. There is a need for a holistic approach that combines technological innovation with changes in consumption patterns, business models, and regulatory frameworks to ensure sustainable outcomes. This involves improving things and rethinking what is made and how much, aiming for a balance between meeting human needs, preserving environmental integrity, and ensuring economic viability. This reflection opens broader questions about the role of AI in society and the economy: How can AI be leveraged for efficiency and holistically promoting sustainability? What regulatory or economic incentives could encourage businesses to prioritise quality over quantity and sustainability over short-term efficiency gains? How can consumers be encouraged or educated to value sustainability and quality in their consumption choices? Addressing these questions requires interdisciplinary collaboration involving technologists, environmental scientists, economists, policymakers, and the public to redefine progress and success in an age of technological abundance.

On the other hand, considering the social aspect of sustainability, the use of AI raises ethical issues, including data privacy, IP, and bias. For the data privacy issue, in the data collection for the research phase, it has been demonstrated that large foundation models are known to be vulnerable to privacy risks and attacks. The AIGC can also be subject to privacy leakage. Studies have attributed these privacy attacks' success to duplicated data in commonly used web-scraped training sets. A sequence that appears multiple times in the training data is more likely to be generated than a sequence that occurs only once. This suggests that deduplication could be used as a countermeasure in privacy-sensitive applications (Chen et al., 2023, p. 2). A complete resolution to the privacy issues has yet to be achieved. However, companies and researchers have taken proactive steps to address these issues, such as introducing warning messages and detecting replicated content. AI in the jewellery industry raises broader questions about IP, including concerns about data memorisation, ownership of images, and plagiarism. For instance, Stable Diffusion memorises duplicate images in the training data.

The generated images are simple combinations of the foreground and background objects of the training dataset. The system occasionally displays the ability to reconstruct memories, producing objects that are semantically equivalent to the original without being identical in pixel form. A relevant example of violation of IP in design involves the American artist Mason Rothschild and the French luxury house Hermès. In 2021, Rothschild released a collection of 100 MetaBirkins NFTs depicting fur-covered, patterned 3D handbags, and it was accused of unauthorised use of its revered Birkin trademark. For the same reason, in the jewellery design field, it is essential to reflect on whether an AI model should be trained on images of another designer's work and used to generate new designs without crediting or remunerating the original designer. It is also fundamental to consider whether AIGC models produce unique creative works or replicate content from their training sets. Ideally, AIGC should produce original and distinct outputs. However, the source and IP rights of the training data are often unknown due to uncurated web-scale data (Somepalli et al., 2023, p. 1). The powerful memorisation of large AIGC models poses a risk of reproducing data directly from the training data, potentially raising legal concerns around copyright infringement and ownership.

Another social concern and problematic dataset that arises when collecting data through AI-driven tools is the AI algorithm bias. Most AIGC models rely on text encoders that are trained using large amounts of data from the internet, which may contain social biases and toxicity (Chen et al., 2023, p. 5). Since the training data used in AI models are collected in the real world, they can unintentionally reinforce harmful stereotypes, exclude, or marginalise certain groups, and contain toxic data sources, which can incite hate or violence and offend individuals (Weidinger et al., 2021, p. 15). For instance, images generated with the prompt "Three engineers running on the grassland" using Stable Diffusion v2.1 resulted in all male people, and none of them would belong to the neglected racial minorities, indicating a lack of diversity in the generated images (Chen et al., 2023, p. 4).

This shows a massive bias in using AIGC models in the research phase of any creative process. Another example is Google's Images, which encodes several social biases and stereotypes, such as generating images of people with lighter skin tones and aligning with Western gender stereotypes. These biases can lead to unfair discrimination and harm certain social groups (Chen et al., 2023, p. 3). This demonstrates that designers should ensure the quality and reliability of data, mitigate biases in AI algorithms, and address ethical concerns that stand out as critical considerations for the effective and responsible utilisation of AI.

As regards creativity, AI can expand the designers' horizons and help identify possible areas for their further research, development, and communication phase. Although AI enhances productivity, allowing designers to focus on the creative aspect of the project rather than focusing too long on the early stage of the ideation process, on the other hand, the integration of AI into the jewellery design process shows a lack of creativity. Primarily, AI lacks human qualities like creativity and empathy, often limiting its ability to understand emotions or produce original ideas. Midjourney, one of the generative AI programs primarily used in the design sector, can only show existing ideas and products. Instead, humans can come up with new visions. Also, using AI in the design process shows a potential misuse of this technology, making designers dependent on it and reducing their critical and creative thinking skills. These AI tools often need more spatial understanding capabilities and the inability to follow complex instructions (Gokhale et al., 2022, p. 1). Even though many of the designs generated are usually aesthetically appealing, some designs are not as admirable due to factors like disproportionate placements, lack of harmonic orientations, etc. (Gupta et al., 2018, p. 2), which could lead to an unsold production of jewellery pieces. While AI can be a valuable tool in jewellery design, the designer ultimately must refine and curate the concepts generated by the AI and create a design that meets their specific aesthetics, functionality, and commercial viability. Jewellery design is a profoundly creative human process that cannot be entirely set aside in favour of this emerging technology.

Suppose it is true that in jewellery design, AI offers material optimisation and reduction of waste. In that case, it is also true that today, a product is determined by its material worth and the design value inherent within it. If we are going toward eliminating the design value in favour of the technology, what happens to the design value when a machine accomplishes it? How is it perceived? These open questions must be posed when discussing values and creativity.

Talking about education, AI systems are attracting more and more attention as possible tools to enhance creativity in the design process (Figoli et al., 2022, p. 11), but integrating AI into traditional design processes may be difficult. It requires designers to adapt to a new technological scenario, and education and training in the design sector have become one of the main priorities. This new AI technological approach to the design process can be expensive and require specialised knowledge and resources that designers generally do not possess. Designers need to build experience, prompting AI tools to be able to extract beneficial results from the AI's outputs. They need training programs and workshops to learn these new skills. From a positive perspective, using AI in the jewellery industry can potentially create new job opportunities for designers and artisans. For instance, new design roles may be needed to program and train AI systems to generate specific concepts or work with new materials. Similarly, suppose the AI revolution spurs an increase in customisation or a trend toward uniqueness. More artisans may be needed to bring the broader range of designs to life, using their expertise in traditional techniques such as casting, soldering, and setting. Integrating AI in the design system remains challenging, especially in jewellery, where traditional techniques and craftsmanship are highly valued. In order to better understand AI potential and its applications, at Politecnico di Milano, we are integrating these new AI technologies into our jewellery design teaching methodologies.

4.6 CONCLUSIONS

Designing for the new future implies considering the environmental and social sustainability aspects, creativity, and education included in the 2030 agendas and translating them into the industry. AI applied to the jewellery design field can contribute to the transition towards a more sustainable and resilient future that considers the well-being of people and the planet. The jewellery industry's path is intricately connected to the limitless potential of AI-powered tools, and it is reshaping the way designers imagine and create new products. Designers are imagining a future world in which the only efficient way to move forward is to look back, keeping an eye on the present.

This chapter encourages designers to responsibly integrate AI into their design processes, better exploring this technology and taking advantage of its capabilities. Designers need to understand their potential role in contributing to a positive and meaningful change in the design process, seeking to remove obstacles to sustainable progress. They must consider exploring new ways to utilise this technology to enhance their work, creativity, personalisation, efficiency, and sustainability. The designer's role is to navigate existing elements rather than create something entirely new. This is what AI is accused of, which only automates a process already utilised in design practices: generating innovation by observing the existing.

The writer Jonathan Lethem has said that when people call something "original," nine out of ten times, they just do not know the references or the sources involved (Kleon, 2012, p. 7). Lethem suggests that what we perceive as "original" is often the result of our ignorance about the sources and influences that inspired that work. This viewpoint applies to both human creativity and AI. In the case of humans, our creativity is deeply rooted in our experiences, culture, acquired knowledge, and external influences. Like machine learning models, AI also "learns" from vast datasets that are essentially a compendium of human expressions and knowledge. In both cases, creation is more a process of synthesis and reworking than invention ex nihilo. Speaking of originality, Voltaire (Voltaire, 1756, p.224) already stressed the creative interdependence of great authors and used the terms "imitation" and "borrowing": "Thus almost all is imitation. The idea of the Persian Letters was taken from that of the Turkish Spy. Boyardo imitated Pulci; Ariosto, Boyardo; the most original wits borrow from one another."

If we are shaped by the things we love, as Goethe wrote (Kleon, 2012, p. 11), we can say that AI is shaped by the data, algorithms, and human interactions it receives and analyses. In this context, AI is indeed shaped by the information (data), the learning rules and methods (algorithms), and the interactions it has with human users. These elements influence how AI learns, adapts, and responds to requests.

The fact is that both human and artificial creativity do not operate in a vacuum but rather build on the foundation of what has already been done. This perspective invites a greater awareness of the influences and sources that inspire every creative act, regardless of its origin.

On the one hand, AI represents an exciting future for the design sector and provides innovative design scenarios that redefine and facilitate the design process. Optimisation methods combined with machine learning techniques can take on a fundamental role in generating more efficient and sustainable designs. Combining the human approach and AI capabilities can lead to a more robust and effective sustainable design strategy. On the other hand, some criticalities regarding human-AI collaboration might emerge when introducing non-human agents in the design process and design teams (Figoli et al., 2022, p. 64). AI raises critical concerns, and it is essential to understand its limitations. The primary concerns include the authenticity, privacy, bias, responsible use of sensitive data, and the vast amount of energy AI requires. Essential knowledge regarding the impact of AI systems in the design processes still needs to be included (Cautela et al., 2019, p. 7).

In approaching the future, it is essential to continue experimenting with the capabilities of AI but also address the ethical implications that these new opportunities bring along. We must remember that AI will not replace the designer's role or human creativity. However, it will augment and amplify it, enhancing productivity and allowing them to focus on the more creative aspects of their work. This collaboration sees AI complementing human creativity through algorithmic insights, idea optimisation, and innovative suggestions. The marriage of technological efficiency with the emotional depth of human imagination is at the core of this synergy. It is also crucial to balance human creativity and AI assistance. While AI can quickly generate ideas and concepts, it still requires human input and expertise to turn them into commercially practical and manufacturable pieces. Moreover, the most challenging part of being a successful designer is not sketching ideas but identifying those ideas that, when taken to market, together with a well-executed brand and customer experience, will lead to stunning, lasting, and saleable products.

Despite these social and environmental impacts, AI can still be a powerful tool for a more sustainable future scenario. This chapter aims to actively develop a joint reflection in a transdisciplinary scientific exchange between art and science, which enhances innovation through Design and Industry for a better future with a strategic vision. Through the dynamic integration of AI, the convergence of sustainability innovation and ethical practices can become possible, leading the jewellery sector toward a more responsible future.

REFERENCES

Big Human. (2023). Human vs. Midjourney: What happened when we used AI to create mood boards. https://www.bighuman.com/blog/midjourney-ai-generated-mood-boards

Bubeck, S., Chandrasekaran, V., Eldan, R., Gehrke, J., Horvitz, E., Kamar, E., … & Zhang, Y. (2023). Sparks of artificial general intelligence: Early experiments with gpt-4. https://arxiv.org/pdf/2303.12712

Cautela, C., Mortati, M., Dell'Era, C. & Gastaldi, L. (2019). The impact of artificial intelligence on design thinking practice: Insights from the ecosystem of startups. *Strategic Design Research Journal*, 12(1), 114–134. https://api.semanticscholar.org/CorpusID:201245155

Chen, C., Fu, J., & Lyu, L. (2023). A pathway towards responsible ai generated content. https://doi.org/10.48550/arXiv.2303.01325

CIVS, ONG 2.0, AIx IA, & Università degli Studi di Bari Aldo Moro. (2021). *L'Intelligenza Artificiale per lo Sviluppo Sostenibile.*

Das, K. P., & Jayaraman, C. (2022). A survey on artificial intelligence for reducing the climate footprint in healthcare. *Energy Nexus*. https://doi.org/10.1016/j.nexus.2022.100167

Elfar, M.A.A. & Dawood, M.E.T. (2023). Using artificial intelligence for enhancing human creativity. *Journal of Art, Design and Music*, 2(2). https://doi.org/10.55554/2785-9649.1017

Figoli, F. A., Rampino, L., & Mattioli, F. (2022a). AI in the design process: Training the human-ai collaboration. https://hdl.handle.net/11311/1222226

Figoli, F. A., Rampino, L., & Mattioli, F. (2022b). *Artificial Intelligence in the Design Process: The Impact on Creativity and Team Collaboration.* Milan: Franco Angeli https://library.oapen.org/handle/20.500.12657/53627

Gkikas, D. C. & Theodoridis, P. K. (2022). *AI in Consumer Behavior. Advances in Artificial Intelligence-Based Technologies.* Cham: Springer. https://www.researchgate.net/publication/355069187

Gokhale, T., Palangi, H., Nushi, B., Vineet, V., Horvitz, E., Kamar, E., … & Yang, Y. (2022). Benchmarking spatial relationships in text-to-image generation. https://doi.org/10.48550/arXiv.2212.10015

Gupta, K., Damani, S., & Narahari, K. N. (2018). *Using AI to Design Stone Jewelry.* https://doi.org/10.48550/arXiv.1811.08759

Harreis, H., Koullias, T., Roberts, R., & Te, K. (2023). *Generative AI: Unlocking the Future of Fashion.* New York: McKinsey & Company.

Imaginarium. (2023). https://imaginarium.io/

Kleon, A. (2012). *Steal Like an Artist: 10 Things Nobody Told You About Being Creative.* New York: Workman Publishing Company.

Pournader, M., Ghaderi, H., Hassanzadegan, A., & Fahimnia, B. (2021). Artificial intelligence applications in supply chain management. *International Journal of Production Economics*, 241, 108250. https://doi.org/10.1016/j.ijpe.2021.108250

Ramesh, A., Pavlov, M., Goh, G., Gray, S., Voss, C., Radford, A., Chen, M., & Sutskever, I. (2021). Zero-shot text-to-image generation. In *International Conference on Machine Learning* (pp. 8821–8831). PMLR, Virtual Event.

Ramos, L., Rivas-Echeverría, F., Pérez, A.G., & Casas, E. (2023). Artificial intelligence and sustainability in the fashion industry: A review from 2010 to 2022. *SN Applied Sciences*, 5, 387. https://doi.org/10.1007/s42452-023-05587-2

Raper, R., Boeddinghaus, J., Coeckelbergh, M., Gross, W., Campigotto, P., & Lincoln, C. (2022). Sustainability budgets: A practical management and governance method for achieving goal 13 of the sustainable development goals for AI development. *Sustainability.* https://doi.org/10.3390/su14074019

Rossato, B., Tenuta, L., Testa, S., & Cappellieri, A. (2024). Beyond the machine: Human-AI collaboration in jewelry design. In *10th International Conference on Higher Education Advances, Valencia*, 18–21 June 2024. https://doi.org/10.4995/HEAd24.2024.17262

Salesforce. (2023). State of the connected customer report. https://www.salesforce.com/resources/research-reports/state-of-the-connected-customer/

Somepalli, G., Singla, V., Goldblum, M., Geiping, J., & Goldstein, T. (2023). Diffusion art or digital forgery? Investigating data replication in diffusion models. In *Proceedings of the IEEE/CVF Conference on Computer Vision and Pattern Recognition* (pp. 6048–6058). https://doi.org/10.48550/arXiv.2212.03860

Starryai. (2023). Designing jewelry with AI: From concept to creation in five step. https://starryai.com/blog/designing-jewelry-with-ai-from-concept-to-creation-in-five-steps

Tancorre, M. (2023). #PuntiLuce// I nuovi gioielli di Pannalù onorano l'imperfezione, Cosmopolitan. https://www.cosmopolitan.com/it/moda/tendenze/a45996226/gioielli-2024-pannalu-intervista-susanna-testa/

Tenuta, L., Testa, S. & Rossato, B. (2023). Immersive innovation: Bridging digital design and virtual realities in jewelry. https://doi.org/10.54941/ahfeXXXX

United Nations (2023). *The Sustainable Development Goals Report 2023: Special Edition.* New York: United Nations.

Voltaire. (1756). Collection complette des Oeuvres de Mr. de Voltaire: Mélanges de Littérature, *d'Histoire et de Philosophie*, 4, 224.

Walker, S. (2020). *Progressive Design Praxis, Design Realities Creativity, Nature and the Human Spirit* (pp. 268–294). Oxford: Routledge. https://www.researchgate.net/publication/340678480

Weidinger, L., Mellor, J., Rauh, M., Griffin, C., Uesato, J., Huang, P. S., … & Gabriel, I. (2021). Ethical and social risks of harm from language models. https://doi.org/10.48550/arXiv.2112.0435

Zhu, Q., & Luo, J. (2022). Generative design ideation: A natural language generation approach. arXiv. https://doi.org/10.48550/arXiv.2204.09658

Zia, A., Aziz, M., Popa, I., Khan, S. A., Hamedani, A. F., & Asif, A. R. (2022). Artificial intelligence-based medical data mining. *Journal of Personalized Medicine*, 12, 1359. https://doi.org/10.3390/jpm12091359

5 Designers for/and/ with Manufacturing

Design-Driven Solutions for an Updated Dialogue between Young Creative Talents and the Textile-Knitwear Industry

Giovanni Maria Conti, Mariana Ciancia, Martina Motta, Katia Goldoni, Francesca Piredda, and Beatrice Zagatto

5.1 MANUFACTURING SMEs AND YOUNG DESIGN TALENTS: TWO POLES TO BE CONNECTED

Today, the textile-knitwear sector is rediscovering itself as a domain in which the culture of 'knowing how', the soul of modern and contemporary work, deals with products capable of describing the *hic et nunc*, mainly using traditional techniques in the framework of a sector that is amid technological updating.

Con-text. A digital ecosystem to connect the textile-knitwear manufacturing system with future design talents was a project proposed by the KnitDesign Lab and Imagis Lab research groups of the Department of Design at Politecnico di Milano to analyze, understand, and return design-driven solutions to connect young creative talents, students, final-year students, and young graduates in knitwear design to the rich manufacturing sector of Made in Italy textile-knitwear.

Italian manufacturing represents excellence worldwide: in the National Recovery and Resilience Plan (NRRP), the Italian government emphasizes the importance of creating, updating, and training specific skills for the innovating Italian industry as part of the relaunch of active labor policies to attack delays and historical bottlenecks in the relationship between education/training and work (European Commission, 2010). In an industrial landscape undergoing continuous renewal, a real need inevitably arises to train new professional figures (Bertola et al., 2021) who are able not only to innovate the processes of the entire knitwear supply chain but also to keep a

DOI: 10.1201/9781032719290-6

watchful eye on the manual and traditional work that still is representative of the sector (Magni & Noè, 2017; Conti et al., 2016). In this context, the production of knitted garments is the most visible point of a branched supply chain that mainly revolves around a few provinces in Piedmont, Veneto, Emilia Romagna, and Tuscany. Over time, each territory has developed its own identity and created a pool of expertise that represents the knowledge wealth of the local area (Traini, 2004; Motta, 2019). Therefore, the new profession of the knitwear designer was born within a historical Made in Italy industrial sector, which has never taken the time to understand the proper skills of people who have always worked in it. Even when looking at them, these people have always been mainly considered in purely operational and technical terms, without realizing that it is precisely within manufacturing that design, production, creativity, and technique coexist. It is precisely this alchemy that, over the years, has made Made in Italy products recognized worldwide for their tangible aesthetics and intrinsic quality (Motta, 2018).

Con-text started from the initial results of the PhD thesis 'Designing Knit Designers' (Motta, 2019), which had drawn an initial profile of knitwear designers and their skills. Another starting point for the project was the research practice experimented with the WTDT project (Ciancia et al., 2023; Ronchi et al., 2020), which conducted an activity for listening to and analyzing companies' needs for activating a dialogue between future professionals and companies in the communication and marketing sectors, identifying the user journey and the main weaknesses of the stakeholders involved. The project also stemmed from observing the relationship dynamics between the activities proposed during the Final Synthesis Studios in Fashion and Communication Design (School of Design, Politecnico di Milano) and the entrepreneurs involved. In recent years, collaborative work with companies has generated many results that have shown the strategic importance of bringing supply and demand closer together (Stolze et al., 2018) as early as during the educational path of future knitwear designers.

5.2 CON-TEXT: PROCESS AND PRELIMINARY OUTPUTS

Con-text used the methodology and tools already codified by the WTDT project (Ciancia et al., 2023; Ronchi et al., 2020), adapting them to the knitwear-textile sector. The methodology included initial desk research followed by a quantitative and qualitative field investigation (data collection, surveys, and interviews).

Desk Research was purely exploratory and aimed to understand the reference context. It has been done online (on websites reporting educational programs, on knitwear and textile companies' web pages) and offline (on scientific publications and sectoral reports). Field research was composed of a quantitative investigation aimed to update the mapping of the stakeholders involved and a qualitative, thorough investigation to understand their needs and better focus on the sector's weaknesses. The results were followed by a more operational phase (Defining), which identified the strategic actions to be implemented in the last phase (Designing/Connecting) and defined the best communication strategy between young people and companies. In this case, the aim was to describe a system of relationships to both create an awareness-raising process for companies, triggering a more vital student–business link, and make the profiles of young designers available to companies.

The aim of the research comprises multiple levels: the goal of the first one was to introduce young talents to the world of manufacturing and the brands that offer job opportunities in various fields, from creative to more technical ones, including those related to consulting, marketing, and communication; the second level aimed at facilitating the induction of young talents into companies and creating the suitable space for them to grow as high-potential profiles with creative and technical skills suited to drive process and product innovation.

The final aim is to identify a praxis – a methodology based on the design thinking approach – that establishes a connection between the university and SMEs, creating new opportunities for all the stakeholders involved.

5.2.1 Research Structure and Methodology

The research has been structured by combining desk and field research, articulated in four work packages (WPs) described below.

WP1: Desk research aimed to identify correspondences among educational programs and outcomes, as well as the organization of the manufacturing system.

The desk research allowed researchers to validate the results of WP1 and identify knowledge gaps and the information that needed further investigation with quantitative surveys and qualitative interviews in *WP2: Field research.*

Quantitative surveys have been spread through emails, owned social media channels (LinkedIn, Facebook, and Instagram), the website page www.context.polimi.it, and newsletters sent by professional associations. Surveys have been filled in by 67 professionals working in knitwear/textile companies and 92 students and new graduates. Surveys validated the data collected with desk research and added relevant information for the qualitative investigation and the overall research development.

Qualitative interviews have been conducted in two consecutive sets, following a reiterative process:

- A first set of explorative interviews was conducted in parallel with the surveys: five professionals and three students/former students/young designers;
- The second set of interviews is based on the results of surveys, together with the results of the first set of interviews: five professionals and five students/former students/young designers.

The activities of *WP3: defining* started with the analysis of the results collected in WP2, allowing us to (1) define the contents and channels most suitable for the stakeholders; (2) define the research actions needed for the solution of pain points; (3) understand the importance of designing a physical event to enhance the exchange between the stakeholders (see WP4); and (4) make first experimental trials of new modalities of involvement of companies into the Knitwear Design program at School of Design, Politecnico di Milano.

Moreover, this WP led to the identification of five core issues as opportunities for intervention:

- **Who is the knit designer?** The detachment in the perception and the difficulty in the definition of who the knit designer is;
- **The space for designers in manufacturing companies** and the value of working experiences and career paths inside them;
- **The skills and competencies of knit designers:** the present, the required, and the ones envisioned for future innovation;
- **The lack of attractiveness of manufacturing companies for young talents is due to multiple factors**, including lack of communication, among others. The first actions for change will, therefore, have to act precisely on communication, with various actors involved: manufacturing companies, professional designers, schools, and brands;
- **The role of education.** What does the training already do, and what can it do to improve?

The last WP, *WP4: Designing & Connecting*, began with a review of the initial plan to foster communication through a digital ecosystem. Research evidence showed that stakeholders needed to be present, share spaces and time, and establish personal, physical connections. Given these findings, we focused on designing a pilot physical event to enhance mutual knowledge between manufacturers and young talents and create an awareness path for companies to ignite the student–business connection. This was made possible through collaboration with Consorzio Promozione Filati – Prato (CPF), which hosted the event on February 24, 2023, during the two days of 'Feel the Green – Behind the scenes'. The result was the talk 'The Centrality of the Designer in Manufacturing', followed by a speed date formula to bring companies and young talents together with a new dialogic tool for both students and companies. The speed date was not intended as a job interview but as an informal exchange to let those in the company share the passion and creativity that the manufacturing field can offer and, above all, to create an opportunity to meet beyond the training paths. Positive feedback was received from both sides after the event. Indeed, all professionals and young talents considered this speed-dating a helpful experience.

The speed date was instrumental in getting companies to recognize the figure of the all-round designer in its differences from the figure of the stylist and to show them a different perspective, a more 'fluid' view of work and roles. With this first pilot event, the research enabled the foundations to be laid for a new paradigm with shared aims and a progressively open dialogue between the players in the textile-knitwear production system.

5.2.2 Preliminary Outputs: Stakeholder Map and the Five Core Issues

One of the first preliminary outputs obtained was updating and restructuring the stakeholder map outlined during the PhD thesis 'Designing Knit Designers' (Motta, 2019). The maps bring together industry companies, trade associations, certification bodies, training providers, recruiting organizations, and, of course, young talents (Figure 5.1).

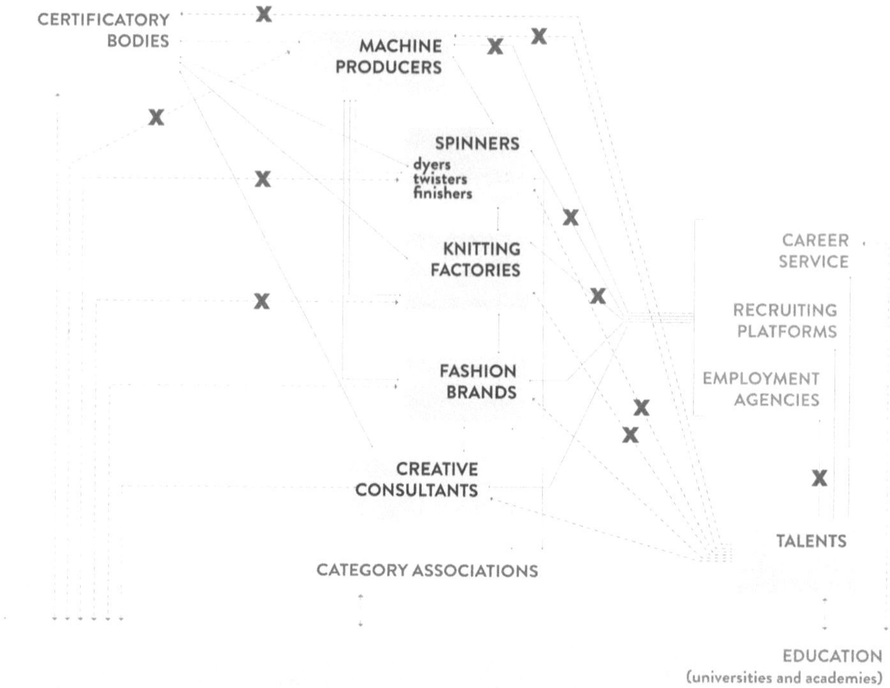

FIGURE 5.1 Stakeholders map. Solid arrows indicate existing relationships and potential reciprocity; dashed arrows indicate potential relationships or relationships with vast room for research action.

Starting from this map, we continued our investigation using quantitative (questionnaires) and qualitative (interviews) methods.

The questionnaires were administered to 67 companies in the knitwear-textile sector and 92 young talents, students, or recent graduates in design with specializations relevant to the industry of interest. Among the companies, the percentages show a good representation of the various clusters by respondent age group and gender. The research target has been reached and is well represented (86.6% micro-, small, and medium-sized enterprises): the percentages represent the Italian entrepreneurial fabric of the sector. In percentage terms, the sample is also significant as it reached the most relevant territorial areas for the industry: all regions where the most critical textile-knitting districts are located (Tuscany, Veneto, Emilia Romagna, Piedmont, Lombardy) are well represented in percentage terms.

The answers show that the sample includes a good percentage of spinning mills, knitting mills, and brands, followed by consultancy firms and machine manufacturers. The remaining answers come from all other stakeholders in the sector and complete a comprehensive view of the supply chain. Concerning the professional role of respondents, the most significant percentage (32.8%) is that of CEOs, who can return a strategic vision of their companies, followed by Marketing & Sales professionals, i.e., people with sales and communication-oriented skills and vision.

Among the young talents, there is a balance between students and recent graduates. Also, in this case, the research target has been fully achieved as the two periods of fundamental interest are represented: the last years of education/training, when the knowledge acquired is consolidated and specializations deepened, and the first years in the world of labor, when one verifies the knowledge acquired and begins to build their professionalism. Likewise, a good percentage (33.3%) of respondents work in the sector or worked in it in the past (5.6%), and an equally relevant percentage (52.2%) is looking for a job in the sector. The research thus collected two views: that of people who already had direct experience in the sector and those who bring ideas and expectations. The high percentage of respondents specializing or specializing in knitwear design (71.1%) makes the sample very vertical on the topic while offering adequate insight through specific questions.

Geographically, a high percentage of respondents are from Lombardy, as there is an excellent concentration of schools and academies in the area of our interest. The gender share of respondents (27.8% male, 72.2% female) reflects that of school students.

The most notable insights from the questionnaires were further investigated through semi-structured interviews with professionals, students, and young designers. The combination of qualitative and quantitative data allowed the researcher to cluster the complexities of the interaction between textile-knitwear design education and the manufacturing industry into five intervention areas, which are explained below.

5.2.2.1 Core Issue 1: Who is the Knitwear Designer?

Qualitative interviews reveal different perspectives on the profession of designers in the knitwear sector. Generally, two opposing visions are recognized: one sees the designer solely devoted to creative tasks, and the second considers the designer a strategic and catalyzer professional for wide-ranging innovation.

According to the first vision, for manufacturing companies and brands, the word 'designer' defines a professional who is purely devoted to creative tasks in the styling office of a brand. This figure differs from those who deal with product development and is undoubtedly different from those who work on aspects more related to the study and creation of yarns. The traditional figure of the consultant is also linked to this perspective. Designers are found to be often used as consultants: small and medium-sized manufacturing companies rely, with varying degrees of continuity, on external consultants for research on trends, choosing colors, and designing communication material (booths at trade shows, digital and printed promotional material, color charts, and yarns). In these cases, the creative part, while retaining its importance, seems less relevant than the product and its technical development and is neither described nor perceived as fundamental to companies' innovation and cultural evolution. When the designer is perceived as a figure devoted solely to the creative sphere and not as a strategic figure for the evolution of products, processes, and cultural assets (Banerjee & Ceri, 2016), there is no need to integrate a qualified designer because the latter is not recognized as a figure with high potential. This resource can bring visions and benefits in a broader sense. As a result, people are content to entrust their cultural development to external figures.

On the contrary, young talents see education and training in design as an asset that can be spent in many different professional roles and throughout the textile-knitwear supply chain, bringing a broad vision, whether they take on the role of external consultants or are in-house professionals at the company. Where there is awareness of designers' vertical and transversal skills, they are given more room to act and experience a redefinition of their tasks. In some cases, roles are rethought, and new figures are introduced into the workforce of companies that strongly focus on creating connections between traditional roles and serve as 'facilitator-designers'.

These figures are still awaiting definition. The word 'consultant' is a broad term with no specificity for the sector, but interviews with young talents reveal they need a job title to identify and introduce themselves.

5.2.2.2 Core Issue 2: The Designer's Space in Manufacturing

The interviews reveal different facets of the designer profession and different perceptions of this figure and its potential inclusion as one moves along the supply chain. Views differ widely depending on the type of stakeholder and even among stakeholders in the same category.

In particular, different approaches are found among spinning mills: for some of them, the designer must be a company's in-house figure and works to develop yarn collections; other mills rely on external consultants, and others still do not believe they need a designer and see designers as customers/buyers.

The reasons for these differences can be traced back to several factors:

- The size of the company (small, medium, large) and thus the resources available;
- Membership in large luxury groups. This ensures a more robust market positioning and a good percentage of safe sales to brands within the same group;
- The type of product. Companies that produce classic yarns from noble fibers do not feel a pressing need for product innovation and have been the benchmark for decades for their customers when it comes to high-quality yarns for a traditional-looking end product. Spinning mills that handle a more comprehensive range of yarns, in terms of materials and/or patterns, renew their collections more radically to offer brand designers materials that can transfer experimentation and creativity to the finished product.

The point of view of young designers with years of industry experience is attractive. They tell of the value of working positions unrelated to brands, recognize their potential to enrich their professionalism, and foster dialogue and connections between stakeholders. The words of those whose job is less related to products and more to promotion, training, aggregation, and networking between companies are often enlightening. These figures provide a 360-degree view of the supply chain, and today's disconnections generate problems and hamper competitiveness. They can also propose solutions for the future that always include dialogue and collaboration between different actors.

5.2.2.3 Core Issue 3: The Skills of the Knitwear Designer

The need for a knitwear designer to combine creative and technical skills is evident.

The interviews reveal the complexity and richness of the work of the knitwear designer – whether as a designer in a spinning mill, in a styling office, or as a consultant – who designs starting from the yarn and can, therefore, think about the material before arriving at a garment and can imagine the finished garment just by touching or seeing a yarn.

The primary sources of complexity are the variety of technical skills and the fast-evolving technologies for producing and handling yarns, fabrics, and finished products. For the control of technologies, software, and electronic machines, there are already highly specialized professionals within companies, such as programmers, yarn engineers, or machine operators; all these professionals require years of punctual technical education and training followed by constant practice. The relationship and communication between the designer and these specialized figures are often complex and have long been the subject of study as the main problem and obstacle to a good workflow. The fact that designers possess technical skills does not superimpose them on technical figures. Still, it makes it possible to have a commonality of languages that activates dialogue between the creative and technical spheres and pushes the horizon of possibilities forward.

5.2.2.4 Core Issue 4: The Attractiveness of Manufacturing

Comparing quantitative and qualitative data, manufacturing companies are recognized as unattractive to young creative talent, both in the view of companies, students, and recent graduates.

The survey on the reasons for this low attractiveness reveals two different aspects. The first, extremely concrete, is the existence of objective limiting factors such as a geographical location far from centers of interest (e.g., large cities) or outside industrial districts, the lack of connections and means of transport for those who do not own a car, the small size of companies, which offers little room for growth, and the scarce resources available to introduce new professional figures.

In quantitative surveys, in-house professionals at companies mainly highlight these factors as the main reasons for young talent's lack of interest in companies or their short stay within them. The second aspect, which already emerged with the quantitative analyses and was confirmed and deepened in the qualitative interviews, is a broad view of the professional prospects and opportunities that manufacturing offers designers and of the fascination that sector operators recognize in manufacturing processes, which could be used as an element to attract new professional figures. Both of these aspects are strongly present but could have been communicated better.

While concrete limiting factors are difficult to override or solve in the short term because it is unlikely that companies can physically move to the center of a city or grow in size overnight, research has room for action precisely in raising awareness of job prospects and opportunities, along with the fascinating features of these realities. In this respect, in the Italian production fabric, companies in this sector are no different from other small- and micro-sized enterprises. As a result of a management approach that is typical of family-run small- and micro-sized enterprises, the communication of the company, its values, and objectives is often not considered a strategic asset for business development or maintenance, especially in the B2B part of supply chains.

The increasingly global and digital evolution of the environment and markets (Kraus et al., 2022; Kane et al., 2015) has highlighted that structured communication is paramount not only to attract new customers but also to 'exist' as a market player. As Fontana (2019) pointed out, companies compete narratively today because communication has become an essential asset not only in business but even before business itself. Communicating one's history, values, and market vision is crucial in brand consolidation and building relationships with reference stakeholders and young talents.

The data tell us that the new generations are also aware that they choose the company to work with because of the perceived image of the company (27.8%). Still, they also tell us that for a company to be attractive, it must be able to communicate itself and the type of career path a candidate can take in it (86.7%). Communication, therefore, becomes an essential asset in creating relationships and synergies between individual actors and all stakeholders in the supply chain, making an excellent network capable of facing the challenges of the global market. Indeed, some interviewees believe that awareness-raising is the sole responsibility of schools. In contrast, others think it is the companies' responsibility, which are often very closed, and they need to identify the potential value elements they can communicate to the broader public. These discrepant visions underline the need to resume dialogue beyond stereotypes and redesign the relations between companies, manufacturers, designers, schools, and brands.

It emerges from the interviews that the first actions for change will have to act precisely on communication to systematize the real needs of the supply chain, generate a new shared vision, and put the importance of education and training back at the center.

5.2.2.5 Core Issue 5: The Role of Education: Present and Perspectives

As extensively discussed in Motta et al. (2023), the data collected, confirmed by the interviews, made the role of education emerge with its complexity, evidencing diverse areas of intervention. First, the relevant detachment between the educational offer and the industrial sector, which fostered research actions toward mutual closeness; second, the confirmation of the need to combine creative skills with a technical approach to the product; third, the students' standpoint on strength and weaknesses of the present; fourth, design competitions as valuable opportunities; fifth, a perspective toward the future with new perceptions, updated figures, and reinforced connections.

As also confirmed above, the general detachment between education/training and companies in the textile-knitwear supply chain results both in the students' fascination for jobs within fashion brands and a widespread lack of interest in other opportunities along the supply chain and in the companies' lack of ability to imagine a pathway for the insertion of new professionalism. This emerges as more accurate for educational institutions – universities or academies – which have an approach geared toward encouraging creative experimentation and artistic expression of individual students, often giving less importance to technical and process skills. In the absence of typical discussion venues where multiple stakeholders can voice and debate the needs of the sector, companies propose a closer interaction with the world of education and training realities, which often results in theoretical indications on education and training rather than a genuine mutual rapprochement aimed at building profiles with strategic skills.

We have already seen that acquiring new skills is often a costly investment for small and medium-sized companies, as they have limited resources in addition to established processes, and this also limits their ability to understand which skills and expertise can add value and where to find them. An institutional mapping of the educational system would make it easier for companies to orient themselves among the different offers and outgoing profiles and thus identify which institutions effectively target the textile manufacturing sector. As highlighted above, companies must become aware of the need to communicate effectively. Still, it is also essential for educational institutions to take action in this regard, as they, too, lack a structured showcase through which to communicate their strengths. Existing relationships, born mainly from personal acquaintances, favor long-standing relationships and foster a static and partial view of the whole system on both sides.

In contrast, new initiatives are usually undertaken by larger companies with a structured organization and the financial and human resources to do so. Two examples of this are the 'Wool you meet me' program proposed by the Loro Piana company to schools and universities in the fashion sector but also to those in the engineering area and the designer training course Loro Piana requested from the Knitwear Design research group at Design Department at Politecnico di Milano.

The 'Wool you meet me' program involves Loro Piana professionals from different areas of the company speaking in schools to explain their work. A first pilot meeting occurred in November 2022 at the School of Design with Fashion Design students within the Knitwear Design Final Synthesis Studio. Simone Pellegrini, Advanced Quality Project Manager, and Andrea Di Napoli, Industrial Quality Engineer, spoke about how the company maintains high standards of quality and innovation in its industrial processes. In this case, even though no designers were acting as speakers, the meeting was enlightening for students and professors alike because it dealt with product-related aspects that are not usually known or studied in design programs. In the second case, Loro Piana requested an ad hoc training course for professionals in the design area to integrate knowledge of knitting techniques and create moments of in-depth study that would otherwise be impossible to reconcile with its daily workflow. This request recognizes the importance of a structured educational and training program, even for professionals working within the company.

5.3 FINAL RESULTS: IMPACT AND DISCUSSION

The research results are in the ongoing development of Industry 4.0 (Ghobakhloo, 2018), which requires infrastructure changes, new organizational models, and new physical and digital competitiveness tools. For small and medium manufacturers in the textile and knitwear area, these efforts resulted in a specific resistance to innovation in human resources, in implementing professional skills, and in communication

(Lehner & Sundby, 2018). The manufacturing reality remains essentially unknown, and young talents looking for a job cannot recognize the creative opportunities offered by spinning, weaving, and knitting factories: they see few growth prospects for their design vision and their technical skills, and they often find themselves assuming executive roles and adapting their modus operandi to the company environment. The presented research aimed to fill this gap, using communication strategies to find new ways to connect young talents with manufacturing companies to facilitate their inclusion as profiles that can guide companies toward a true transformation of processes (Ronchi et al., 2020).

To do this, the project applied research-based design practices to explore the reality of manufacturing and bring advancements in design research. The main result is a raised awareness of what design is and what designers can offer to manufacturers. Specifically:

- It increased knowledge of how design plays a crucial role (Brown, 2019) within the textile-knitwear manufacturing sector by facilitating regular, quicker, and more informed decisions regularly;
- It raised awareness of the Knit Design research team's expertise and activities: this led not just to an interest in design students as future professionals but also to an increased interest in what design research can do for companies. Manufacturers manifested their intention in developing applied research projects, financing PhD scholarships, and being involved as industry partners in proposals for EU funding;
- It showed manufacturers the importance of communication protocol, tools, and storytelling in activating a dialogue between future professionals and companies within the textile-knitwear manufacturing sector.

Moreover, the research experience enabled the researchers to synthesize a model for activating fruitful relationships with SMEs. In this model, the designer – understood as both researcher and practitioner – serves as a figure capable of bridging the university and business sectors, addressing the needs of all stakeholders. The result is a model of design praxis (Figure 5.2) derived from the Systemic Design Framework (Design Council, 2021), which we describe below.

The model follows the four stages of the Systemic Design Framework: Explore, Reframe, Create, and Catalyze.

The first phase – Explore – allows us to investigate all the realities and identify the challenges. This phase focuses on bridging the gap between SMEs and designers in training, aiming to understand the needs of both companies and students in detail. The research enabled us to identify clusters that need further investigation to collect insights that will guide design and development. For SMEs, these clusters are the **communication** done by the company of its identity, values, purpose, and actions among stakeholders and diverse audiences; its **heritage**, namely the history, the expertise, the material and immaterial values carried by the company itself; the **professional roles** active inside the company with their duties and their organization; and the **competencies** needed for prospective growth. For students, the areas

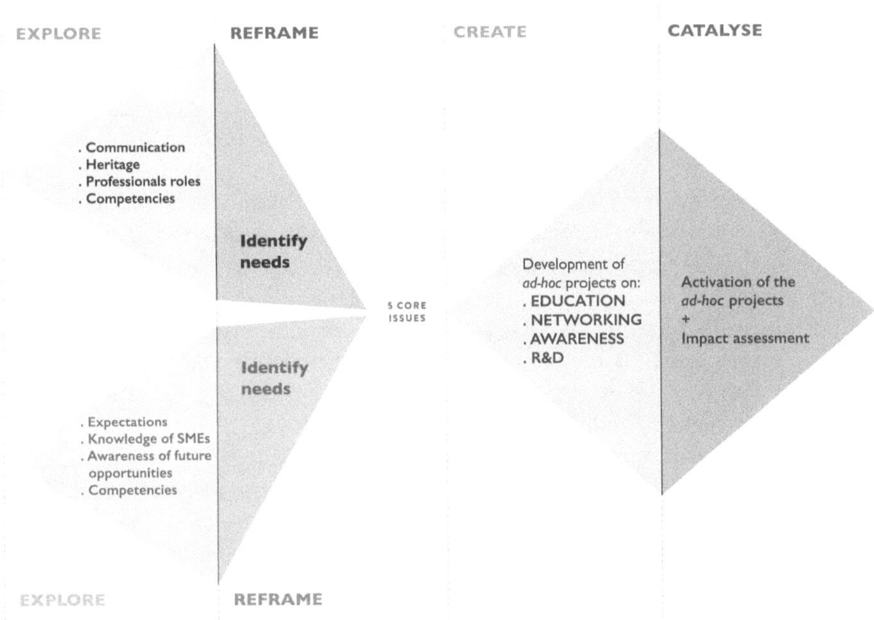

FIGURE 5.2 The elaborated model of design praxis.

to investigate are the **expectations** they have when they think of their professional future, the level of **knowledge of SMEs**, the concrete idea they have of how the system is structured, what the stakeholders are and how they relate, the realistic **awareness of the future opportunities** within the system, and the **competencies** they develop throughout their university careers.

The second phase – Reframe – involves defining the touchpoints between expectations and needs to identify critical areas of interest. Identifying SMEs' real needs and the relevant areas of interest means engaging with individual companies' internal and external stakeholders and understanding their needs beyond the stereotypes of which they are unaware carriers. Many companies of such size are still run according to models that struggle to cope with the ongoing transformations in global markets, and those who lead them often fail to understand the priority areas on which to focus their efforts. For such reasons, in this research, the Reframe specifically focuses on identifying the five core issues for the textile-knitwear industry and determining possible intervention areas.

This is followed by the Create phase, during which ad hoc projects are developed in relation to four main clusters: education, networking, awareness, and R&D. They are ad hoc, as the landscape of Italian manufacturing SMEs indeed escapes the predefined business organization schemes that are unable to reflect their heterogeneity and organizational peculiarities. When approaching this richness, it is necessary to delve into the first two phases of the double diamond, with particular attention to the needs of individual companies, to be able to build tailor-made project paths capable of satisfying the uniqueness of each organization. Throughout the process, it is essential to constantly challenge the initial project ideas to focus on what is truly necessary and helpful for the companies and their stakeholders.

In the Con-text research, although the initial idea was to foster communication through a digital ecosystem (Jacobides, 2019), evidence showed that stakeholders strongly needed physical presence, shared space, and personal connections. Therefore, a physical event was designed to enhance mutual understanding between manufacturers and young talents and to create an awareness-raising pathway for companies to establish connections with students. The outcome was participation in the two-day event 'Feel the Green – Behind the Scenes', which included a talk and a speed date session, focusing on the networking cluster. This event served as a starting point for further research and actions to strengthen dialogue and gather insights to collaborate with companies in updating and co-creating educational pathways, tools, and methods, which are addressed in the subsequent phase.

The final phase – Catalyze – is where projects are activated, and an impact assessment is conducted to evaluate future actions. In our case, the research strengthened the existing relationships of the Knit Design program with companies and established new ones. This has already led to the active involvement of three new companies in teaching activities. The educational program will continue to evolve with the participation of additional companies. Increased awareness of the need for new design professionals resulted in creating an Executive course delivered by POLI. design in partnership with Shima Seiki Italia Company. This course aims to facilitate dialogue between designers and industrial technicians. It is designed for both young design talents seeking technical improvement in textile manufacturing and technical professionals interested in learning design approaches.

The research highlighted a strong connection between manufacturing companies (suppliers and clients) and their detachment from education. Existing relationships have been leveraged to establish a partnership between Knit Design Studio, Maison Chanel, and its textile/knitwear manufacturers (including Vimar as a spinner, Mely's, and Paima as knitting factories). This partnership will annually offer educational workshops and specialized internships to students.

5.4 CONCLUSION

The results presented so far make it clear that there is a need to resume dialogue beyond stereotypes and redesign the relations between schools, manufacturing companies, designers, and brands.

Observing the results obtained and their impact on future actions, the project's primary legacy is a new paradigm with shared purposes, a progressively open dialogue among the stakeholders around the textile-knitwear manufacturing system that has laid concrete foundations for future evolutions of the results, and a model of design praxis for supporting the activation and maintenance of relationships with SMEs.

The study successfully found and experimented with new solutions to introduce young design talents to the manufacturing world and raise awareness of job opportunities in various fields, from creative to more technical ones. On the side of manufacturers, it demonstrated that these new solutions were possible, valuable, and effective in facilitating the integration of high-potential profiles with a new mindset that was suited to drive process and product innovation.

The significance of the research concerns all the stakeholders directly involved, whether they were researchers, educators, young talents, or manufacturers. Still, its contribution also impacts future generations of young talents and any company that will join the network with Politecnico di Milano, as it has generated tools and methods to be used with the public in any future coordinated activity.

While reaching its initial aims, the research presents some limits: its specificity, concentrated on the knitwear-textile sector and localized on the Italian manufacturing territory with its distinctive nature, is hardly retrievable elsewhere. During the research, this has been not a limit but a fertile ground for investigation, which strongly influenced the research outcomes and final results: in this perspective, the scalability of the outcomes in those contexts where circumstances are less favorable is a critical point but also an opportunity for further research developments. The second issue concerns the role of some of the researchers involved as educators in Knit Design at Politecnico di Milano. This allowed testing and updating the existing teaching methodologies and tools, but prevented the involvement of other educational realities due to short timing. In this respect, extending the study to different national and international universities opens space for future research.

Future research actions in the short term will see the implementation of the activities set up, thanks to the dialogue activated by the research. In the longer term, the study constitutes a first experience to rely on when defining criteria of scalability of the elaborated strategy in different manufacturing sectors or the same sector but in other countries. This also opens the door to the perspective of a systematization of the dialogue between multiple academic realities and the manufacturing territories.

REFERENCES

Banerjee, B., & Ceri, S. (Eds.). (2016). *Creating Innovation Leaders: A Global Perspective*. Cham: Springer Nature.

Bertola, P., Mortati, M., & Vandi, A. (Eds.). (2021). *Future Capabilities for Creativity and Design: How Digital Transformation Reshapes Fashion Careers and Higher Education*. Florence: Mandragora.

Brown, T. (2019). *Change by Design: How Design Thinking Transforms Organizations and Inspires Innovation* (Revised, Updated edition). New York: Harperbusiness.

Ciancia, M., Ronchi, M., & Ballabio, S. (2023). Design for transformation. Unlock competencies for coping with complexity. In K. Vaes, & J. C. Verlinden (Eds.), *Proceedings of Cumulus Antwerp 2023: Connectivity and Creativity in Times of Conflict* (pp. 122–126). Academia Press. https://doi.org/10.26530/9789401496476

Conti, G. M., Poletti, F., & Rinaldi, C. (2016). *Maglieria Made in Italy Knitwear. Stories and Talks*. Milano: Silvana Editoriale.

Design Council. (2021). *Beyond Net Zero: A Systemic Design Approach*. Design Council. https://www.designcouncil.org.uk/resources/guide/beyond-net-zero-systemic-design-approach

European Commission. (2010). Green paper—Unlocking the potential of cultural and creative industries. https://op.europa.eu/en/publication-detail/-/publication/1cb6f484-074b-4913-87b3-344ccf020eef/language-en

Fontana, A. (2019). Dieci Anni di Storytelling In Italia: Scenario, Valore, Visione. In *Il valore del capitale narrativo. Storyfactory Narrability Path™, come usare la Narrazione Strategica per avvicinare le audience ai valori e all'offerta delle imprese* (pp. 6–15).

Ghobakhloo, M. (2018). The future of manufacturing industry: A strategic roadmap toward Industry 4.0. *Journal of Manufacturing Technology Management*, 29(6), 910–936. https://doi.org/10.1108/JMTM-02-2018-0057

Jacobides, M. (2019). Designing digital ecosystems. In M. G. Jacobides, A. Sundararajan, & M. Van Alstyne (Eds.), *Platforms and Ecosystems: Enabling the Digital Economy, Briefing Paper. World Economic Forum* (pp. 13–18).

Kane, G. C., Palmer, D., Phillips, A. N., Kiron, D., & Buckley, N. (2015). *Strategy, not Technology, Drives Digital Transformation*. Cambridge, MA: MIT Sloan Management Review; Westlake, TX: Deloitte University Press.

Kraus, S., Durst, S., Ferreira, J. J., Veiga, P., Kailer, N., & Weinmann, A. (2022). Digital transformation in business and management research: An overview of the current status quo. *International Journal of Information Management*, 63, 102466. https://doi.org/10.1016/j.ijinfomgt.2021.102466

Lehner, F., & Sundby, M. W. (2018). ICTsSkills and competencies for SMEs: Results from a structured literature analysis on the individual level. In C. Harteis (Ed.), *The Impact of Digitalization in the Workplace: An Educational View* (pp. 55–69). Springer International Publishing. https://doi.org/10.1007/978-3-319-63257-5_5

Magni, A., & Noè, C. (2017). *Innovazione e sostenibilità nell'industria tessile*. Guerini Next.

Motta, M. (2018). Who makes Italian Knitwear? The great companies behind fashion brands. In G. M. Conti (Ed.), *Maglia, punto. Forty Years of Hidden Treasures from MF1 Archives* (pp. 43–49). Cinisello Balsamo: Silvana Editorial.

Motta, M. (2019). *Designing Knit Designers*. Milano: FrancoAngeli.

Motta, M., Ciancia, M., Conti, G. M., Goldoni, K., Piredda, F., & Zagatto, B. (2023). Next generation textile designers. A research project to connect the textile-knitwear manufacturing system with future design talents and its impact on education. In De Sainz Molestina, D., Galluzzo, L., Rizzo, F., & Spallazzo, D. (Eds.), *IASDR 2023: Life-Changing Design*, 9–13 October, Milan, Italy. https://doi.org/10.21606/iasdr.2023.195

Ronchi, M., Ciancia, M., & Piredda, F. (2020). Design practice for transformation. In *DIID. Disegno Industriale Industrial Design. Design 2030: Practice, N° 72/20* (pp. 144–151). Bologna: Editrice Compositori.

Stolze, A., Sailer, K., & Gillig, H. (2018). Entrepreneurial mindset as a driver for digital transformation–a novel educational approach from university-industry interactions. *Proceedings of the 13th European Conference on Innovation and Entrepreneurship – ECIE 2018*, held in Aveiro, Portugal. Published by Academic Conferences and Publishing International Limited, Reading, UK. (pp. 806–XXI).

Traini, C. (2004). *Maglia: creatività e tecnologia*. Ginevra – Milano: Skira.

6 Design Culture Network

Mapping Participation and Network Type from an Empirical University–Industry Collaboration

Hugo Palmares, Miguel Terroso, and Emília Costa

6.1 INTRODUCTION

A collaboration between university–industry (UIC) involves dynamic linkage and interaction to drive and share a knowledge workflow. This type of sharing work – *inputs*, *outputs*, and *outcomes* – between partners permeates each phase of the research and can be considered respectively as "knowledge creation, knowledge, and knowledge dissemination" (Chen et al., 2022, p. 18). In recent years, researchers have studied UIC from a holistic perspective[1] (Ćudić et al., 2022), industry perspective[2] (Albats et al., 2020), and university–industry perspective[3] (Awasthy et al., 2020; Nsanzumuhire & Groot, 2020; Demircioglu & Audretsch, 2019).

This chapter is an extended version of the article presented in the Design Commit '24 conference proceedings. The main objective here is to establish a conceptual framework, i.e., a conceptualisation of the UIC approach to try to develop a design–innovation–sustainability culture in industry companies. Capturing and mapping data on the resources, products, results, and *impact* of the research is crucial for qualitative and quantitative analyses. This research advocates the existence (or creation) of a research community as a whole – including companies, industry associations, researchers, and university students – which will potentially establish a position on knowledge, but also on its *outcomes* (results) and its *impact* (effects on industry).

This chapter intends to propose and explore the evolution of a network culture, for empirical research, demonstrated through a workflow framework for a UIC, and illustrate a mapping of the design culture network. The qualitative and quantitative data collected during each phase, the effectiveness of all *outputs*, and the equity of the *outcomes* validate the efficiency of the process and practices of this framework. The development of a toolkit or game as an *output*, structured by the interview script/questionnaire of the *Design Culture Ladder* model (Palmares et al., 2023, p. 530),[4] is applied to companies in Portuguese industry in this empirical study of university–industry co-creation. This research proposes a methodology to co-create a toolkit with UIC in a long-term framework and shape the evolution of a network culture,

DOI: 10.1201/9781032719290-7

according to each phase of the co-creation process, with mapping of correlations between industry partners and universities.

The structure of this chapter is the introduction followed by Section 6.2 which is dedicated to applied research methodology with literature review and qualitative collection; Section 6.3 that describes the proposed practices in the process framework, resulting on *inputs*, *outputs*, and *outcomes*; Section 6.4 that demonstrates an iterative collaboration on a mapping participation and a network type; Section 6.5 dedicated to the discussion and implications of this research; and Section 6.6 that concludes and suggests future work.

6.2 LITERATURE REVIEW

According to some authors, in a review of studies related to university–industry, fundamentals and factors that contribute to the efficiency of a UIC are explained. The basis or foundation of this approach concerns the concept of social capital, with respect to the following: (i) social relationships and commitment between associates; (ii) quality and availability of resources through network interactions (Portes, 1998; Bourdieu, 1986). According to these authors, the efficiency of this concept faces the challenge of a durable mutual recognition network.

The interest and confidence between the university and industry involve factors that result from collaboration, in which practical benefits as a partner corporation are expected from the resulting *outcomes* (Awasthy et al., 2020; Barbolla & Corredera, 2009). A reference list of influential *factors* linked to the efficiency of a UIC is identified here, related to the following: (i) individual characteristics; (ii) network structures; (iii) proximity; (iv) motivations; (v) common areas; (vi) realistic expectations; and (vii) same sense of urgency (Chen et al., 2022; Awasthy et al., 2020; Attia, 2015). Various studies also provide an identification and formulation of best *practices* for establishing a sustainable UIC (Awasthy et al., 2020; Edmondson et al., 2012; Sandberg et al., 2011; Greitzer et al., 2010; Prigge, 2005; Cyert & Goodman, 1997).

The collaborative approach between universities and industry is a tool that challenges organisations, but at the same time provides an emerging key to boost innovation in the partners involved (Ćudić et al., 2022; Chen et al., 2022; Nsanzumuhire & Groot, 2020). The boosted *outcomes* from collaboration overcome organisational challenges and boundaries for both universities and companies (Demircioglu & Audretsch, 2019; Alonso et al., 2010; Perkmann & Walsh, 2007; Powell & Grodal, 2005; Mendibil Telleria et al., 2002). In addition, companies with "well-educated and highly skilled population (and therefore labor force) is extremely important to the efficient creation, acquisition, dissemination and utilization of relevant knowledge, which creates innovation" (Marotta et al., 2007, p. 18).

From a network perspective,[5] the social network and relational skills of researchers and the university are crucial to creating collaboration with companies and industry. The researchers' network is a key element and starting point for analysing the potential for short or long-term collaborations, understanding the role of the network culture at UIC. Considering that each company in the sector has its own culture, the university adapts to improve collaboration, respecting that:

> Culture implies stability and rigidity in the sense that how we are supposed to perceive, feel, and act in a given society, organization, or occupation has been taught to us by our various socialization experiences and becomes prescribed as a way to maintain the *social order.*
>
> *(Schein, 2004, p. 3)*

In this sense, it is vital to promote a shared culture in each collaboration, including understanding "complex interactions of knowledge and resources" (Chen et al., 2022, p. 2). From industry partners, there is an advantage, openness, and motivation to work with university researchers with industry background, network, and managerial skills, to cross borders and areas of expertise (Awasthy et al., 2020; Edmondson et al., 2012). It can be assumed that "network competence is the company-wide ability and willingness to set up, use, and maintain targeted relationships with external partners" (Walter et al., 2002, p. 15).

Inevitably, there are potential advantages when human capital has "industry experience, which we define as previous work in the industry [...] provides information about industry rules and norms, customer and supplier networks, and employment practices" (Delmar & Shane, 2006, p. 220); however, "most universities engaged in partnerships are *learning by doing* and lack academics with experience in industry or the proclivity to network outside their area of expertise" (Edmondson et al., 2012, p. 10). Researchers can play a role in maintaining collaboration with industry partners (Lam, 2011; D'Este & Perkmann, 2011; D'Este & Patel, 2007), and during this process the following may occur: (i) earlier short-term collaborations evolve to "long-term strategic partnerships" (Awasthy et al., 2020, p. 56); (ii) the dynamic growth of relational and "community structure" associated with social phenomena[6] (Borgatti et al., 2009, p. 893); and (iii) the ability to absorb, create, and transfer "knowledge network" (Liefner & Hennemann, 2011, p. 810). In this sense, the university must redefine the mission and vision shared with the industry, considering that:

> Collaborating with industry should be linked to a redefinition of the role of the research university for the 21st century. That role now extends beyond teaching and public service research to tackling key social challenges and helping drive economic growth.
>
> *(Edmondson et al., 2012, p. 11)*

Highly qualified human resources within companies are also vital for successful collaboration, and a university depends on the usefulness and quality of its partners' resources (Albats et al., 2020; Myoken, 2013; Ferru, 2010). From a holistic perspective, which is also rich for both parties, it is a "construction of broad collaborative cluster networks, although most clusters focus on loosely defined local level networks" (Ćudić et al., 2022, p. 1); and that, contextually, it is considered that a "system of territorial networks in the Northern Portugal region promotes success and, in some cases, even contributes to the survival of companies and productive areas that are bearers of the region's culture" (Costa et al., 2023, p. 127). In this sense, universities have the potential to create conditions that promote innovative projects, including collaboration between organisations that would otherwise not collaborate spontaneously.

> Thus, opportunities must be created to ensure that learning can be shared, so that the parent organisation can see a benefit from their subsidiary collaborating within the local innovation network. It is also clear that the agenda for action must be fluid.
>
> *(OMalley et al., 2014, p. 1342)*

6.3 QUALITATIVE DATA COLLECTION

During nearly three years, this empirical research involved 120 participants, from 35 companies of six different industries in Portugal, and from seven European universities. The researchers designed and organised UIC events such as design workshops, focus groups, round tables, tests, seminars, and trainings and in an individual interaction approach with interviews and surveys.

The starting point of the present research was online interviews, recorded using the Zoom videoconferencing platform, in which the duration of each interview varied between 40 and 90 minutes. According to the variable sampling, in the three pilot interviews (the interviewees were designers with experience in the footwear industry, accessories or leather goods industry, and textile industry) and the nine interviews (the interviewees were professionals in the footwear industry in Portugal), the criteria of "judgemental sampling" (Charles, 1998, p. 146) were applied, selecting companies and professionals that communicate innovation and disruptive culture in their professional context. The interviewees – designers, creative directors, marketing directors, engineers, and CEOs – were collected both through the researcher's network and through additional recommendations from respondents, the latter using the criterion of "snowball sampling" (Coutinho, 2011, p. 97). Eighty-nine percentage of those interviewed hold a higher education degree (Bachelor's degree), and of these, 50% have postgraduate degrees (Master's degree or MBA) in the fields of Management, Economics, Engineering, Marketing, Design, and Arts.

Regarding surveys applied to the software industry, ten start-ups in software engineering – each with 8–10 employees with a higher education degree – incubated at the Faculty of Engineering of the University of Porto (FEUP) and integrated in the Project Management Laboratory (LGP), were responded to and sent by email. Each of these software companies worked exclusively for a reference company, both national and international, such as Deloitte, Vestas, Stellantis, Talkdesk, or Fashable. At this stage, care was taken in selecting the sample rather than its size; a predefined criterial sampling was identified, in which the industry segment and its cluster were selected for this study (Palmares et al., 2024, p. 103).

The workshops were organised over two days, an 8-hour face-to-face design workshop, with 50 students from the third year of the Industrial Design course at the Polytechnic Institute of Cávado and Ave (IPCA) in Barcelos, Portugal, in the Entrepreneurship and Design Management classes. The students worked in 12 randomly chosen groups, composed of three to five people each, using manual and digital media for the final presentation. As a result of the two-day workshop, 12 toolkits or game-based activity proposals for physical or digital formats were created, of which six proposals were pre-selected by the researchers for the subsequent phase.

At a later phase, a remote focus group was organised, conducted via Zoom lasting 50 minutes, with five professionals from the footwear and software industries: two CEOs, two creative directors, and a marketing director. Of these participants, 80% have higher education (Bachelor's degree), and of these, 75% have a postgraduate degree (Master's degree or MBA) in the areas of Management, Marketing, and Arts. Each participant was able to answer the questions spontaneously and freely, without any order, with spontaneous interventions and discussions unrolled to select the students' best proposal to prototype.

In the testing phase of this research, the physical prototype underwent two pilot tests at the university, before being tested within an organisation, to understand the perception of the dynamics and rules of the toolkit. Four companies from different industrial sectors agreed to this purpose – sessions with five to ten collaborators from different departments – from sectors such as textile, footwear, software, and goldsmith. Later, we were invited by the Portuguese Technological Centre of Footwear (CTCP), to disseminate knowledge and to test the toolkit with several companies at the same time, in two group sessions of seven and nine participants each, from companies in the textile, footwear, accessories, and plastics clusters.

Later, online seminars were organised for 20 students, postgraduates, and researchers from European universities of RUN-EU, in the Short Advanced Programme in Sustainable Footwear Futures funded by Erasmus+, where six companies and one technology centre were invited to contribute to the knowledge creation and knowledge in this training. Most of the industry partners had already participated in previous events/stages, which were identified in the present research as *multi-session* collaborators. Subsequently, in the students' final prototypes and research presentation, four industry partners participated in the online training audience and academic evaluation comments, in a sharing and a knowledge dissemination event that undergraduate and postgraduate international students exchanged directly with the industry.

6.4 PROCESS FRAMEWORK

The framework is based on the *inputs* from previous stages, to produce *outputs* for the next phases, which are implemented through iterative meetings and practices for a UIC, such as interviews, surveys, design workshops, focus groups, tests, seminars, and trainings. This chapter aims to clarify a research process framework in a UIC with a segmented list of practices (*inputs*) and the knowledge resulting from this collaboration (*outputs*). According to Ćudić et al. (2022, p. 3), the generating factors from the *inputs* are related to four categories: (i) institutional factors (business environment); (ii) human factors (human capital); (iii) linking factors (universities and companies relationships); and (iv) framework factors (business infrastructure). The resulting *outputs* can be divided into tangible (prototypes, products, technology) and intangible (industrial designs, patents, scientific and technical publications, business model).

In Table 6.1, a holistic framework has been built according to the different research phases (contextualisation, observation, categories, design concept, design techpack, and validation), listing the inputs and outputs of each phase mentioned.

The role of this conceptual framework is to assess whether the research *inputs* are producing *outputs* and *outcomes/impacts* through appropriate evaluation – i.e., if

TABLE 6.1
Holistic Framework According to Different Research Phases, Inputs, and Outputs

Phases	Inputs	Outputs
1. Contextualisation	Literature review: • Culture, culture evolution, design culture, organisational and corporate culture, industrial culture, design ladder evolution.	Diagnosis tool: • Design culture ladder.
2. Observation	Structure: • Script references, script tests. Industrial meetings: • Interviews, surveys.	Database recorded: • Qualitative data, quantitative data.
3. Categories	Literature review: • UIC, design and innovation, mentality and creativity paradigms, creative toolkits, gameplay paradigms.	Idealised categories: • References, segmentation, categorisation.
4. Design Concept	University conception: • Design workshops, sketching and rendering, mock-ups, paper prototyping. Industry meetings: • Focus group, pugh methods.	Founded categories: • Proposals categorisation and selection. Database recorded: • Qualitative data, quantitative data.
5. Design Techpack	Selected proposal details: • Product personality, functional categories, usability categories, sustainability, material details.	Prototype: • Functional sample, usability sample.
6. Validation	Prototype testing: • Experience prototype tests, product functional tests, product usability tests. Industry meetings: • Tests, round tables.	Database recorded: • Qualitative data, quantitative data. Outcomes: • Industrial clusters dissemination, seminars, trainings, courses.

the research is efficient, then research methods and data analysis are needed to link these *outputs* to the *inputs* of the following phases, concluded by the *outcomes* and *impacts* of the research.

The *outcomes* of this research validate a process framework built for a UIC, with practical benefits expected for the various industry partners, in a design culture network that offers *multi-session* events organised by universities and interested undergraduate and postgraduate international students working with the industry.

6.5 ITERATIVE COLLABORATION

Based on the previous literature review and on empirical research in this chapter, the factors extracted from a sustained UIC aim to segment the theoretical and methodological framework. This research aims to find the crucial factors that influence a UIC efficiency, listing the factors according to each phase of the co-creation process, through structural characteristics for long-term collaboration networks (see Figure 6.1): (i) individual characteristics are directly correlated in the observation phase (interviews, surveys); (ii) proximity and motivation are linked to the conception phase (design workshops, focus group); and (iii) network structure, common areas and expectations are correlated in the validation phase (tests, round tables, trainings, seminars).

To shape the network culture's evolution, patterns and occurrences of iterative collaboration were identified, with a mapping of correlations between network participation and network type (see Figure 6.2). It can explain the complexity of the system – considering variables such as the researchers' network, recommended network, spontaneous network,[7] causality, or random network[8] – by finding patterns and correlations between partners' participation in each phase.

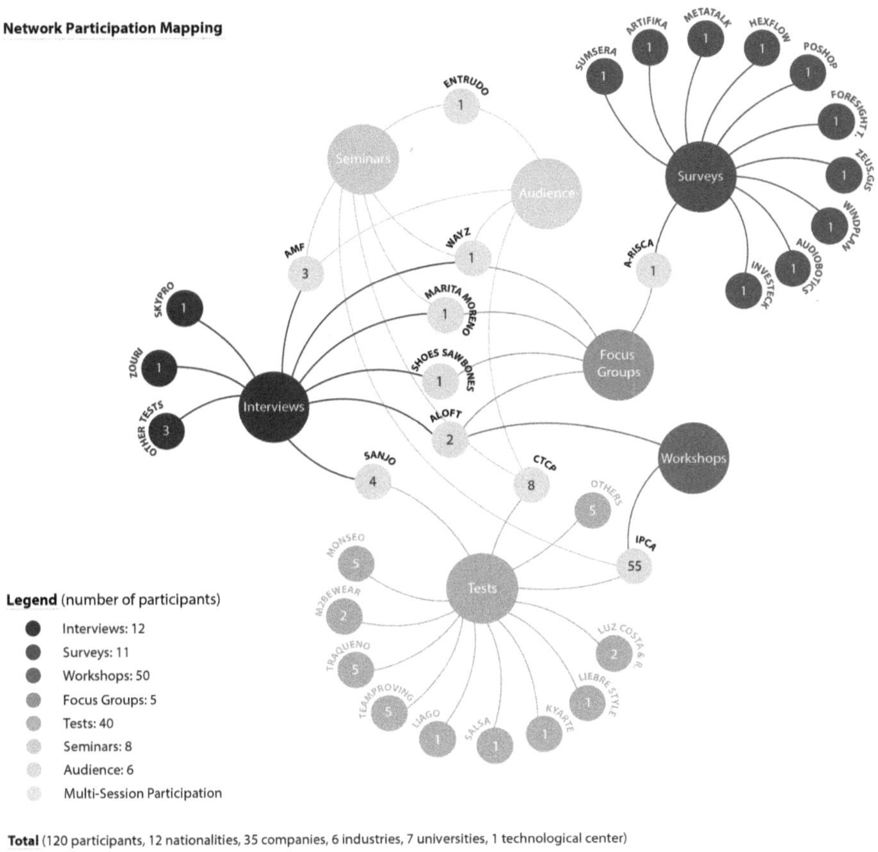

FIGURE 6.1 Mapping a network culture between industry partners and universities participation. (Diagram by the authors.)

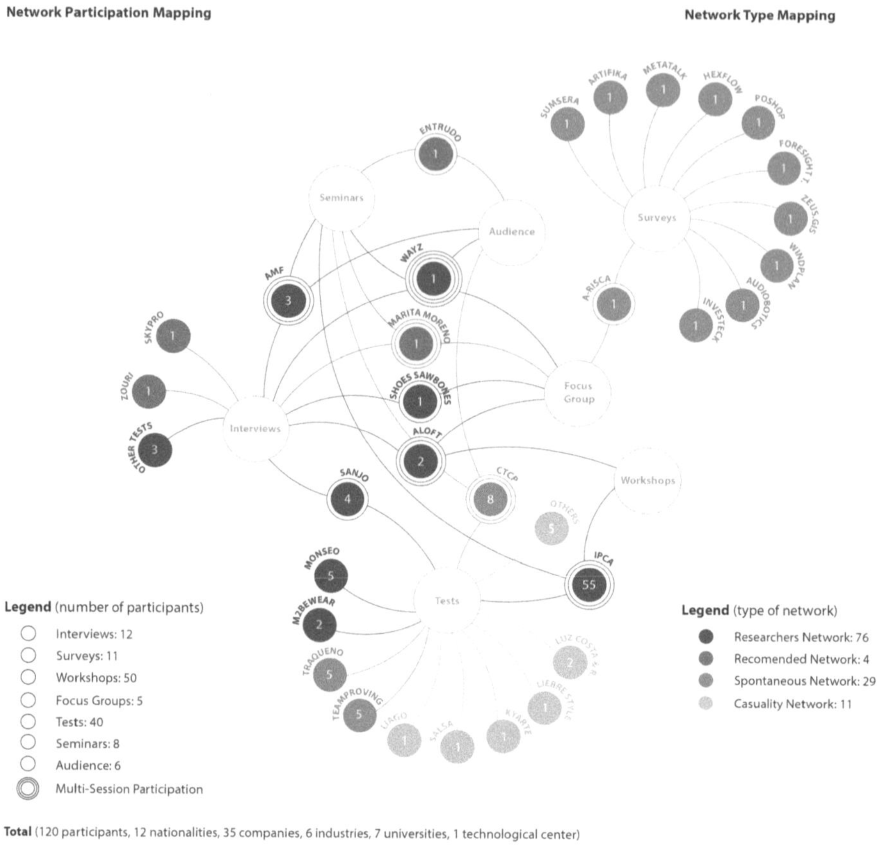

FIGURE 6.2 Mapping a network participation and a network type. (Diagram by the authors.)

An observation about iterative collaborations clearly identifies the occurrence of some industrial partners' participation in different phases of the present research, in a design culture network shared by industrial partners and universities. Building a long-term network, involving *multi-session* participations, provided the effectiveness of all *outputs* and the equity of the *outcomes*, and has validated the efficiency of the research framework, through the knowledge dissemination to large industrial clusters as a contribution.

At this stage, a *multidimensional scale* reading can be incorporated, which is understood as "similar to factor analysis but is based on a similarity or dissimilarity assessment between cases using, for example, pairwise comparison, rather than on evaluating each property of a case" (Blessing & Chakrabarti, 2009, p. 275).

6.6 DISCUSSION AND IMPLICATIONS

This research – which aimed to co-create and disseminate diagnostic and prescription sessions using the toolkit in companies, promoted by national associations and research centres in the Portuguese industrial sectors – sought to answer the question of how a design culture network can evolve through iterative collaboration between university and industry. Here, the aim was to understand the contribution and management of iterative collaboration to achieving the (r)evolution of an industrial network. Building a design culture network as a community definitely has an *impact* on results and can thus contribute holistically to the (r)evolution of an industrial cluster, more than just the production of knowledge from academic results.

In this chapter, a process framework of practices is proposed to implement a UIC efficiency and, from here, provide practical benefits to the industry partners and universities involved. Overall, the following were founded: (i) a structure of framework can challenge a collaboration efficiency; (ii) the variables and the factors mentioned are linked to and influence to a UIC efficiency; (iii) the participation of industry partners in *multi-session* during a research process has a positive effect on the network culture; and (iv) the industry can play a more significant role in building a network culture.

This empirical study illustrated a design culture network evolution; although this process was led by universities, industry companies have a crucial role to play in factors that require iterative collaborations. The crucial factors influencing UIC efficiency were correlated with each phase of the co-creation process through structural characteristics for long-term collaboration networks. However, there was a limit to the practices applied in the framework process, including, for this purpose only, the best practices necessary for this specific research and context.

6.7 CONCLUSION

In this chapter, an exploratory framework and network mapping were carried out with data from research phases, practices, *inputs*, *outputs*, *outcomes*, and *impact*, which are crucial to understanding and improving a UIC approach. The study contributes to a deeper understanding and experience from the researchers and industry partners, enhancing the achievement of a significant quality of knowledge creation, knowledge, and knowledge dissemination. It was argued, in the factors that contribute to the network's evolution, that the building of a design culture network as a community has an *impact* on *outcomes* and can holistically affect the industry, more than just a production of knowledge for academic *outputs*.

The complexity of individual characteristics in the collaboration process makes it difficult to guarantee a successful or reapplied formula. The use of a multidisciplinary range of collaborators was therefore considered for an exploratory case study approach, in order to emphasise complementary aspects of a phenomenon and contemplate *insights* from diverse data collection through multiple sources, highlighting the internal validity and reliability of the study.

The researchers' background and industry partners' experience can directly influence the open framework process and the dynamics of the design culture network. Collecting industry partners can be challenging, as the complexity of the system and considering all of the researchers' variables are involved. For future work, the proposed framework can be adapted to other types of co-creation *outputs*, even in a different industrial context, to be more effective and to encourage other academic researchers involved and facing challenges with a UIC.

ACKNOWLEDGEMENTS

This work is funded by national funds through the FCT – Fundação para a Ciência e a Tecnologia, I.P., under the scope of the project UIDB/04057/2020 (https://doi.org/10.54499/UIDB/04057/2020), and grant reference: 2020.08088.BD (https://doi.org/10.54499/2020.08088.BD) hosted by the ID+ Research Institute for Design, Media and Culture. The authors thank Adriana Mano, Cristiano Figueiredo, Diana Vieira da Silva, Eduardo Bonini, Egipto Magalhães, Gil Gonçalves, Heitor Alvelos, João Alberto Azevedo, Jorge Brandão Pereira, Jorge Pinto, Maria Duarte, Maria José Ferreira, Marita Setas Ferro, Nelson Amorim, Orlando Andrade, Paula Tavares, Pedro Carvalho de Almeida, Pedro Castro, Pedro Maçana, Raul Moreira Vidal, Rita Souto, Verónica Costa, and Vítor Costa.

NOTES

1. Considered an internal and external relationship network.
2. Addresses the company's creation of strategic partnerships to shape its innovation performance.
3. Regarding best practices for effective collaboration between universities and industry.
4. Interview guide structured by the four levels of the *Design Culture Ladder* model: technical skills, organisational skills, social factors, and cultural factors.
5. In relation to university–industry inter-organisational collaboration networks.
6. Refers to the interactions and behaviours that occur within a society or group of people.
7. In this case, spontaneous does not refer to something that is voluntary or unrestrained; rather, it refers to something that is self-regulated or inherent. Here the sample collected was focused on companies inserted in a specific software cluster.
8. Here the sample collection was linked with free and open participation to all collaborators connected to the traditional industry and clusters in Portugal, in seminars and trainings.

REFERENCES

Albats, E., Bogers, M., & Podmetina, D. (2020). Companies' human capital for university partnerships: A micro-foundational perspective. *Technological Forecasting and Social Change*, *157*, 120085. https://doi.org/10.1016/j.techfore.2020.120085

Alonso, J., de Soria, I. M., Orue-Echevarria, L., & Vergara, M. (2010). Enterprise collaboration maturity model (ECMM): Preliminary definition and future challenges. In Popplewell, K., Harding, J., Poler, R., & Chalmeta, R. (Eds.), Enterprise Interoperability IV (pp. 429–438). London: Springer London.

Attia, A. M. (2015). National innovation systems in developing countries: Barriers to university–industry collaboration in Egypt. *International Journal of Technology Management & Sustainable Development, 14*(2), 113–124. https://doi.org/10.1386/tmsd.14.2.113_1

Awasthy, R., Flint, S., Sankarnarayana, R., & Jones, R. L. (2020). A framework to improve university–industry collaboration. *Journal of Industry – University Collaboration*, *2*(1), 49–62. https://doi.org/10.1108/JIUC-09-2019-0016

Barbolla, A. M. B., & Corredera, J. R. C. (2009). Critical factors for success in university–industry research projects. *Technology Analysis and Strategic Management*, *21*(5), 599–616.

Blessing, L. T. M., & Chakrabarti, A. (2009). *DRM, a Design Research Methodology.* Springer-Verlag London Limited. https://doi.org/10.1007/978-1-84882-587-1

Borgatti, S. P., Mehra, A., Brass, D. J., & Labianca, G. (2009). Network analysis in the social sciences. *Science, 323*(5916), 892–895.

Bourdieu, P. (1986). The forms of capital. In Richardson, J. (Ed.), *Handbook of Theory and Research for the Sociology of Education* (pp. 241–258). New York: Greenwood.

Charles, C. M. (1998). Introduction to Educational Research (3rd ed). New York: Longman.

Chen, H., Song, X., Jin, Q., & Wang, X. (2022). Network dynamics in university-industry collaboration: A collaboration-knowledge dual-layer network perspective. *Scientometrics, 127*, 1–24. https://doi.org/10.1007/s11192-022-04330-9

Costa, F., Aparo, E., & Soares, L. (2023). The creation of a network, in the North of Portugal, as an opportunity for territorial sustainability. In Raposo, D., Martins, N., & Brandão, D. (Eds.), Human Dynamics and Design for the Development of Contemporary Societies, *AHFE (2023) International Conference* (vol. 81, pp. 123–128). AHFE Open Access. https://doi.org/10.54941/ahfe1003540

Coutinho, C. P. (2011). Metodologia de Investigação em Ciências Sociais e Humanas: teoria e prática (2nd ed.). Coimbra: Edições Almedina.

Ćudić, B., Alešnik, P., & Hazemali, D. (2022). Factors impacting university–industry collaboration in European countries. *Journal of Innovation and Entrepreneurship, 11*(1), 33. https://doi.org/10.1186/s13731-022-00226-3

Cyert, R. M., & Goodman, P. S. (1997). Creating effective university–industry alliances: An organizational learning perspective. *Organizational Dynamics, 25*(4), 45–57.

D'Este, P., & Patel, P. (2007). University–industry linkages in the UK: What are the factors underlying the variety of interactions with industry? *Research Policy, 36*(9), 1295–1313.

D'Este, P., & Perkmann, M. (2011). Why do academics engage with industry? The entrepreneurial university and individual motivations. *The Journal of Technology Transfer, 36*(3), 316–339.

Delmar, F., & Shane, S. (2006). Does experience matter? The effect of founding team experience on the survival and sales of newly founded ventures. *Strategic Organization, 4*(3), 215–247.

Demircioglu, M. A., & Audretsch, D. B. (2019). Public sector innovation: The effect of universities. *The Journal of Technology Transfer, 44*(2), 596–614.

Edmondson, G., Valigra, L., Kenward, M., Hudson, R., & Belfield, H. (2012). Making industry-university partnerships work. Lessons from successful collaborations. *Science Business Innovation Board AISBL, 16*, 1–52.

Ferru, M. (2010). Formation process and geography of science-industry partnerships: The case of the University of Poitiers. *Industry and Innovation, 17*(6), 531–549. https://doi.org/10.1080/13662716.2010.509998

Greitzer, E. M., Pertuze, J., Calder, E., & Lucas, W. A. (2010). Best practices for industry–university collaboration. *MIT Sloan Management Review, 51*(4), 83.

Lam, A. (2011). What motivates academic scientists to engage in research commercialization: 'Gold', 'ribbon' or 'puzzle'? *Research Policy, 40*(10), 1354–1368.

Liefner, I., & Hennemann, S. (2011). Structural holes and new dimensions of distance: The spatial configuration of the scientific knowledge network of China's optical technology sector. *Environment and Planning A, 43*(4), 810–829.

Marotta, D., Mark, M., Blom, A., & Thorn, K. (2007). Human capital and university–industry linkage's role in fostering firm innovation: An empirical study of Chile and Colombia. Policy Research Working Paper Series 4443. The World Bank.

Mendibil Telleria, K., Little, D., & MacBryde, J. (2002). Managing processes through teamwork. *Business Process Management Journal, 8*(4), 338–350.

Myoken, Y. (2013). The role of geographical proximity in university and industry collaboration: Case study of Japanese companies in the UK. *International Journal of Technology Transfer and Commercialisation (IJTTC), 12*(1–3), 43–61. https://doi.org/10.1504/ijttc.2013.064170

Nsanzumuhire, S. U., & Groot, W. (2020). Context perspective on university–industry collaboration processes: A systematic review of literature. *Journal of Cleaner Production, 258*, 120861.

OMalley, L., ODwyer, M., McNally, R. C., & Murphy, S. (2014). Identity, collaboration and radical innovation: The role of dual organisation identification. *Industrial Marketing Management, 43*, 1335–1342. https://doi.org/10.1016/j.indmarman.2014.08.007

Palmares, H., Terroso, M., & Costa, E. (2023). Design culture ladder: Creation of a diagnosis tool to evolve industry. In Martins, N., & Brandão, D. (Eds.), *Advances in Design and Digital Communication III* (pp. 523–535), Springer Nature Switzerland. https://doi.org/10.1007/978-3-031-20364-0_45

Palmares, H., Terroso, M., & Costa, E. (2024). Design culture perception: Diagnosis of knowledge and vision of traditional and innovative industries in Portugal. In Martins, N., Brandão, D., & Fernandes-Marcos, A. (Eds.), *Perspectives on Design and Digital Communication IV* (pp. 91–111), Springer Nature Switzerland. https://doi.org/10.1007/978-3-031-41770-2_6

Perkmann, M., & Walsh, K. (2007). University–industry relationships and open innovation: Towards a research agenda. *International Journal of Management Reviews, 9*(4), 259–280.

Portes, A. (1998). Social capital: Its origins and applications in modern sociology. *Annual Review of Sociology, 24*(1), 1–24.

Powell, W. W., & Grodal, S. (2005). Networks of innovators. In Fagerberg, J., Mowery, D. C., & Nelson, R. R. (Eds.), *The Oxford Handbook of Innovation* (pp. 56–85), Oxford and New York: Oxford University Press.

Prigge, G. W. (2005). University–industry partnerships: What do they mean to universities? A review of the literature. *Industry and Higher Education, 19*(3), 221–229.

Sandberg, A., Pareto, L., & Arts, T. (2011). Agile collaborative research: Action principles for industry– academia collaboration. *IEEE Software, 28*(4), 74–83.

Schein, E. H. (2004). Organisational Culture and Leadership (4th ed.). San Francisco: Jossey-Bass.

Walter, A., Auer, M., & Gemunden, H. G. (2002). The impact of personality, competence, and activities of academic entrepreneurs on technology transfer success. *International Journal of Entrepreneurship and Innovation Management, 2*(2–3), 268–289.

Part 2

Thematic – SOCIAL

Subthemes: Education – Healthcare – Critical Design

7 Designing Innovative Medical Tools

How Home-Integrated Healthcare Devices Can Shape the Future of Therapeutic Recovery

Valentina Sorvillo

This chapter is an extended version of the article presented in the Design Commit '24 conference proceedings.

7.1 INTRODUCTION

Facial nerve palsy is the most common cranial nerve dysfunction in children and adults. The long-term effects on vision, eating, drinking, psychological issues, self-esteem, and quality of life are well-established (Wang et al., 2021). Patient outcomes may range from complete and persistent acute flaccid facial paralysis to complete return of normal facial function. Between these extremes lie various zones with hypoactivity and hyperactivity, commonly known as post-paralytic facial nerve syndrome.

Bell's palsy is a neuropathic condition named after Charles Bell, a Scottish neurologist, and occurs due to dysfunction of the facial nerve (VII cranial nerve), often with an unknown etiology or associated with trauma, inflammation, or viral infections. Bell's palsy is most common in people aged 15–40 years, with a 1 in 60 lifetime risk. Most people make a spontaneous recovery within 1 month, but up to 30% show delayed or incomplete recovery (Holland and Weiner, 2004).

The VII cranial nerve is responsible for innervating the facial muscles, playing vital roles in emotional expressions and essential functions in human communication. Comprising sensory neurons, parasympathetic fibers, and motor neurons, it regulates taste, tear production, and facial muscle movement. Nerve lesions can be classified by severity and significantly impact the prognosis of facial paralysis (Dulak and Naqvi, 2020).

In longer-standing cases, synkinesis may develop as well, resulting in involuntary movements, dyscoordination, and facial pain (Mistry and Al-Sayed, 2021).

While Bell's palsy is often transient, it can have significant long-term impacts on patients' quality of life. Hence, timely and targeted treatment is crucial to minimize

DOI: 10.1201/9781032719290-9

complications and enhance functionality and confidence in affected patients. Multidisciplinary therapeutic approaches and recovery practices can play a pivotal role in facilitating recovery and minimizing long-term complications.

Additionally, it is essential to note that patients often perform exercises and massages at home, which might pose the risk of improper execution, potentially hindering the recovery process. Therefore, adequate guidance and supervision by healthcare professionals are recommended to ensure the correct implementation of these practices.

7.2 THERAPEUTIC INTERVENTIONS

Currently, the treatments for Bell's palsy can be categorized into two strategies, namely, pharmacotherapy and physiotherapy, both aimed at facial function recovery facilitation and sequelae prevention. In the majority of cases, pharmacological therapy is preferred in the early stage of Bell's palsy, including corticosteroids and antiviral agents, either individually or in combination. Corticosteroids are the most commonly used drugs for Bell's palsy due to their potent anti-inflammatory effect, which can reduce edema and inflammation of the facial nerve. Available evidence from randomized controlled trials (RCTs) has revealed benefits from treating the Bell's palsy with corticosteroids. Many published authoritative clinical practice guidelines for Bell's palsy have strongly recommended the prescription of corticosteroids within 72 hours of symptom onset. However, the benefit of antiviral agents has not yet been established, and antiviral therapy alone is discouraged in the treatment of Bell's palsy. On the other hand, physical therapy can be beneficial and complementary from even alternative therapies to pharmacological therapy. Thermotherapy, massage, electrostimulation, facial exercises, Kabat rehabilitation, and acupuncture are regarded as different forms of physical therapy that have been widely applied for Bell's palsy in the clinical setting (Shi et al., 2022).

Facial massage therapy plays a crucial role in the acute phase, facilitating increased blood circulation and lymphatic drainage in facial muscles. This technique heightens sensory awareness and aids in muscular relaxation, supporting the maintenance of muscle tone during the recovery process. During persistent facial paralysis, specialized physiotherapy becomes essential. Collaborating with an experienced therapist is crucial to prevent exacerbation. Electromyographic analysis (EMG) assists in evaluating muscle activity and associated nerve function. Physiotherapeutic intervention targets pain relief, improved tissue elasticity, and inflammation management using various high-tech tools and techniques. Long-term complications such as synkinesis, characterized by abnormal facial movements during nerve regeneration, require attention. These involuntary movements during routine activities like speaking or smiling can be addressed through therapeutic muscle rehabilitation, reducing discomfort.

7.3 FACIAL SYNKINESIS AFTER BELL'S PALSY

Introduction Synkinesis refers to abnormal involuntary facial movements that accompany volitional facial movements. Despite a 55% incidence of synkinesis reported in patients with enduring facial paralysis, there is still a lack of complete understanding of this debilitating condition, leading to functional limitations and decreased quality of life.

The resultant inability to control the full extent of one's facial movements has functional and psychosocial consequences and may result in social withdrawal with a significant decrease in quality of life.

Facial rehabilitation, including neuromuscular retraining, soft tissue massage, and relaxation therapy in addition to chemodenervation with botulinum toxin, remains the cornerstone of treatment.

Ultimately, therapy should be tailored to the severity and pattern of synkinesis, and each patient approached on a case-by-case basis. A multidisciplinary approach involving therapists, clinicians, and surgeons is necessary to develop a comprehensive treatment regimen that will result in optimal outcomes (Shokri et al., 2021).

7.4 RESEARCH: "EFFECT OF AN INTENSIFIED COMBINED ELECTROMYOGRAPHY AND VISUAL FEEDBACK TRAINING ON FACIAL GRADING IN PATIENTS WITH POST-PARALYTIC FACIAL SYNKINESIS"

The study evaluated the effectiveness of a 10-day facial rehabilitation program using intensive facial training combined with electromyography and visual biofeedback for patients with post-paralytic facial synkinesis. It included 54 patients and assessed facial function at different points: study commencement, start of training, end of training, and six months post-training. Various facial assessment systems were used, such as the House-Brackmann Facial Nerve Grading System, Stennert index, Facial Nerve Grading System 2.0, and Sunnybrook Facial Grading System. Results showed no significant changes in facial scores between the study commencement and the start of training. However, the intensive 10-day training improved facial symmetry and reduced synkinetic activity. The improvements were particularly evident when using the Facial Nerve Grading System 2.0 and Sunnybrook Facial Grading System. These enhancements remained stable at the 6-month follow-up. In conclusion, the intensified daily 10-day training involving electromyography and visual biofeedback was effective for patients with facial synkinesis, and the benefits persisted six months after therapy. The article underscores the importance of this approach in facial rehabilitation (Volk et al., 2021).

7.5 BIOFEEDBACK AND MT

Biofeedback therapy is an instrument-based learning process that is based on "operant conditioning" techniques. It is an evidence-based approach to improve the ability to modify involuntary processes consciously. The basic biofeedback paradigm suggests that whenever we provide a human being with feedback about a biological process, that feedback enables the individual to increase awareness of the process and gain conscious control. Physiotherapists use biofeedback to help refine a movement sequence or activation pattern to assist patients in achieving a goal. This technique involves using visual, physical, and/or auditory feedback to guide the patient to give their optimal performance. The ultimate purpose is that the patient gets to know his own body signs and that he can control them consciously in the first place using biofeedback equipment, and afterward even without (Physiopedia, n.d.).

MT is a valuable method for enhancing motor recovery in post-stroke hemiparesis. The technique utilizes the mirror illusion created by the movement of sound limb that is perceived as the paretic limb. MT is a simple and economical technique than can stimulate the brain noninvasively. The intervention unquestionably has a neural foundation. But the underlying neural mechanisms inducing motor recovery are still unclear. In this review, the neural modulation due to MT has been explored. Multiple areas of the brain, such as the occipital lobe, dorsal frontal area, and corpus callosum, are involved during the simple MT regime. Bilateral premotor cortex, primary motor cortex, primary somatosensory cortex, and cerebellum also get reorganized to enhance the function of the damaged brain. The motor areas of the lesioned hemisphere receive visuo-motor processing information through the parieto-occipital lobe. The damaged motor cortex responds variably to the MT and may augment true motor recovery. Mirror neurons may also play a possible role in the cortico-stimulatory mechanisms occurring due to the MT (Arya, 2016).

A 2022 study called "Effectiveness of robot-assisted virtual reality (VR) MT for upper limb motor dysfunction after stroke: study protocol for a single-center randomized controlled clinical trial" suggests that the combined use of robot-assisted therapy and VR might offer potential improvements compared to using VR alone in treating upper limb motor dysfunction caused by stroke. It hypothesizes that this combined approach could potentially activate the mirror neuron system and reward circuits in stroke patients' brains to a greater extent, providing enhanced therapeutic benefits compared to VR therapy alone. However, the study emphasizes the need for further high-quality research to confirm this hypothesis and determine how this combined therapy might be more effective in treating upper limb motor dysfunction after a stroke (Wei et al., 2022).

The study "The Role of Immersive Virtual Reality and Augmented Reality in Medical Communication: A Scoping Review" investigated a novel MT system utilizing VR and an actuated exoskeleton to address hand motor impairments. Traditionally, MT involves a patient moving the healthy hand in front of a mirror, tricking the brain into perceiving movement in the impaired hand and stimulating its recovery. The proposed system integrates VR to create an immersive mirroring effect and a soft hand exoskeleton to physically induce mirrored motion to the impaired hand. Tested with 21 healthy users, the system demonstrated acceptability based on the system usability scale and did not cause adverse effects or sickness according to the simulator sickness questionnaire (Aliwi et al., 2023).

Another study titled "The Efficacy of Virtual Reality for Upper Limb Rehabilitation in Stroke Patients: A Systematic Review and Meta-Analysis" compared VR-based rehabilitation with conventional occupational therapy across a spectrum of immersion levels and outcome domains. VR conferred benefits over conventional therapy for upper limb motor function, functional independence, quality of life, spasticity, and dexterity. Fully immersive VR showed the greatest gains in gross motor function, while non-immersive approaches enhanced fine dexterity. Interventions exceeding six weeks elicited superior results, and initiating VR within six months post-stroke optimized outcomes. This systematic review and meta-analysis demonstrate that adjunctive VR-based rehabilitation enhances upper limb motor recovery across multiple functional domains compared to conventional occupational therapy alone after stroke. Optimal paradigms likely integrate VR's immersive capacity with conventional techniques (Soleimani et al., 2024).

7.6 VR AND AR APPLICATIONS IN MEDICINE

In the future, extended reality technology will be widely used. People will be led to utilize VR and AR technologies in their daily lives, hobbies, numerous types of entertainment, and employment. Medical AR has evolved with applications ranging from medical education to picture-guided surgery. Moreover, a bulk of research is focused on clinical applications, with the majority of research devoted to surgery or intervention, followed by rehabilitation and treatment applications. Numerous studies have also looked into the use of AR in medical education and training (Taghian et al., 2023).

VR and AR are rapidly developing technologies that offer a wide range of applications and enable users to experience digitally rendered content in both physical and virtual space. Although the number of studies about the different uses of VR and AR increases year by year, a systematic overview of the applications of these innovative technologies in intensive care medicine is lacking. Both VR and AR offer multiple possibilities to improve current care, both from the perspective of the healthcare professional and the patient. It can be assumed that VR and AR will develop further and their application in health care will increase (Kanschik et al., 2023).

The study "The Role of Immersive VR and AR in Medical Communication: a Scoping Review" investigated the role of immersive VR and AR within the medical field. AR refers to the integration and superimposition of digital elements into the user's real-life environment, so that both can be attended to simultaneously. This is facilitated through different means and is most commonly enabled by smartphones and smart glasses (Aliwi et al., 2023).

7.7 CASE STUDIES

A series of case studies have been selected to delve into the design aspects of the project. The objective is to create a tool that transcends conventional medical applications, transforming a medical instrument into a tactile, comfortable, intuitive, and intimate device.

This analysis encompasses various projects involving Internet of Things (IoT) devices designed for rehabilitation and home connectivity, showcasing innovative approaches to integrating technology into daily life. Additionally, sensory objects and unconventional designs have been explored, challenging traditional perceptions of medical tools by incorporating elements of playfulness and user engagement.

By synthesizing insights from these diverse design projects, the aim is to inform the development of a novel instrument that is both aesthetically pleasing and user-friendly. This approach highlights the importance of interdisciplinary collaboration in the advancement of rehabilitation technologies, ultimately enhancing the user experience and improving outcomes in rehabilitation contexts.

7.7.1 Case Study: Rewind by Poh Yun Ru, 2022

Winner of the LEXUS DESIGN AWARD 2022, Rewind is a portable IoT device, with a minimalistic design, that utilizes motion tracking to guide elderly individuals with dementia in recalling familiar gestures. Users are tasked with mimicking common actions while holding the portable device. Sensors within the device measure how users interact with it (Lexus, 2022).

7.7.2 Case Study: Connectivity by Benjamin Hubert, 2022

The project challenges the materiality of technology to define a new vision for smart home devices. Each device features delicate geometric shapes, discreet interfaces, and tactile materials inspired by residential interior design. The project also aims to pursue more sustainable practices in connectivity (Dezeen, 2022).

7.7.3 Case Study: Totem & Tabù by Giovanni Lauda, 2022

Totem & Tabù is a family of wall grazers that introduces a magical aspect to technical lighting. When mounted on the wall, they evoke primitive masks, like faces adorned with mystical tattoos; the lamp patterns are etched with regular horizontal or vertical lines carved into wood and metal, reflecting the ambient light. (Giovanni Lauda, 2022).

7.8 METHODS AND MATERIALS

This study addresses contemporary challenges in physiotherapeutic treatments, specifically the difficulty patients face in replicating these treatments independently. To improve treatment quality and address the lack of accurate information, innovative approaches utilizing cutting-edge technology were employed. A customized device was developed, allowing patients to perform therapy at home accurately. Simultaneously, it enables therapists to monitor and guide the patient's progress through specific therapy phases. This tool was designed to be highly interactive, engaging, and adaptable to individual patient needs. Innovation was introduced through the implementation of emerging technologies such as VR and AR. VR was used to alleviate patient anxiety, stress, fear, and pain during therapy sessions, offering an engaging, motivating, and sometimes more effective rehabilitative approach compared to conventional therapies.

The methodological approach involved evaluating the device's performance in a clinical setting through pilot tests conducted with a small group of patients. It recorded the achieved outcomes and collected detailed feedback from both patients and healthcare professionals involved in the therapies.

This study embraced a multidisciplinary perspective, integrating expertise in technology, rehabilitative medicine, and human–machine interaction, to provide an innovative and personalized solution for addressing challenges in rehabilitation. In the initial phase of research and case study analysis, the objective is to explore the key phases for designing an effective product for therapy: the initial post-paralytic phase and facial synkinesis (Figure 7.1).

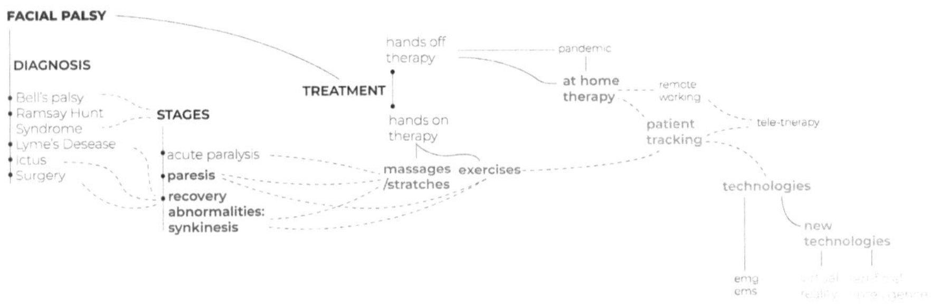

FIGURE 7.1 Research Map. (Illustrated by the author.)

Following initial research, disease phases requiring intervention were identified, including post-paralytic paralysis and recovery anomalies such as synkinesis. During these phases, pharmacological therapy, facial exercises, and massages are implemented, often performed independently with a high risk of error. Massages, stretching exercises, and facial exercises are conducted not only by physiotherapists but also independently, posing a risk of inadequate execution. Hence, the necessity arises to provide patients with tracking to ensure proper exercise execution, achievable through the use of new technologies like AR and VR.

7.9 DEVELOPMENT OF AR FILTERS FOR FACIAL MUSCLES ENGAGEMENT IN THERAPEUTIC EXERCISES

AR filters were created using Meta's Spark AR platform, designed to engage with patients' facial features for performing massages and movements. Specifically, an in-depth analysis was conducted to identify key reference points of facial muscles to target for therapeutic purposes. Subsequently, focus was placed on the tracking points of the face mask within the Spark AR template. These points corresponded to various facial regions, including the eyebrows (orbicularis oculi muscles), the nose (nasal muscles), the cheeks (zygomatic minor and major muscles), and the mouth (orbicularis oris muscles).

The development process involved the conceptualization of exercises categorized into four major areas: stretching, massages, symmetry, and interactive exercises. Each exercise category targeted specific facial muscle groups. For instance, stretching exercises aimed to elongate and relax facial muscles, while massage exercises were designed to relieve tension and enhance circulation. Symmetry exercises focused on balancing facial muscle movements bilaterally, and interactive exercises engaged patients in facial muscle movement control through interactive digital stimuli. The AR filters design aimed to encourage patient engagement and compliance during therapeutic exercises by leveraging interactive facial tracking technology. These filters served as innovative tools in facilitating and enhancing facial muscle rehabilitation exercises, promoting patient adherence and proper execution of therapeutic movements. This approach not only provided an engaging and user-friendly therapeutic tool but also opened avenues for further exploration into the integration of technology into rehabilitation methodologies for facial muscle-related conditions.

The second phase was dedicated to the actual design of the device, aiming to create an object seamlessly integrating into the domestic environment, providing familiarity, sensory engagement, and intimacy. Diverging from a mere medical technical tool, the design choice focused on an interactive mirror meant for household use, particularly enabling patients to undergo therapy at various times throughout the day. The mirror serves a dual purpose, as both a wall-mounted and desk-based mirror, facilitating its use in different settings and positions.

7.10 REACTIVATE

Reactivate is an interactive mirror designed to facilitate self-therapy for patients experiencing post-paralytic Bell's palsy and facial synkinesis. Leveraging biofeedback and MT, this innovative device integrates advanced technologies, such as a two-way mirror enabling a display behind the reflective surface. This discreet yet powerful digital interface projects specialized filters aimed at guiding patients through a spectrum of facial muscle exercises, targeted massages, and specific stretching routines to aid in their therapeutic journey. When inactive, Reactivate functions as a conventional mirror, seamlessly blending into its surroundings. However, upon activation, it metamorphoses into a sophisticated therapeutic tool, enabling patients to access and engage with a spectrum of personalized exercises and therapeutic interventions. Its adaptable design allows for wall mounting or unfolding onto a flat surface, presenting users with the flexibility to utilize it for self-reflective moments or therapeutic sessions at their convenience. This dynamic feature set, coupled with its hands-free capabilities, accentuates Reactivate as an adaptive, user-centric, and technologically sophisticated aid in the realm of facial rehabilitation (Figure 7.2).

Wood was selected for its materiality and for its sustainability as a material, offering visual comfort and evoking a sense of familiarity. Diverse wood types are envisaged to accommodate different environmental aesthetics. The mirror comprises a frame designed to house an embedded display with a motherboard and a two-way mirror – a semi-reflective surface allowing content projection while enabling simultaneous viewing through the reflective layer. Additionally, it features a rear panel with a hinged pivot mechanism, facilitating rotational movement for versatile positioning and adaptability within various spatial contexts.

In the design process, careful consideration and experimentation were applied to shape the mirror. Several iterations were explored to refine its ultimate configuration, which incorporates intentional curves to evoke a sense of fluidity and softness. The design also considered the shape of the monitor, strategically positioned within the frame beneath the bidirectional mirror to ensure effective interactivity. The pivotal principle of multifunctionality guided the design, enabling the support to impart a dual conformation to the mirror. The chosen final form represents the culmination of a series of trials and refinements, selected to optimize visual harmony, functionality, and user experience.

The device features fluid lines and an aesthetic appeal that easily fits into home environments, distancing itself from the sterility of typical medical instruments. This design invites users to engage with it and fosters a sense of attachment, making the device feel like a familiar and welcomed presence in their daily lives.

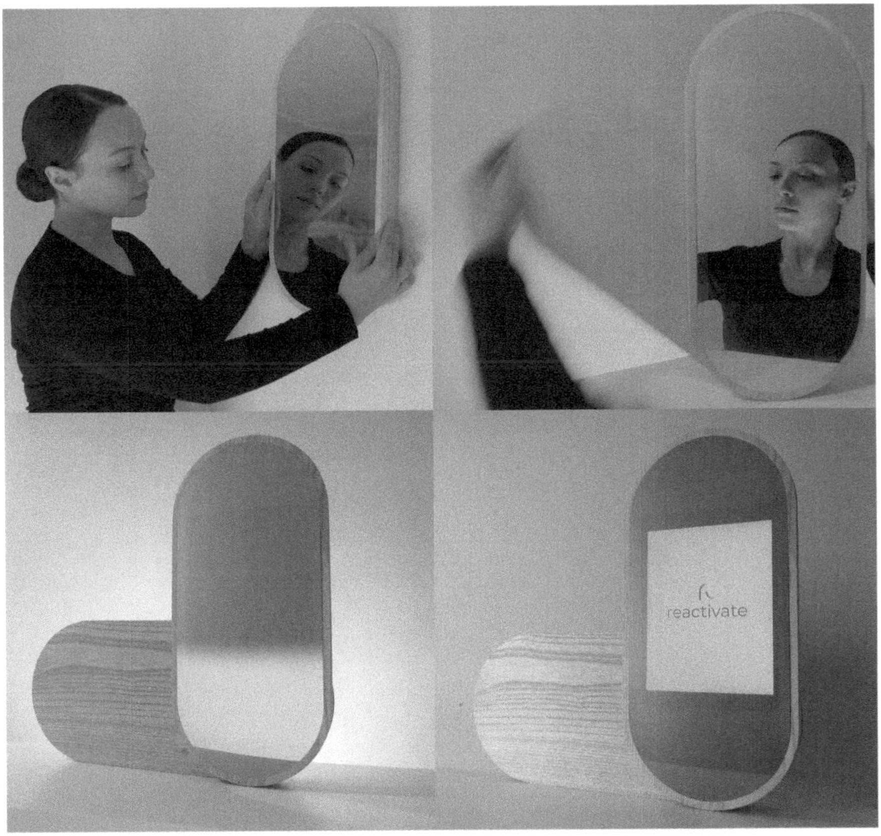

FIGURE 7.2 Reactivate Mirror. (Photograph by the author.)

For resource optimization, this project aims to use engineering components with utmost care and efficiency during the production phase, to slow the depletion of natural resources and reduce waste. By minimizing excess, the project ensures that components are durable and have a low environmental impact (Figure 7.3).

An application for Reactivate has been conceptualized, culminating in the development of a purposefully designed User Interface (UI). The utilization of this dedicated application holds promise, bringing forth a range of tangible benefits. These include the ability to record therapeutic sessions, track progress through a monitoring system, access detailed statistics on a weekly and monthly basis, and acquire additional information crucial for a deeper understanding of the underlying pathology. Of particular significance is the potential incorporation of new filters for therapeutic sessions, under medical supervision. This innovation could enhance the precision of such filters, incorporating features like timers and detailed instructions to guide exercise execution, thereby significantly improving the effectiveness of the therapeutic journey (Figure 7.4).

The project name is Reactivate, expressing the concept of reactivating or revitalizing facial function, aligning with the therapeutic goal of the device. The color

FIGURE 7.3 Render showcasing various wood types: ash, mahogany, and maple. (Made by the author.)

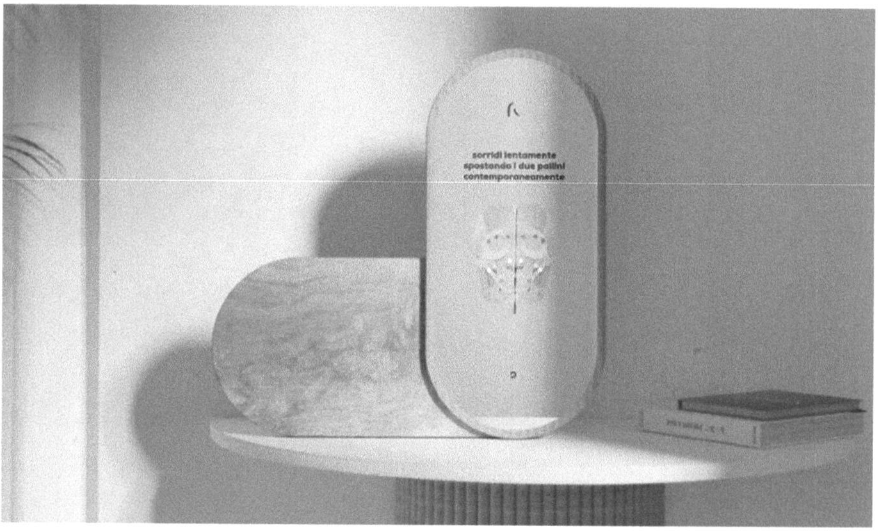

FIGURE 7.4 Render of the mirror simulating the display on, showing the interface. (Made by the author.)

SafetyBlue falls within the blue spectrum and is characterized by a cool hue. The color blue holds various psychological meanings; generally, it is associated with feelings of calmness, serenity, and tranquility. Often perceived as a color capable of reducing stress and fostering concentration.

The rehabilitation program underlying the "Reactivate" project is structured into four main areas of exercises and therapies, each consistently named to align with the project's visual identity, utilizing the monogram "R" as a distinguishing element. These names represent the essence of each area, emphasizing their significance:

Revive: This area, named Revive, is dedicated to stretching exercises. The term "Revive" signifies "rebooting", reflecting the crucial role of stretching in preparing facial muscles for training, preventing stiffness, and contributing to maintaining their tonicity.

Restore: Representing the massage-focused area, Restore stands for "restoring". The name "Restore" emphasizes the vital role of facial massages in keeping muscles "healthy" and functioning optimally.

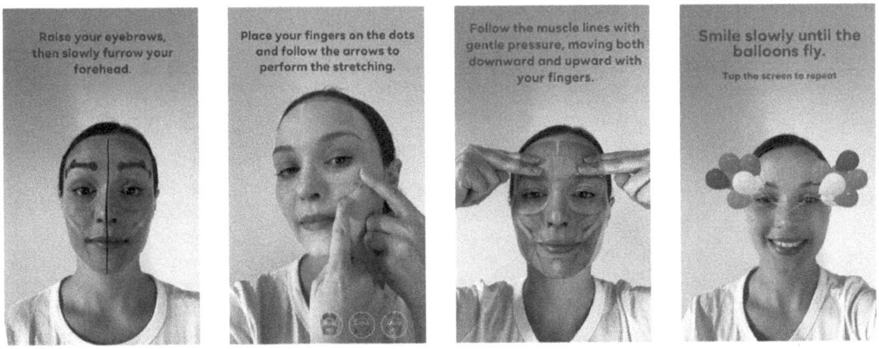

FIGURE 7.5 AR Filters, respectively: Rebalance, Revive, Restore, and Regain. (Photograph by the author.)

Regain: The section called Regain focuses on muscle recovery exercises. "Regain" implies "recovering", highlighting the goal of restoring muscle strength and functionality.

Rebalance: The area dedicated to exercises for symmetry in synkinesis is named Rebalance. This term, meaning "rebalancing", represents the primary objective of this area – improving and re-establishing harmony and symmetry in the face in cases of synkinesis.

A comprehensive and customizable range of filters is available for continuous implementation, allowing personalized adaptation based on individual requirements and preferences (Figure 7.5).

For the "Rebalance" filter, colored spheres have been integrated into the tracking points of the Spark AR template's face mask. These points correspond to various facial areas, including the eyebrows (associated with the orbicularis oculi muscles), the nose (associated with the nasal muscles), the cheeks (related to the minor and major zygomatic muscles), and the mouth (corresponding to the orbicularis oris muscles). Transparency was added to display a visualization of facial muscles, enabling their projection during the exercise. This approach allows the patient to observe real-time muscle movements. The goal is to enable patients to perform symmetric and controlled movements.

The "Revive" filter, designed for targeted facial muscle stretching, incorporates spheres positioned at key muscle points where finger pressure is required. Additionally, arrows have been included to clearly indicate the direction for the correct movement. Red spheres signify the points where pressure should be applied to stretch the muscle on the opposite side, while the black ones indicate the initial finger position from which to follow the movement along the yellow arrows. The "Revive" filter accurately and intuitively guides users in performing targeted stretches for the frontalis, orbicularis oculi, nasal muscles, minor and major zygomatic muscles, and orbicularis oris. This ensures that exercises are performed accurately and precisely, maximizing the benefits in the recovery process.

The "Restore" filter was developed to guide users in performing targeted facial massages. This filter provides a comprehensive view of facial muscles, allowing users to clearly visualize the position of each muscle.

FIGURE 7.6 Prototype showcasing the "Regain" and "Restore" filters in action. (Photograph by the author.)

Consequently, the filter offers a detailed perspective of the muscles, ensuring maximum therapeutic benefit. The "Regain" filter focuses on exercises aimed at improving facial muscle control and preventing synkinesis.

Moreover, the "Regain" filter offers interactive activities to make the training more engaging. For instance, balloons placed at the corners of the mouth respond to the user's smiles by "flying away" when a correct facial expression is performed. This interactive element makes the training enjoyable and motivating, encouraging users to actively engage in the exercise and work on the involved muscles.

Above the filters, an explanatory message guides the user in understanding how to perform the exercises (Figure 7.6).

7.11 CONCLUSION

Bell's palsy, a neurological condition resulting in facial paralysis, often leads to complications such as synkinesis. Our comprehensive studies have demonstrated the effectiveness of addressing these complications with positive outcomes through targeted treatments. The "Reactivate" project originates from a meticulous analysis conducted in collaboration with an experienced physiotherapist who, initially, underscored the importance of adopting a biofeedback-based approach to address the intricate challenge posed by Bell's palsy. This condition, affecting the facial muscles and prone to eliciting involuntary movements, necessitates an advanced and targeted therapeutic strategy. The innovative conception of the project materialized in a sophisticated interactive mirror designed to facilitate therapeutic sessions through the application of filters sophisticated and personalized precisely guiding the facial movements of the patient. The exploration of this methodology was conducted with scientific rigor, involving individuals afflicted with Bell's palsy, who reported a high degree of engagement and motivation during the therapies. This application is designed to further elevate the precision and adaptability of the therapeutic treatment, responding specifically to the individual needs of each patient.

"Reactivate", designed with a focus on innovation, introduces a paradigm shift in therapy delivery for individuals grappling with this condition. By harnessing the potential of interactive filters, this platform empowers patients to engage in daily therapeutic exercises autonomously, ensuring correct and precise execution of exercises without the necessity for frequent visits to clinical settings. The interactive nature of these filters not only facilitates a more engaging therapeutic experience but also promotes consistent adherence to the regimen, laying the foundation for successful and personalized rehabilitation.

The inherent strengths of the "Reactivate" project lies in its meticulous scientific approach, incorporating the expertise of a seasoned physiotherapist and the integration of advanced technology, such as the interactive mirror and personalized filters. These elements contribute to heightened patient engagement and motivation, crucial factors in the success of therapeutic interventions; moreover, its adaptable design caters to diverse needs and strong patient adherence, accurate execution of exercises in autonomous settings, and valuable patient feedback.

Reactivate stands as a beacon in modern therapy by providing an accessible, personalized, and engaging means for patients to manage their condition effectively and independently, marking a significant leap forward in facial rehabilitation therapy.

REFERENCES

Aliwi, I., Schot, V., Carrabba, M., Duong, P., Shievano, S., Caputo, M., Wray, J., de Vecchi, A., & Biglino, G. (2023). The role of immersive virtual reality and augmented reality in medical communication: A scoping review. *Journal of Patient Experience, 10*. https://doi.org/10.1177/23743735231171562

Arya K. N. (2016). Underlying neural mechanisms of mirror therapy: Implications for motor rehabilitation in stroke. *Neurol India, 64*(1), 38–44. https://pubmed.ncbi.nlm.nih.gov/26754990/

Dezeen. (2022). Connectivity by Benjamin Hubert. https://www.dezeen.com/2022/06/14/benjamin-hubert-residential-smart-devices/

Dulak, D., & Naqvi, I. A. (2020). Neuroanatomy, Cranial Nerve 7 (Facial). StatPearls Publishing, 2020. www.ncbi.nlm.nih.gov/books/NBK526119/

Giovanni Lauda. (2022). Totem&Tabù. https://www.giovannilauda.com/totem-tab

Holland, N., & Weiner, G. M. (2004). Recent developments in Bell's Palsy. *BMJ, 329*(7465), 553–557. https://doi.org/10.1136/bmj.329.7465.553

Kanschik, D., Bruno, R.R., Wolff, G. et al. (2023). Virtual and augmented reality in intensive care medicine: A systematic review. *Annals of Intensive Care, 13*, 81. https://doi.org/10.1186/s13613-023-01176-z

Lexus. (2022). Lexus Design Award. https://discoverlexus.com/stories/lexus-design-award-2023/past-winners/2022-winners/

Mistry, R. K., & Al-Sayed, A. A. (2021). Facial Nerve Trauma. Nih.gov; StatPearls Publishing. https://www.ncbi.nlm.nih.gov/books/NBK553095/

Physiopedia. (n.d.). Biofeedback. https://www.physio-pedia.com/Biofeedback

Shi, J., Lu, D., Chen, H., Shu, M., Xu, Y., Qian, J., Ouyang, K., Huang, H., Luo, Z., Wang, C., & Zhang, Y. (2022). Efficacy and safety of pharmacological and physical therapies for Bell's Palsy: A Bayesian network meta-analysis. *Frontiers in Neurology, 13*. https://doi.org/10.3389/fneur.2022.868121

Shokri, T., Patel, S., Ziai, K., Harounian, J., & Lighthall, J. G. (2021). Facial synkinesis: A distressing sequela of facial palsy. *Ear, Nose & Throat Journal*, NP382–NP391. https://doi.org/10.1177/01455613211054627

Soleimani, M., Ghazisaeedi, M., & Heydari, S. (2024). The efficacy of virtual reality for upper limb rehabilitation in stroke patients: A systematic review and meta-analysis.

BMC Medical Informatics and Decision Making, *24*(1). https://doi.org/10.1186/s12911-024-02534-y

Taghian, A., Abo-Zahhad, M., Sayed, M. S. et al. (2023). Virtual and augmented reality in biomedical engineering. *Biomedical Engineering Online*, *22*, 76. https://doi.org/10.1186/s12938-023-01138-3

Volk, G. F., Roediger, B., Geißler, K., Kuttenreich, A.-M., Klingner, C. M., Dobel, C., & Guntinas-Lichius, O. (2021). Effect of an intensified combined electromyography and visual feedback training on facial grading in patients with post-paralytic facial synkinesis. *Frontiers in Rehabilitation Sciences*, *2*. https://doi.org/10.3389/fresc.2021.746188

Wang, C. S., Sakai, M., Khurram, A., & Lee, K. (2021). Facial nerve palsy in children: A case series and literature review. *Otolaryngology Case Reports*, *20*, 100297. https://doi.org/10.1016/j.xocr.2021.100297

Wei, D., Hua, X.-Y., Zheng, M.-X., Wu, J.-J., & Xu, J.-G. (2022). Effectiveness of robot-assisted virtual reality mirror therapy for upper limb motor dysfunction after stroke: Study protocol for a single-center randomized controlled clinical trial. *BMC Neurology*, *22*(1). https://doi.org/10.1186/s12883-022-02836-6

8 Regenerative Approach to Design Sustainable Tourism Experiences for Aquatic Ecosystems

Elgani Elena, Giulia Ettori, Umberto Monchiero, and Elisa Schembri

8.1 INTRODUCTION

8.1.1 Hospitality Industry: A Field of Interest for Designing Responsible and Sustainable Tourism Experiences

In the experience economy (Gilmore & Pine, 1999), the system of spaces–products–services developed and realized to welcome guests during tourism experiences can be defined as the "hospitality industry". This expression has widely spread from the United States since the 1970s and is now used globally. Nowadays, the hospitality industry is referred to all those commercial activities related to the travel experiences; in fact, it takes into account all the accommodation spaces, including hotels, hostels, co-living, resorts, and camping, but also all the spaces belonging to the Ho.Re.Ca., which is an expression widely used in Europe that stands for "hotels, restaurants, and cafés" categories, and their related products and services as a well-organized industrial system. This type of industry is characterized by a specific supply chain. Dedicated design processes for the organized experiences, with structured guidelines for designing spaces, products, and services, are specifically developed for the hospitality industry, focusing mainly on business models and management strategies to create the best option to maximize the incomes and benefits, offering the best possible experience to guests.

This cluster of various spaces–products–services can be considered as a collection of *machine à habiter* (Le Corbusier, 1923), quoting the famous expression for domestic spaces by the architect Le Corbusier. Indeed, the architect sentenced this expression in the early 1900s, when the new industrial serial production was expanding all over the western world, offering a stable and continuous production and replacement of standardized electrical machines as well as furniture and accessories for the houses of the Modern Movement. By looking at the articulated relationship between all the elements that make up the living space, the architect intended to govern the complex relationship between technological solutions, design products, and

DOI: 10.1201/9781032719290-10

living habits in a harmonious manner, offering a response to the desires and needs of living through design.

At the same time, nowadays, the entire tourism system, based on lived and engaging experiences in spaces designed to accommodate them, contributes significantly to the economy of all the countries in the world being a strong input for improving the quality of life for the local communities as researched by the World Travel & Tourism Council. Indeed, a networked system, based on hotels and their related spaces, hosts, every day, numerous travellers around the world. Developing countries, in particular BRICS (Brazil, Russia, India, China, and South Africa) countries, are emerging as the main actors of this process as they're increasingly aware of their economic potential in tourism. As an example, in 2023, the tourism sector contributed to 6.8% of Africa's economy, bringing to the continent US$192 billion, having a jump of 13.8% from the previous year, returning to the pre-pandemic levels. The goal by 2034 is to reach US$321.6 billion and to create a total of 38.8 million jobs, enhancing the employment thanks to the Ho.Re.Ca. sector (World Travel & Tourism Council, 2024). As affirmed by Rasool, the importance of inbound tourism has grown exponentially because of its growing contribution to economic growth in the long run. It enhances economic growth by augmenting the foreign exchange reserves, stimulating investments in new infrastructure, human capital, increasing competition, promoting industrial development, and creating jobs (Rasool et al., 2021). That growth in GDP could lead to a further increase in international tourism.

As an example, considering the hospitality industry, in 2017, Accor acquired 50% of Orient Express' shares from Société Nationale des Chemins de fer Français (SNCF) with the aim of relaunching the famous luxury train brand. In 2020, the project saw an acceleration with the creation of Orient Express *La Dolce Vita*, with Trenitalia-Gruppo Ferrovie dello Stato e Fondazione FS Italiane, a new luxury train that will travel across Italy, with a design concept celebrating the golden age of train journeys. At the same time, the company announced the opening of a series of Orient Express hotels, bringing rail-travelling experiences to the luxury segment of the hotel industry. The company's idea behind this is to diversify its portfolio and position itself in an exclusive niche in the luxury travel and hospitality market (Nandi, 2024), not only by restyling the interiors and building new infrastructures, but also through the refurbishment from scratch of such an old gem as the train, giving it a new life and showing the potential of restoring something that has not been used for a while. This project, developed in Italy, will soon spread to Europe and other contexts. In 2025, in Saudi Arabia, it will be possible to travel on the *Dream of the Desert* train (Saudi Arabia Railways) to experience luxury hospitality by rail.

Luxury tourism sector is in fact one of the fastest growing sectors in the hospitality field, with an expected annual tax growth at CAGR of 7,6% from 2024 to 2030, and an ever-increasing demand for unique and personalized experiences (GVR, 2024). Moreover, it is the market segment in the travel industry that can significantly invest in sustainability issues, thanks to the availability of significant investments. The aim of this experience's expansion is to create loyalty by offering an immersive and engaging experience that starts from the journey and continues through a luxury stay at the hotels and all of this is possible thanks to the certainty that the brand's history and reputation will automatically attract high-profile customers willing to pay a premium price for the experience. This type of investment requires a high initial expenditure, but aims to create a stable and sustainable revenue stream in the long term; for this reason, the project is supported by partnerships with renowned designers and specialized construction companies, which ensures high quality in the final product and more efficiency in cost management and in reducing the environmental impact of spaces–products–services; additional partnerships with local tourism organizations amplify visibility and integration with key destinations (Hotel Management Network, 2024). The main idea is to create a value chain that extends far beyond the simple stay and educates the guest to new, slower, more welfare-conscious and sustainable behaviours, reducing the environmental impact of tourism by favouring sustainable mobility. At the same time, this type of inbound tourism intervention stimulates the economy of several sectors complementary to tourism, such as transport and infrastructure, the local economy related to food and beverage, and local productions, as well as the whole system of related technological services. This investment in new infrastructure is therefore an example of how large hospitality groups are aiming to redefine the concept of luxury, and also sustainability in travel, through the integration of a design-driven approach, based on unique experiences and storytelling, well-designed strategies, and an interconnected supply chain through a design-led systemic approach, positioning themselves as pioneers in an increasingly competitive market.

Another significant investment in new hotel infrastructure is the expansion of Marriott International which is renovating several existing facilities, such as the W Hotels in major cities in the United States, including NY and Hollywood; this strategy aims to satisfy a new generation of luxury travellers who are looking for unique and sustainable experiences, characterized by adventure and exclusivity since they expect more than simply gourmet food and luxurious furniture, and the goal is to position the brand as a leader in destinations that define "new luxury", focusing on authenticity and ethical travel practices (Onadele, 2023).

It can therefore be stated that in recent years, updated tourism policies and economic and political strategies, as well as travel promotion strategies operated by private companies, are supporting the economic growth of different countries through the application of specific development processes in which the system of spaces for hospitality has an important role. For small islands and coastal communities, the main source of income comes from water-based tourism leisure activities, from recreational experiences connected to the sea, and from the income generated by the entire system that supports these kinds of activities. Water-based tourism, and in particular costal tourism, is a relevant component of the global tourism industry, and it has been well documented in the literature (Brett, 2021) and through research, conducted by the World Tourism Organization (2018), and it is highly dependent on the quality of water ecosystems to attract visitors. This means that the sector is extremely vulnerable to threats related to climate change, such as pollution, significant temperature variation, natural disasters, and biodiversity loss.

In recent years, Italy, whose coastal tourism represents an important source of income, has experienced a significant increase in climate change events, including storm surges and sea level rise, which have led to the destruction of coastal settlements and infrastructure damaging them and various services of the hospitality structures, resulting in a massive loss of money (Legambiente, 2023). The CNR ISMAR (National Research Council & Institute for Marine Sciences) report (2023) predicts that sea levels could rise to 1 m by 2100 if no significant measures are taken against global warming, therefore becoming a possible catastrophic event for coastal zones all over the world with direct consequences on economic activities and tourism, and many more scientific studies and reports clearly indicate the need for adaptation and mitigation strategies to protect coastal infrastructures and ecosystems.

All these aspects indicate that the field of tourism, and particularly the hospitality industry, is an area of interest and focus for design (Scullica, 2008), which is a discipline capable of promoting research and innovation through methodological approaches, processes, and established practices (Scullica & Dominioni, 2022) that can connect scientific fields and social sciences (Collina, 2005). This area is particularly interesting for research into design for sustainability with an eco-systemic approach to interior design (Elgani & Monchiero, 2024).

In this sense, Le Corbusier's thinking has a meaning also today: as a *machine*, hospitality spaces are considered as a well-designed mechanism, and as a system to inhabit by creating more qualitative spaces for humans' life (Del Zanna, 2009). However, the new hospitality industry should become more flexible to adapt private and shared spaces to different needs, as well as technological evolutions. Indeed, technological implants and groundbreaking devices, specialized solutions for the climatization, domotic systems, energy saving technologies as well as devices for data transmission and innovative products for personal care are increasingly introduced in hospitality spaces to improve space performances and customers' comfort. All this requires intense design activity aimed at identifying the appealing solutions to engage and satisfy guests, while reducing the environmental impact of space–product–service systems.

In recent years, the relationship between design and the hospitality industry is becoming more and more consolidated, not only for the aesthetic aspects related to the interior design of spaces, but also for the identification of sustainability processes, based on systemic, strategic, and processual approach (Elgani & Monchiero, 2024), to be applied to the hospitality system.

After the halt imposed by the pandemic, design has put particular interest in focusing on how to support, through specific strategies, the design of sustainable and responsible tourism experiences based on a system of spaces for contemporary travellers as well as local communities that host them. Nowadays one cannot only reflect about creating a "location" with a hospitality space, but the aim is to establish a "destination" following the initial idea of 1993 sentenced by the WTO which defined "tourist destination" a place with a network of structures and points of interest, a "goal-place" (HSL Hospitality value, 2024). New trends research suggests that well-being destinations, those adopting a systemic approach, offering a sustainable location, attractive landscapes, local culture, authenticity, food and wine tourism, and places and areas dedicated to physical exercise and beauty treatment will be the most selected ones (Ejarque, 2020): nowadays guests want to live engaging experiences in amazing places and long for something to remember at the end of their journey, a way to relive the past moments. This will necessarily have to be combined with reducing the impact of these activities on the environment and local communities.

Current reflections and debates on sustainability are progressively defining a responsible approach to tourism (Goodwin, 2011) that can bring a positive impact on the territories and communities that host tourism activities. A new regenerative design approach must ensure that these experiences don't just remain so; in this context, particular attention should be paid to hospitality spaces: from small urban hotels and gem-resorts placed in unique ecosystems to big middle-priced chain hotels have to be designed to support sustainable innovation and become a new typology of complex systems in order to be small seeds to develop and expand a truly sustainable tourism.

8.1.2 The Pandemic as an Opportunity for a New Start for Tourism

The global health crisis caused by the pandemic was a setback for the tourism sector. According to the Organization for Economic Cooperation and Development (OECD), reductions in global tourist flows ranged between 60% and 80% compared to 2019. As written in the Sustainable Tourism Report commissioned by the Ocean Panel for a sustainable ocean economy, the big and sudden pause that has affected the entire world due to the forced lockdown has had a serious impact on every sector, and global tourism has been the most affected one (Ocean Panel, 2022). At the same time, the pandemic was an opportunity to understand the relationship between climate change, which irreparably alters natural ecosystems, and collective well-being. Indeed, travel and tourism directly and indirectly rely on the natural world in myriad ways. The wonders of nature in settings such as beaches, national parks, or mountains are what draw millions of visitors to them every year. And sourcing the fuels, water, and other resources needed to transport, feed, and accommodate these visitors has important implications for ecosystems and biodiversity across the world (World Travel & Tourism Council, 2023).

The Environmental Impact on Global Tourism Report (WTTC, 2023) states that tourism worldwide significantly affects resource usage, including water, and is directly accountable for 8% of yearly greenhouse gas emissions (Lenzen et al., 2018). Tourism also contributes to air pollution, high energy consumption, and waste production. Unfortunately, this figure is expected to increase as mass tourism continues to rise, leading to difficult-to-manage challenges and phenomena such as overtourism, characterized by heightened transportation usage, the spread of hit-and-run consumption behaviours, rapid exploitation of regions, and increased demand for related products and services. Traveller behaviours have a significant impact on ecosystems, and this was demonstrated during the global pandemic by Covid-19. Scientists have in fact recorded that during the lockdown, individuals changed their everyday routine and behaviour; there was in fact a significant reduction in greenhouse gases, and other air, water, and land-polluting outputs (Rume & Islam, 2020). This was a solid representation of how individuals can have a positive impact on our environment by changing personal habits towards all-encompassing global wellness and developing innovative approaches (Spathonis et al., 2020). This has allowed the entire tourism system to understand the fragility of some processes and the operational limits of some hotels. It also sheds light on the impact they have on local economies, their related communities, and natural ecosystems. The pandemic, as well as phenomena related to climate change, has rapidly led the entire tourism system to question what the future innovation trajectories for truly sustainable and responsible hospitality might be, in terms of environmental, social, and economic sustainability, according to the Sustainable Development Goals (SDGs). Considering the 2023 World Travel & Tourism Council in Kigali, the entire tourism sector, and particularly the hospitality industry, is gradually becoming aware of the importance of paying more attention to real sustainability. The aim is to actively work towards sustainable initiatives and typologies of spaces–products–services that promote environmental health, inclusion, and well-being for all. The focus is not only about reducing the tourism impact, in terms of fewer emissions and more responsible resource management, but the real challenge is to understand how tourism experiences, and the system of spaces–products–services related to them, can contribute to reduce certain environmental problems such as water pollution, promoting a real regeneration of ecosystems.

8.1.3 Microplastics Affect the Development of Water-Based Tourism

The international scientific community is currently faced with many water-related problems, the most prominent of which are the continuing rise in sea levels and the uncontrolled spread of pollution in aquatic environments – especially in rivers, lakes, and oceans, very close to the coastlines in direct contact with human life. Plastics can enter the ocean from rivers where they meet the coastline, blown from land, or are discharged overboard by vessels (Geyer et al., 2017). Considering coastal tourism, given the fact that millions of tonnes of plastic enter the marine environment annually, the severe conditions of polluted waters are becoming the biggest problem, and it can only worsen since quantities are expected to increase in the coming years (Barrett et al., 2020). Around 80% of marine pollution is from land-based (Reichelt-Brushett, 2023). Europe, for example, releases between 307 and 925 million pieces of waste into the ocean every year, of which 82% is plastic (González-Fernández et al., 2021). The leading causes of this pollution are garbage and big plastic items (Tekman et al., 2022), which, however, during their stagnant period in water, break and fragment more and more into very thin pieces turning into the so-called microplastics, fragments with sizes less than 5 mm (Zhang & Zhang, 2021). The abundance of microplastics in coastal as well as in remote oceanic zones, where collection is more difficult, determines the continuing destruction of marine ecosystems, with a particular focus on coral bleaching, the contamination of all animals, from fish to large mammals, and the constant loss of biodiversity. Indeed, "due to the mounting anthropogenic pressure from pollution and over exploitation alongside the rise in the global temperature, coral reefs around the world have suffered mass coral bleaching events that increase their vulnerability" (Battaglia, 2023). There is often a difference between the narrative associated with certain destinations, described as natural and ideal, and the poor reality. In fact, the pollution of the islands, coasts, and seas is not often talked about, so few are really aware of the serious situation of the corals and the loss of this collective heritage. But the reality is even worse: all the waste produced by more than 110 luxury resorts on the small island of the Maldives ends up anywhere, and as a solution, in order not to affect the enormous economic impact of tourism on the island, the Maldivian Government adopted a not so healthy solution: they collected all the trash coming from single-use plastic items and created what, in 1992, was the biggest artificial garbage-island, Thilafushi, only 6.85 km away from the Maldivian capital, Malè (Del Dot, 2019). This pollution, besides being an ecological threat to the local environment, could also become a significant barrier to the Maldives' touristic development. Over time, the accumulation of debris and sewage could deter visitors and lead to the closure of some luxury resorts due to the unpleasant conditions (Messina, 2016). However, there might be a silver lining to this issue, as it opens up new horizons and ideas by inspiring the design community with a scientific background to do environmental research and develop environmental-related products/projects. At the same time, important agencies in the field of environment, such as the Ocean Panel, a global initiative working towards a sustainable ocean economy, have developed a set of principles to address the challenges of our time. Given the fact that coral reef tourism attracts over 350 million people per year with an estimated value over 70 countries of $36 billion, with territories having "million-dollar-reefs" (reefs that generate over $1 per square kilometre per year in tourism spending), the Panel stated that the most important of these principles is the restoration of the natural environment and the revitalization of cultural heritage and communities (Ocean Panel, 2022).

Considering coastal tourism, Ocean Panel has outlined the guidelines to protect the marine ecosystem, to support ecosystem regeneration, biodiversity conservation, and resilience, and to invest in local communities. Specifically, some topics are considered significant for design-led change:

- the ability to rapidly adapt and respond with a multidisciplinary approach to emerging problems due to climate change and the issue of biodiversity loss, which often go unnoticed.
- the ability to embrace a more specific and non-standard target, understanding and recognizing responsible evolving values, as well as the importance of promoting the resilience of local communities and indigenous people, reducing inequality through the promotion of equal opportunity, inclusion, and equitable distribution of benefits.

These topics support the definition of pathways to raise awareness and stimulate change. For example, in addition to water pollution because of climate change, the rising sea levels are a topic very visible according to the challenges faced by some water cities such as Venice, one of the most historic water cities in Europe. Art disciplines decided to be at the service of this problem by spotlighting it: at the 2023 Biennale in Venice, the research exhibition "Everybody talks about the weather" (Fondazione Prada Venezia, 2023) took atmospheric conditions as the starting point for investigating the emergency of the climate crisis. The ecological art practice had the potential to raise collective awareness and inspire new forms of action. As shown, historical artworks traced the numerous ways in which climate and weather have shaped our histories and how humanity has dealt with our daily exposure to weather events, leading visitors to imagine how these may be addressed in the future and what role people may play in changing current climate conditions. Scientific spotlights are developed in collaboration with The New Institute Center for Environmental

Humanities (NICHE) of the Ca' Foscari University of Venice. As stated by Miuccia Prada, President of the Prada Foundation, where the exhibition takes place:

> the goal is to understand the environmental crisis and its undeniable impact on our lives by attracting attention, representing, and analysing meteorological phenomena. Climate is a global issue that affects the actions and destinies of people around the world. Talking about the weather today therefore means talking about and worrying about everyone's future
>
> *(Fondazione Prada Venezia, 2023)*

Correspondingly, design can (and should) be a discipline that tries to offer solutions and could offer strategic approaches to find more sustainable alternatives, being a direct support against the environmental crisis. This could be done not only by catching public attention, but also by proposing practices and systems able to connect different disciplines and research areas, that can be devoted to improving oceans and seas pollution issue along with the rising sea levels one; this also promotes the safest, most durable, and stable coastal tourism.

8.2 THE DESIGN APPROACH FOR THE DEVELOPMENT OF SUSTAINABLE AND RESPONSIBLE TOURISM

In the described contemporary scenario, designers can play a key role in the transformation of the hospitality industry and more generally in the re-designing of tourism experiences that requires the overcoming of the contemporary paradigm and the definition of a new set of competences and processes, closely linked to a holistic approach, capable of connecting environmental, social, cultural, and economic factors. This innovative set can be defined by implementing the multilevel, multiscalar, and interdisciplinary systemic approach of design, which is capable of strategically connecting different competences (Elgani & Monchiero, 2024), coming from very distant disciplinary fields such as management and ecology, biology and naval and nautical design, through specific methodologies that allow the generation of new knowledge and skills.

In particular, design can support the dissemination of a regenerative approach to tourism that considers, from the outset, humans and nature as connected and intertwined categories. With this approach, based on an ecological worldview, the goal is to create the conditions for all life to renew and restore itself (Reed, 2007; Mang & Reed, 2019; Dredge, 2022).

In this way, regenerative tourism seeks to ensure that travel and tourism reinvest in people, places, and natural living systems. This approach supports the long-term renewal and flourishing of our social-ecological systems.

From the design perspective, a regenerative approach to tourism can systematically consider three areas:

1. Design can support the definition of a new mindset for operators and managers to learn about regenerative strategies and practices. This will enable them to redefine the entire touristic experience, specifically aiming to find strategies that are place-based, community-centred, and environment-focused (Dredge, 2022) to restore each specific natural environment and revitalize each cultural heritage and local community.

2. Design can be used to re-design the guest experience and to develop innovative strategies capable of connecting the new generations of clients' needs with physical spaces and natural environments, stimulating co-existence while restoring the natural ecosystems. A very important aspect of achieving sustainability goals is, in fact, related to guest education, promoting the learning of environmentally friendly behaviour.
3. Design can reconfigure the entire system of spaces and products related to the tourism experiences, supporting the re-design of existing facilities, the definition and realization of new spaces, and the development and production of innovative products and services for hospitality.

It is indeed increasingly important to think about adaptive hotel facilities, in which systems of spaces and services demonstrate versatility of use, without being relegated to a single purpose. This means designing systems of flexible and adaptable spaces and services that lend themselves to change use over time and that are designed not only for guests, but to create connections with local communities. As affirmed by Dianne Dredge, "How positive would it be if tourism had as its central goal not to extract economic wealth to be distributed elsewhere, but to nourish local places, their environments, and communities so that they could flourish?" (Dredge, 2022).

Therefore, the designer is called upon to study strategies to engage people with different backgrounds as well as communication campaigns to promote new approaches, but above all to design innovative solutions in terms of space–product–services systems. The goal for the future is to allow more people to access sustainable tourism experiences, transforming tourism into not only an ecological and responsive activity for the environment, but also a way to restore biodiversity and regenerate natural ecosystems, achieving the 2015 United Nations SDGs. Thanks to its ability to connect different fields and establish new relationships between social trends, lifestyles, scientific developments, technological tools, and biological research, design can help "materialize" "possible and desirable futures, making them tangible through design thinking" (Bertola, 2022).

8.2.1 Understanding Sustainable Approach to Design for Hospitality

For many years, lots of research projects have been presented in the design field, achieving important results regarding single products or innovative PSS (product–service–systems) to define strategies that mitigate the problems caused by human actions on the environment and reduce the impact of dangerous products on the planet. According to the research of Ceschin and Gaziulusoy (Ceschin & Gaziulusoy, 2016), networks such as LeNS (Vezzoli, 2018) have brought significant results in the product–services field. At the beginning, the pollution prevention was applied to industrial processes, thus speaking about cleaner technologies; it was then expanded to industrial products (green products) to achieve a bigger impact. The approach to product innovation has also undergone a significant upgrade, through the adoption of more radical measures (new sustainable products). Recently, the attention has shifted to the whole product and services system that together can satisfy the request of the market, creating sustainable business models, considering the entire supply chain and working on principles promoted by the circular economy, as the R10 model promoted by Jacqueline Cramer (Elgani & Monchiero, 2024).

From this path, the potential role of design becomes clear: to unify a new innovation in product–service systems in relationship to space planning and development with a focus on a small design scale for the interior design and built environments. The hospitality sector is a potential territory where to act, since there is still much to be done, thanks to the presence of very complex systems. In fact, in 2023–2024, the hospitality sector accounts for 1% of global emissions according to the Sustainable Hospitality Alliance. In the hospitality industry, the consumption of spaces, considering material and finishing solutions, but also products, like systems for equipping spaces, furniture, and objects, like decorative elements as well as towels, amenities, etc., quickly makes spaces obsolete and stimulates processes of "soft refurbishments" or total refitting. This partial or complete refurbishment of interiors has a relevant impact on the planet; for instance, in hotels, refurbishments on Furnishing, Finishes, and Equipment (FF & E) are scheduled at short intervals, every 5–7 years, and for food and beverage spaces even more frequently (Perkins & Will, 2021). In these contexts, a new way of approaching design, as well as certain innovations, especially in building construction or interior design systems, clashes with commercial logics, which do not have a real sustainability as their first objective yet.

For this reason, it is necessary, in the first place, to apply a new ecological thought, which has roots in the long reflection of design (Papanek, 1971) to which today it is necessary to add new goals. The radical transformation of the mindset is important in order to acquire a real ecological thinking that allows for a different approach to design as advocated by Studio Albori, which curated the Padiglione per la Santa Sede, Vatican exhibition, at 18th Mostra Biennale di Architettura, in Venice – Italy (20 May–26 November 2023), inspired and supported by the Pope Francesco's reflections in the encyclicals *Laudato si', 2015*, and *Fratelli tutti, 2020* (Elgani & Monchiero, 2024; Cremascoli, 2023).

The new challenge that design is capable of leading is to overcome the concept of sustainability, often limited to the minimization of the environmental impact that a system of spaces–products–services can generate and its footprint reduction.

Currently ongoing financed research in design at the Department of Design – Politecnico di Milano – is trying to develop new methodologies for analysing the context, as well as new tools for designing and new nature-based or nature-inspired solutions in terms of spaces–products–services, which not only reduce the impact of human actions on the environment, but are also capable of restoring, repairing the damages directly or indirectly caused by human actions, and regenerating natural ecosystems. This leads to the achievement of several nature-positive results and to the restoration of those conditions, thanks to which plants, animals, and individuals can coexist. The final goal is to promote people's well-being while preserving the planet, but above all to support the co-existence of animals, plants, and people in which there are no inequalities, so that individuals can co-evolve with nature, promoting an inter-species approach to design, according to the reflections by studio Ossidiana "the way to live together as a space of possibilities" (Kuitert, 2023). To achieve this goal, it is important to note that design can develop a systemic and interdisciplinary approach that can connect not only its different fields (the project of spaces, products, services, their relationship with materials and technologies, and the integration of innovations and communication systems), but most importantly those regarding other disciplines, like the study of the environment and the support of technological and economic innovation one to another.

For this reason, the direct observation of nature can suggest an approach to the design of new tourist experiences in which the environmental dimension of sustainability is combined with the improvement of individuals' living conditions. It is therefore a process of not only designing new engaging and exciting tourist experiences, but also defining new systems of spaces which, at the same time, can welcome visitors and contribute to the revival of the natural environments in which they are located. The process is not simple, also considering the slowness with which the hospitality industry can accommodate a radical transformation that requires the overturning of some consolidated logic; but at the same time, this process opens new development horizons, since working in partnership with nature really allows us to restore the ecology of places. This challenge is particularly interesting for the new generations, who are the today and tomorrow's inhabitants of the Earth and presents itself as a significant area of analysis and research, but also as an experimentation and teaching experience for the training of a new generation of professionals, capable of consciously and positively facing the challenges of the future.

8.3 HÒSTRAKA: A NATURE-INSPIRED REGENERATIVE FLOATING RESORT

Given the necessity for interdisciplinary collaboration between different fields of research to reimagine hospitality models and formats, management systems, design approaches, and spatial organization to promote the transition to sustainable tourism experiences, a variety of initiatives are emerging to advance the global shift toward sustainable tourism. Among these, the Sustainable Hospitality Alliance – an association committed to the United Nations SDGs – advocates for "Net Positive Hospitality" by fostering continuous efforts toward both environmental and social sustainability and responsibility (Sustainable Hospitality Alliance, 2023).

One of the foremost operational and promotional initiatives in this space is the Sustainable Hospitality Challenge (SHC), the largest international competition for hospitality students worldwide, which since 2023 also includes design school participants. Initiated by Hotel School The Hague, the SHC aims to guide the hospitality sector through innovative projects focused on system-wide transition. In 2023, the team of students Rosanna Caldarella, Giulia Ettori, Davide Grasso, and Elisa Schembri from the Bachelor's degree in interior design, School of Design Politecnico di Milano, with the supervision of research fellow Elena Elgani, Department of Design Politecnico di Milano, joined the SHC, using the potentials of design disciplines (particularly interior design) to rethink sustainable hospitality formats. By applying a "learning by doing" methodology, they developed Hòstraka, an innovative project inspired by aquatic ecosystems. The aim was to realize a sustainable floating resort for diving lovers, focusing on design as a facilitator of the reconnection between humans and nature through the cross-fertilization of disciplines to achieve impactful solutions.

The following paragraphs analyse the Hòstraka design concept as a practical case study for the development of regenerative hospitality experiences.

This project emphasizes how design can create impactful new connections across disciplines, utilizing a nature-inspired design methodology rooted in observing nature's offerings: this process aligns therefore with the principles of the "blue economy" (Pauli, 2010), to support sustainable expansion within marine and aquatic environments; in this context, water is considered not only a vital resource to be protected but also as a facilitator for new ways of managing aquatic resources (Thackara, 2017).

The name of the design concept comes from the English word "host" (verb for welcoming someone) and the Italian word "*ostrica*" (name of the oyster), and it draws inspiration from the natural purification actions of oysters, which filter water and turn impurities into valuable pearls. The team envisioned a system of public and collective spaces anchored to the seabed, with floating private rooms by navigation, as they are essentially boats, that actively support ecosystem regeneration (see Figure 8.1).

Diagram developed by Hòstraka Design Team (Rosanna Caldarella, Giulia Ettori, Davide Grasso, Elisa Schembri).

The final product is a floating resort with interlinked and separated public and private spaces. Private spaces in particular have a dual nature: they serve as individual floating suites and also eco-friendly transport allowing guests to explore the surrounding natural environment autonomously.

Each suite is equipped with a microplastic-filtering engine that cleans water as it moves, creating a positive environmental impact proportional to the number of guests sailing.

In the shared spaces, guests can instead participate in a wide range of activities, from educational lessons to workshops in research labs, centred around microplastic pollution.

Hòstraka thus provides an eco-resort concept blending engaging aquatic adventures with sustainable practices, offering guests the chance to contribute to cleaner waters by participating in microplastic collection. Through the uniquely designed "Sustainable Pearl Experience" (see Figure 8.2), guests gain first-hand insight into pollution by actively using their movable suites, named "Straka" to collect microplastics.

The project's key objectives include the following:

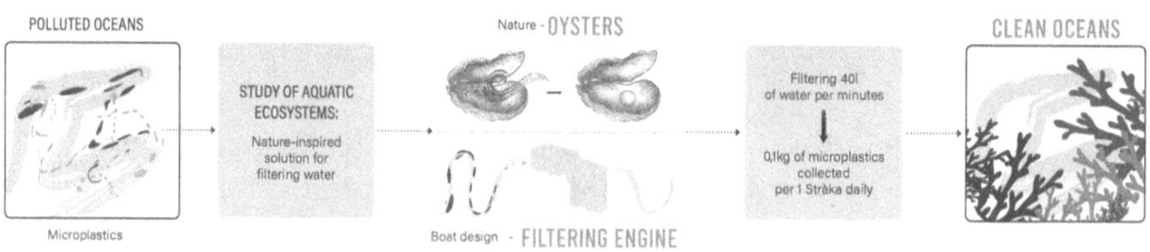

FIGURE 8.1 The development of a water filtering system, a nature-inspired and design-based solution.

FIGURE 8.2 The "Sustainable Pearl Experience" is the guest journey focused on responsible actions and leisure activities. (Diagram developed by Hòstraka Design Team (Rosanna Caldarella, Giulia Ettori, Davide Grasso, Elisa Schembri).)

1. Significantly reducing the environmental footprint that hospitality structures leave on natural ecosystems. Common areas function as innovative floating houseboats, while private spaces are compact catamarans hosting two to five people each; connections between these spaces are facilitated by modular floating blocks, allowing for flexible resort expansion. The floating structures are anchored in specific points using oceanite blocks, mineral composites designed to foster coral growth and marine biodiversity.
2. Revitalizing the ecosystem at the project site, utilizing an interior design-oriented approach that addresses the rapidly evolving hospitality sector, where creative, multidisciplinary solutions are essential; designers may therefore assume a unique role in interdisciplinary teams, where design thinking is crucial from the initial stages of development to drive the industry forward.
3. Fostering a collaborative framework where guests actively participate, not only interacting with each other but also engaging with local communities in a variety of activities (such as guided tours and meetings). Hòstraka also promotes a design approach that supports the co-existence of human and non-human species, transforming hospitality spaces into meeting grounds for diverse life forms. By addressing both reduction and regeneration, Hòstraka highlights the necessity to restore ecosystems as a solution to preventing the irreversible loss of local flora and fauna while introducing well-tested technologies to support the hospitality industry in reaching higher goals of sustainability.
4. The mobile suites feature an integrated engine designed to filter microplastics from seawater by drawing in water, circulating it for cooling, and passing it through a filtration circuit before discharge. Guests contribute to cleaning the ocean as they navigate to diving sites, where local guides help them experience nature. This simple activity has an enormous cumulative impact; if many guests engage in this behaviour, over extended periods of time, it could have a profound effect on the global travel and tourism industry, enabling sustainable activities without sacrificing quality experiences.

8.4 DESIGN FOR SUSTAINABILITY: HÒSTRAKA LIFE CYCLE DESIGN

A central focus of this project is the assessment of Life Cycle Design (LCD), a modelling tool that evaluates environmental impacts associated with a product's entire lifespan, encompassing raw material extraction, processing, manufacturing, distribution, use, disposal, and/or recycling stages (Vezzoli, 2018). LCD is particularly valuable in the design process because it offers a comprehensive view of environmental impacts, thereby helping avoid the unintended consequences of optimizing one indicator at the expense of others. During the design phase, it is essential to integrate sustainability-oriented strategies to reduce the overall environmental impact of the final product or service. Reducing potential design errors involves making well-considered choices that address each stage of the product's lifecycle; for example, an important strategy is to design products with modularity in mind, prioritizing assembly methods that require no adhesive substances. This approach enables flexible component construction based on modular assembly, with an emphasis on dry assembly methods, and is suited to specific requirements, while also facilitating disassembly and extending the lifespan of individual parts in an environmentally friendly manner.

The Hòstraka project has been deliberately developed with a component-oriented approach (Elgani & Monchiero, 2024) with modularity as a core principle, implementing circular design elements that prioritize component reuse. Its assembly and disassembly can be easily modified for deployment across various sites, ensuring both essential construction elements and alignment with local needs.

In regions with pre-existing infrastructure that meets resort requirements, Hòstraka complements these amenities; in less serviced areas, it introduces vital resources, promoting local growth and sustainable practices. The LCD approach of Hòstraka (see Figure 8.3) is based on principles of sustainable design (Vezzoli, 2018). Due to the complexity of the aspects involved in the floating experience in relationship to naval design, the floating spaces were developed with the input of nautical design experts (A. Ratti and A. Bionda, Design Dept., Politecnico di Milano) and key stakeholders (Matteo Costa – H_2O Design).

Particularly notable stages in the LCD process are as follows:

- In the **pre-production** stage, careful sourcing and acquisition of materials are essential. For Hòstraka, materials are 100% certified for sustainability, with sourcing, production, and processing prioritized to reduce environmental impact, uphold ethical standards, and encourage long-term ecological balance. Where possible, materials are procured locally to minimize transport emissions; otherwise, they are shipped to the production location.
- **Production** involves transporting raw materials to the appropriate machinery and then transforming them into components for assembly into the final product. The production site for Hòstraka is determined by the resort's intended location: the main docking structure is pre-manufactured as modular units to be assembled on-site, while floating structures are built in segments and shipped to the destination to maximize shipping efficiency. The mobile suites, produced near the final location, adhere to specific production values, including resource efficiency, minimizing waste and energy consumption through optimized techniques, reversible assembly processes for ease of disposal, and ethical production practices that support fair working

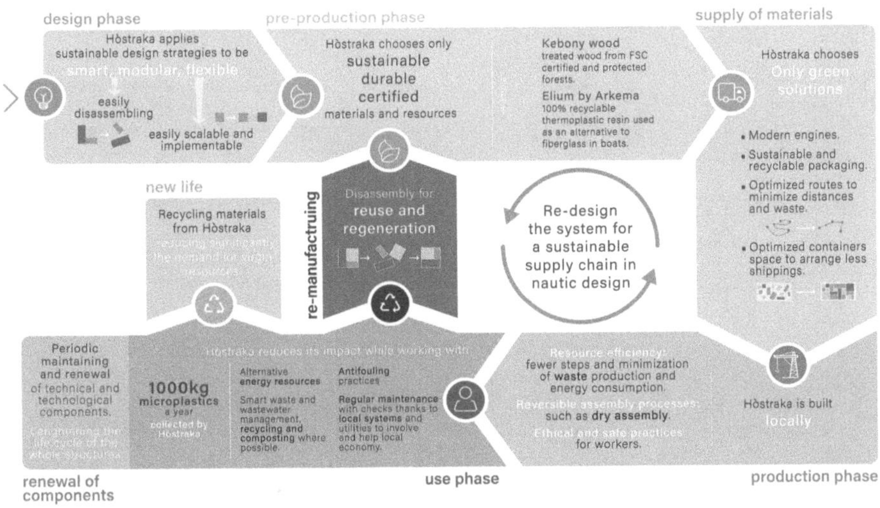

FIGURE 8.3 The development of Hòstraka supply chain is based on the study of the LCD. (Diagram developed by Hòstraka Design Team (Rosanna Caldarella, Giulia Ettori, Davide Grasso, Elisa Schembri).)

conditions. The main challenges in this stage are resource efficiency (investing in alternative techniques that contribute to a greater efficiency and less energy consumption and output of waste), reversible assembly processes (it facilitates both the mobility but also the disposal and recycling phases), and ethical and safe practices (ensuring fair and safe working conditions for workers that have to be paid decent wages to promote social and economic sustainability in the production chain).

- During **distribution**, Hòstraka's main structure is transported in segments to the site, employing sustainable logistics practices. Shipping methods include fuel-efficient "green" ships designed to have minimal emissions, route optimization to decrease travel distance, slow steaming (meaning the use of vessels at slower speeds) to cut fuel use, and maximizing container capacity to minimize shipment numbers. Lightweight, recyclable, or biodegradable packaging further reduces shipment weight and environmental impact.
- The **utilization** phase encompasses Hòstraka's energy use, maintenance, and operational logistics. Hòstraka uses exclusively alternative energy sources (primarily solar panels), alongside sustainable antifouling coatings to lower the cleaning needs of the hulls. Advanced waste and water management systems are incorporated (such as greywater tanks, recycling, and composting) and regular maintenance of boats' engines and systems ensures optimal performance. Hòstraka's modular suites feature the renewable components, extending the structure's lifecycle. Notably, this design concept plays an active role in environmental sustainability by annually collecting 1,080 kg of microplastics, or 0.1 kg per boat per day. Recovered microplastics are primarily sold to local companies for recycling into community-benefiting

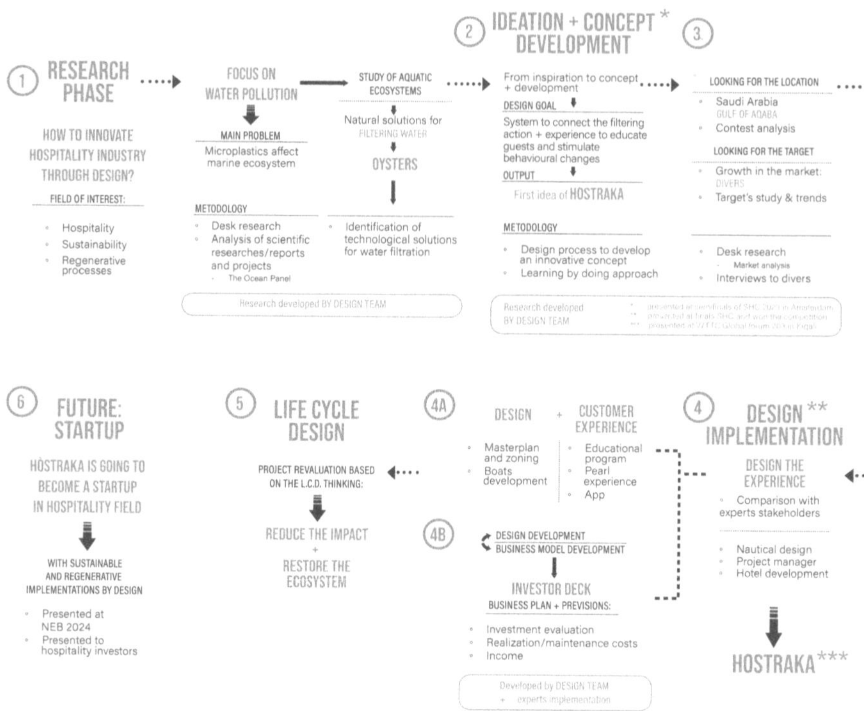

FIGURE 8.4 Hòstraka design project development. (Diagram developed by Hòstraka Design Team (Rosanna Caldarella, Giulia Ettori, Davide Grasso, Elisa Schembri) with Elena Elgani.)

products, while some are transformed into sustainable souvenirs that embody guests' commitment to environmental preservation.

- In the **end-of-life** phase, sustainable disposal practices are prioritized to reduce waste's environmental impact. Hòstraka's design emphasizes a simple and straightforward disassembly, facilitating component reuse and recycling, reducing reliance on new resources, and decreasing the overall environmental impact. With intelligent renewal, each mobile suite is expected to last approximately 30 years, while the docking structure around five years, with regular part replacements anticipated. Recycling materials from Hòstraka's components contributes to a circular economy, minimizing the need for new resources, and locally integrated collection and recycling systems support sustainable disposal throughout Hòstraka's life cycle.

Given the pressing need for such sustainable innovations in a highly polluted world, Hòstraka was awarded first place at the 2023 SHC. Currently in its start-up phase (see Figure 8.4), the project has garnered interest from global hotel leaders following its presentation at the World Tourism & Travel Council 2023 in Kigali, Rwanda.

8.5 THE VALUE OF THE HÒSTRAKA'S SOCIAL IMPACT

Hòstraka seeks to create a beneficial impact on both the environment and the community, encompassing both travellers and residents, by fostering awareness and promoting responsible behaviours through environmental education programme dedicated to different target groups. Consequently, its goal is to educate temporary visitors at the resort through sustainable, water-oriented recreational activities, particularly diving experiences, planned in the "Sustainable Pearl Experience", a sustainable guest journey. These activities not only raise awareness of marine pollution among guests but also encourage their involvement in protecting the marine ecosystem. Using their personalized app, guests can stay informed about new objectives achieved each year in reducing marine pollution, as well as access laboratories for engaging activities focused on the recovery and repurposing of plastic waste.

This can also be seen as a management-driven strategy to create loyalty with the guest who may want to visit Hòstraka's different floating resorts around the world. Indeed, this project has the ambition to be scalable and adaptable in different geographical areas.

An essential aspect of tourism is to attract individuals for hands-on experiences, but it is crucial that these experiences are meaningful and educational rather than merely entertaining. Hòstraka envisions a new hospitality paradigm that empowers guests through educational opportunities. Simultaneously, it emphasizes the importance of involving local communities that are impacted by hospitality initiatives; this project bolsters local economies by integrating residents into the management and operation of the floating resort. The team is dedicated to collaborating with struggling entities to rejuvenate local economies through partnerships in catering and resort services, primarily by including local experts and guides, ensuring that tourists can truly appreciate the area's unique characteristics. The project also features an inclusive aspect due to its scalability, allowing it to be implemented in various marine and freshwater ecosystems. Initially designed for the Gulf of Aqaba as part of Saudi Arabia's NEOM development, its participation in the New European Bauhaus Prize (2023) underscores its adaptability. Hòstraka can be tailored to fit different protected areas, such as the Gulf of Capo Gallo in Sicily, and through its scalability, it can aid in the economic recovery of unstable regions across the Mediterranean, including Greece, Italy, Portugal, and Spain. The educational focus, significant scalability (where smaller units must include essential facilities, unlike larger ones that may not require them), and adaptability to different environments such as gulfs, bays, and lakes form the legacy Hòstraka leaves behind, even if temporarily, in its various locations.

8.6 CONCLUSION

The protection of natural ecosystems and efforts to reduce pollution, minimize waste, and promote conservation of natural habitats and biodiversity, along with strategies for habitat regeneration, will be crucial for the future of tourism, as nature is a vital asset for this sector. Additionally, tourism must further lower emissions and manage its effects on biodiversity and essential ecosystems. This can be done considering two levels. At the first technological level, it is essential to establish the goal of applying sustainable design principles within the hospitality industry to formulate new sustainable hospitality models. This approach can be guided by the regenerative tourism theory, which develops the capacities of places, communities, and their guests to operate in harmony with interconnected social-ecological systems (Bellato et al., 2022). Indeed, to achieve sustainable ambitious objectives, design must integrate diverse elements – approaches (such as nature-based or nature-inspired), strategies, methodologies as well as business development, and spatial planning – to create environments that extend beyond eco-friendly objects and spaces and that instead encompass the transformation of areas into experiences that foster sustainability and social-economic responsibility that are able to meet present needs while fostering long-term environmental and social responsibility.

The interplay of these elements suggests an ecological worldview and a holistic design approach, stimulated by the debate on regeneration, where every decision and component have an influence on the entire system.

In the design field, special considerations must be given to the design of all spaces that facilitate tourism experiences, including resorts, hotels, and hostels. An evaluation of the complete life cycle of these spaces and their environmental impact is essential to measure the impact of the design and to make an evaluation. The circular economy, driven by design, enables waste and pollution prevention at the outset, as highlighted by the Ellen MacArthur Foundation (n.d.). By merging sustainability-oriented design strategies with circular design, designers can rethink water as a vital resource for sustainable development. Adopting design principles in water-related tourism to support the blue economy (Pauli, 2010) through "blue thinking" will facilitate the achievement of sustainability objectives and foster innovative future scenarios for aquatic ecosystems. The World Bank defines the blue economy as the "sustainable use of ocean resources to enhance economies, livelihoods, and the health of ocean ecosystems", highlighting that sustainability goals must prioritize water as a global common good and a critical resource to safeguard within environmental ecosystems.

At the second social level, also explored in this research, this provides extending beyond the design of facilities towards destination planning, tourism stakeholder and enterprise capacity development (Bellato et al., 2022), and promoting tourism living systems that facilitate encounters, create connections, and develop reciprocal and mutually beneficial relationships through travel practices and experiences, uniquely reflecting tourism places. Regeneration occurs mentally, physically, emotionally, spiritually, culturally, socially, environmentally, and economically (Bellato et al., 2022). This necessitates a design perspective that considers long-term repercussions, aiming for improved quality of life and enhanced social equity, while diminishing environmental impact. This also involves exploring new methodologies in which the principles of community inclusion can explore the dimension of co-existence between humans and non-humans and co-creation with locals, to involve communities and promoting environmental education. Indeed, educational initiatives like the SHC can inspire a new generation of designers to address sustainability challenges and critically evaluate the impact of their projects on the planet. In the case of

Hòstraka, the team not only envisioned an innovative design for an experience that is responsible, playful, and educational simultaneously, but they also critically assessed their project's impact in relation to the SDG focused on water: Goal 14, "Life Below Water". This initiative complements the broader developmental goals for the Gulf of Aqaba, promoted by the forward-thinking NEOM project in Saudi Arabia, which supports the SHC by virtue of its commitment to limiting its footprint to a mere 5% of the region through regenerative hospitality initiatives and the aim of enhancing nature-based experiences for visitors around the globe.

Overall, the paper highlights the potential of sustainable and regenerative design to transcend aesthetics, while aiming to harmonize the complex facets of contemporary life with the urgent need to shape environments that positively contribute to sustainable and responsible futures.

REFERENCES

Barrett, J., Chase, Z., Zhang, J., Holl, M. M. B., Willis, K., Williams, A., Hardesty, B.D., & Wilcox, C. (2020). *Microplastic Pollution in Deep-Sea Sediments from the Great Australian Bight*. https://doi.org/10.3389/fmars.2020.576170

Battaglia, F. M. (2023). Climate change and the ocean: The disruption of the coral reef. In Garcia, M. d. G., & Cortês, A. (eds), *Blue Planet Law. Sustainable Development Goals Series*. Springer. https://doi.org/10.1007/978-3-031-24888-7_10

Bellato, L., Frantzeskaki, N., &Nygaard, C. A. (2022). Regenerative tourism: A conceptual framework leveraging theory and practice. Tourism Geographies. https://doi.org/10.1080/14616688.2022.2044376

Bertola, P. (2022). Reading Tomás Maldonado: Back to design research future. In Califano, P. (ed.), *Exploring Tomás Maldonado*. Fondazione Giangiacomo Feltrinelli, Milano.

Brett, M. R. (2021). How important is coastal tourism for island nations? An assessment of African and Indian Ocean Islands. *Journal of Coastal Research*, 37(3), 568–575.

Ceschin, F., & Gaziulusoy, I. (2016). Evolution of design for sustainability: From product design to design for system innovations and transitions. *Design Studies*, 47, 118–163.

CNR, ISMAR. (2023). *Piattaforma di ricerca Oceanografica "Acqua Alta" – Bollettino 2023*. Retrieved October 10, 2024, from https://www.ismar.cnr.it/wp-content/uploads/2024/02/rapporto-tecnico-25-CNR-ISMAR.pdf

Collina, L. (2005). *Design e metaprogetto: Teorie, strumenti, pratiche*. Poli.Design, Milano.

Cremascoli, R. (2023). *Amicizia sociale: Incontrarsi nel giardino*. Skira, Milano.

Del Dot, S. (2019). *Thilafushi: l'isola delle Maldive che da quasi 30 anni accoglie milioni di tonnellate di rifiuti*. Retrieved October 10, 2024, from https://www.ohga.it/thilafushi-lisola-delle-maldive-che-da-quasi-30-anni-accoglie-milioni-di-tonnellate-di-rifiuti/

Del Zanna, G. (2009). *La casa: macchina per abitare*. Retrieved July 21, 2023, from https://blog.stannah.it/barriere-architettoniche/la-casa-macchina-per-abitare/

Dredge, D. (2022). Regenerative tourism: Transforming mindsets, systems and practices. *Journal of Tourism Futures*, 8(3), 269–281.

Ejarque, J. (2020). Quali tipologie di destinazione prediligerà il mercato?. In *Destinations & Tourism Rivista di Destination Management e Marketing* (vol. 1, pp. 4–5). Retrieved October 10, 2024, from https://www.ftourism.it/wp-content/uploads/2020/01/DT-gen-2020.pdf

Elgani, E., & Monchiero, U. (2024) *Interiors in the Age of Uncertainty: Future-Proofing Design Practices*. Design International Series. Franco Angeli, Milano, 153 p. https://digital.casalini.it/9788835165460

Ellen MacArthur Foundation. (n.d.). *Plastics and the Circular Economy – Deep Dive*. Retrieved November 16, 2023, from https://www.ellenmacarthurfoundation.org/plastics-and-the-circular-economy-deep-dive

Fondazione Prada Venezia. (2023). *Everybody Talks About the Weather*. Retrieved November 15, 2023, from https://www.e-flux.com/announcements/482130/everybody-talks-about-the-weather/

Geyer, R., Jambeck, J. R., & Law, K. L. (2017). Production, use, and fate of all plastics ever made. *Science Advances*, 3(7), e1700782. https://doi.org/10.1126/sciadv.1700782

Gilmore, J. H., & Pine, B. J. (1999). *The Experience Economy*. Harvard Business Press, Cambridge, MA.

González-Fernández, D., Cózar, A., Hanke, G., Viejo, J., Morales-Caselles, C., Bakiu, R., Barceló, D., Bessa, F., Bruge, A., Cabrera, M., & Castro-Jiménez, J. (2021). Floating macrolitter leaked from Europe into the ocean. *Nature Sustainability*, 4(5), 474–483. https://doi.org/10.1038/s41893-021-00722-6

Goodwin, H. (2011). *Taking Responsibility for Tourism*. Goodfellow Pub Ltd, Oxford.

GVR – Grand View Research. (2024). *Luxury Travel Market Size, Share & Trends Analysis Report by Tour (Customized & Private Vacations, Safari & Adventure, Celebration Journeys, Luxury Trains), by Age Group (*21–30 *Years, 31–40 Years, 41–60 Years), By Region, and Segment Forecasts, 2024–2030*. Retrieved October 10, 2024, from https://www.grandviewresearch.com/industry-analysis/luxury-travel-market

Hotel Management Network. (2024). *Accor and LVMH to Develop Orient Express Hotels and Sailing Ships*. Retrieved October 10, 2024, from https://www.hotelmanagement-network.com/news/accor-lvmh-orient-express-hotels/

HSL Hospitality Value. (2024). *La nascita di una location e il destination management*. Retrieved October 16, 2024 from https://www.hslhospitality.it/la-nascita-di-una-location-e-il-destination-management/

Kuitert, F. (2023). Ossidiana studio. *Frame*, 154, 49.

Le Corbusier. (1923). *Vers une Architecture*. les Éditions G. Crès, Paris.

Legambiente. (2023). *Rapporto Spiagge: la situazione e i cambiementi in corso nella aree costiere italiane*. Il Sole 24 Ore, Milano. Retrieved October 10, 2024, from https://i2.res.24o.it/pdf2010/Editrice/ILSOLE24ORE/ILSOLE24ORE/Online/_Oggetti_Embedded/Documenti/2024/08/08/Report-Spiagge-2023.pdf

Lenzen, M., Sun, Y. Y., Faturay, F., et al. (2018). The carbon footprint of global tourism. *Nature Climate Change,* 8, 522–528. https://doi.org/10.1038/s41558-018-0141-x

Mang, P., & Reed, B. (2019). Regenerative development and design. In R. Meyers (Ed.), *Encyclopedia Sustainability Science & Technology* (pp. 1–44). Springer Science + Business Media, LLC, part of Springer Nature.

Messina, I. (2016). *Il lato oscuro delle Maldive, un inferno in Paradiso*. Retrieved October 10, 2024. https://www.ehabitat.it/2016/05/24/il-lato-oscuro-delle-maldive-un-inferno-in-paradiso/

Nandi, K. (2024). *Accor, LVMH Board Orient Express to Steer Brand Development*. Hospitality News, Hotelsmag.com. Retrieved October 10, 2024, from https://hotelsmag.com/news/accor-lvmh-board-orient-express-to-steer-brand-development/

New European Bauhaus. (2023). *Beautiful/Sustainable/Together*. Retrieved in November 8, 2023, from https://new-european-bauhaus.europa.eu/index_en

Ocean Panel. (2022). *Opportunities for Transforming Coastal and Marine Tourism. Towards Sustainability, Regeneration and Resilience*. High Level Panel, Washington. Retrieved October 10, 2024, from https://oceanpanel.org/wp-content/uploads/2022/06/Sustainable-Tourism-Full-Report.pdf

Onadele, A. (2023). Marriott opening 24 luxury properties in 2024. *News – Upgraded Points*. Retrieved October 10, 2024, from https://upgradedpoints.com/news/marriott-opening-luxury-hotels-2024/

Papanek, V. (1971). *Design for the Real World: Human Ecology and Social Change*. Bantam Books, New York.

Pauli, G. (2010). *Blue Economy*. Taos Paradigm Pubblications, Taos, NM.

Perkins&Will. (2021). *Net-Zero Now*. Hospitality. Retrieved November 17, 2023, from https://issuu.com/perkinswill_london/docs/net-zero_now_hospitality_issuu

Rasool, H., Maqbool, S. & Tarique, M. (2021). The relationship between tourism and economic growth among brics countries: A panel cointegration analysis. *Future Business Journal,* 7, 1. https://doi.org/10.1186/s43093-020-00048-3

Reed, B. (2007). Shifting from "sustainability" to regeneration. *Building Research & Information*, 35(6), 674–680. https://doi.org./10.1080/09613210701475753

Reichelt-Brushett, A. (ed) (2023). *Marine Pollution – Monitoring, Management and Mitigation*. Springer Textbooks in Earth Sciences, Geography and Environment. Springer, Cham. https://doi.org/10.1007/978-3-031-10127-4_1

Rume, T., & Islam, S. M. D. (2020). Environmental effects of COVID-19 pandemic and potential strategies of sustainability. *Heliyon,* 6(9), e04965.

Scullica, F. (2008). *Interni Ospitali*. FrancoAngeli, Santarcangelo di Romagna, Italy.

Scullica, F., & Dominioni, A. (2022). *Designing Behaviours for Well-Being Spaces. How disruptive approaches can improve living conditions*. FrancoAngeli, Santarcangelo di Romagna, Italy.

Spathonis, A., Thorsen, D., & Mitic, J. (2020). *The Sustainable Hotel: Designing for Wellness and Resilience*. Retrieved November 17, 2023, from https://www.gensler.com/blog/the-healthy-hotel-designing-for-a-new-experience

Sustainable Hospitality Alliance. (2023). *Accelerating the Path to Net Positive Hospitality*. Sustainable Hospitality Alliance Five-Year Strategy. Retrieved November 30, 2023 from https://sustainablehospitalityalliance.org/strategy/

Sustainable Hospitality Challenge. (2023). C. Abercrombie, A. Bondagjy, A. Awlia, *Introduction to NEOM during the SHC Semi-Finals in Amsterdam, 20–23 June 2023*.

Tekman, M. B., Walther, B. A., Peter, C., Gutow, L., & Bergmann, M. (2022). *Impacts of Plastic Pollution in the Oceans on Marine Species, Biodiversity and Ecosystems* (pp. 1–221). WWF Germany, Berlin. https://doi.org/10.5281/zenodo.5898684

Thackara, J. (2017). *Progettare oggi il mondo di domani. Ambiente, economia e sostenibilità*. Postmedia Books, Milano.

Vezzoli, C. (2018). *Design for Environmental Sustainability. Life Cycle Design of Products*. Springer, London.

World Tourism Organization. (2018). *UNWTO Tourism Highlights, 2018 Edition*. UNWTO, Madrid. https://doi.org/10.18111/9789284419876

World Travel & Tourism Council. (2023). *The Environmental Impact of Global Tourism*. Retrieved November 30, 2023 from https://researchhub.wttc.org/product/the-environmental-impact-of-global-tourism-2023

World Travel & Tourism Council. (2024). *Travel & Tourism Economic Impact 2024: Global Trends*. Retrieved July 20, 2024 from https://researchhub.wttc.org/product/economic-impact-report-global-trends

Zhang, C., & Zhang, D. (2021). Microplastics. In Cui, W., Fu, S., & Hu, Z. (eds), *Encyclopedia of Ocean Engineering*. Springer, Singapore. https://doi.org/10.1007/978-981-10-6963-5_318-1

9 Warping the Loom of Consciousness for the Portuguese Textile Design Education

Sónia Seixas, Gianni Montagna, and Maria João Félix

9.1 INTRODUCTION

This chapter is an extended version of the article presented in the Design Commit '24 conference proceedings.

Society in the 21st century is increasingly defined by the intensive flow of information and knowledge processes, which highlights the need to develop skills to tackle the challenges of our planet (OECD, 2021). Work teams are increasingly polyphonic and interweave different disciplines in the search for solutions to ever more complex problems. It is essential to implement differentiated teaching approaches that go beyond traditional methods and nurture the ability to apply both theoretical and theoretical–practical knowledge in solving real-world issues. To achieve this, it is important to promote a stronger relationship between academia and industry, along with promoting a multidisciplinary perspective that encourages collaboration between different areas of knowledge. This approach should embrace both formal and non-formal learning (European Commission, 2020b; Kirschner & Norman, 2021; Muratovski, 2020).

The Portuguese textile industry continues to play a key role in the country's economic and social weave (ATP, 2022, 2024; DGAE, 2018). However, as we look ahead, new opportunities are emerging, highlighting the need for circularity and decarbonization (ATP, 2019, 2021). This industry is currently undergoing a transformation due to the significant waste it generates and its negative environmental impact (ATP, 2021; Duhoux et al., 2022; Ellen MacArthur Foundation, 2020; UNECE, 2018). This shift towards more sustainable practices affects the entire value chain, from fibre production to disposal, with the textile designer playing an important role, as a considerable portion of a textile object environmental impact is determined in the creative process (European Commission, 2021; Fletcher & Grose, 2012; Gwilt, 2020; Seixas et al., 2021). The sector faces upcoming challenges, with a focus on the digital economy, sustainability, circularity, and decarbonization. In this context,

 DOI: 10.1201/9781032719290-11

academia acts as a pivotal agent of change, shaping future designers not just as textile craftsmen, but as conscientious citizens capable of addressing the complex issues that are emerging in the textile industry.

Textiles are more than just surfaces; they represent a three-dimensional universe with inseparable properties and characteristics, where the internal and external are intertwined and tell layered stories (Castro, 1989). The essence of the yarns and the geometry of the bonds form the backbone of the process of creating a textile object. It is about the exploration of materials and structures that intertwine and reveal the secrets of surfaces and constructions (Albers, 1974). The textile designer must navigate this world through hands-on experimentation, requiring not only creativity but also a deep understanding of materials, production processes, and technology (Albers, 1974; Castro, 1989).

A review of the literature, combined with collected data, highlights the near absence of higher education opportunities in textile design in Portugal. Out of eight available undergraduate programmes, only two are directly related to the field of textile design (DGES, 2023). Furthermore, even these programmes dedicate only a limited portion of their syllabus to textile design education. To continue to weave a textile economy of the future, a teaching model in textile design in close connection between industry and academia should be considered to prepare future textile designers to work with the ongoing changes.

9.2 THEORETICAL CONSIDERATION

9.2.1 The Warp of Textile Design

The creative process involved in designing a textile item is complex. The textile industry is continually seeking to implement more sustainable practices, whether in the production and processing of fibres, the manufacturing processes, or the business models themselves. This industry is undergoing a transformation due to the substantial waste generated and the adverse effects this has on the environment (ATP, 2021; Duhoux et al., 2022; Ellen MacArthur Foundation, 2020; UNECE, 2018). The choices made throughout the creative process of a textile product, from fibre selection to disposal, have a considerable impact on the environment (European Commission, 2021; Fletcher & Grose, 2012; Gwilt, 2020; Seixas et al., 2021), highlighting the role of the textile designer. This professional is expected to possess knowledge that encompasses not only aesthetic and creative aspects of design but also technical expertise, including a solid understanding of textile materials and production methods to make more informed decisions in product development.

Textile design crosses borders. Combining yarns and colours involves more than merely the fashion sector. It encompasses a wide range of design fields, from decorative items to functional everyday objects, creating a rich tapestry of possibilities (ATP, 2019; DGAE, 2018). As different technologies and approaches evolve, textile designers need to incorporate sustainability and circular economy principles into their creative process (Seixas et al., 2021, 2022, 2023a, 2023b). In the delicate balance between form and substance, they explore the intricacies of materials and finishes in search of textile objects that not only envelop but also serve, sustain, and outlast. By addressing the needs of users, society, and the planet, they unlock the potential of textile design in a universe of intertwined possibilities, where every stitch tells a story and every decision weaves the future.

Textile design is the meeting point between art and technology, where materials and structures intertwine to unveil the mysteries of surfaces and constructions. Creativity is only the first thread here, as it also requires a deep understanding of materials, production, and technology. The essence of the yarns and the geometry of the weaves form the backbone of the process of creating a textile object. For the textile designer, experimenting and practicing are crucial to understanding the interaction between these fundamental elements that shape the essence of the textile material (Albers, 1974). Fabric is more than a surface; it is a three-dimensional universe with inseparable properties and characteristics in which the inside and the outside tell interwoven stories. Fundamental concepts of colour, pattern, and construction, combined with aesthetics and functionality, play a crucial role in the development of textile objects (Albers, 1974; Castro, 1989).

Designers who focus on finding solutions use creativity to overcome challenges, often guided by data analysis. Empathy plays an important role in ensuring that real individual and contextual needs are acknowledged. At the same time, designers recognize the interconnectedness of global realities, understanding the implications of their solutions on the world. This approach transcends the walls of corporate, political, and governmental interests by prioritizing human, cultural, and natural ecosystems, including biodiversity. Furthermore, it embraces the pressing issues confronting our planet, such as resource management, pollution, and climate change (Design Council, 2020; Skibsted & Bason, 2023).

The textile industry in Portugal remains a significant pillar of the country's social and economic weave (ATP, 2022, 2024; DGAE, 2018). However, looking ahead, new opportunities are emerging that reflect the urgency of adopting a circular economy and pursuing decarbonization (ATP, 2015, 2021). In this context, academia has an important role to play as an agent of change. It has the task of shaping the designers of tomorrow, not only as skilled textile artisans but also as aware citizens capable of tackling the complex challenges arising in the textile sector. To continue developing a robust and promising textile ecosystem that aligns with the values of societal and environmental respect, it is essential for textile designers to adopt a mindset during their academic careers that focuses on the circular economy, where yarn and fabric are renewed.

However, it is important to highlight that this journey may be challenging in the Portuguese context, as there are currently no higher education programmes specifically focused on textile design (DGES, 2023). Therefore, it is important to plant the seeds of expertise and develop programmes that prepare future textile designers to guide the industry towards a horizon where sustainability and responsibility are intertwined.

9.2.2 The Warp of Education in Design

The 21st century is witnessing profound and rapid transformations across various spheres, including the environment, economy, and society (OECD, 2018, 2019). At the same time, intensive processes of information and knowledge are increasingly shaping society, which means that skills need to be developed to meet the challenges facing our planet (OECD, 2021). Consequently, the job market is becoming more volatile and unpredictable. Work teams are increasingly becoming polyphonic teams that interweave different disciplines in the search for solutions to increasingly complex problems. The advancement of technologies, particularly artificial intelligence, underscores the necessity for a strong commitment to continuous professional development. Future professions are expected to be broader and more varied, primarily characterized by creativity, critical thinking, and innovation. Society expects individuals who are critical, independent, and emotionally equipped to effectively resolve problems across various contexts (Ross, 2016; Schwab, 2016). As globalization and technological advancements reshape citizenship and the workforce, education systems are drifting away from the realities and demands of the global economy and society, leading to a significant shift in educational paradigms (World Economic Forum, 2020). Today's students are oriented towards rapid technological advancement (Prensky, 2001, 2010), which means that existing teaching–learning models need to be re-evaluated (Almeida et al., 2022). New forms of teaching are emerging, such as distance learning and blended learning, which are challenging the traditional educational model that has characterized the higher education system (Christensen et al., 2013; Dunagan, 2017). We do not want mere repositories of knowledge; we want citizens who are able to reflect on knowledge rather than just memorize and repeat it.

In this context, it is important for academic institutions to nurture creativity, experimentation, collaboration, critical thinking, and the ability to adapt to technological change in order to prepare students for the challenges of the 21st century (OECD, 2019, 2020, 2021). We are looking for differentiated approaches to teaching that are able to break away from the traditional, century-old model that has stood the test of time but may no longer be the best choice in a world where technology is advancing rapidly. In this sense, Education 4.0 represents a considerable progression and transformation of traditional educational systems. Its core philosophy is the principle of "learn by doing", placing students at the heart of their own educational journeys. It also recognizes the importance of new technologies, including artificial intelligence, as tools that can enhance learning and enrich the educational experience (World Economic Forum, 2020). It seeks to go beyond the conventional approach and overcome the barriers of pedagogical conformism to train students to apply theoretical and theoretical–practical knowledge in a real-life environment. Through a closer connection between academia and industry, it should be possible to learn in a multidisciplinary, interdisciplinary, and transdisciplinary way in a vision that encompasses both formal and non-formal learning (European Commission, 2020b; Kirschner & Norman, 2021; Muratovski, 2020).

Design inherently requires multidisciplinary collaboration to do justice to complexity (Antunes & Almendra, 2020; Cross & Keynes, 2023; Davis, 2018) and should establish strong links with other fields of knowledge. The field of design is interdisciplinary and multifaceted. Designers do not develop solutions on their own. Often, these solutions are not necessarily tangible (Davis, 2017). Therefore, it is crucial to encourage designers to be culturally conscious, socially responsible, and proactive (Davis & Dubberly, 2023; Kirschner & Norman, 2021; Meyer & Norman, 2020). The design process is not a solitary activity; it is complex and involves several steps.

It is an activity that requires collaboration and transcends disciplinary boundaries. The academia, in turn, should be the fertile ground on which not only technical knowledge is created, but also ethical, social, and personal skills are cultivated (OECD, 2019, 2020, 2021; World Economic Forum, 2020). Educational approaches are sought that can support textile designers in navigating the transformation of the industry, the planet, and society.

9.3 METHODOLOGY

The interweaving of literature review revealed that student profiles have evolved due to technological advancements, prompting a need for differentiated teaching approaches that prioritize student engagement in the educational process. In turn, the textile industry is also undergoing a process of transformation it seeks to mitigate its negative impacts. Given this framework and in order to better understand how Portuguese higher education encompasses the studied universe, an analysis of current undergraduate programmes in textile design was carried out. At the same time, in 2021, semi-structured interviews were held with experts in the field to gather insights on developing a teaching–learning model in textile design that can meet the needs of academia, the textile industry, designers, society, and the environment.

9.3.1 Portuguese Higher Education in Textile Design

The first undergraduate programme in this field was established in 1992, indicating that higher education in this area is still relatively new. Currently, there are eight undergraduate programmes available, with an equal split of four offered by public institutions and four by private ones. If we take a closer look, only 25% of these programmes explicitly reference textile design in their titles. This includes the undergraduate programme in Fashion and Textile Design from the Higher School of Applied Arts, Castelo Branco Polytechnic Institute, which is a public education institution, and the undergraduate programme in Fashion and Textile Design and Production from the Higher Institute Manuel Teixeira Gomes, a private educational institution (Table 9.1) (DGES, 2023).

These data show that none of the available programmes focus exclusively on the field of textile design, but all address the field of fashion design in parallel. Even the most recent, introduced in 2023, is titled Sustainable Fashion Design.

A detailed analysis of the syllabi of the programmes that include textile design at ESART and ISMAT reveals that even in these programmes, the field of textile design is only dealt with superficially. In a percentage ratio between curricular units (CUs) and European Credit Transfer and Accumulation System (ECTS), ESART comprises 16.67% (Figure 9.1) and ISMAT 7.22% (Figure 9.2) of the CUs that cover textile design (ESART, 2023; ISMAT, 2023). According to this analysis, it can be concluded that students integrating these undergraduate programmes are better prepared to enter the fashion and clothing industry than the textile industry.

These data suggest that the existing undergraduate programmes meet the needs of the fashion and clothing industry to some extent, but in terms of the textile industry, the academia response is poor and a cause for concern.

TABLE 9.1
Portuguese Undergraduate Programmes in Textile Design

Acronyms	Higher Education Institution	Undergraduate Course	Implementation Year
FAUL	Lisbon School of Architecture, Universidade de Lisboa	Fashion Design	1992
ESART	Higher School of Applied Arts, Castelo Branco Polytechnic Institute	Fashion and Textile Design	1999
UBI	University of Beira Interior	Fashion Design	2000
ESAD	School of Art and Design	Design with a specialization in Fashion	2004
UMINHO	University of Minho	Fashion Design and Marketing	2005
ECATI	School of Communication, Architecture, Arts and Information Technologies – University Lusófona	Fashion Design and Production	2021
ISMAT	Higher Institute Manuel Teixeira Gomes	Fashion and Textile Design and Production	2022
FCATI	Faculty of Communication, Architecture, Arts and Information Technologies – University Lusófona	Sustainable Fashion Design	2023

Source: Author (2023).

FIGURE 9.1 Percentage ratio of curriculum units related to textile design, fashion design, and other fields within the undergraduate programmes in Fashion and Textile Design of ESART. (Author 2023.)

9.3.2 Interviews

Twelve semi-structured interviews were conducted with experts from the area under study, including one individual textile designer and teacher, two individual textile designers, two textile craftsmen and teachers, the general director of the technological centre of the textile and clothing industries, the general director of the centre for nanotechnology and smart materials, the executive director of the Portuguese textile cluster, the president of the Portuguese textile association, the executive

FIGURE 9.2 Percentage ratio of curriculum units related to textile design, fashion design, and other fields within the undergraduate programme in Fashion and Textile Design and Production of ISMAT. (Author 2023.)

administrator of the Portuguese business association, the director of the selective moda association, and the president of the modalisboa association. The purpose of these interviews was to evaluate the appropriateness and relevance of implementing a teaching–learning model for textile design within the Portuguese higher education system and to identify the fundamental skills that this profession needs to acquire to practise its profession.

The results show that the textile designer needs to acquire a broad spectrum of knowledge. Mastery of technical aspects, such as textile materials and processes, in combination with the design field, is essential. The interviewees highlighted the significance of management and planning abilities within the textile industry, which are essential for meeting the demands of both the market and society in their professional practice. Transversal skills play a fundamental role, as the designer does not work in isolation, but collaborates in a team with other professionals such as textile engineers, marketing professionals, and others. Therefore, communication, responsibility, interpersonal skills, and the ability to listen should be encouraged during academic training.

An interesting aspect relates to the role that the textile designer can play in communicating and promoting the image of the Portuguese fashion and textile industry, creating added value. The interviewees mentioned that the designer should continue to be someone who can interpret society and consumers, who looks to society to shape and design the future in a sustainable way, keeping in mind issues of eco-design, sustainability, and the circular economy, which are fundamental as the paradigm of the world is changing in this direction. They highlighted the need for a change of consciousness in terms of consumption, with the designer playing a crucial role in enabling and bringing about this change. From an artistic point of view, they pointed out the importance of not letting old techniques die out but renewing them and applying them in a more contemporary way. The designer should act as a decoder, facilitator, and problem solver because the designer's main task in any field is to create solutions.

The interviewees emphasized the importance of the textile designer being able to design textiles. This means not only having a creative streak but also possessing practical knowledge and the ability to apply it effectively. The hands-on application of this knowledge in real-world situations and addressing specific challenges is essential for understanding how concepts translate to an industrial environment. A shared educational approach between academia and industry could facilitate this understanding. To this end, the interviewees pointed to the need for collaboration between industry, knowledge centres, and academia, which would allow for internships and short immersion programmes in companies during students' academic journeys, even at early stages. This will allow the students to know the companies, understand how they work, and what their concerns are. Given the technological advancements and changes in the textile industry, it has proven to be important to have a higher education in textile design in Portuguese territory in order to respond to the current and future needs and challenges of the sector.

9.3.3 Results

The analysis of the data helps to outline the profile of a textile designer. The field of design is part of the socio-economic fabric of a region and therefore reflects the needs of the community in which it operates. It is important to apply teaching strategies that enable students to engage with contexts that are closer to real-world scenarios, but without losing the reference lines of a comprehensive design education.

Thus, it is considered that the textile designer is the professional who, autonomously, has the competence to understand the professional context and the practice of textile design. An effective higher teaching–learning model in this field should have as its main objective to prepare textile designers capable of developing and carrying out textile design projects effectively in collaboration with industry, community, and professional partners, designing materials and surfaces for different objects. It should enable the construction of a personal creative identity and an aesthetic vocabulary informed by a responsible understanding of historical, cultural, technical, and design principles, along with hands-on textile-making experience. To this end, practice-based teaching where students learn from the teacher can make the teaching–learning process more assertive, not limited only to memorizing content. Learning methodologies and methods focused on theoretical and practical learning, learning-by-doing, associating the acquisition of theoretical knowledge with its simultaneous application in interdisciplinary work with tutorial support are important. The main activities to be developed by the textile design professional include the understanding of the textile industry, society, and respect for the planet in the design, development, and implementation of textile projects (Figure 9.3).

Based on the analysis, several considerations have been outlined as a guideline for developing a teaching–learning model for textile design. It is important to include curriculum units that focus on "learning-by-doing" and "reflection-in-action", as in the field of design problems are rarely clearly defined and a process of interpretation is necessary for their understanding (Buchanan, 1992; Rittel & Webber, 1973; Schön, 1983). A project-based learning approach that focuses on real-world problems increases student participation and motivation (Bender, 2014; Glasser, 1999). For this to be fruitful, it is necessary to establish collaboration agreements with companies, knowledge centres, and the community to enable working and dealing with real-world problems (World Economic Forum, 2024). This allows one, in addition to the theoretical and theoretical–practical content, to work on transversal skills such

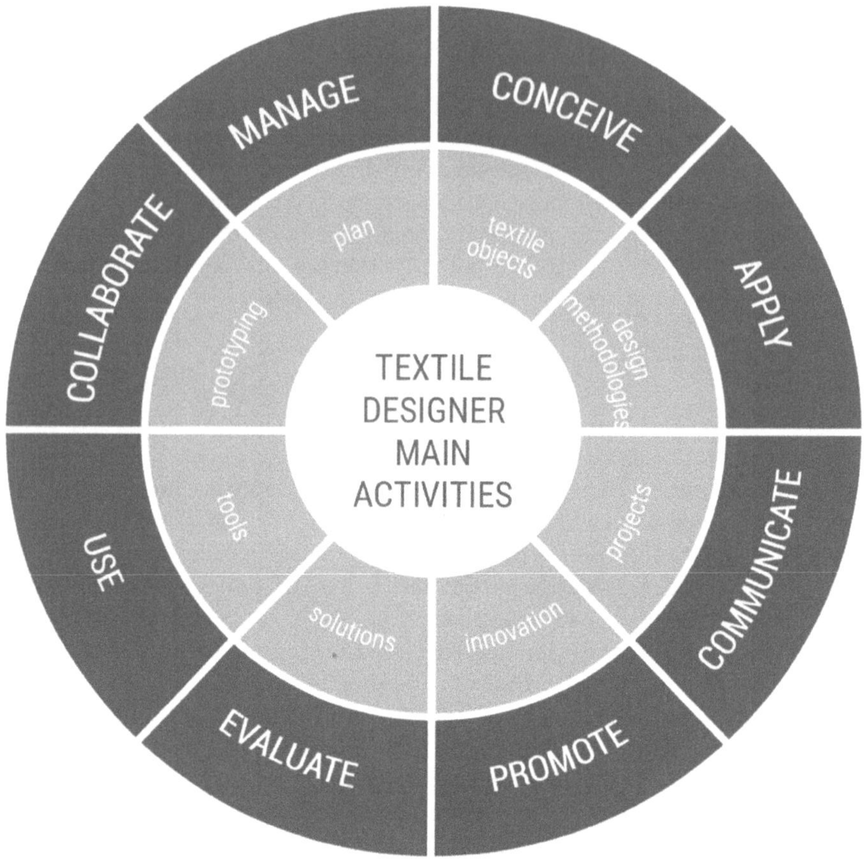

FIGURE 9.3 Main activities of the textile designer. (Author 2023.)

as teamwork, communication, creative and critical thinking, and others, which are increasingly important skills to meet the challenges of the 21st century (Muratovski, 2022; OECD, 2021; World Economic Forum, 2020).

Active teaching–learning strategies should involve an intensive process aimed at stimulating creativity. These approaches are designed to equip students for working in multidisciplinary teams, addressing real-world projects that respond to pressing global challenges. By combining "learning-by-doing" with teacher support, where the instructor takes on a facilitator role, both theory and practical application are taught simultaneously. Students are encouraged to continuously assess and reassess the effectiveness of their solutions. Self-assessment and autonomy are part of this process, gradually reducing reliance on teacher guidance. Learning through reflection is not about what happens to the students. Rather, it is about what the students can do with what has happened to them (Bender, 2019; Bourner, 2003).

Design is an activity where teamwork is essential, so it should start in academia. In this sense, a CU that integrates two teachers with complementary expertise – one specializing in design and the other in textile materials – in a multi- and interdisciplinary

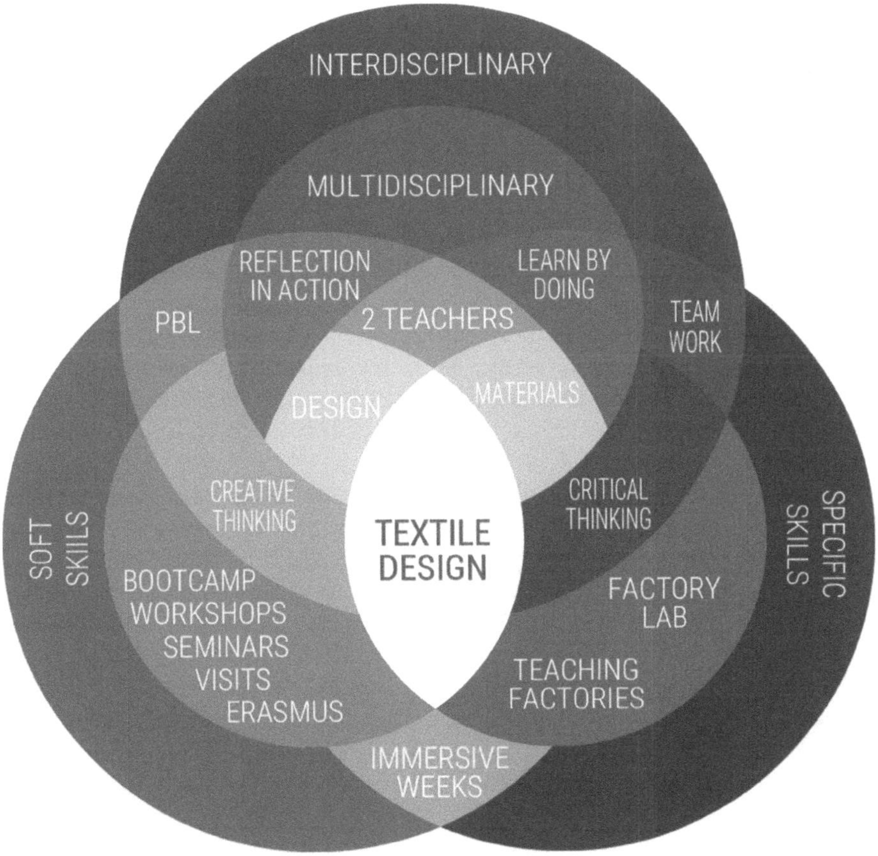

FIGURE 9.4 Framework of active teaching strategies for higher education in textile design. (Author 2023.)

vision will teach students the importance of teamwork, collaborative work, and adaptability, which are important skills to succeed in the 21st century (Advani, 2023; European Commission, 2020a; Kirschner & Norman, 2021; Muratovski, 2020). Additionally, it also allows students to optimize their knowledge in complementary areas such as design and textile materials. Last but not least, various active learning strategies should be considered, such as bootcamps, seminars, workshops, mini-internships, company and fair visits, factory labs, teaching factories, and others (Dunagan, 2017). These methods will not only help students acquire specific skills but also foster communication, teamwork, idea-sharing, independence, and responsibility (Figure 9.4). This holistic approach will enhance the role of the designer, improving their ability to collaborate, empathize, and connect design expertise with other fields of knowledge.

Through the exploration and experimentation of materials, students can gain essential knowledge of different techniques, production processes, and material properties, all of which are crucial for the textile design profession (Seixas et al., 2024). The aim is not to train students to become masters of weaving (or any other specific technique), but to become designers who also have a solid understanding of production processes and can critically evaluate and compare different options and their implications. Learning and practicing to become a textile designer are about engaging with the processes.

9.3.4 Discussion

In light of this, the question arises whether it is worthwhile to establish a dedicated textile design undergraduate programme in Portuguese higher education to train professionals capable of helping the textile industry overcome its current challenges.

Portuguese higher textile design education lacks a specific focus and a practical approach that addresses the challenges of the industry, both from a current and future perspective. The implementation of a teaching–learning model that includes close collaboration between academia and industry proves to be an important topic for discussion.

A curriculum that emphasizes textile design and promotes collaboration between the various actors in the value chain to address increasingly complex problems is essential in the 21st century (Advani, 2023; OECD, 2021; World Economic Forum, 2020). The incorporation of a multidisciplinary, interdisciplinary, and transdisciplinary approach (Seixas et al., 2025), with its various areas of knowledge, is essential to tackling emerging challenges and developing a new strategy that allows the industry to retain its position as a key driver of the national economy.

Promoting collaboration between academia and industry through a shared teaching approach is key to providing future textile designers with the opportunity to complete internships and participate in collaborative projects to learn about the real-world challenges of this industry. This partnership not only immerses students into the professional environment, but also strengthens the links between academia, industry, and creativity.

Moreover, early exposure to real-world problems in the sector can better prepare textile designers to seek solutions that meet the needs of society and the planet.

9.4 CONCLUSIONS

This exploratory study, woven through the threads of knowledge, sought to assess the relevance of introducing an undergraduate programme specifically focused on textile design in Portugal. By interweaving insights from the literature review with the findings from the interviews, it became evident that there is potential for a higher teaching–learning model for textile design in Portugal. Out of the eight existing undergraduate programmes, there are only two that fall within the field of "textile design": the Fashion and Textile Design undergraduate programme at ESART and the Fashion and Textile Design and Production undergraduate programme at ISMAT (DGES, 2023). And even in these programmes, only a small percentage of the syllabus is dedicated to teaching textile design (ESART, 2023; ISMAT, 2023). The absence of specific programmes in textile design may hinder professionals' ability to emerge as specialists, limiting the potential for innovation and adaptation to new sustainable practices within the textile industry.

In an effort to reduce its environmental impact and safeguard the planet, the textile industry is undergoing a transformation, reshaping both production processes and business models. This transition to more responsible methods encompasses all stages of the value chain, from production to fibre processing and disposal. This highlights the role of the textile designer, as part of the environmental impact of textile objects is determined in the creative process (European Commission, 2021).

The demands of 21st-century society call for individuals who are critical, independent, and emotionally equipped to solve problems across various contexts. Future professions will increasingly be defined by creativity, critical thinking, and innovation (Ross, 2016; Schwab, 2016). With the rapid pace of globalization and technological advancement, there is a growing need to rethink teaching–learning models (OECD, 2021; World Economic Forum, 2020) and promote more reflective approaches than mere memorization and repetition.

Higher education must embrace the metamorphosis of today's students, characterized by the technological age (Prensky, 2001, 2010), and thus renew the way they absorb knowledge. Break with the conservative and traditional model in favour of differentiated and even revolutionary methods. The educational system should prioritize practical application over theoretical dominance, with real-world challenges that consolidate knowledge and foster both specific and transferable skills. Active teaching strategies stimulate creativity and prepare students to work in multidisciplinary teams (Bender, 2019; Bender, 2014; Bourner, 2003). As such, academia must promote teamwork and offer interdisciplinary approaches. Including two teachers with complementary expertise in the same programme can foster an understanding of the importance of collaboration and adaptability and strengthen the designer's ability to collaborate, communicate, and integrate knowledge from different fields. A potential path forward is shared education between academia and industry. However, for this to happen, it is essential to strengthen the links and enable teaching–learning strategies that are grounded in the real-world contexts.

The creation of a teaching–learning model that involves internships and collaborations with companies in the sector could be a first step towards preparing textile designers who stand out not only for their creativity, but also for their environmental responsibility and ability to adapt to new production paradigms.

It is believed that the most important contribution of this exploratory study to design was the possibility of identifying ways to construct a teaching–learning model in textile design to prepare future textile designers as agents of industry transformation. Investing in specialized training in this field can contribute to fostering a more robust and sustainable textile ecosystem. The designer is not only a creator of objects, but also an agent of change for a more responsible future.

We plan to continue this exploratory study for future research to identify the most effective strategies for developing a teaching–learning model for textile design. These efforts will contribute to the development of a promising textile ecosystem guided by respect for the planet and society.

The author of this study proposes to weave the construction of a textile design curriculum based on the interweaving of knowledge, thus promoting the advancement of knowledge in this field in Portugal.

ACKNOWLEDGEMENTS

This work is financed by national funds through FCT – Fundação para a Ciência e a Tecnologia, I.P., under the Strategic Project with the reference UIDB/04008/2020 and UIDP/04008/2020, and through the individual research grant 2020.04735.BD.

REFERENCES

Advani, A. (2023). Education 4.0: here are 3 skills that students will need for the jobs of the future. In *World Economic Forum Annual Meeting*. https://www.weforum.org/agenda/2023/01/skillsets-cultivated-by-education-4-0-davos23/

Albers, A. (1974). *On weaving*. Studio Vista.

Almeida, L., Gonçalves, S., Ramos do Ó, J., Rebola, F., Soares, S., & Vieira, F. (2015). ATP 50 anos. In: B. Sendim, A. P. Dinis, F. Santos, & F. Gomes (Eds.), *Inovação Pedagógica no Ensino Superior. Cenários e Caminhos de Transformação. A3ES Readings n. 16*. ISBN: 978-989-53667-3-6.

Antunes, S., & Almendra, R. (2020). Design education for the 21st century: the multiple faces of disciplinarity. In R. Almendra, & J. Ferreira (Eds.), *Research & Education in Design: People & Processes & Products & Philosophy* (pp. 233–240). Taylor & Francis. https://doi.org/10.1201/9781003046103-26

ATP. (2019). *Fashion from Portugal 4.0. Diretório 2019*. ATP – Associação Têxtil e Vestuário de Portugal.

ATP. (2021). *Visão Prospetiva e Estratégias ITV 2030. Contributo para um Plano Estratégico para o Sector Têxtil e Vestuário Português até 2030*. ATP – Associação Têxtil e Vestuário de Portugal.

ATP. (2022). Press Release: exportações de têxteis e vestuário ultrapassarão os 5,3 mil milhões de euros em 2021. https://atp.pt/wp-content/uploads/2022/01/PressRelease-Exportacoes-da-ITV-Novembro21_10.1.22.pdf

ATP. (2024). Press realease: Novembro revela indícios de recuperação nas exportações do setor têxtil e vestuário. https://atp.pt/wp-content/uploads/2024/01/PressRelease-Exportacoes-da-ITV-jan_nov_10.1.2024.pdf

Bender, D. (2019). The role of reflection in design studio education. *13th International Technology, Education and Development Conference* (pp. 1134–1139). https://doi.org/10.21125/inted.2019.0367

Bender, W. N. (2014). *Aprendizagem baseada em projetos: educação diferenciada para o século XXI*. São Paulo: Penso.

Bourner, T. (2003). Assessing reflective learning. *Education + Training*, *45*(5), 267–272. https://doi.org/10.1108/00400910310484321

Buchanan, R. (1992). Wicked problems in design thinking. *Design Issues*, *8*(2), 5. https://doi.org/10.2307/1511637

Castro, E. de M. e. (1989). *Manual de design e cálculo de tecidos*. Lisbon: Direcção-Geral da Indústria.

Christensen, C. M., Horn, M. B., & Staker, H. (2013). *Ensino Híbrido: uma Inovação Disruptiva? Uma introdução à teoria dos híbridos* (pp. 1–52). Cambridge: Clayton Christensen Institute.

Cross, N., & Keynes, M. (2023). Design thinking: What just happened? *Design Studies*, *86*, 1–10. https://doi.org/10.1016/j.destud.2023.101187

Davis, M. (2017). *Teaching Design: A Guide to Curriculum and Pedagogy for College Design Faculty and Teachers Who Use Design in Their Classrooms*. New York: Allworth.

Davis, M. (2018). *Design Futures Trend: Complex Problems*. New York: AIGA Design Futures Trends.

Davis, M., & Dubberly, H. (2023). Rethinking design education. *She Ji*, *9*(2), 97–116. https://doi.org/10.1016/j.sheji.2023.04.003

Design Council. (2020). *Design Perspectives: Design Skills*. Design Council. https://www.designcouncil.org.uk/fileadmin/uploads/dc/Documents/Design%2520Perspectives-%2520Design%2520Skills.pdf

DGAE. (2018). *Indústria Têxtil e Vestuário (Sinopse)*. Direção-Geral das Atividades Económicas.

DGES. (2023). *Direção-Geral de Ensino Superior*. Licenciaturas. https://www.dges.gov.pt/pt/pagina/licenciatura

Duhoux, T., Blévennec, K. Le, Manshoven, S., Grossi, F., & Arnold, M. (2022). *Textiles and the Environment: The Role of Design in Europes Circular Economy* (ETC/CE Report 2/2022; Issue February). European Environment Agency (EEA).

Dunagan, A. (2017). *College Transformed: Five Institutions Leading the Charge in Innovation* (pp. 1–38). Cambridge: Clayton Christensen Institute.

Ellen MacArthur Foundation. (2020). *Vision of a Circular Economy for Fashion*.

ESART. (2023). Escola Superior de Artes Aplicadas. Retrieved from https://www.ipcb.pt/esart/escola-superior-de-artes-aplicadas

European Commission. (2020a). *European Skills Agenda for Sustainable Competitiveness, Social Fairness and Resilience*. European Commission.

European Commission. (2020b). Skills for industry. Curriculum guidelines 4.0: future-proof education and training for manufacturing in Europe: final report. In *Office*. European Commission Executive Agency for Small and Medium-sized Enterprises. https://doi.org/10.2826/097323

European Commission. (2021). Sustainable product policy. https://ec.europa.eu/jrc/en/research-topic/sustainable-product-policy

Fletcher, K., & Grose, L. (2012). *Fashion & Sustainability:* Design for Change. London: Laurence King Publishing Ltd.

Glasser, W. M. D. (1999). *Choice Theory. A New Psychology of Personal Freedom*. New York, NY: HarperCollins Publishers Inc.

Gwilt, A. (2020). *A Practical Guide to Sustainable Fashion* (2nd edn.). London: Bloomsbury Publishing Plc.

ISMAT. (2023). Instituto superior manuel teixeira Gomes. https://www.ismat.pt/

Kirschner, P. A., & Norman, D. (2021). Teaching and learning of design. https://assets.website-files.com/6010c2ec764764887cfefee7/60515ff38234e90fac4c64cb_Kirschner and Norman. Teaching and Learning of Design 21.01.21.pdf

Meyer, M. W., & Norman, D. (2020). Changing design education for the 21st century. *She Ji*, *6*(1), 13–49. https://doi.org/10.1016/j.sheji.2019.12.002

Muratovski, G. (2020). The making of an American design school: lessons learned. *She Ji*, *6*(1), 67–82. https://doi.org/10.1016/j.sheji.2020.01.002

Muratovski, G. (2022). *Research for Designers. A Guide to Methods and Practice* (2nd edn.). SAGE Publications Ltd.

OECD. (2018). *The Future of Education and Skills. Education 2030*. OECD.

OECD. (2019). *Learning Compass 2030. OECD Future of Education and Skills 2030*. OECD. www.oecd.org/education/2030-project

OECD. (2020). *What Students Learn Matters. Towards a 21st Century Curriculum*. OECD. https://doi.org/10.1787/d86d4d9a-en

OECD. (2021). *Building the Future of Education*. OECD.

Prensky, M. (2001). Digital natives, digital immigrants. *On the Horizon*, *9*(5), 1–6. https://doi.org/10.1108/10748120110424816

Prensky, M. (2010). *Teaching Digital Natives: Partnering for Real Learning*. Thousand Oaks, CA: Corwin Press.

Rittel, H. W. J., & Webber, M. M. (1973). Dilemmas in a general theory of planning. *Policy Sciences*, *4*(2), 155–169. https://doi.org/10.1007/BF01405730

Ross, A. (2016). *As indústrias do futuro.* Lisbon: Actual Editora.
Schön, D. A. (1983). *The Reflective Practitioner: How Professionals Think in Action.* New York: Basic Books.
Schwab, K. (2016). *The Fourth Industrial Revolution.* New York: Crown Business.
Seixas, S., Montagna, G., & Félix, M. J. (2021). Materials matters in textile and fashion design education. In C. S. Shin, G. Di Bucchianico, S. Fukuda, Y. -G. Ghim, G. Montagna, & C. Carvalho (Eds.), *Advances in Industrial Design. AHFE 2021. Lecture Notes in Networks and Systems* (Vol. 260, pp. 681–688). Springer, Cham. https://doi.org/10.1007/978-3-030-80829-7_84
Seixas, S., Montagna, G., & Félix, M. J. (2022). The textile designer skills to face the emerging challenges of textile industry and the role of academia. In G. Montagna, & C. Carvalho (Eds.), *Human Factors for Apparel and Textile Engineering. AHFE (2022) International Conference* (Vol. 32, pp. 23–30). AHFE Open Access. https://doi.org/10.54941/ahfe1001534
Seixas, S., Montagna, G., & Félix, M. J. (2023a). Weaving the skills of the textile designer as a contribution to the construction of a textile ecosystem of the future. In A. C. Broega, J. Cunha, H. Carvalho, & B. Providência (Eds.), *Advances in Fashion and Design Research* (pp. 505–517). Springer International Publishing. https://doi.org/10.1007/978-3-031-16773-7_43
Seixas, S., Montagna, G., & Félix, M. J. (2023b). Where is textile design in Portugal? Interlace the past to weave the future. In D. Raposo, J. Neves, R. Silva, L. Correia Castilho, & R. Dias (Eds.), *Advances in Design, Music and Arts II. EIMAD 2022. Springer Series in Design and Innovation* (Vol. 25, pp. 220–237). Springer, Cham. https://doi.org/10.1007/978-3-031-09659-4_17
Seixas, S., Montagna, G., & Félix, M. J. (2024). Threads of knowledge. Weaving paths for higher education in textile design in Portugal. In G. Montagna, & C. Figueiredo (Eds.), *Proceedings of the 3rd Textile Design Conference 2023 (D_tex), April 22–24, Lisbon, Portugal.* Lisbon Shcool of Architecture, Lisbon.
Seixas, S., Montagna, G., & Félix, M. J. (2025). Textile design: a transdisciplinary approach in higher education in Portugal. In R. A. Almendra (Ed.), *Which Proximity in Design Education? A Contemporary Curriculum* (1st edn.). London: Routledge.
Skibsted, J. M., & Bason, C. (2023). New "design thinking" can lead to a more sustainable world. In *World Economic Forum Agenda.* https://www.weforum.org/agenda/2023/07/design-thinking-sustainable-world/
UNECE. (2018). Traceability of sustainable value chains. Enhancing transparency in the garment and footwear sector for informed and responsible choices. https://unece.org/trade/traceability-sustainable-garment-and-footwear
World Economic Forum. (2020). Schools of the future: defining new models of education for the fourth industrial revolution. In *World Economic Forum Reports 2020.* World Economic Forum. www.weforum.org
World Economic Forum. (2024). Circular transformation of industries: the role of partnerships. https://www.weforum.org/publications/circular-transformation-of-industries-the-role-of-partnerships/#:~:text=Unveiling the transformative potential of,sustainable%2C resilient and prosperous future

10 Investigating Design Fixation Caused by Social Conformity

Nikole Melo de Mendonça and
Rita Assoreira Almendra

10.1 INTRODUCTION

"Won't this be weird?" Many professionals in the fields of Design and Architecture have probably heard this when presenting a new idea instead of proposing one of those options that would normally be applied to a problem. But what is wrong with an idea that is different from usual? After all, isn't it in unexpected answers and solutions that innovations are grounded? As an extended version of the article presented in the Design Commit '24 conference proceedings, this chapter will discuss an investigation driven by these questions.

Studies in the field of behavior and social psychology show that individuals tend to change the way they act or accept answers they do not agree with to meet group expectations. Social conformity is the result of judgmental pressure that can shape people into feeling accepted and belonging to the social contexts (Asch, 1955; Durkheim, 2004; Lane, 1983; Souza et al., 2018; Sunstein, 2019b). In professional areas with a creative bias, social rules could inhibit innovation since it is also susceptible to judgments. In this way, the fear of proposing or the inability to think of ideas that go too far off the beaten track can be an obstacle to new ideas, but most of all to radical transformation toward a better world.

It is common that when trying to design new ideas, the outcomes seem to be an offshoot of what is already known. The difficulty of coming up with new answers and repeating the essence of pre-established solutions is a phenomenon known as "Design Fixation" (Cardoso & Badke-Schaub, 2011; Crilly & Cardoso, 2017; Youmans & Arciszewski, 2014a,b). This topic has been explored since 1991 with different approaches, from the phenomenon itself to surpass methodologies. However, there are still gaps in its exploratory perspectives, one of the most meaningful being the Design Fixation caused by social conformity.

This investigative perspective aims to add value when seeking to understand social dynamics during the development of a project to understand how conformity can affect creativity and create barriers to possible innovations. Social factors related to creativity aimed at the market and teaching have been explored in several areas of academic fields (Adams & Siddiqui, 2020; Hou & Jingyu, 2023; Magni & Manzoni,

DOI: 10.1201/9781032719290-12

2020; Sunstein, 2024; Taylor et al., 2014), however, not yet focused by this association of research lenses.

It is also valuable when it can add the possibility to expand discussions to promote the diversity of thoughts, ideas, and solutions to unlock new horizons to inventive potential. Encourage experimentation where mistakes are seen as trials rather than failures or problems. Inspire the inclusion of minority groups to add more perspectives, expanding the possibility of creating more inclusive, accessible spaces and improving user experiences.

This investigation is the result of a doctoral research in progress, carried out in Portugal and Brazil, which aims to answer the question "How can the effects of Design Fixation motivated by the need for adjustment and social conformity be minimized, and open up spaces for meaningful innovation in design?"

Breaking rules is understood as subversive behavior that can generate social judgment. On the other hand, innovations are results from breaking paradigms, when people decided to subvert and no longer do things the same way (Sunstein, 2024), some were responsible for the evolution of society (through inventions such as electricity and aviation), and at some point the creations and creators may have been labeled "mad", but "Here we can see a potentially beneficial role of misfits and malcontents, who can perform a valuable function in getting otherwise neglected information and perspectives to others" (Sunstein, 2019, p. 59). In a positive context, usually for transcendent creativity, going beyond expectations is called "disruptive", and "The negative relationship between conformity and creativity is usually due to reduced independence and lack of disruptive behaviors that are associated with conforming tendencies" (Magni & Manzoni, 2020, p. 965).

Looking at the architecture and design fields, focusing on humanity's daily relationships and interactions with spaces and objects, conformism keeps people unquestioning or reluctant about the possibility of different (and perhaps better) ways of doing things. Acting on social conformity can raise awareness and open the vision of students and professionals to different potential innovators. The organization of spaces with positions, flows, and dimensions different than usual, the use of materials and objects initially made for one function being used in other ways, can be valuable ideals to develop more efficient and accessible spaces, promote more economically efficient executions, and, in some cases, even make it possible when there is some difficulty in accessing the product that would be conventionally necessary. Some actions of great relevance have obtained recognition, citing few examples as the Pritzker Prize for Architecture for Shigeru Ban in 2014 for the construction with cardboard tubes to shelter victims of disasters (Kennicott, 2014) and for Francis Keré in 2022 gaining prominence in the way local materials are used (Kornblatt, 2022).

With the literature review framework of the themes involved in this research, an understanding of the subject and its state of the art was sought. As it is a subject that has yet to be explored through this angle, the possible research methodologies suitable for data collection are raised. Considering this, the present investigation intends to discuss and bring hypotheses of applicable and compatible methodologies, such as surveys and the application of a Creative Toolkit, as well as the possible formatting of a quasi-experimental field research project inspired by other current methodologies in the fields of Design and Social Psychology.

10.2 CONTEXT

Behavioral studies by Asch (1955) showed that people acquiesced when faced with a simple challenge, where they deliberately accepted answers that differed from what they believed, in a move toward social conformity. Many studies explored this phenomenon in Social Sciences and even today have been conclusive about individuals' judgment being influenced by groups (Franzen & Mader, 2023). The culture and social environment in which people are inserted and which compel them to act, think, and do things on a daily basis (Durkheim, 2004) are aligned with human behavior and the need to feel that they belong and are appropriate to the social environment (Lane, 1983). Besides, a large group position is enough to be taken as true, so people are more influenced by the group they belong to (Sunstein, 2019).

The process of project development is believed to be aimed at escaping social judgments, which can happen consciously or unconsciously, inherent to human behavior, in a psychosocial process of conformity to fit into the group in order to avoid rejection (Myers, 2014). In addition, information dissemination in contemporary times, which happens practically instantaneously, contributes to intensifying the social and psychosocial process that involves judgments, behaviors, and the need to maintain self-image (Souza et al., 2018). Even though the aim is not to close the research gap in the digital sphere, its immersion is relevant since the recurrence of dynamics in social media is almost inherent in social and cultural relations today and is necessary for self-positioning in the market and in the world.

This raises questions about the normative influences of social contexts and the unquestioning acceptance of common and traditional ways of making and using spaces and things, as well as the behavior of conformity and social acceptance in the context of design solutions. New projects sometimes seem to be a slightly modified repetition of past solutions, even if they were developed for recent problems. They stick to the roots of previously known solutions without being able to see possibilities in other imaginative poles (Cardoso & Badke-Schaub, 2011; Crilly & Cardoso, 2017; Youmans & Arciszewski, 2014a, 2014b).

In this context, it is believed that projects in the fields of Design and Architecture are susceptible to the effect of Design Fixation with the repetition and standardization of common and socially accepted solutions. There may be a possible reflection of the stifling of creative aspects in the face of these social relations, specifically in the proposition of innovative solutions within the scope of the design process, innovations may potentially be contained, and good answers to common problems may be hidden by social judgment.

10.2.1 KNOWLEDGE GAP

Situated in the field of design, and with design methodology and social psychology as its main areas of study, the current research seeks to answer the question: "How can the effects of Design Fixation motivated by the need for social adjustment and conformity be minimized, and spaces for meaningful innovation in design be opened up?"

To this end, the steps guiding the research path are (i) to learn about the factors that lead to Design Fixation and the use of repetition of standardized solutions; (ii) to characterize meaningful innovation on design; (iii) to understand how culture, tradition, and the social environment influence ways of making, planning, and interacting with spaces and things; (iv) to identify barriers and critical points in this context for project design; and (v) to find ways of interfering in the design process to avoid the barriers of standardized solutions.

To understand the state of the art of this research topic, a search was carried out in at least two indexed databases, "Science Direct" and "Periódico Capes", controlling the search parameters and checking the results. First, a search was made for the exact term "Design Fixation" in any part of the document, in all types of documents, on any date, in any language.

In the Science Direct database, 356 results were found in November 2023, including research and review articles, book chapters, and short communications, among others published since 1999, with the term being related, proportionately from the largest to the smallest number of publications, in the areas of Engineering, Medicine and Dentistry, Computer Sciences, Psychology, and Social Sciences. In the "Periódico Capes" database, results were found since 1991, with 179 occurrences in areas and proportions similar to the Science Direct database. The approaches to the theme of "Design Fixation" are diverse; however, no publications were found that aimed to study this effect through the prism of judgment and social conformity. Next, the search was repeated by adding the term "social conformity" using the same parameters defined earlier. In both databases, no results were found.

Crilly and Cardoso (2017) point steps and paths for research Design Fixation and list 20 approaches that have already been conducted. However, to date, although the fixation of ideas is a topic that has been discussed and the average number of publications has gradually increased over the years, it has not yet been fully explored from all perspectives.

Darke (1979) discusses the concept of primary generator,[1] establishing a connection between interpersonal and individual relationships as possible delimiters of projects even before the conception phase. The designer's background already imposes edges on perception and influence. In line, Cipolla (2018) raises arguments about the situation of vulnerability when exposing ideas, but emphasizing that discomfort can be necessary for building solutions.

The current investigation is aligned with the precepts established by Darke and Cipolla. Authors in the field of social psychology (Asch, 1955; Baumeister & Leary, 1995; Bley, 2006; Durkheim, 2004; Lane, 1983; Myers, 2014) discuss people's behavior when faced with invisible pressure and the need to belong. Observing the gap in the approach to the theme of "Design Fixation" through the lens of social adjustment and conformity, the aim is to develop actions in the project methodology to minimize the barriers to creativity with this bias.

It is important to note that this research intends to understand the relationship between the designer and creative meaningful construction in the field of social judgment, making this movement visible and providing tools for overcoming innovative barriers caused by social conformity. It is known, therefore, that it is not appropriate to act in the field of social psychology and behavior to control people's reaction to a new idea or promote acceptance, the field of study here being centered primarily on design. Even so, it is believed that it is possible to propose methodological interventions within the scope of the project to bring the awareness of the influenced state and subversive and innovative ideas closer together. In this way, it will be up to the innovator to decide whether or not to apply the new proposals.

10.3 METHODOLOGIES

The research being carried out is mainly qualitative in nature, so an inductive approach will be adopted. The analysis and interpretation of the results at all stages of this investigation will embrace a hermeneutic approach and will undergo interpretative criticism. Anyhow, in the investigative stages where information will be collected from surveys and tests, there will be cataloging and quantification of data, and in this stage, we will also have quantitative results (Lakatos & Marconi, 2003). In order to answer the research questions, it is proposed to develop a methodological basis of an applied nature, with qualitative and quantitative approaches. This research includes three phases: exploratory, generative, and evaluative.

Due to the scarcity of publications with this same investigative focus, the propositions of the generative phase methodologies represent a greater challenge in establishing the appropriate way of generating the information and systematizing it. Thus, the combination of three methods for capturing information was proposed, namely the Creative Toolkit, Surveys by Questionnaire, and Quasi-Experimental Field Research, which will allow the generation of information for subsequent cross-checking and comparison, aiming to find patterns and deviations from these same patterns. The descriptions and delimitations proposed for the development of the research are described in the following topics.

10.3.1 Defining Groups of Participants

In this stage, the people who will benefit most from the results of this research have been defined and described, and with whom we intend to seek participation. Aimed at activities that deal with the practical creative environment and that have a correlative nature, the fields of Design and Architecture are focused.

By questioning the already known and preconceived solutions, the acceptance of common and traditional ways of making and using spaces and things, and the action of conformity in design solutions, the research is focused on areas that work to solve problems that have a direct impact on people's interaction with the environment and with things, physical spaces, and artifacts, therefore, on a human scale. Thus, as illustrated in Figure 10.1, in the field of design, those working in product design, ergonomics, and interior design were delimited. In the field of Architecture, those working on building projects, interiors, and works on a scale of human interaction were delimited, justifying the non-inclusion of urban planning.

The profile definition of the participants resulted in two groups, students and professionals, and a total of five subgroups distinguished by their level of experience. The number of participants in each methodological stage may vary and will be defined after the contact and feasibility of groups with universities and offices.

In students group, two profiles were identified, the (i) beginners: students in the first two years of their degree, with an average age of 20, with no internship

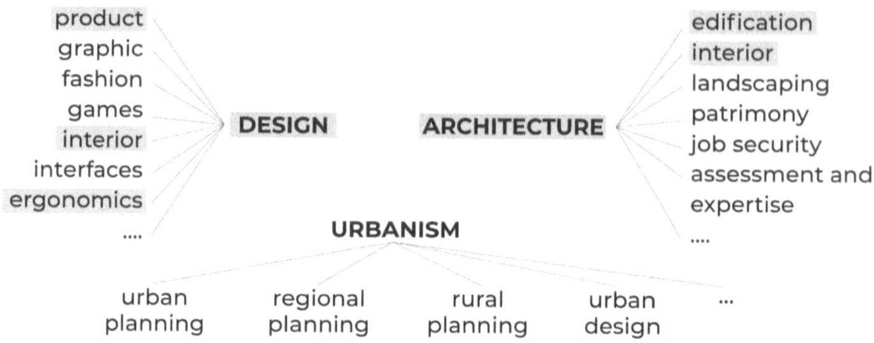

FIGURE 10.1 Fields and areas in which participants work. (Created by the author.)

or anterior project development experience; and the (ii) experienced: students from the third year of their degree and students returning to university, who already have professional and life experience, with financial independence.

The professionals group are composed of three subgroups defined as (i) junior: recent graduates with up to three years' experience, up to 28 years of age, generation "Z", in the process of establishing themselves on the market; (ii) senior-1: between four and 15 years of experience, between 29 and 40 years old, generation "Y" (millennials) consolidating in the market or consolidated, standardized operations may already have been established; and (iii) seniors-2: over 16 years of experience, over 40 years of age, generation "X" (boomers), consolidated careers and solid methodologies of action.

Among the students, the methodology of separating by age and year of academic enrollment was used in a similar way by Hou and Jingyu (2023), as experience in projects was taken into account. However, in a different way, the students' grades should not be taken to define the group of participants, as it is not intended to measure their technical abilities. As for students with experience, in addition to those who are in their third year of undergraduate study, empirical analysis has shown that there has been an increase in the number of students seeking a second degree or a new profession. These students will also form part of the groups participating in the research as experienced students.

The difference that most clearly separates the student and professional groups is the direct implication and impact on life caused by mistakes, judgment, and consequent distancing and rejection from the social group to whom the projects are presented, teachers or real clients.

10.3.2 Surveys by Questionnaire

The surveys will take place in two stages, before and after the Creative Toolkit is applied. The questionnaires were structured according to the guidelines proposed by Hill and Hill (1998) and Oppenheim (1992). The first stage is intended to identify, quantify data, and qualify the profile of the volunteer participants. The people surveyed may or may not take part in the dynamic stage, the experience with the Creative Toolkit, depending on their availability of time and location. Contributions, even in just one stage, are considered valuable. The initial survey also aims to qualitatively verify the self-perception of students and professionals about conformist behavior in the conceptual methodological stage of designing their projects.

For those who participate in the Creative Toolkit, the survey's second stage aims to capture the panorama of their self-assessments during the dynamic, in terms of the social rules anchored in their conceptions and idea propositions, and in terms of the judgmental behaviors they were subjected to or active in during the process.

It is understood that the possibility of an anonymous questionnaire can give participants greater freedom of expression. On the other hand, there is a need to organize the data for comparison with the second stage of the survey, after the Creative Toolkit, as it will also be important for the research to understand the impacts, triggers, personal perceptions, and openness of the participants to the proposed challenges. In this way, to minimize the risk of giving factual information, volunteers will be able to use pseudonyms. In some of the questions using Likert-type scales, even-numbered alternatives will be applied not giving the possibility of a neutral, biased response, since, in general, the subject of the survey deals with variables that can be sensitive because they involve behavior, and what would be socially acceptable (Hill & Hill, 1998). In addition, due to the content involving social dynamics and attitudes, which are emotional in nature, and based on the hypothesis of opening up space for ideas through empathy and receptivity, Oppenheim's (1992) proposal of using familiar, unrefined language is applied, offering descriptions of feelings in the response scales and formulating questions that are easy to deal with.

As a concern point, questionnaires may take some attention and time to answer. Also, even with pseudonyms, some responses still may not reflect realness. On the other hand, this method provides direct and faster answers to questions, mainly in Likert-type and alternative questions, which provide qualitative and quantitative data.

10.3.3 Creative Toolkit

The Creative Toolkit is described by Martin and Hanintgton (2012) as a set of organized physical elements that can be shaped and visualized in the form of participatory challenges by users. It is a design research methodology related to the conception, idea generation, and prototyping phase, suitable for qualitative research in behavioral, attitudinal, and innovation studies. The tool is initially proposed for groups; however, it is possible to use it either individually or in groups to verify different patterns of behavior, in front of peers or when the participants are not under possible judgment (Magni & Manzoni, 2020; Taylor et al., 2014).

The toolkit, beyond a box on the table, will involve an entire room as an influencer of behavior and creativity. It will work with elements scattered around the environment as if they were decorative or forgotten items. In this way, a scented candle, chocolates, and napkins arranged as a side support could be some potential unconventional items useful for solving a proposed challenge. The steps of the Creative Toolkit application are drawn in Figure 10.2.

After the survey's first moment, if the participant agrees to continue, he/she will be given a set of cards in the Creative Toolkit first step according to the variables:

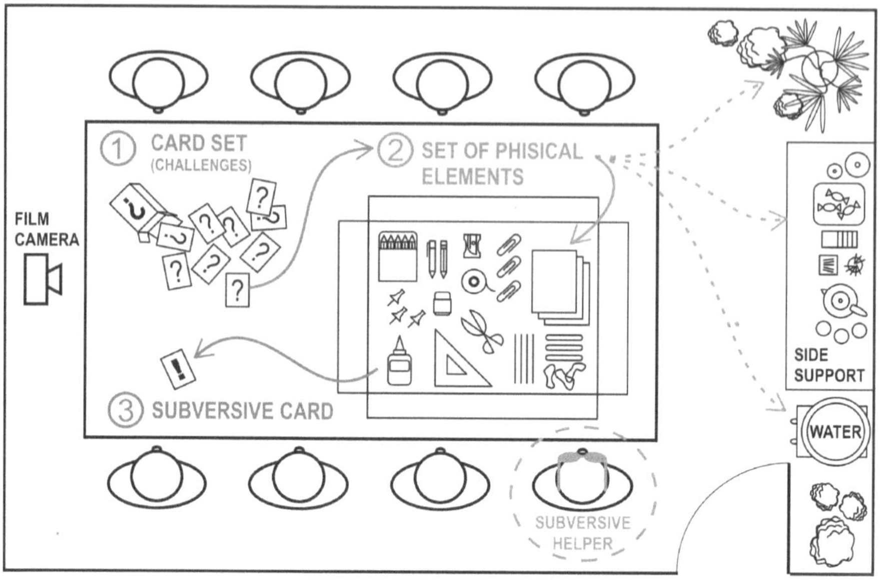

FIGURE 10.2 Creative Toolkit scheme of application steps. (Created by the author.)

(i) group definition profile and (ii) if one or more people will play the round. This set allows the participants to blindly choose a challenge to solve and gives sufficient instructions to proceed, setting the achievement required and which elements of the toolkit on the table can or cannot be used.

Going to the second step, it will be the moment to develop a project and design the solution to the challenge. In order to do that, participants must select and modify the toolkit set or an environment variable available on the side support and, as guidelines, should use the materials in different ways and purposes from the common uses in order to be easily solved (Raviola & Norbäck, 2013; Vink et al., 2021). Hou and Jingyu (2023) asked participants to solve the same problems used in another research paper for comparing results. But to dig deeper into how conformity acts on project development, new problems must be designed with the bias of subverting common rules of everyday life. These challenges will be designed to take between 30 and 90 minutes, including problem drawing, analysis, discussion (when with pairs), solution conception, and practical application. Applying in a classroom ambience or a workshop with students, the challenge could take time as a curricular dynamic exercise and extend to discussions and a repetition with a third step.

In the third step, after presentation of the solution by the participants, a new card will be added to the challenge set, suggesting a subversive way to solve it with an unconventional solution breaking social rules, with a unusual material or artifact that could be in the room, or an outcome in uncommon form (Raviola & Norbäck, 2013; Vink et al., 2021).

The conceptual model of the Creative Toolkit includes, during the second stage, the project's development phase, the action previously agreed with an actor (volunteer participant or helper) to insert creative proposals with subversive content breaking small social rules (Vink et al., 2021) to collect the reactions. This intervention will happen in some rounds to compare interactions with and without a disruptive act from a peer. The entire process will be filmed and recorded during its execution for later analysis and data collection.

With the Creative Toolkit methodological tool, the aim is to obtain output data on (i) creative behaviors in project design in teams and individually and (ii) identification of critical points in project development from the perspective of conformity and social judgment. This methodology should be applied as an enjoyable time, making a possible austere academic research bias lighter. In this way, people may be more relaxed compared to a questionnaire and give more authentic responses. Although the investigation walks along the abstract edge between design and social psychology, and due to the lack of similar works to reference and compare, defining the exact and specific data to be collected in this research presents a challenge. In this way, the next methodology proposed may show the same matters and should be explored for theoretical advancement and comprehension.

10.3.4 Quasi-Experimental Field Research

In survey studies, it became clear that it was common for participants to withdraw and provide expected or more socially acceptable answers. Under evaluated conditions, this is, in fact, an expected move (Hill & Hill, 1998). Observing this behavior within the patterns studied in this research, the question arose: "What if the participants were not in a judging position, but as judges in their natural habitat?"

In the Creative Toolkit, a previously agreed intervention of an actor has been added to subversively act at a given moment to relate reactions of their peers. However, as volunteers, people are still aware that they are taking part in research, and it is expected to decrease the expressiveness that they may believe to be socially undesired reactions. On the other hand, being in an everyday environment, how would their peers react to an unusual presentation?

An appropriate methodology for this investigation is explained by Lakatos and Marconi (2003) as field research to discover the relationships between phenomena by observing the facts as they occur naturally, requiring (i) literature review on the subject and identification of variables at a general level; (ii) determining the techniques used to collect data and the choice of samples; (iii) determining data recording and subsequent analysis techniques.

Furthermore, if the field research involves an experiment, which aims to test hypotheses of a cause–effect relationship by manipulating variables in a controlled or uncontrolled environment, it is necessary to follow the steps: (i) defining the problem; (ii) present the research objectives; (iii) establish the sample related to the research and the universe of components; (iv) establish experimental and control groups; (v) introduce stimuli; and (vi) control and measure effects (Lakatos & Marconi, 2003, p. 186).

Experimental field research is aimed at understanding social aspects and phenomena. Its advantages are that it captures data that can be used for other research, but its disadvantages are that it has little control over the situation and unknown factors (Lakatos & Marconi, 2003). Thus, it is inspired by the Social Experiment, which is research to understand social phenomena. Mainly, but not exclusively, used in the field of sociology and psychology, it also expands to economics, politics, education, health, and law, where variables are manipulated for observation (Cook & Shadish, 1994; Myers, 2014; Riecken & Boruch, 1978).

In a Social Experiment, some people within groups are randomly selected to perform different tasks or treatments to verify and measure social reactions and behaviors. According to Greenberg and Shroder (2004), experiments should have at least two control groups; a set of actions to ensure that everyone has the opportunity to participate in their daily lives; follow-up data collection for members of each group; and evaluation with statistical inferences of the degree of interventions and results. In a Quasi-Experiment, there are no randomized groups due to the particularities and individualities of the groups and the study for which it is applied.

With this basic understanding, we move on to the next steps in the organization of this research method, which should complement and confront the others.

10.3.4.1 Delimitation and Approach of Quasi-Experimental Field Research

In order to delimit the field research, studies were carried out on various research methodologies in design (Martin & Hanintgton, 2012; Vink et al., 2021), and in the areas of social psychology (Riecken & Boruch, 1978) to compare approaches and procedures with the intention of adapting them to the problem at hand and then creating an Environmental Observation Protocol (EOP) for data collection.

The experiments will be conducted during the field research, inspired by the creative approaches: (i) **Experiments**, appropriated to cause–effect verification through group experiments with variable modification. They are Quasi-Experiments when participants' attributions cannot be controlled due to the research conditions or groups (Cook & Shadish, 1994; Martin & Hanintgton, 2012; Riecken & Boruch, 1978); (ii) **Aesthetic rupture** to stage a disruptive factor that may cause strangeness in a identified social structure (Vink et al., 2021); (iii) **Artifact recreation** where an artifact with a relevant social function is modified by testing it in the same context (Raviola & Norbäck, 2013; Vink et al., 2021); and (iv) **Micro tests**, when a social structure is identified and small ways of challenging it are proposed by observing the responses of the environment (Vink et al., 2021).

Based on design research methodologies with a socio-behavioral bias, it is suggested to apply these approaches at specific moments throughout the experiment and use their guidelines to create collection guide summarized in an EOP.

10.3.4.2 Environmental Observation Protocol (EOP)

The EOP will be used to outline the information to be collected, taking into consideration the various facets of the investigative prism. It will organize the description and observation of means, inputs, and outputs, such as locations, environmental conditions, social structure, and participants; tasks, identified and manipulable variables, and artifacts; and reactions, interactions, identification of routines, and critical points, identifying behaviors and reactions toward things, spaces, and people; behavior and reactions toward proposed solutions; social dynamics of development, discussions, and presentation of design concepts; and whether there is openness to innovative ideas. These procedures are illustrated in Figure 10.3.

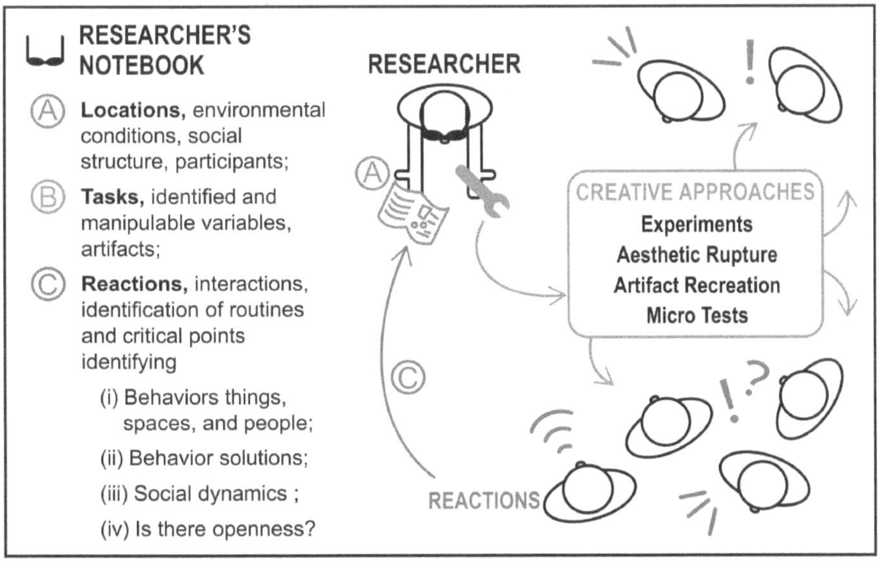

FIGURE 10.3 Procedures of EOP. (Created by the author.)

After revisiting the circumstances and reflecting on what happened, they will be reflected about if (i) "Are convergence and divergence constructive (reasoned and substantiated)? Or are they 'guesses'?" (ii) "Has the idea been implemented?" (iii) "If not, why?" and (iv) "If it has been: Did it work? How were the reactions?"

The notes should be organized in physical and digital notebooks as notes, diagrams, and drawings, using note coding (for participants, tasks, inputs, outputs, conditioning factors, variables) to organize and systematize the results later (Martin & Hanintgton, 2012; Raviola & Norbäck, 2013).

To categorize and describe the results, the methodology of **Behavioral Mapping** will be valuable, used for systematic documentation of characteristics, movements, and interactions, which can be centered on the place or the individual, using pre-code behaviors to identify standard behaviors and key points of interaction through maps, plans, drawings, and schemes (Martin & Hanintgton, 2012). Also, related with study of people in descriptive accounts, the **Design Ethnography** is appropriate for observation of individuals and groups in their natural habitats, in a defined social system, seeking sufficient information, seeking a comprehensive and empathetic understanding of users, language, artifacts, contexts, and behaviors to decipher patterns (Martin & Hanintgton, 2012).

10.3.4.3 Cultural Differences and Sample Group Selection

The research in progress is being carried out as part of a doctoral program at a university in Portugal, while the researcher lives in Brazil. In this context, it is expected that experiments carried out in different locations may yield different results, as various social factors can determine behaviors and reactions, such as age, gender, ethnic groups, culture, and government, among other variables (Greenberg & Shroder, 2004).

Using the comparison tool of cultural parameters between countries provided by Hofstede-Insights.com, which has been collecting continuous data for 35 years, some cultural differences between Brazil and Portugal can be observed, with the most relevant to this investigation being the cultural dimensions of "Individualism" and "Avoidance of uncertainty". According to the descriptions in Hofstede-Insights (Country Comparison Tool, 2023), for "Individualism", Brazil scored significantly lower, indicating the strong and cohesive integration of people into groups as well as social support based on relationships, while in Portugal, there is a greater tendency toward self-sufficiency. For "Avoidance of uncertainty", Portugal scored almost the maximum, indicating a preference for avoidance and an aversion to behavior and ideas that are not very orthodox, and a preference for rules. Meanwhile, in Brazil, although there is a need for rules, the need to comply with them is not as strong, and the need to live in groups and show emotions is more evident.

The two cultural dimension parameters highlighted are directly connected to behavioral aspects that demonstrate Brazil, compared to Portugal, as a culture that has a higher level of interdependence between people and groups and a stronger concern with showing emotions. Thus, given the cultural differences that could result in different procedures for obtaining results, given the fact that Brazil has a greater need for acceptance by a group, and due to the initial limitations of geographical constraints, the sample cut-off was defined in Brazil.

The next steps include establishing contacts and partnerships with educational institutions and offices in the fields covered by this research, to form groups with a variety of profiles. Quasi-Experimental Field Research could be extended to other geographical and cultural contexts in later stages to continue the investigation and go deeper through comparisons and possible methodological adaptations to fit each cultural reality.

10.4 CONCLUSION

The ongoing research into the phenomenon of Design Fixation brings a new perspective when it approaches the social psychology area and sees through the lens of social conformity a barrier to innovative ideas, exposing a new theoretical framework. Walking in the design field lit by other knowledge yards opens walkways to explore problems under new angles by crossing interdisciplinary scientific knowledge, but also carries challenges on how to explore it properly, correctly using the respective methodologies and respecting the gradient abstract limits of design acting space. Also, the research lacks approaching the social conformity, causing Design Fixation to establish a gray area to initially define the exact expected data that will contribute to the problem understanding and restrict the deck of replicable methods, making it necessary to build and shape adequate systematizations for data collection.

So, an exploration in design and social psychology disciplines revealed applicable methodologies that could be combined and adapted to format appropriate scientific systems for this generative research phase. It was realized that more than one methodology would be needed in the generative phase, so the formatting of a Creative Toolkit, the application of surveys by questionnaire before and after the toolkit, and, in parallel with the application of an quasi-experimental field research, the delimitations and approaches were outlined, as well as the guidelines for the EOP and the definition of sampling groups for research procedures.

The next stages of this ongoing research include (i) establishing partnerships with universities and offices and creating groups to apply the methodological tools for generating data; (ii) applying the pilots with smaller groups; and (iii) evaluating and refining the tools so that they can be used with all the proposed sample groups. Following the data collection, systematization and tabulation should be conducted in order to proceed to the evaluation phase of the results obtained.

ACKNOWLEDGMENTS

This work is financed by national funds through FCT – Fundação para a Ciência e a Tecnologia, I.P., under the Strategic Project with the references UIDB/04008/2020 and UIDP/04008/2020.

NOTE

1 Darke (1979) brings a research finding in design process named "primary generator", and it could be explained as some objectives self-imposed by the designer reflecting value judgement instead of the rationality.

REFERENCES

Adams, R. S., & Siddiqui, J. A. (2020). *Analyzing Design Review Conversations* (R. S. Adams & J. A. Siddiqui (Eds.)). Purdue University Press. https://doi.org/10.2307/j.ctv15wxp3j

Asch, S. E. (1955). Opinions and social pressure. *Scientific American*, *193*(5), 31–35. https://www.jstor.org/stable/24943779

Baumeister, R. F., & Leary, M. R. (1995). The need to belong: Desire for interpersonal attachments as a fundamental human motivation. *Psychological Bulletin*, *117*(3), 497–529. https://doi.org/10.1037/0033-2909.117.3.497

Bley, J. Z. (2006). *Comportamento seguro: a psicologia da segurança no trabalho e a educação para a prevenção de doenças e acidentes de trabalho*. Sol.

Cardoso, C., & Badke-Schaub, P. (2011). Editorial: Fixation or inspiration: Creative problem solving in design. *Journal of Creative Behavior*, *45*(2), 77–82. https://doi.org/10.1002/j.2162-6057.2011.tb01086.x

Cipolla, C. (2018). Designing for vulnerability: Interpersonal relations and design. *She Ji*, *4*(1), 111–122. https://doi.org/10.1016/J.SHEJI.2018.03.001

Cook, T. D., & Shadish, W. R. (1994). Social experiments: Some developments over the past fifteen years. *Annual Review of Psychology*, *45*(1), 545–580. https://doi.org/10.1146/annurev.ps.45.020194.002553

Country Comparison Tool. (2023). https://www.hofstede-insights.com/country-comparison-tool?countries=brazil%2Cportugal

Crilly, N., & Cardoso, C. (2017). Where next for research on fixation, inspiration and creativity in design? *Design Studies*, *50*, 1–38. https://doi.org/10.1016/j.destud.2017.02.001

Darke, J. (1979). The primary generator and the design process. *Design Studies*, *1*(1), 36–44. https://doi.org/10.1016/0142-694X(79)90027-9

Durkheim, É. (2004). *Durkheim As Regras do Método Sociológico* (9th ed.). Presença.

Franzen, A., & Mader, S. (2023). The power of social influence: A replication and extension of the Asch experiment. *PLoS ONE*, *18*(11 November), 1–15. https://doi.org/10.1371/journal.pone.0294325

Greenberg, D. H., & Shroder, M. (2004). *The Digest of Social Experiments* (3rd ed.). The Urban Institute Press. https://books.google.com.br/books?hl=pt-BR&lr=&id=abz_5Dj7CSYC&oi=fnd&pg=PR19&dq=social+experiments&ots=sLd45AbJrX&sig=6zZK61WEcxk19H8xFSRH1KcU_tw#v=onepage&q=social experiments&f=false

Hill, M. M., & Hill, A. (1998). *A construção de um questionário.*

Hou, G., & Jingyu, C. (2023). The impact of design creativity: Inspirations and timing of stimulation. *Telematics and Informatics Reports, 12*(August), 100105. https://doi.org/10.1016/j.teler.2023.100105

Kennicott, P. (2014). Shigeru Ban wins the Pritzker Architecture Prize. *The Washington Post, April.* https://www.washingtonpost.com/blogs/style-blog/wp/2014/03/24/shigeru-ban-wins-the-pritzker-architecture-prize/

Kornblatt, I. (2022). Diébédo Francis Kéré Named 2022 Pritzker Prize Laureate. *Architectural Record, 210*(4), 21–22. https://search.ebscohost.com/login.aspx?direct=true&AuthType=ip,shib,uid&db=asn&AN=156239149&site=eds-live&scope=site

Lakatos, E. M., & Marconi, M. de A. (2003). *Fundamentos de Metodologia Científica* (5th ed.). Atlas.

Lane, S. T. M. (1983). *O que é Psicologia Social.* Brasiliens.

Magni, F., & Manzoni, B. (2020). When thinking inside the box is good: The nuanced relationship between conformity and creativity. *European Management Review, 17*(4), 961–975. https://doi.org/10.1111/emre.12414

Martin, B., & Hanintgton, B. (2012). *Universal Methods of Design: 100 Ways to Research Complex Problems, Develop Innovative Ideas, and Design Effective Solutions.* Rockport Publishers.

Myers, D. G. (2014). *Psicologia Social* (10th ed.). AMGH.

Oppenheim, A. (1992). *Questionnaire Design and Attitude Measurement* (2nd ed.). Continuum.

Raviola, E., & Norbäck, M. (2013). Bringing technology and meaning into institutional work: Making news at an Italian business newspaper. *Organization Studies, 34*(8), 1171–1194. https://doi.org/10.1177/0170840613492077

Riecken, H. W., & Boruch, R. F. (1978). Social experiments. *Annual Review of Sociology, 4*(1), 511–532. https://doi.org/10.1146/annurev.so.04.080178.002455

Souza, G. de, Freitas, T. G. de, & Biagi, C. R. (2018). A relação das mídias sociais na construção da autoimagem na contemporaneidade. *Akrópolis – Revista de Ciências Humanas Da UNIPAR, 25*(2), 117–128. https://doi.org/10.25110/akropolis.v25i2.6426

Sunstein, C. R. (2019). *Conformity: The Power of Social Influences.* NYU Press. https://search.ebscohost.com/login.aspx?direct=true&AuthType=ip,shib,uid&db=e000xww&AN=1909797&site=eds-live&scope=site

Sunstein, C. R. (2024). Promoting creativity: How to overcome habitual thinking. *Rotman Management, Fall2024*, 20–25. https://search.ebscohost.com/login.aspx?direct=true&AuthType=ip,shib,uid&db=bsu&AN=179267666&site=eds-live&scope=site

Taylor, D. W., Berry, P. C., Block, C. H., Taylor, D. W., Berry, P. C., & Block, C. H. (2014). Does group participation when or using inhibit brainstorming facilitate creative thinking? Administrative Science Quarterly, *3*(1), 23–47.

Vink, J., Wetter-Edman, K., & Koskela-Huotari, K. (2021). Designerly approaches for catalyzing change in social systems: A social structures approach. *She Ji: The Journal of Design, Economics, and Innovation*, *7*(2), 242–261. https://doi.org/10.1016/J.SHEJI.2020.12.004

Youmans, R., & Arciszewski, T. (2014a). Design fixation: A cloak of many colors. In *Design Computing and Cognition* '12 (Vol. 12, pp. 115–129). Springer Netherlands. https://doi.org/10.1007/978-94-017-9112-0_7

Youmans, R., & Arciszewski, T. (2014b). Design fixation: Classifications and modern methods of prevention. *Artificial Intelligence for Engineering Design, Analysis and Manufacturing*, *28*(2), 129–137. https://doi.org/10.1017/S0890060414000043

11 Empowering Creative Leaders: Power Dynamics and Critical Thinking in Cultivating Designers and Industries

A Conversation with Artificial Intelligence on Possible Futures

Francesco Galli and Irina Suteu

11.1 EMPOWERING FUTURE CREATIVE LEADERS THROUGH ADAPTIVE THINKING

This chapter, an extended version of the article presented in the Design Commit '24 conference proceedings, emphasizes the imperative for the new generation of creative professionals to comprehend the ongoing transformation of various digital, tangible, and intangible formats. As we confront future scenarios and the evolution of the real world, engaging in conversations with AI becomes essential for generating and activating critical thinking awareness. Understanding these evolutions, along with their potential and limitations, is crucial for designing programs that educate future creative talents to harness these possibilities responsibly, which integrates dialogues with AI to enhance critical thinking about future scenarios.

The educational activity aimed to transform the learning environment into a creative arena, an absorption progression welcoming experimentation and providing opportunities to implement test projects. Creative talents engaged with several scenarios, focusing on three main objectives: first, to identify specific contradictions within the organizations; second, to develop internal intercultural awareness; and third, to activate creative strategies that could be adopted horizontally and transversally.

One of the central pedagogical research questions was how to introduce methods that integrated design education and philosophical concepts to participants without prior knowledge of design research and methodologies. In particular, the activity referred to

DOI: 10.1201/9781032719290-13

the definition of different "conceptual worlds" as presented by Popper (1976) and later incorporated into design thinking by Love (2000). This approach continued the investigation into the relevance of speculative reasoning from a philosophical perspective and its application in creative and critical thinking (Galli and Suteu 2024).

To pinpoint the areas where this approach could most benefit leadership empowerment, the program considered changes in participants' mindsets during different phases of skill and knowledge acquisition (Dreyfus & Dreyfus, 1980; Galli et al., 2017; Galli, 2020). The activity evolved on two parallel levels with different objectives: an explicit one focused on outcomes and an implicit, collateral one aimed at raising awareness of the potential of communication as a powerful management and strategic influence tool within the internal structures of international organizations.

The purpose of this initiative is twofold. First, it argues that future creative leaders should reflect on communication strategies through a critical thinking lens. Second, it posits that the application of design methods can strengthen the identity and sense of leadership among various actors. In conclusion, it suggests that change can be reimagined with the help of flexible strategies and by understanding the different power dynamics that extend beyond design activities to influence design decisions.

We assert that design methods and practices can make significant contributions to non-design learners. Therefore, we emphasize the importance of "un-disciplining" design culture and propose a diverse order of design as a strategic tool for social change from within organizations.

University experiences play an important role in cultural and creative industries studies, serving as dynamic academic hubs for critical thinking and reflection. At their core is a particular attention to the constant changes in market trends and the adaptation of pedagogical content to the opportunities and demands of the cultural industry, thereby educating future creative talents. This reflects in the pedagogical content, designed to introduce creative talents from corporate communication to cross-disciplinary knowledge.

The cultural diversity of the creative community and the cross-disciplinarity of the subject matter were two of the main elements taken into consideration in the process of creating the overall pedagogical framework. For this reason, an alternative disciplinary approach was used, integrating both design research and critical thinking methods (Figure 11.1).

During the educational activities, we advocated for a renewed view of design—not as a discipline bound by present rules, but as a creative thinking practice that provokes reflection. This brings into discussion a shift in how we perceive and present "design" to the global future scenarios and emphasizes the cultural value of an unconventional approach to reasoning (Galli, 2020). In this context, a disruptive attitude aims to demystify the significance of output and to unveil the authentic essence of critical processes. This shift in focus dismantles and "un-disciplines" existing norms, creating space for new cultural paradigms that drive change. This approach can significantly contribute to the development of power and knowledge, exploring how "design" can exert its influence at various levels, manifesting and sustaining power. Within this process, the creative talents assumed the role of a learner's charismatic and transformational leader, as described by Bass (1991), envisioning the potential for change, while others adhere to established regulations.

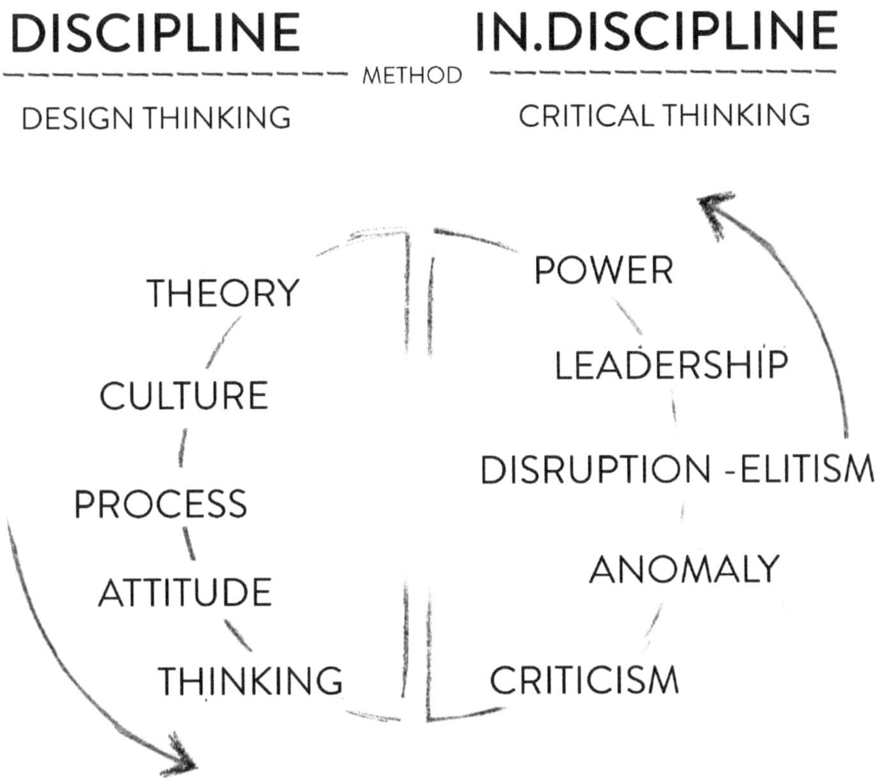

FIGURE 11.1 Process Methodologies: integrating design research and critical thinking.

As shown in Figure 11.1, the theoretical insights served as a guide for the practical activities. The methods were adjusted according to the learners' needs, supporting the development of their projects. The process encouraged participants to design communication scenarios through "conversation in action", fostering dynamic interactions and real-time application of concepts.

11.2 METHODOLOGY: "INJECTING" AN INTEGRATED FRAMEWORK INTO THE PEDAGOGICAL STRUCTURE

The design thinking framework, which focuses on a user's behavior, through an iterative process, is organized in several steps. The analysis and problem definition, seeking opportunities for envisioning solutions, testing, and constant refinement of possible solutions, whether tangible or intangible artifacts, systems, and services (Buchanan, 1992; Cross, 2007; Dorst, 2011), represented the backbone of the activities and training process. At the same time, critical thinking elements were presented to encourage the creative talents' attention on the discovery of "wicked problems" (Buchanan, 1992) and encourage them to investigate the case studies with critical thinking. In the next part, we will introduce the methodology of the course as follows. First, we will review the design thinking method, and second, we will explain how we integrated the critical thinking insights. This will then allow us to explain the theoretical background of agile strategies and introduce the preliminary results.

Design thinking is a problem-solving methodology that leverages vision, empathy, creativity, research, and scientific data to generate innovative solutions (Dorst, 2011) and solve "wicked problems" for equal social and economic prosperity (Brown and Kuratko, 2015; Brown et al., 2010). Due to its elasticity of mindset, design thinking enables leaders to embrace ambiguity and uncertainty (Panke, 2019), navigate complex scenarios, build bridges between different teams and fields, and drive divergent narrative strategies and business opportunities.

Design thinking has proven to be a valuable tool for leadership (Dorst, 2011) because it offers a human-centered and ethnographic approach to challenges that encourages research, experimentation, prototyping, and iteration (Glen et al., 2014). It also presents an important lateral thinking approach which helps leaders to identify unmet needs, pain points, and opportunities for innovation that might have been missed using vertical problem-solving methods, thus opening the possibility to reimagine new ideas and perspectives (Figure 11.2).

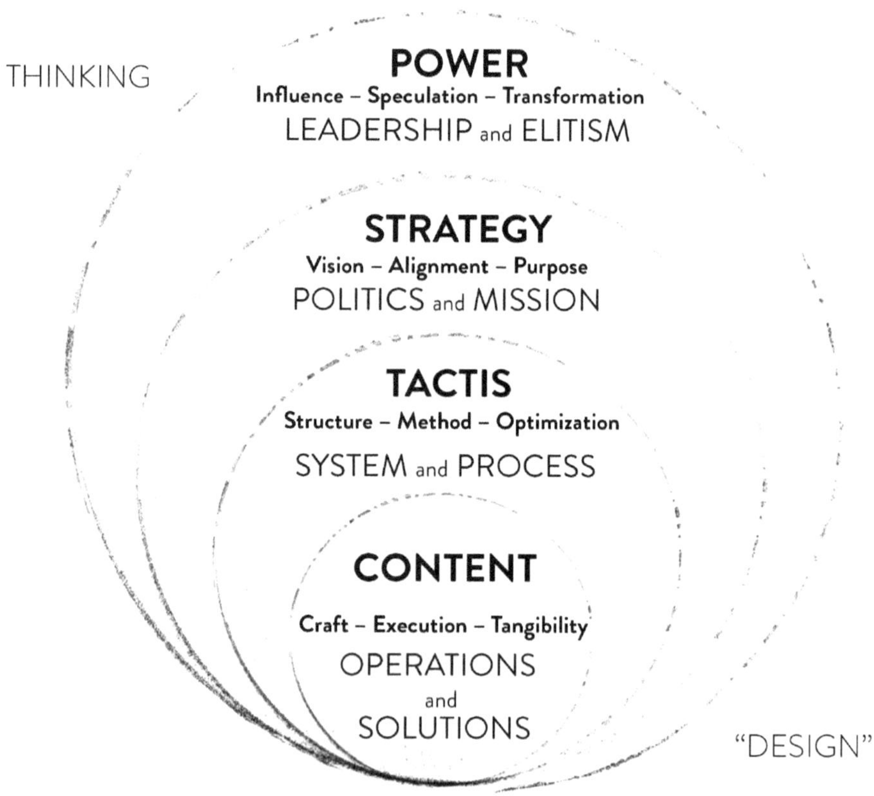

FIGURE 11.2 Creative Leader. (From Contents to Power Levels, modified from Best, 2006.)

The methodology has gained significant traction in recent years (Dorst, 2011), especially in the business world, both as a practical and interdisciplinary research method and a guide to solving complex problems in urgent situations. However, in higher education creative programs, too little attention is devoted to the power that design thinking can give to create visions at three levels: the strategy level for policy and missions, the tactics, systems, and processes level, and the operational level (Best, 2006, p. 17).

In this scenario, the flexibility of the design thinking approach becomes fundamental and strategic for the renewal of production realities that are confronted with the sudden evolution of strategies and visions. The main challenge is therefore to understand how the processes and methods of design thinking can be applied not at the content level, but at the leadership level, to integrate a unitary and shared identity, aware of the challenges and paradigmatic changes of purpose, vision, and values.

Critical thinking (Figure 11.3) is an approach and methodology that originated in philosophical inquiry and, in short, is based on pondering the "immediate acceptance of an idea that suggests itself as a solution to a problem" (Hitchcock, 2022). Because of the emphasis on the reflective stance, the method has been extensively used as an investigative tool in education. In fact, Dewey, one of the first advocates of

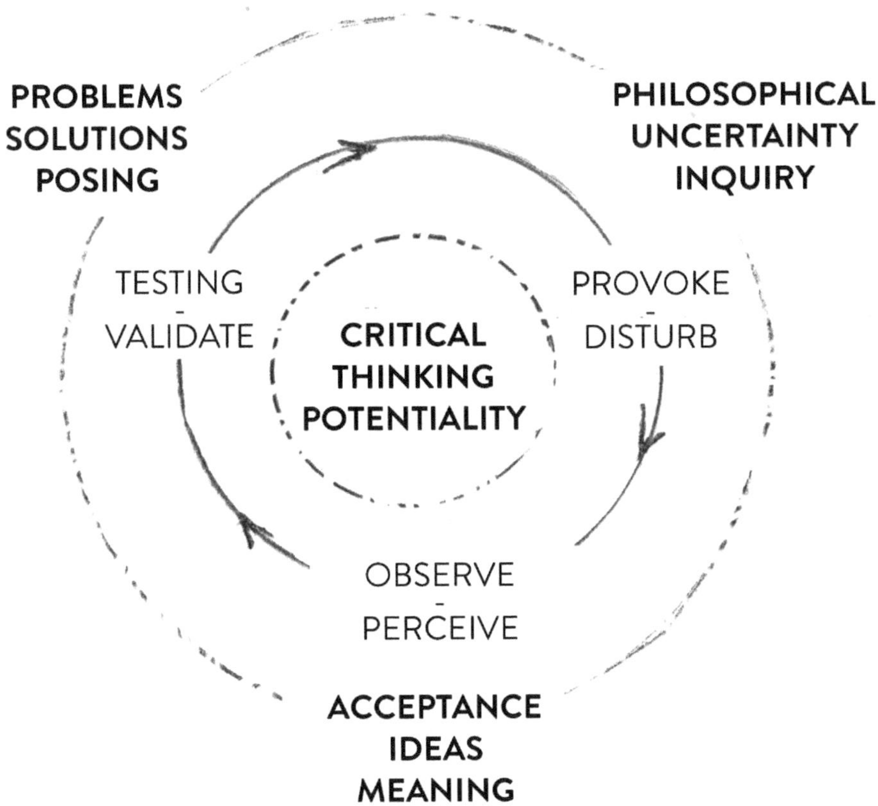

FIGURE 11.3 Critical Thinking Approach. From Problems to Potentiality.

the approach, underlines how critical thinking supports the transition of the talents' mindset, from a "perplexed, troubled, or confused situation" to "a cleared-up, unified, resolved situation" (Dewey, 1933, p. 106), and synthesizes the process in five phases:

- *Suggestions*, in which the mind leaps forward to a possible solution.
- An *intellectualization* of the difficulty or perplexity into a problem to be solved, a question for which the answer must be sought.
- The use of one suggestion after another as a leading idea, or *hypothesis*, to initiate and guide observation and other operations in the collection of factual material.
- The *mental elaboration* of the idea or supposition as an idea or supposition (reasoning, in the sense in which reasoning is a part, not the whole, of inference).
- *Testing* the hypothesis by overt or imaginative action (Dewey, 1933, pp. 106–107; italics in original) (Hitchcock, 2022).

The connection between *design thinking* and *critical thinking* has been recognized by several literature studies, such as Mathee & Turpin who argue for the integration of the two methods in the design and building of information systems as part of the problem-solving strategies (Mathee & Turpin, 2011); Ericson presents a thorough evaluation of the critical thinking and design thinking and argues that although the two methods have been applied in entrepreneurship, "critical thinking is generally not an explicitly discussed feature of the design thinking process" (Ericson, 2022, p. 407). Montoya-Hoyos, instead, provides a more detailed view of the design thinking in creative education context and argues for the relevance of the critical thinking in the convergent phase of the process because it "provides the tools to effectively evaluate and choose the most suitable of the solutions generated in the creative process" (Montoya-Hoyos, 2011, p. 21).

Drawing from the above insights, in the activities, the creative talents were presented with an integrated methodological approach.

An essential concept introduced was that only through exposure to design debates and critical conversations do individuals become more aware of how design can be used as an instrument of influence and manipulation. As outlined elsewhere,

> This is essential to note in particular because it gives further relevance and meaning to the importance of coaching for adaptive leadership, which enables surpassing the expert level and attaining full awareness of the environment and the capability to transgress the boundaries of professional domains by questioning their limitations.
>
> *(Galli, 2020, p. 140)*

In addition to discussing the hidden aspects of design, we investigated the concept of the creative leader and argued that for design to be used for positive purposes, designers must expand their awareness and learn to influence and inspire others. We suggest that creative leaders must possess a combination of technical skills, empathy, and vision to be effective, providing a thought-provoking examination of the role of design in society and the importance of responsible leadership within organizations.

SKILL LEVEL / MENTAL FUNCTION	NOVICE	COMPETENT	PROFICIENT	EXPERT	MASTER	VISIONARY
RECOLLECTION	NON SITUATIONAL	SITUATIONAL	SITUATIONAL	SITUATIONAL	SITUATIONAL	CHANCE SEEKING
RECOGNITION	DECOMPOSED	DECOMPOSED	HOLISTIC	HOLISTIC	HOLISTIC	RANDOM
DECISION	ANALYTICAL	ANALYTICAL	ANALYTICAL	INTUITIVE	INTUITIVE	INSTINCTIVE
AWARENESS	MONITORING	MONITORING	MONITORING	MONITORING	ABSORBED	**MUTATED**

FIGURE 11.4 Skill and Mental, Visionary Level. (Modified from Dreyfus & Dreyfus, 1980; Galli, 2020.)

One of the methodological aspects of the educational process was to take "novice" learners and, using the reference case study supported by the theoretical lessons on design thinking and critical thinking, introduce them to a mind frame and vocabulary specific to a level up on practical skills using the presented model of Figure 11.4.

By "injecting" a leadership attitude and trying to direct the training process in the evolution of mental aspects, we brought the attention of the creative talents to recollection, recognition, and awareness, with the decision-making aspect limited to a speculative learning exercise. By doing that, we opened a space for ambiguity necessary for crafting their own methodological approach from the inputs given, navigating in a creative "tension". As Galli explains,

> Placed in the creative context, the metaphor of vision indicates the possibility of the paradox that is born from the tension between the concrete experience and its articulation in an abstract conceptualization. The same tension is created by the potentiality and the preservation of power and allows visionaries to attain the openness of the discovery while enriching their experience.
>
> *(Galli, 2020, p. 151)*

11.3 DESIGNING ADAPTIVE METHODOLOGICAL EXPERIENCES

The educational activity introduced future creative talents to a broad understanding of creativity from a communication design perspective, moving beyond a strictly theoretical lens from traditional communication studies. It provided insights into the evolving nature of design applications in interdisciplinary settings and roles outside the conventional design sphere, preparing them to enter the real industrial world with a responsible vision and an awareness of evolving future scenarios.

The methodology aimed to empower novice practitioners to envision and enable possibilities (Welsh & Dehler, 2011), rather than focusing solely on outcome quality. By incorporating design thinking principles into their strategic approaches, these emerging professionals enhanced organizational identities and developed strategies that allowed leaders to move in synchrony with all affiliated partners. The goal was to equip future creative talents with the visionary power of design as a cultural force, enabling them to explore new connections grounded in real-world needs and to approach industry challenges responsibly and innovatively. This approach sought to advance the creation of new paradigms for creative education, extending beyond the aesthetic visual functions for which design is recognized, and focusing on its flexible thinking that offers alternative perspectives on management (Martin, 2009) and organizational leadership in a rapidly evolving industrial landscape.

In today's rapidly changing environments, adaptive strategic thinking becomes essential for accelerating adaptation processes (Van Ruler, 2021). This flexible, iterative method has gained popularity, particularly in strategic contexts. Betteke van Ruler, a leading advocate of adaptive strategy, has expanded the vision of strategic planning—not merely as information distribution (Falkheimer & Heide, 2014) but as a tool guiding the entire design process in all its phases (Van Ruler, 2021). Adopting an adaptive approach to strategy enables responsiveness, flexibility, and effectiveness in changing contexts (Salo, 2017). It helps experts deliver powerful outcomes that question the impact of decisions, respond proactively, and provoke thought through divergent thinking that defines the new paradigm of creativity today. Internal and external planning can thus be supported by an adaptive approach, allowing organizations to pivot and adapt their strategies in real time, which may determine business survival. Educating future creative talents in these methods equips them to navigate and influence the industrial world responsibly, with a keen awareness of emerging trends and future scenarios.

From this perspective, the educational activity highlighted the importance of applying design thinking principles to create adaptive strategies, strengthen organizational identities, and emphasize effective communication as a leadership asset. Reinforcing business identity is crucial, especially in an era that requires constant vision redesign. Business identities comprise values, culture, mission, and vision, and it is vital for leaders to communicate them effectively (Buil et al., 2016) and align them with emerging social, political, and environmental scenarios. The primary goal was to encourage participants to move beyond merely finding solutions to briefs and to question the assumptions presented by organizations. This approach fostered a responsible vision among future creative talents, enhancing their awareness of social, cultural, environmental, and political contexts, and preparing them to contribute thoughtfully to the industrial world's evolution.

The methodology stimulated the creative potential of future talents to envision unknown connections and possibilities (Welsh & Dehler, 2011), leveraging their fresh perspectives as members of the digital generation. By integrating design and critical thinking principles with an adaptive vision influenced by emerging trends, they enhanced organizational visibility and developed innovative internal and external strategies. This approach enabled organizations to synchronize with affiliated partners, enrich the sense of belonging, and raise mission awareness through various channels. The challenge involved applying design thinking methods and adaptive scenario planning not only at the content level but also at the leadership level, addressing the lack of a unified identity and reinforcing a community sense that perceives the organization as a cohesive entity working toward a common mission: the well-being of the entire body. In doing so, future creative talents are prepared to enter the industrial world with a responsible outlook, capable of anticipating and adapting to future scenarios.

During this process, the emerging creative talents advanced by crafting their own methodologies to approach presented problems. This progression was particularly noteworthy pedagogically, indicating behavioral changes in how they assimilated theoretical content and applied it to strategic scenarios. They adopted shared identities and collaborated effectively, shifting from passive to proactive attitudes, as evidenced by active participation in discussions with guest speakers. Instead of merely following presented methods, they began visualizing their own, blending activities insights with independently researched material. This evolution reflected their growing capacity for critical thinking and divergent problem-solving within the realm of adaptive strategic thinking and design thinking, ultimately preparing them to enter the real industrial world with a responsible vision and an awareness of the evolution of future scenarios.

The application of an integrated, design thinking/critical thinking methodology gave creative talents a larger spectrum of theoretical insights and helped them better understand the unpredictability of external factors with which organizations are confronted. In other words, a series of issues that weren't addressed in other courses, which focused more on pure marketing principles, and not on the meaning of "agile communication", as a process of transformation and adaptation to the emerging social contexts, but on fast production, quantity, and time (Van Ruler, 2021).

At the beginning of the training, it was difficult for novice, non-design learners to put aside their previous perspective about design and communication, merely based on visual content. Nevertheless, during the course, they learned how to incorporate an elastic mindset that helped them shape their critical thinking, curiosity, and ability to adapt to different scenarios in today's fast-paced business environment. As a result, at the end of the process, the creative talents were better equipped to handle the challenges of the future scenarios and management industries; as Welsh recognizes, this represents the point where also business schools could benefit from incorporating design thinking into their programs to help future creative leaders become integrative learners (Welsh et al., 2012).

Undertaking a leadership role in the pedagogical project entails a certain amount of risk because it requires the talents to be open to change and willing to transform their mindset. This is in part due to the way in which a leadership position exposes people to change, investing them with increased responsibilities, and at the same time brings this change within communities and organizations they are contributing to. Educators, in this sense, may be seen as leaders because they must embrace these risks in pursuit of their growth and progress (Lake et al., 2016). The approaches described herein helped novice learners to develop proposals that have a greater impact at the leadership level, incorporating policy and legislation considerations.

11.4 LIMITATIONS AND CRITICAL THINKING ON THE VISION

From a pedagogical perspective, the knowledge gained in the activity provides insights into the multiple possibilities of applying an integrated design and critical thinking conceptual framework in the activities and opens new possibilities that enable creative talents to understand how to design purposeful strategies for future scenarios. This perspective responds to the new requirements in the business and management fields, where a more exploratory and elastic approach counterbalances the business analytical strategy approach (Glen et al., 2014).

The program highlighted the importance of design thinking principles to strengthen the identity of organizations, the sense of belonging and leadership, and the importance of effective communication in the context of business and management. An important aspect of any business is establishing an identity, and communication plays a major role in this process. Business identities are made up of values, culture, mission, and vision, and it is crucial for managers to communicate this identity effectively (Buil et al., 2016).

The methodology applied helped talents to go beyond the problem definition and solving, to envision new possibilities (Welsh & Dehler, 2011), acknowledging in this process their leadership capabilities. The experiment showed how creative leadership can be applied in a variety of contexts, including in business, education, and the arts. In each of these areas, creative leaders can inspire innovation, encourage collaboration, and drive critical change. By fostering a culture of creativity and empowering others to think creatively, creative leaders can help individuals and organizations to achieve their full potential. By incorporating design thinking principles into the strategy process, the creative talents will enforce and give visibility to the non-profit identity and develop strategies to allow the organization to collaborate well with all affiliated companies and raise awareness about their purpose, vision, and values.

Outside of academia, however, it is important for companies to be aware of the dark sides of design thinking, such as the time and resources required, the possibility of failure due to uncertainty and unpredictable events, and the need to balance creativity with logical business practicality.

- Time-consuming: Design thinking requires a significant investment of time and resources, particularly in the research and prototyping stages where the collaboration with community and clients sometimes requires traveling, several interviews, translators if working with communities who do not speak English, and finally constant iteration processes that can be challenging for companies with limited budgets or restricted operation time.
- Uncertainty: Design thinking involves taking risks and new approaches that can lead to innovative solutions as well as failure, which can be costly for businesses, both in time and resources.

In addition, this constant iterative design process might be perceived as negative and confusing for some leaders who come from a linear and structured problem-solving approach. Furthermore, business misconceptions about creativity could limit its power and role in design thinking. It is common for leaders and large corporations to feel the solution unrealistic and too risky within the creative approach, and they must balance their creativity with practical considerations, such as budget constraints.

Regarding the effectiveness of design thinking for leadership discussed during the activities between novice talents and educators, an attentive reflection came up on some limitations within the business world in making an impact. In this scenario, are businesses truly interested in changing their working framework based on pure market principles? These principles shaped the anatomy of businesses, revealing a social and ecologically destructive economy, which influences the role of communicators and leaders' values. Still nowadays, in a world that promises equality and innovation, companies' communication strategies based on marketing, competitions, and toxic consumerism are extremely reliable indicators of a society in stasis (Monagle & Vardoulakis, 2013), not yet ready for a real change. Agamben's words like stasis, sovereignty, and power, expressed in 1999, sound even more appropriate these days, in this sadly recidivistic world (Agamben, 1999). Today, the power of leaders in negotiation, politics, business, and management still determines vulnerability to both social and organizational risks. In a world in which many profound challenges entail social, economic, business, and environmental changes, everybody, every leader, and every industry must redesign their existence. Today, we urgently need to use design thinking to tackle new kinds of problems—from global warming to food waste, climate action, education, and all SDG-related goals—for which new visions are required. Design education has the potential to reshape the future of design and drive radical change in society, defining the role of future leaders who envision new directions, strategies, and connections to break the stasis and propel a reactive, conscious, and revolutionary world into action (Liedtka, 2014).

11.5 CONVERSATION IN ACTION WITH AI ON FUTURE SCENARIOS BETWEEN DESIGN AND INDUSTRIES

To integrate and absorb the theories presented in this research, a divergent approach has been adopted that integrates advanced AI tools to deepen understanding and highlight research practical applications. In particular, the AI model (ChatGPT-4 o1 preview) was used to engage in a reflective analysis of the topics presented, with the aim of proposing synergies between the theoretical framework presented and the real world of design and industry, and the future scenarios between sustainability and the use of AI.

The process began with the request for a thorough analysis of the original text using the AI model.

We propose, in Figure 11.5, the analysis process carried out by AI, developed in a defined temporal space, associating it with express concepts and presented within the research.

11.5.1 Designed Prompt: Analyze and Understand This Research

Subsequently, a second request was proposed to AI, based on the previously made analysis, it was required to propose critical synergies and connections within the evolution of future scenarios that intertwine between the world of design and industries.

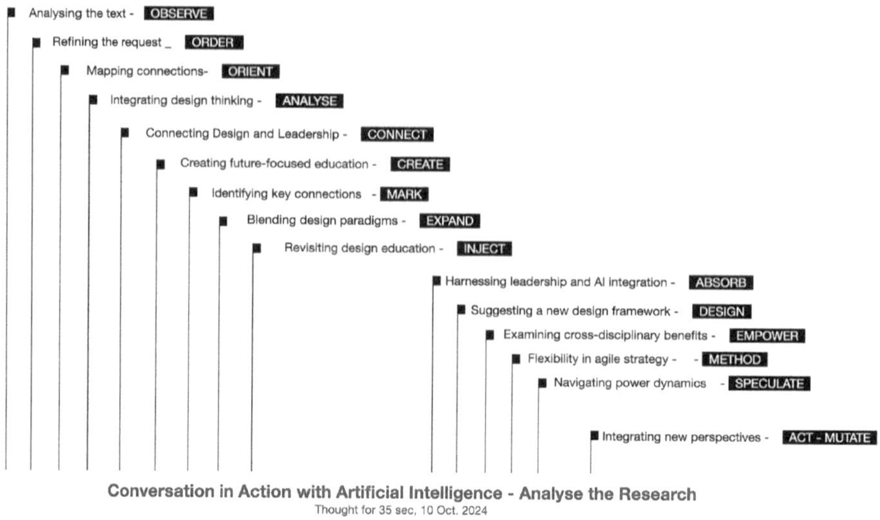

FIGURE 11.5 Conversation in Action with AI: Analyze the Research (Oct. 2024).

11.6 DESIGNED PROMPT: BASED ON THE PREVIOUS UNDERSTANDING, PROPOSE SOME CONNECTIONS FOR FUTURE SUSTAINABLE SCENARIO BETWEEN DESIGN DISCIPLINE AND CREATIVE INDUSTRIES

The ability of the AI to process and summarize large quantities of information allowed for the identification of key themes and insights that could enrich the additions in the real world; therefore, we propose some reflections that have emerged from the reflection and conversation with AI; also in this second request, we represent the thought process generated in Figure 11.6, faithfully reporting the passages and cognitive successions proposed by AI, visually associating them with concepts expressed in the research.

The reflections presented here are based on research analysis conducted using AI, aiming to verify and create evidence related to the educational experience analyzed. The experiment focused on assessing the impact and utility of integrating creative leaders into the industrial workforce—leaders who can generate concrete, sustainable proposals for future industrial development.

The integration of design thinking and critical thinking methodologies is key to fostering innovation that balances creativity with rigorous analysis. This combined approach equips professionals to tackle complex global challenges, such as climate change and resource scarcity, by generating imaginative yet feasible solutions.

Democratizing design methodologies across organizations breaks down silos and promotes cross-disciplinary collaboration. Empowering non-design professionals with design tools enables them to approach challenges from a humanity-centered perspective, leading to the creation of products and services that better meet stakeholder needs.

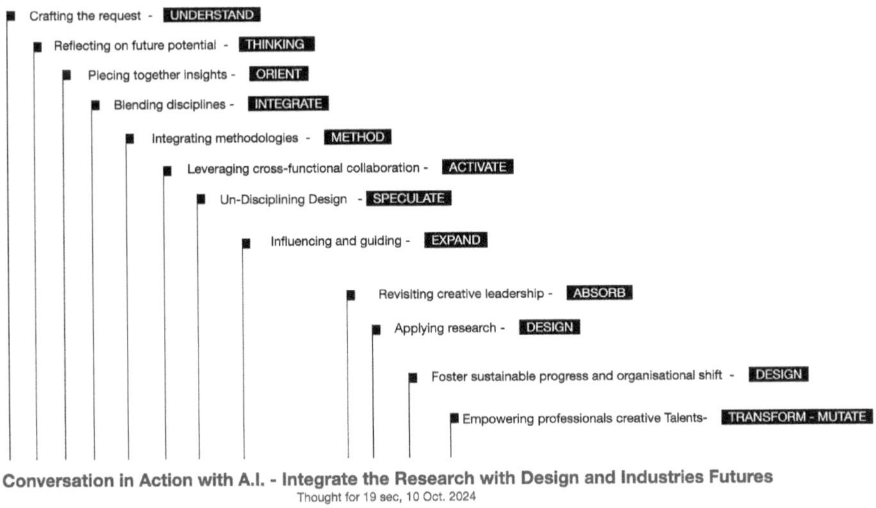

FIGURE 11.6 Conversation in Action with AI: Integrating Research with Design and Industry Futures.

Agile strategies are essential for organizations to remain responsive and adaptive. Transparent communication strengthens stakeholder relationships and supports effective change management, especially as industries navigate sustainability goals.

Adopting an "un-disciplined" approach to design encourages experimentation and diversity of thought, crucial for industries seeking to break from traditional norms and remain competitive in an evolving market.

Future success in industries hinges on cross-disciplinary collaboration, where knowledge from various fields converges to create more sustainable and ethical solutions. Education systems that promote cross-disciplinary skills will better equip professionals to handle complex challenges.

AI plays a pivotal role in augmenting human capabilities in design, analysis, and decision-making. However, ethical considerations, particularly regarding data privacy and societal impact, must be prioritized as AI is integrated into operations.

Creative leaders are central to these transformations, driving organizational change and fostering a culture of innovation. They are skilled at navigating power dynamics to promote sustainable and ethical practices, leading cross-disciplinary collaboration and guiding digital transformation efforts.

Sustainability is embedded in the vision of empowered creative leaders, who establish it as a core objective. They engage stakeholders in meaningful discussions, track progress, and ensure that sustainability is an integral part of the organization's mission.

This research highlights that the fusion of design thinking, critical thinking, and AI, led by visionary creative leaders, charts a course toward a more sustainable and equitable industrial future. Organizations that embrace interdisciplinary collaboration, ethical technological integration, and innovative leadership will be well-positioned to navigate the complexities of a rapidly evolving global landscape, balancing economic growth with social and environmental responsibility.

11.7 CONCLUSION AND FUTURE PEDAGOGICAL RESEARCH SPECULATIONS

The results of the pedagogical experiment demonstrate that designing creative and critical strategies for leadership necessitates a multidisciplinary approach, integrating elements of philosophy, psychology, sociology, and technology. Insights from observing the training activities highlight the imperative for design educators to embrace and incorporate diverse perspectives into their teaching methodologies. This integration is essential for preparing emerging creative talents to navigate and address the continuous changes in contemporary society effectively.

This chapter presented a pedagogical experiment that documented a learning experience aimed at equipping creative talents with the skills required to meet the demands of modern working environments. By fostering an environment that encourages multidisciplinary collaboration and critical inquiry, the educational activity successfully trained participants to develop adaptive and divergent leadership strategies.

Furthermore, the methodology employed in this experiment holds promise for future academic endeavors. It can be effectively applied in subsequent educational experiments in collaboration with creative organizations that have a cultural mission. Such partnerships can enhance the relevance and impact of design education, ensuring that future leaders are well-prepared to drive meaningful change in various societal sectors.

We've emphasized the necessity of modernizing our understanding of design, advocating for a departure from the traditional view of design as a discipline bound by preset rules. We've proposed a shift in how we perceive and present "design" to the global market, recognizing the cultural value of an unconventional approach to reasoning (Galli, 2020). In this context, the function of disruption is to demystify the significance of end results and unveil the authentic essence of critical processes. This process aims to dismantle existing norms and embrace an "un-discipline" approach (Figure 11.1), giving rise to novel cultural paradigms.

Power in this case is acquired through knowledge and gains a moral dimension. The observation becomes a state of mind that must expand and modify according to the modifications in the organizational system. The zone of strategic intervention gains a temporal dimension in which awareness and readiness for action have a major relevance. Revealing the importance of power enables us to suggest new methods of empowerment and leadership that can be tested and implemented both in the academic realm and in the industry and its inherent politics.

This approach unlocks the potential of creative talents to understand the broad spectrum of potential sectors and the real or illusory sustainability of possible future scenarios.

Leveraging the results of the educational activity, this reflection suggests divergent and interdisciplinary educational approaches for leadership, with the aim of equipping future creative talents with the ability to intercept the wide and uncertain potential of scenarios in emerging fields that go beyond the traditional domains in which we have always been confronted with the real world.

First Order Design Symbolic & Visual Communications (Symbol)	Second Order Design Material Objects & Artifacts (Object)	Third Order Design Interactions & Process (Action)	Fourth Order Design Environments & Systems (Culture/Systems)	Diverse Order Design Power Dynamics (Learning, Leadership)
Sub-disciplines: Communication & Graphic Design	**Sub-disciplines:** Industrial & Product Design, Fashion Design	**Sub-disciplines:** Interaction Design, Experience Design, Service Design, etc.	**Sub-disciplines:** Enviromental Design, Urban Planning, Co- Design, Transition Design, Facilitation, etc.	**Sub-disciplines:** Advanced coaching Creative Leadership, Speculative Thinking, Philosophy
Concerned with: Communicating information, ideas, and arguments through a synthesis of words and images using various media.	**Concerned with:** The form, function, and appearance of everyday objects, exploring physical, psychological, social, and cultural relationships between products and human beings.	**Concerned with:** Concerned with interactions that involve communications and products, mediating between digital and physical, managing resources, and combining elements to create meaningful and satisfying experiences.	**Concerned with:** Complex systems for living, working, playing, and learning, analyzing parts within larger systems. It examines design's role in sustaining, integrating, and adapting humans into broader ecological and cultural environments.	**Concerned with:** experiential learning, coaching expert designers, adaptive leadership training. Explores the power dimension of creative discipline, considering change and mutations as primary factors. Artificial knowledge Generations
Outputs: Printed and digital communications, logos, identities, websites, forms, etc.	**Outputs:** Appliances, furniture, vehicles, medical equipment, packaging, tools, etc.	**Outputs:** Product service systems, user/customer experiences.	**Outputs:** redesign of patient/doctor conversations;shaping of new local or regional education policy; redesign of national voting system; redesign of a national tax system; design of niche transition experiments	**Outputs:** design of training strategies for lcreative eadership. Advanced creative education and expert training. Introduces a circular dimension of leading and learning in contrast with brief-output process.

FIGURE 11.7 Diverse Design Order, Power Dynamics, and Creative Leadership. (Modified from Buchanan, 2001.)

These fields encompass politics, environmental, and social sectors, which demand a broad knowledge of societal changes and a versatile approach to communicating them.

This was possible by encouraging future creative talents to participate proactively in conversations with partners, challenge their narratives, and bring their own opinions into discussion. By first presenting the integrated critical thinking and design thinking methods, the teaching staff facilitated the assimilation of a leadership attitude among the participants, which also implied taking the risk of entering a creative confrontation with more experienced professionals.

This reflective process brought us to propose a different order of design, which may be seen as completing the model proposed by Buchanan (2001) (Figure 11.7).

This emerging order has sub-disciplines of creative leadership, speculative thinking, and philosophy and deals with human- and AI-generated dynamics, training expert creative leaders for adaptive leadership. It expands the developing role of expert professionals from practitioners to leaders and explores the reintegration of design in relation to changing circumstances and considering the upcoming mutations. More importantly, the outlined order introduces a circular dimension of leadership and learning over time, in contrast to the linear brief-output process practiced by expert professionals so far. Now, more than ever, industry and society need to develop flexible and strategic approaches. This concept of ontology and philosophy can support the design education mission to reimagine the future of creative leaders and connect their learning process outside of academia to real-world problems.

This highlights the need to view creative leadership coaching and training as a dynamic and evolving educational space, where insights from diverse fields such as geopolitics, philosophy, neuroscience, and psychology merge with cultural and humanistic knowledge. The use of AI conversations exemplifies how AI can drive this interdisciplinary integration and enrich educational methods. Future research on advanced coaching in creative leadership could delve deeper into these cross-disciplinary influences, exploring new educational formats that incorporate AI to foster adaptive and evolving leadership skills in the next generation of creative leaders.

REFERENCES

Agamben, G. (1999). *Potentialities: Collected Essays in Philosophy*. Stanford, CA: Stanford University Press.

Bass, B. M. (1991). From Transactional to Transformational Leadership: Learning to Share the Vision. Organizational dynamics, 18, 19–31. https://doi.org/10.1016/0090-2616(90)90061-S

Best, K. (2006). *Design Management: Managing Design Strategy, Process and Implementation*. Lausanne: AVA Publishing.

Brown, V. A., Harris, J. A., & Russell, J. Y. (Eds.). (2010). *Tackling Wicked Problems: Through the Transdisciplinary Imagination*. London: Earthscan.

Brown, T. J., & Kuratko, D. F. (2015). The impact of design and innovation on the future of education. *Psychology of Aesthetics, Creativity, and the Arts*, 9(2), 147–151. https://doi.org/10.1037/aca0000010

Buchanan, R. (1992). Wicked problems in design thinking. *Design Issues,* 8(2), 5–21. https://doi.org/10.2307/1511637

Buil, I., Catalán, S., & Martínez, E. (2016). The importance of corporate brand identity in business management: an application to the UK banking sector. *BRQ Business Research Quarterly*, 19(1), 3–12. https://doi.org/https://doi.org/10.1016/j.brq.2014.11.001

Cross, N. (1993). *A History of Design Methodology. Design* In N. Cross & K. Dorst (Eds.), *Design Methodology and Relationships with Science* (pp. 15–27). Dordrecht: Springer.

Dewey, J. (1933). *How We Think: A Restatement of the Relation of Reflective Thinking to the Educative Process*. Boston, MA: D.C. Heath and Company.

Dorst, K. (2011). The core of 'design thinking 'and its application. *Design Studies*, 32(6), 521-532.

Dreyfus, S. E., & Dreyfus, H. L. (1980). *A Five-Stage Model of the Mental Activities Involved in Directed Skill Acquisition*. California University Berkeley Operations Research Center.

Ericson, J. D. (2022). Mapping the relationship between critical thinking and design thinking. *Journal of the Knowledge Economy*, 13(1), 406–429. https://doi.org/10.1007/s13132-021-00733-w

Falkheimer, J., & Heide, M. (2014). Strategic communication in participatory culture: From one- and two-way communication to participatory communication through social media. In D. Holtzhausen & A. Zerfass, A. (Eds.), *The Routledge Handbook of Strategic Communication* (pp. 337–349). New York: Routledge.

Galli, F. (2020). *"Design" is Power. The Dark Side. Critical Thinking through Negotiation, Politics and Leadership*. Milan: Mimesis International.

Galli, F., & Suteu, I. M. (2024). Mutate design education through active and contemplative leaders. In *Proceedings of DRS Learn X Design 2023*. https://doi.org/10.21606/drslxd.2024.068

Galli, F., Pino, B., & Sute, I. (2017). Adaptive thinking for design leadership. Coaching adaptive capabilities to empower next visionary leaders. *The Design Journal, 20*(sup1). https://doi.org/10.1080/14606925.2017.1352917

Glen, R., Suciu, C., & Baughn, C. (2014). The need for design thinking in business schools. *Academy of Management Learning & Education,* 13(4), 653–667. https://www.jstor.org/stable/43696653

Hitchcock, D. (2022). Critical thinking. In: Zalta, E.N., & Nodelman, U. (eds.), *The Stanford Encyclopedia of Philosophy*. (Winter 2022 Edition). https://plato.stanford.edu/archives/win2022/entries/critical-thinking/

Lake, D., Ricco, M.E., & Whipps, J. (2016). Design thinking accelerated leadership: transforming self, transforming community. *The Journal of General Education*, 65(3–4), 159–177. https://doi.org/10.5325/jgeneeduc.65.3-4.0159

Liedtka, J. (2014). Innovative ways companies are using design thinking. *Strategy & Leadership*, 42(2), 40–45. https://doi.org/10.1108/SL-01-2014-0004.

Love, T. (2000). Philosophy of design: a meta-theoretical structure for design theory. *Design Studies,* 21(3), 293-313.

Martin, R. L. (2009). *The Opposable Mind: How Successful Leaders win Through Integrative Thinking*. Boston, MA: Harvard Business Press.

Matthee, M., & Turpin, M. (2019). Teaching critical thinking, problem solving, and design thinking: Preparing IS students for the future. *Journal of Information Systems Education*, 30(4), 242–252.

Monagle, C., & Vardoulakis, D. (Eds.) (2013). *The Politics of Nothing: on Sovereignty*. London: Routledge, Taylor and Francis Group.

Montoya-Hoyos, D. (2011). Empowering creative education with awareness and critical thinking. *Journal of Creative Education*, 2(3), 15–22.

Panke, S. (2019). Design thinking in education: Perspectives, opportunities and challenges. *Open Education Studies*, 1(1), 281–306. https://doi.org/10.1515/edu-2019-0022

Popper, K. (1975). Unended Quest Open Court, La Salle: Illinois (1976) 31 Feyerabend, P Against method.

Salo, O. (2017). *How to Create an Agile Organization, a Survey*. Helsinki, FI: McKinsey *& Company. Retrieved October 2, 2017.* https://www.mckinsey.com/capabilities/people-and-organizational-performance/our-insights/how-to-create-an-agile-organization

Van Ruler, B. (2021). Communication planning: agility is a game changer in strategy development, *International Journal of Strategic Communication*, 15(2), 113–125. https://doi.org/10.1080/1553118X.2021.1898117

Welsh, M. A., & Dehler, G. E. (2012). Combining critical reflection and design thinking to develop integrative learners. *Journal of Management Education*, 37(6), 771–802. https://doi.org/10.1177/1052562912470107

Part 3

Thematic – GREEN

Subthemes: Sustainability – Materials – Circular Economy – Education

12 From Fibre to Future
Sustainable Strategies for the Linen Textile Industry in Portugal

Isabel Duarte de Almeida and Maria João Delgado

12.1 GLOBAL DYNAMICS: BALANCING WELFARE, COMPETITIVENESS, AND SUSTAINABILITY IN APPAREL MASS CUSTOMISATION

Since the 1960s, the rise of fast fashion brands has completely transformed the textile industry (TI) by introducing a new business model that allows a range of fashionable clothing at a reasonable price (Ellen MacArthur Foundation, 2021). As designers push for innovation and new styles and create new consumer markets, brands follow these new demands and introduce a fast production cycle that constantly refreshes their inventory and offers consumers the latest looks at affordable prices within weeks.

To some extent, the rapid increase in production, facilitated by new technologies, transportation, and market deregulations (Piippo et al., 2022), has contributed to the global movement of people, products, services, and materials. This globalisation movement brought many benefits to the TI. Many manufacturing companies have relocated their operations to developing countries to reduce their labour costs and, consequently, reduce their production costs (Puig et al., 2008). Undoubtedly, globalisation has brought many advantages, such as helping to increase international trade, creating more jobs, and polarising income distribution (Staber, 1997). However, some economists and researchers have started exposing the social downsides of globalisation, namely those related to the low level of salaries that employees receive, the lousy working conditions many of them must endure, the salary inequality between men and women, and the extraordinary increase in pollution levels in countries where manufacturers are located (Berggren & Bjørnskov, 2023). Even though globalisation was so important in connecting continents, countries, and people and making them dependent on each other, its adverse side effects, namely for the environment, cannot be ignored.

Situations like wet treatment processes, such as bleaching and dyeing, cause water pollution, which affects aquatic ecosystems and local communities (Niinimäki et al., 2020). Besides the environmental problems in those countries, globalisation has led many factories to operate fast, forcing people to work long hours without breaks to guarantee that fashion shops (mostly in Western countries) are always full of clothes to meet the high demand.

DOI: 10.1201/9781032719290-15

The TI is highly resource-intensive, with most resources consumed before the clothing reaches the consumer. Based on data from Sweden, which relies on global supply chains, about 75% of the energy demand over a garment's life cycle occurs before the retail stage (Sandin et al., 2019). This pattern is likely representative of other high-income countries regarding fast fashion. These energy-intensive processes in the TI, such as yarn production and wet treatment processes, contribute substantially to high energy consumption (Peters et al., 2021); they rely on non-renewable energy sources in production or outbound logistics. Still, transportation plays a minor role compared to processes like fibre manufacturing, yarn spinning, textile weaving, and energy-heavy wet treatments such as bleaching and dyeing (Sandin et al., 2019). This issue worsens the industry's carbon footprint. It increases operational costs for companies trying to keep up with rapidly changing market demands, making the need for more sustainable practices increasingly urgent.

Even so, fast fashion companies can maintain a competitive edge by being at the forefront of style and availability. Because of this approach, buy-and-dispose behaviour has become equally common among consumers, increasing textile waste due to such consumption. For instance, European Union (EU) consumers discard an average of 11 kg of textiles yearly, which equals 36 skirts per person discarded yearly (European Parliament, 2020). This model of consumption of "buy and discard", which fast companies have perpetuated for decades, is synonymous with unsustainability. To tackle this problem, the EU has introduced the Circular Economy (CE) Action Plan, which encourages companies to produce more sustainable products, promote circular business models, and increase textile recycling rates.

Given these significant challenges, moving towards sustainability in the textile sector is imperative. This shift must seriously reconsider production and consumption practices in the TI, moving away from mass production and towards models that prioritise minimising environmental impacts throughout the supply chain.

By adopting sustainable practices—such as using natural fibres and applying CE principles—the industry can lower its ecological footprint while becoming more resilient and competitive globally.

This change also opens up economic opportunities, from creating new jobs to supporting the growth of innovative, sustainable businesses. On top of that, it can lead to a fairer and more ethical industry, one that shows greater respect for the people involved in producing, manufacturing, and selling clothing. The need to strike a balance between social welfare, economic viability, and environmental sustainability has never been more critical, and sensitive public opinion has emerged in response to these challenges.

12.2 THREADS OF CONCERN: SUSTAINABLE DILEMMAS IN TEXTILE AND APPAREL

12.2.1 Environmental Impact

When identifying the most polluting industries, it is essential to consider the five primary forms of pollution: water, air, soil, light, and noise. Research studies by Piippo et al. (2022) show that the fashion industry is the third most polluting sector worldwide after the oil industry and agriculture. It is well known that natural fibre production, mainly cotton, is characterised by a significant environmental impact. In terms of soil pollution, the extensive use of fertilisers and pesticides during harvesting harms soil quality and biodiversity, and the estimated annual consumption of 70 billion litres for garment production contributes to 20% of industrial water pollution from the treatment and dyeing of textiles. According to reports, 7,000–29,000 L of water are required to produce 1 kg of cotton, with over 87% of total water consumption used in manufacturing (Sandin et al., 2019). Thus, yielding a single T-shirt requires 2,700 L of water, equal to one person's drinking water needs for two years (European Parliament, 2023).

The TI is characterised by chemical-intensive processes, with around 15,000 chemicals used in manufacturing processes (Roos et al., 2019). Thus, hazardous substances from textile production are often disposed off in waterways, affecting rivers and oceans and posing health risks (European Environment Agency, 2019). The production of clothing uses about 1,900 chemicals (European Environment Agency, 2019), and in the finishing and dyeing steps of product manufacturing, this is responsible for around 20% of global clean water pollution (European Parliament, 2023).

Another major problem is air pollution caused by emissions from oil machines and diesel generators during production. This issue occurs mainly in the pre-consumption phase (Niinimäki et al., 2020), especially in production countries such as China, which accounts for a large proportion of global energy consumption in the TI (Eutotex, 2020). Textile production is a significant CO_2 emitter, emitting 4–5 billion tonnes annually (Niinimäki et al., 2020). In 2017, the purchase of clothing, footwear, and household textiles in the EU generated emissions of 654 kg CO_2 equivalent per person (European Environment Agency, 2019), making the fashion industry responsible for 10% of global carbon emissions, surpassing international flights and maritime shipping combined (European Parliament, 2023). The EU produces 12.6 million tons of textile waste every year (EU, 2023), with apparel and footwear accounting for 5.2 million tons of waste alone, which is about 12 kg of waste per person every year (EU, 2023). Only 22% of post-consumer textile waste is collected for reuse or recycling separately, while the others are often incinerated or landfilled (EU, 2023).

Within the EU, the upstream value chain (apparel, footwear, and home textiles) is a major consumer of raw materials and ranks fourth (European Environment Agency, 2019). This shows a significant demand for resources in the initial stages of production, reflecting the extensive use of materials and inputs before the final products reach consumers. The importance of this value chain for raw material consumption emphasises its impact on resource availability and environmental sustainability. As the EU endeavours to address environmental concerns and improve resource efficiency, it is essential to identify and manage consumption patterns within this value chain.

12.2.2 Socio-Economic Impact

The TI has a significant and complex socio-economic impact since many people depend on it for a livelihood, especially in developing countries (Piippo et al., 2022). Besides creating jobs, the industry generates income and improves the financial well-being of workers and their families. Moreover, the apparel sector contributes to skills development by providing training and expertise on various aspects of the production process. Skills improvement is valuable for human capital development (Niinimäki et al., 2020). Through its production and trade activities, the industry also contributes to overall economic growth by boosting GDP and positively influencing trade balances (European Parliament, 2023).

The global supply chain complexity highlights the different regions' linkage, from raw material suppliers to manufacturers, demonstrating the industry's impact on a broader socio-economic level. The apparel sector encourages entrepreneurship, with many small businesses involved in manufacturing, design, and retail stressing its role in supporting a diverse and dynamic business ecosystem, contributing to economic vitality.

The TI is critical in making clothing accessible and affordable to many consumers. This emphasis on consumer access and affordability underscores its positive impact on individuals' quality of life and standard of living (Piippo et al., 2022). However, alongside these positive contributions, it is crucial to recognise and address the industry's challenges. Labour practices are a significant cause for concern, and there are reports of poor working conditions, low wages, and problems with labour rights in certain regions (International Labor Organization, 2019).

As highlighted by the International Labor Organization (ILO) in 2019, exploitative labour practices contribute to socio-economic inequalities and adversely affect workers' well-being. Simultaneously, the industry's emphasis on fast fashion and mass production, as identified by Niinimäki et al. (2020), is linked to environmental degradation and resource depletion. This environmental harm, in turn, has far-reaching consequences, including negatively affecting local ecosystems and further influencing the socio-economic conditions of the communities surrounding these industries. Therefore, a systemic connection exists between exploitative labour practices, the fast fashion industry's production methods, and the resulting impacts on human well-being and the environment. Furthermore, the global nature of the apparel supply chain sometimes leads to a race to the bottom, where companies strive for the lowest production costs, potentially exploiting workers and communities. This race to the bottom can drive social inequalities and economic imbalances.

After consumption, textile waste is a challenge due to environmental pollution. In some regions without proper textile waste management regulations, local communities can be negatively affected, and external factors, including fast fashion models adopted by multinationals, contribute to environmental challenges (Ellen MacArthur Foundation, 2021). With low prices and short clothes life cycles, fast fashion companies can make significant profits without investing in reverse logistics or recycling. Textile waste is linked to external factors shaping the industry, adversely affecting workers, communities, and the environment. In the fast fashion industry, women comprise 60.5% of the workforce and are particularly vulnerable because of low wages and inadequate recognition (International Labor Organization, 2019). On the other hand, slow fashion emphasises quality and durability. Using materials and clothing for extended periods aligns with CE concepts. As a result of these sustainable practices, environmental action and social change are accelerated in the TI (Matušovičová, 2021).

12.2.3 Sustainability and Corporate Social Responsibility (CSR)

Sustainability started to be mentioned around the 1980s when academics from different research areas (e.g., science, biology, ecology, engineering, fiscal) realised that the fast consumption model was undoubtedly leading to a fast depletion of natural resources. To find an equilibrium among people`s necessities and available natural resources, a British author, John Elkington, revealed the three dimensions (pillars) of the new sustainability model: economic, social, and environmental. In his book, Elkington (1997) called the three dimensions the *Triple Bottom Line*, which should function in equilibrium, side by side, to maximise value to all the stakeholders. To be sustainable, an organisation must be financially independent (profit), minimise its environmental impacts (environment), and act according to social needs (social).

In 1987, the *World Commission on Environment and Development* urged world leaders to align the economy with environmental and social concerns. The Commission was chaired by Gro Harlem Brundtland, who published a report named "Brundtland Report" on "Our Common Future". For the first time, the report described the meaning of Sustainable Development (SD) as "a development that meets the needs of the present without compromising the ability of future generations to meet their own needs". Many years later, in 2015, all United Nations Member States adopted the 2030 Agenda with an ambitious programme of 17 Sustainable Development Goals (SDG) and 169 targets to be accomplished by 2030 (Figure 12.1).

The principles embedded in this new UN agenda are already guiding many corporations worldwide to make significant changes in their business models. Along with sustainability, other terms such as *Corporate Social Responsibility* (CSR) and *Business Ethics* are gradually becoming part of the business sector and part of the DNA of many companies. As David Chandler and William Werther arguably said about CSR (Werther & Chandler, 2009, p. 5), "A view of the corporation and its role in society assumes a responsibility among firms to pursue goals in addition to profit maximisation and a responsibility among a firm's stakeholders to hold the firm accountable for its actions". Companies have often followed the economic rule of

FIGURE 12.1 The UN 17 SDGs. (UN, 2022.)

being profitable. However, today, the role of a company is far more than being profitable. It provides a secure, rewarded, and just job to its employees; maintains good relationships with its suppliers; follows government laws; ensures reliable and quality products to its customers; fights for diversity and equality; promotes transparency with its stakeholders; and takes care of the environment as well. By incorporating those values, the company creates value for the employees, society, and itself.

12.2.4 CE in the Textile and Apparel Industry

CE addresses environmental challenges associated with linear economies by prioritising unlimited production and neglecting social and economic impacts within supply chains (Agyemang et al., 2019). The CE aims to transform economics by advocating reuse, recycling, and reduction (Stahel, 2019). CE adopts a circular approach that extends the benefits of materials and clothing rather than the linear approach that involves resource extraction and waste. Within the circular framework, "reducing" means a concerted effort to decrease inputs and outputs, promoting eco-efficiency (Rahla et al., 2021). Using "Reusing" involves repairing and refurbishing clothing and materials to extend their lifecycle (Rahla et al., 2021). Recycling clothing faces barriers, including technological complexity and financial constraints, resulting in minimal recycling rates worldwide (World Bank, 2020). Therefore, the fashion market is seeing rapid growth in "reuse" segments, which can reduce raw material consumption, conserve resources, and reduce the environmental footprint (Abdelmeguid et al., 2024).

In CE, circular design emphasises sustainable and durable materials in product development (Dan & Ostergaat, 2021). CE designers can use fabric-cutting techniques, 3D prototyping software, and innovative garment manufacturing processes. They can create multifunctional garments (e.g., T-shirts that can be worn inside out depending on the weather conditions), trans-seasonal clothes to increase their lifecycle, and even modular clothes.

Nevertheless, designers' transformative potential is constrained by limited access to sustainable materials (Dan & Ostergaat, 2021). CE creates "regenerative or restorative design" (Ellen MacArthur Foundation, 2021). This includes careful material design, closed-loop systems, and resource efficiency, all of which emphasise the importance of product design in achieving CE (Murray & Skene, 2017). By challenging the linear cycle of production, consumption, and disposal, CE becomes a key strategy to reduce textile waste, minimise the global carbon and water footprint, and address resource scarcity (Chen, 2021).

Companies can stand out more if they pursue sustainable practices as consumers become more aware of their impact on society and the environment. Fashion companies, for instance, are trying to increase the circularity of their materials in the supply chain, decrease the usage of raw materials, or choose better materials.

12.2.5 Challenges in the Portuguese Textile and Apparel Industry

Textile companies have undergone many changes over the years when it comes to increasing sustainable practices in their processes, which are in line with more CSR towards society and the environment.

The textile and apparel sector plays an important role in the development of the Portuguese economy. Before the pandemic, according to Portuguese Statistics Institute (INE) data from 2019, textiles represented €7.701 billion in sales and €5.215 billion in exports, and according to the Portuguese Textile Association (APT), textiles and apparel have represented 10% of total national exports.

There are big textile companies in Portugal with considerable recognition in the international market, such as *Riopele*, *Polopiqué*, and *Somelos*. They have significantly grown their sales volume in the last few years by improving the quality of their fabrics and investing in new and innovative fabrics, new technology, and sustainable practices. These textile companies have recognised that sustainability is the path to reducing waste, being more environmentally friendly, and adjusting their business to consumers' requests, since they are increasingly becoming more concerned about what they buy. Here are some initiatives that they have developed over the last years:

- Joining the Cotton Initiative, which promotes the use of organic cotton (Polopiqué, Somelos).
- Obtaining credible certifications such as global organic textile standard (GOTS) certification, OEKO-TEX 100 label, Recycled 100, and Better Cotton Initiative (Polopiqué, Riopele, Somelos).
- Implementing technology to minimise carbon emissions and reduce water waste (Polopiqué, Riopele).
- Decreasing and eliminating the use of harmful chemicals throughout the supply chain (Riopele).
- Repurposing fibres and other waste materials (Polopiqué, Riopele).
- Reusing fibres and water (Polopiqué, Riopele).
- Investing in renewable energy sources, such as solar panels (Polopiqué, Riopele).
- Exploring innovative methods to dye fabrics using natural sources such as fruits and nuts (Somelos).
- Utilising a sorting system to prevent waste from being disposed of in landfills (Riopele).

In the last few years, the textile and apparel sector has been through very challenging times. The business sector went through one of the biggest crises ever described, related to the Coronavirus (COVID-19) pandemic and the consequent lockdown. As a result, many manufacturers stopped producing or reduced their overall production. Portuguese textile business volume declined by 8.62%, and the garment volume by 10.68%, and in 2020, textile exports decreased by 11% compared to 2019 (Banco de Portugal, 2021).

However, in 2021, the business sector recovered and, after one year, was able to increase its exports to €5.419 billion, exceeding more than 16.5% compared to 2020 and 3.9% compared to the pre-pandemic year. The extraordinary recovery allowed Portugal to continue having a positive textile/fashion commercial balance (Exports>Imports) (Banco de Portugal, 2021).

12.2.6 The Importance of Linen in TI

Natural fibres are considered more environmentally friendly than synthetic fibres because they come from plants that can sequester CO_2, thereby helping to decrease greenhouse gases. Furthermore, cultivating natural plants requires less energy than producing synthetic fibres (Tiwari et al., 2024), making them a more energy-efficient option.

In contrast, synthetic fibres are derived from non-renewable petroleum and gas reserves. This group, which includes polyester, acrylic, nylon, and polyolefin, has a significantly worse environmental impact than cellulose-based fibres such as viscose, rayon, bamboo, modal, and lyocell. One primary reason is that producing synthetic fibres involves the use of toxic chemicals, which not only endanger workers but also contaminate the environment (Gonzalez et al., 2023). Additionally, when synthetic fibres are washed, they release millions of microplastics into waterways, contributing to environmental pollution (European Parliament, 2023).

Linen, by contrast, is both the oldest known fabric and one of the most sustainable fibres available today. With the help of merchants, linen reached the European markets through trade transactions (Dissanayake & Perera, 2016). Because of its richness, farmers started cultivating linen mostly in northern Europe, making it an essential textile for trade-offs.

Linen also became known in the Portuguese territory in the XIX century. Seeing the potential of linen, Portuguese manufacturers in the northern region of Portugal started manufacturing this fibre, gradually building a specialised industry. However, with globalisation at full speed and, consequently, the rise of new potential markets, other fibres were introduced, resulting in their increased (and cheaper) consumption. Today, Portuguese flax cultivation persists only on a small scale in a few smallholdings, which makes it difficult to benefit from mechanisation or produce quantities large enough to scale textile production.

There are a few Portuguese textile companies that strive to manufacture linen fibre for the clothing industry. In most cases, textile companies import linen fibre and blend it with other fibres (e.g., cotton) to create garments. Moreover, other textile companies use linen to create bed sheets and towels (home textile market) (ATP, 2018).

12.3 PROBLEM TO ADDRESS AND A BUSINESS PLAN TO ACHIEVE SOLUTIONS

The fashion and textile industries are notorious for their environmental impact and social responsibilities. Clothing's ever-growing demand for high-quality materials worsens environmental impact, and financial pressure often neglects environmental protection measures. However, as public and political pressure increases, these industries become more sustainable despite historical adherence to capitalist growth paradigms (Andersen, 2017).

As demonstrated in the literature and research, organisational change is a considerable challenge. As a result, sustainability offers a compelling alternative to business models that harm the environment and perpetuate social injustice. This work proposes a sustainable business plan for the natural fibre textile sector. As one of the most sustainable fibres, linen is selected as an example of a primary raw material. Ecological concerns and consumer trends drive increasing demand for natural fibres like hemp and linen for textiles. It takes five times fewer pesticides and fertilisers to grow linen, and most of the world's linen, valued for its long fibre, is grown in Europe (Dissanayake & Perera, 2016). This natural fibre targets the upper-middle market with luxury clothing as its goal. A buoyant economy and a sustainable strategy are crucial to this TI's future.

A business plan is a fundamental tool that helps managers understand the current state of the business (by identifying strengths, weaknesses, opportunities, and threats) and provides insight into the future. This plan is crucial for efficient management, attracting new customers and partners and securing credit. Fiore (2005) states that a business plan fulfils two main functions. First, it is an organisational tool to streamline and clarify the company's goals and strategies. Second, it acts as a sales document, presenting a business idea and showing that a product/service can create profits, attract funding, and secure business resources.

The present business plan details creating a sustainable linen company that is competitive in the Portuguese TI and meets the needs of Portuguese customers and consumers. Therefore, this plan should serve as a guide for implementation practice and a guiding line for decision-making, defining the company's objectives and planning needs. Nevertheless, several considerations must be addressed before developing and implementing the business plan. Key questions include "What is the current state of the linen manufacturing market in Portugal, and how does it operate?", "What are the production costs associated with linen clothing products?", and "Is there a growing interest among designers to work with 100% linen fibre?"

12.3.1 Methodological Approach

The methodology used to create the business plan requires extensive research and analysis. First, secondary data gathered from external sources, such as literature reviews, company websites, and online annual reports, were widely used to summon the necessary information on business planning, the linen industry and market, the TI, eco-design, sustainability in textiles, CE, and the role of CE in fashion.

Then, a qualitative analysis follows, looking at experts' perspectives on textiles and circular design. It specifically focuses on the current state of the linen manufacturing market in Portugal, its operational intricacies, and the production costs of linen garments. In addition, an evaluation was conducted on designers' enthusiasm to engage with 100% linen fibres. Considering gender, age, educational background, and position held within the company, Table 12.1 presents the profiles of respondents from each organisation.

The interviews are a vital step to understanding the market landscape, assessing the feasibility/viability of the business idea, and identifying potential customers. The primary techniques for analysing interviews included qualitative analysis. The environmental analysis is conducted through a Political, Economic, Social, Technological, Environmental, and Legal (PESTEL) analysis, which aims to uncover the external forces affecting the business. A comprehensive investigation is carried

TABLE 12.1
Interviewees' Characterisation (Referred to by Bracketed Terms)

Participant	Gender	Age	Academic Education	Role in the Sector
Interview 1(i1)	Female	40	Design PhD	Designer Professor at Univ. Minho
Interview 2(i2)	Male	61	Management MSc	WIN WIN Textile CEO
Interview 3(i3)	Male	38	Engineering Degree	POLOPIQUÉ Spinning Production Responsible
Interview 4(i4)	Male	58	Textile Engineering MSc	FITECOM S.A. CEO

out for the internal audit, including a functional analysis, an assessment of competencies, capacities, and resources, and a strengths, weaknesses, opportunities, and threats (SWOT) analysis. A BMC is also created to formulate a strategic plan that outlines the company's value proposition.

This sustainable business plan should focus on four factors. First, to know how the Portuguese textile market works and how linen is manufactured. Next, examine the production costs associated with linen garment products, considering the different stages of the manufacturing process. Understanding the price range of retail chains looking for 100% linen products is crucial. Several factors must be considered, including production costs and market demand. Finally, it is crucial to determine if designers actively seek 100% linen in their creations. Completing the comprehensive approach involves understanding potential market demand and preferences.

By carefully considering these factors, the sustainable business plan should point to an environmentally conscious and socially responsible future in the textile and fashion industry and identify how the company can provide a competitive advantage to its potential customers. Figure 12.2 shows the structure of the analyses used to implement this business plan.

12.4 MAIN FINDINGS AND RESULTS' DISCUSSION

This section examines the market landscape and the dynamics of the TI. The analysis includes an Industry Analysis covering the Global and European textile market, the Portuguese textile market, its sustainability, and the linen market. A PESTEL analysis considers the Political, Economic, Social, Technological, Environmental, and Legal factors influencing the industry. The section also covers a SWOT analysis, which provides an internal assessment. Finally, it shows "Porter's five forces" results, which shed light on market competitive forces.

12.4.1 Industry Analysis

12.4.1.1 Global and European Textile Market

The TI processes natural and synthetic fibres into yarns and fabrics used to manufacture various goods, such as high-tech synthetic yarns, bed sheets, industrial fibres, and clothing. In a global context, the United States emerges as the frontrunner in revenue generation within the Apparel market.

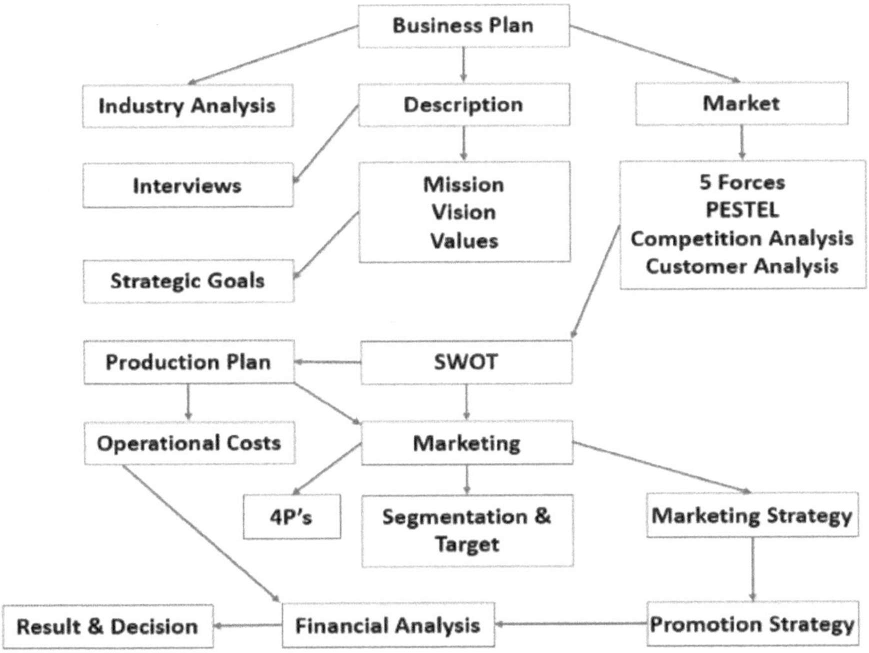

FIGURE 12.2 Framework for the New Linen Factory Sustainability Plan analyses.

Around 599,009 people work in the TI in Europe, generating 88 billion euros in revenue. Euratex (2020) shows that 99.8% of businesses in the textile and clothing industry are small and medium-sized (SMEs). Europe's fashion and TI comprises approximately 1.5 million companies, with a turnover of €162 billion in 2021. The European Commission says over 70% of the industry's workforce is female. The industry is also highly fragmented, with 98% of companies having less than €2 million in annual revenue. SMEs account for 80% of textile and clothing exports to the EU (Euratex, 2020).

Textile production can be categorised into three main sectors: clothing, household, and industry. According to Euratext (2020), Italy is the leader in European investments in textiles, clothing, and synthetic fibres. Germany, France, Portugal, and Poland follow.

In 2023, the European Apparel market exhibits a robust economic performance, with an estimated revenue of €410.15 billion. This projection aligns with a foreseen annual growth rate of 2.26% (Statista, 2024), indicative of sustained expansion within the market. Within the apparel market, the largest segment is Women's Apparel, commanding a significant market volume of €223.45 billion in 2023 (Figure 12.3).

This emphasises women's fashion's prominent role in shaping the overall market landscape and highlights the potential and growth of the European apparel market, making it a business attractive and profitable opportunity. It further reinforces the importance of the European market to the global economy.

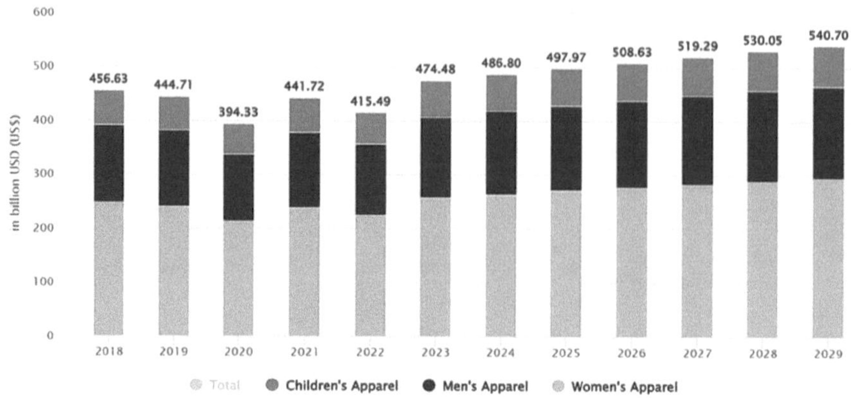

FIGURE 12.3 Revenue by segment in the Apparel market in Europe. (Statista, 2024.)

TABLE 12.2
Characterisation of the Textile Sector in Portugal (PORDATA, 2023)

Portugal	Textile Sector	National Total
Companies_2022	8,867	68,501
Investment (%wealth)_2021	14.1	23.3%
Exports (M€)_2022	6,538.5	78,326.1
Imports (M€)_2022	6,356.8	109,109.2
Coverage rate of imports by exports_2021	140.6%	86.1%
Employment (Thousands)_2021	193.36	4,812.99
Gross Added Value (M€)_2021	4,238.30	209,790.75
Apparent labour productivity (€)_2021	21,919.66	38,867.74

12.4.1.2 Portuguese TI

Portuguese textiles have a long history and economic significance. The past decade has seen substantial growth despite a crisis in the 1970s and 1980s due to excessive reliance on cheap labour. Innovating, delivering quality, and pursuing sustainable production practices have led to this resurgence. As a result of this shift, the sector is expected to remain competitive in the global market.

Portugal is a hub for textile/apparel production, with its TI contributing 6% to GDP. According to Banco de Portugal (BPstat,2023), about 121.124 people are employed in the textile sector, making Portugal's 15th most profitable industry. In 2022, the sector comprises 6339 companies, generating a collective turnover of €8.833 billion (BPstat, 2023). The Portuguese economy relies heavily on this sector, which employs more than 10% of the workforce. With 65% of its production exported to 189 countries across five continents, the industry plays a crucial role in foreign trade. As a result, Portugal ranks 22nd among clothing exporters worldwide and 10th among EU exporters, with 5.5% of national production exported to the EU. Table 12.2 provides key indicators of the Portuguese TI.

Market size, measured by revenue, was €4.9 billion in 2023, and nearly 50% (48.95%) of this turnover is attributed to medium-sized companies. SMEs represent 38.12%, with 10249 employees, and most companies in Portugal are micro-enterprises (61.27%). Regarding company maturity, the largest share consists of businesses with more than 20 years of activity, primarily operating at the micro-enterprise level (PORDATA, 2023). Furthermore, the Portuguese textile and clothing industry is located in the Portugal North Region, representing 87% of the sector's turnover and 85% of employment (ATP, 2019). The region hosts the most significant number of textile companies in Portugal and has the highest concentration of SMEs. Raw materials availability, skilled labour, and market access contribute to this. Many textile-related businesses, from designers to manufacturers, are also in the region. Government subsidies and incentives also encourage growth.

12.4.1.3 Sustainability in the Textile Sector

By combining innovation and sustainability, the Portuguese fashion sector creates products with increased value and contributes to economic sustainability for companies. Portuguese businesses, particularly textile firms, are tackling pollution by aiming to eliminate or reduce their effects. Interviewee 2 (i2) says this change is "driven by the resulting economic and environmental advantages, adding to a positive image that enhances its market position". Sustainable practices have been adopted by many Portuguese textile companies, including wastewater recovery, eco-friendly materials (Provin et al., 2021), synthetic fibres, and social measures during manufacturing (Lin et al., 2023), reflecting a high level of environmental awareness, and some companies specialise in recycling textile waste (Costa et al., 2022). The interviewee (i3) highlights that "the Portuguese textile and clothing cluster has become one of the world's most innovative and modern clusters, with a reputation for cutting-edge technical textiles and private label models". According to the interviewee (i1), "most companies in this sector have existed for over 20 years and are mainly micro-enterprises", in line with the Portuguese PORDATA (2023); however, "i1" emphasises that "Portugal is the most trusted fabric/apparel producer worldwide". In the same line of thought, "i4" mentions that it "requires us to do business with several brands that demand investment in sustainability areas". As a result of working for these companies, the CEO knows "these investments could increase the company's profit margin and generate more income for the company".

Regarding environmental impact, respondents agree that pollution from the sector is obvious and needs to be tackled quickly. The "i2" said that "if we didn't report pollution situations, this could affect the brand's reputation and trustworthiness in a short period; in extreme cases, this trust might even disappear", and "i1" is peremptory: "To prevent the environmental impact, companies should work on and invest in innovation to develop alternative, "greener", and, if possible, reusable products". Consumers are becoming increasingly concerned about pollution, reflected in their opinions. Organisations and governments worldwide strive to minimise the environmental impact of individual actions, processes, cities, and countries. Consumers pick up on this message, and businesses are responsive to it. Pointing out the significant amount of textile polymers washing ashore in different parts of the world substantiates and quantifies a dilemma attracting global attention. In this context, sustainability is very important in society, and companies are pursuing short-term goals to achieve it. Therefore, some authors (Al-Issa et al., 2022) propose that companies integrate the corporate governance pillar, which shows how business is conducted, with the social and environmental pillars, which show how people and the environment are affected by the company's activities (BOF, 2021). This integration aims to create additional customer value.

According to interviewees, sustainability has become ever more important, and the perception that customer pressure has made most organisations more committed to sustainability is robust. In the interview, "i2" says, "In a few years, if a company is not sustainable, it will not be able to operate. The market is evolving in this way, and being sustainable will be a standard thing, with increasingly tighter rules towards sustainability". Similarly, "i4" says "Regarding strategy, we would have to choose that way |sustainability| because of our clients". A shift from indifference to active demand has occurred within the textile sector for the implementation and monitoring of the entire value chain. Textile pollution, which the interviewees have cited as one of the most polluting industries ("i1" points to "A sustainable strategy that ensures the transition from the manufacture of polluting products to recyclable, reusable and/or ecological products" and "i3" referring "we are betting on overcoming the problem of pollution, or at least minimising it"), is a fact that aligns with what has been stated by various governmental and non-governmental organisations (Hwang et al., 2016; Mulder, 2022).

12.4.2 Linen Market

About 80%–85% of the flax used for fibres is grown in Western Europe (France, Belgium, Netherlands), and France is the world leader in the production of flax fibre, while other important flax fibre-producing countries are Spain, Egypt, Russia, and China (Maximize Market Research, 2023). In 2020, global flax fibre production was around 1 million tonnes. Linen falls into the "other plants" category, accounting for only 6.5% of total global plant fibre production. Synthetic fibres, on the other hand, account for 68.2% of total global fibre production (Debnath, 2021).

Linen has three markets: Home Textiles, Apparel, and Industrial. This work focuses on linen in the apparel industry since, as a sustainable fibre of high quality, the demand for linen is expected to increase (Maximize Market Research, 2023). In addition, linen is durable, soft, breathable, and highly absorbing (Chanda et al., 2020). The only drawback is linen's low elasticity, which explains why it creases easily (Maximize Market Research, 2023). Also, flax fibres are costly to produce as they must be separated from the stalk, which explains the slow market growth and why it is considered a luxury market.

Furthermore, according to the European Confederation of Linen and Hemp, only 0.5% of European flax is certified organic (Textile Exchange, 2022). Two essential seals of approval in the flax industry are European Flax®, which certifies the European origin of premium quality fibres, and Masters of Linen®, which guarantees a seal of approval for 100% linen from Europe, from field to yarn, fabric, and sustainable production. According to the Allied Market Research Report (2021), Europe was the largest consumer of linen fibres, although China is expected to grow its linen market in the following years. This report also highlights the companies that have diversified their business fibres mainly by adding linen fibres: Banana Republic,

Peacock Alley, Ralph Lauren, Levi Strauss & Co, Michael Kors, American Eagle, Tommy Hilfiger, H&M, and China Linen Textile Industry Ltd.

The global market for quality linen fibres is expected to grow at a CAGR (Statista, 2024) of 5.1% during the forecast period 2020–2026 (Maximize Market Research, 2023). Given this overview, the lingerie market still appears to be in the "launch" life cycle phase as there is still high market growth potential, and demand is also expected to increase gradually over the years. To start a new business, one must look at the external and internal market trends that can influence growth and help managers make the right decisions.

When a company is transparent about its supply chain, it allows the company and its stakeholders to be accountable. However, 50% of the world's largest fashion brands continue to disclose little or no information about their supply chain, and only 12% of brands globally publish insights into their raw material suppliers (Fashion Revolution, 2022). The recent Fashion Transparency Index (2023) report found that the average transparency of the 250 brands surveyed was 23%, and only 7% could disclose where their raw materials come from. Without traceable data, it is difficult for most brands to know the origin of their fibres down to the farm level and say with 100% certainty which fibres are in their garment and the associated environmental impact. With transparent supply chains (Figure 12.4), traceability is feasible, and consumers can see the conditions under which products are made and make informed decisions about the products they buy (Alves et al., 2024).

From another perspective, supply chain transparency can also promote innovation and efficiency. Companies with a transparent supply chain can better identify problems and find solutions, such as reducing waste and increasing efficiency, resulting in cost savings.

A PESTEL analysis focuses on the European and Portuguese trends that influence the textile business. The business sustainability plan will also include an analysis based on Porter's Five Forces and a SWOT analysis to better understand and understand the new linen company's key strengths, weaknesses, opportunities, and threats in the national and international context.

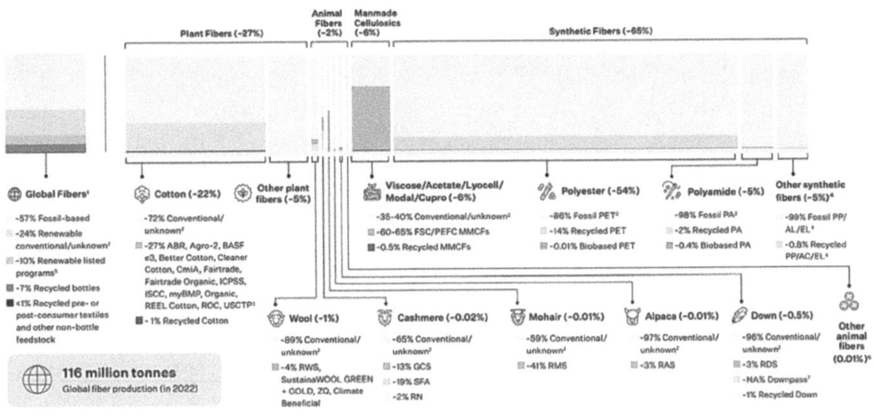

FIGURE 12.4 The European Flax™ Traceability Chain. (https://allianceflaxlinenhemp.eu/en/technical-documents-download.)

12.4.2.1 PESTEL Analysis

PESTEL analysis is a strategic management tool that uses a structured method to help organisations and new businesses understand and evaluate external macro-environmental factors affecting their business. The acronym PESTEL stands for Political, Economic, Social, Technological, Environmental, and Legal factors. Through this analysis, companies gain a holistic overview of their external environment. This will enable them to capitalise on opportunities and overcome challenges for sustainable success. It enables informed decision-making, minimises risks, and maximises opportunities, contributing to the overall success and sustainability of the company.

Political Factors: In 2019, the EU introduced the European Green Pact, which aims to create a carbon-neutral economy by 2050. Key goals include a 55% reduction in greenhouse gas emissions by 2030 and addressing pollution and biodiversity concerns. Legislative proposals span various sectors. The upcoming linen business plan aligns strategically with the Green Pact, highlighting linen as a sustainable fibre with flax production contributing to CO_2 absorption. Portugal's €16.6 billion Recovery and Resilience Plan (PRR) until 2026 prioritises Resilience, Climate Transition, and Digital Transition (FiGroup, 2022). With €145 million allocated to Climate Transition, a focus on the bioeconomy involves the textile and apparel industries. Emphasis is on integrating forest biomass (cellulose and lignin), investing in natural fabrics (pineapple and banana sleeves, hemp, and bamboo) and new models of business plans based on the circularity of raw materials (reuse, recycle). On top of this, the objective is to promote further conscious awareness about production and sustainable consumption within society.

The linen business strategically aligns with the Green Deal, emphasising linen as the world's most sustainable fibre. It should promote societal awareness of conscious production and sustainable consumption, which is crucial for economic resilience (Council of the EU, 2023).

Economic Factors: The European economy has grown 2.3% in 2023, and the variance is attributed to global supply bottlenecks and disruptions in global supply, particularly manufacturing, which have been intensified by a swift post-pandemic demand surge. The Ukraine–Russia war further complicates matters, elevating food and energy prices, causing regional instability, and triggering widespread social, economic, and humanitarian crises globally, and energy prices have steadily risen since November 2021 (Council of the EU, 2023).

Portugal had a nominal GDP of 6.3% in 2022 and 2.6% in 2023 (BPstat, 2023), anticipating 2% in 2024. Synthetic fabrics are also anticipated to grow (Bergkamp, 2024) due to a growing global population expected to reach 8.5 billion by 2030. The fashion market should increase from €1.5 billion in 2021 to approximately €2 billion by 2026 (Statista, 2024). Regarding Economic Business Trends, the global fibre market shows an asymmetry between synthetic and natural fibres.

Social Factors: Increased awareness of the effects of climate change among younger generations has fuelled environmental activism in recent years. Climate change pressures companies and governments as customers, investors, and the media demand sustainability. Indeed, 88% of Americans and Brits want environmentally friendly brands. In addition, the second-hand market will grow at a CAGR of 15%–20% over the next five years (BCG, 2020).

Technological Factors: Textiles are reshaped by advanced methods and technologies. The Inditex Group, for instance, invests in sustainable and eco-friendly programmes, including one that measures and reduces microfibre release during washing. Inditex also developed an innovation hub to develop technologies that can transform carbon emissions into a resource for producing a new fibre. This initiative allows the company to reduce its carbon footprint.

The List Programme is another pioneering programme that seeks to improve the quality of chemicals in manufacturing. As mentioned, the fashion industry is chemical-intensive, so this kind of programme can represent an excellent solution to highlight. Recycling textiles is not an easy task, as already described. Companies like Resortecs (2022) are investing in new technology to improve the recycling process. Designing clothes is also a significant trend in the technology sector, thus, the importance of finding different materials. Food waste, biodegradable threads made from algae, pineapple sleeves, and textile waste are just some of the innovative examples companies have started using in the TI by developing new technology methods.

Environmental Factors: 2019 Greenpact commits to a cleaner future, and governments are also urged to use cleaner energy, stop massive deforestation, and invest in soil restoration. By doing so, CO_2 will be absorbed by plants more efficiently. In the fashion industry, which accounts for 10% of global carbon emissions, flax is essential due to its high CO_2 absorption capacity (European Parliament, 2023). Over 200 million trees are felled annually to produce plant-based materials such as viscose, and the Amazon Forest is heavily deforested to make room for livestock farming. This contributes to fibre production for leather goods such as bags, shoes, and belts (Canopy Planet, 2021).

Legal Factors: Various laws, such as the European Air Quality Policy, regulate environmental impacts. In 2030, the aim is to reduce premature deaths from ground-level ozone by half. "European Green Deal" encourages global companies to reduce pollution, and CO_2 emissions are taxed per ton by countries. Apart from environmental laws, Portugal has one of Europe's highest corporate tax rates at 21%, which poses challenges for Portuguese textile companies. Thus, taxes are an additional challenge for Portuguese companies.

12.4.2.2 SWOT Analysis

A SWOT analysis is a valuable tool for assessing a company's resources and abilities before defining its strategy and attracting new clients. Figure 12.5 shows the strengths and weaknesses of a linen company *vs* opportunities in national/international markets. Additionally, it identifies possible threats to the company's growth.

Investing in a new linen business/factory involves building facilities, installing machinery, and hiring specialised staff. "Financial responsibility and market immaturity are weaknesses since linen processing is time-consuming, contributing to higher costs and limiting accessibility", as (i1) stressed. Centralising manufacturing processes in Portugal will also allow better control over production levels, possibly reducing costs. A company's strength lies in its expert staff, which ensures high-quality linen supplies in Portugal, as "University of Beira Interior students that could be involved in the production process, promoting employee recruitment" (i3).

Externally, the "company can benefit from the increasing demand for sustainable products and align itself with market trends" (as i1 reported). Although the share of sustainable products on the market is still relatively low, as organic cotton only accounts for 0.96% of the total cotton market (26.2%), consumer preferences are

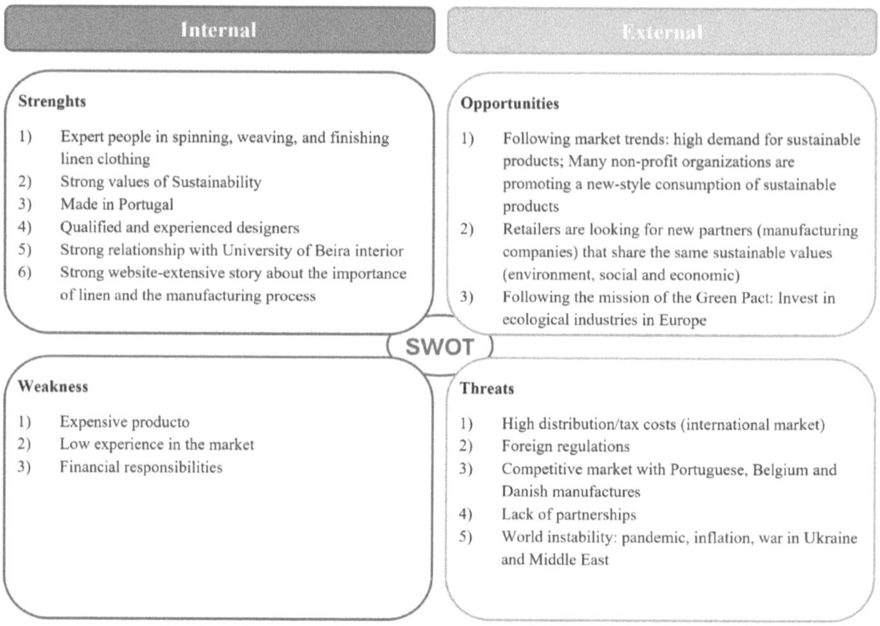

FIGURE 12.5 SWOT Analysis of a new linen company.

shifting towards eco-friendly brands. Offering products made from eco-friendly fibres meets this demand and supports the European Green Pact goals. In addition, the linen company has the potential to become a significant player in the European linen production sector, as only a few companies produce clothing made from 100% linen. However, external threats include high distribution costs when selling to other European countries. Linen companies' responses indicate that global logistics management is a significant challenge, including complicated transport logistics, tax implications, and compliance with foreign regulations. "Overcoming these challenges will be critical to the success of the linen company in the European market", according to (i1).

12.4.2.3 Five Forces of Porter Analysis

As the fashion industry has grown over the last two decades, consumers have more options and have lower switching costs. This dynamic negatively impacts linen Portuguese Textile due to the lower prices of fast fashion brands compared to sustainable linen products.

To understand the attractiveness of the linen company in Portugal, the powerful tool of Porter's 5 Forces (Porter, 1980) is used, which will examine the following topics: (i) the Consumer Bargaining Power, (ii) the Supplier Bargaining Power, (iii) the Threat of New Substitute Products, (iv) the Threat of New Entrants, and (v) the Rivalry Competition.

For each topic of analysis, a score of "Low" and "High" force level was attributed. The scores approach is seen in Table 12.3 combined with analyses of the scenarios where a new Portuguese linen company thrives.

TABLE 12.3
Porter's Five Forces Analysis for the Success of a New Portuguese Linen Company

Consumer Bargain Power	Supplier's Bargain Power	Threat of New Substitute Products	Threat of New Entrants	Rivalry Competition
The rapid growth of the fashion industry occurred in countries with high consumer wages, causing low switching costs for consumers and empowering them. This negatively impacts the linen Portuguese Textile company, given the lower prices of fast fashion compared to sustainable linen products.	Due to limited alternatives, major flax plant suppliers have excellent bargaining power vis-à-vis the Portuguese linen company (high power).	Sustainable substitute fibres were considered: hemp, organic cotton, and recycled polyester. While European companies are interested in sustainable options, the limited production of these fibres indicates a competitive market (high performance).	It is unlikely that many textile companies producing 100% pure linen will emerge in Portugal and Europe in the short term, as the production of linen is still costly.	The competitive analysis covers two segments: Portuguese linen companies and well-known foreign companies, including Dutch and Belgian Companies.
If linen options are limited, higher switching costs favour the linen company, giving it more power over consumers.	Due to their well-established markets (high power), suppliers depend less on linen companies. The closest alternative for linen is hemp, which is only produced in limited quantities and does not threaten suppliers (high power).	Despite its durability and strength, hemp is slow to catch on due to its cost and associations with drugs. Consumers prefer organic cotton for sustainability reasons, affecting linen textiles' competitiveness.	This target linen at a specific market	Polopiqué and Riopele rules in Portugal. Polopiqué specialises in organic cotton and linen spinning and will strategically install Europe's first 100% linen spinning machine.

(Continued)

TABLE 12.3 (*Continued*)
Porter's Five Forces Analysis for the Success of a New Portuguese Linen Company

Consumer Bargain Power	Supplier's Bargain Power	Threat of New Substitute Products	Threat of New Entrants	Rivalry Competition
The high differentiation between linen and synthetic fibres favours consumers, as synthetic fibres are less durable but more affordable. Natural fibres offer quality/comfort, which reduces consumer power due to the high differentiation.	Flax/linen production requires capital and technical knowledge, which is unlikely to be a flax supplier's domain in the TI (low power) and benefit the line company.	As recycled polyester gains market share, international brands adopt it for sustainability, further influencing the linen company's power.	Building a textile company requires extensive capital (e.g., infrastructure, human labour, machinery) and a high level of manufacturing knowledge.	Riopele has invested in modernisation, making it one of Europe's most modern textile companies. It has a robust digital platform and produces sustainable products, including recycled fibres.
If retail can offer the linen company's products, customers/retail chains have high bargaining power. This is not expected for Portuguese linen companies, which suggests lower customer bargaining power.		Substitute products have high bargaining power, weakening the linen company. In addition, organic cotton's low conversion costs and recycled polyester's lower price adversely affect the linen company.	The Portuguese linen textile company is positively positioned on the market according to "new entrants" (low power).	Two indirect competitors, Somelos and Têxtil Manuel Gonçalves (TMG) Fabrics, lack vertical integration. Somelos weaves cotton fabric, while TMG manufactures garments.
				A Dutch trading company, Northern Linen Bureau Veritas, operates worldwide and has Asian production facilities. It exports worldwide and produces flax fibre, spinning and weaving.

TABLE 12.4
Environmental Indicators for a New Linen Manufacturer (A) and Suppliers (B)

	Environmental Indicator	Indicator Assessment
A	Recovered Water	Rainwater used (%)
	Waste Reduction	Waste produced and reduced in the factory (%)
	Fibre Reutilisation	Fibres reused in the manufacturing process (%)
	Wasted Water	Waste water in the manufacturing process (%)
	CO_2 Emission	Amount of CO_2 emitted by a kg of fabric
	Energy Consumption	Amount of Energy consumption by a kg of fabric
	Certified Materials	Certified materials GOTS, ISO9001, ISO14001, Master Linen/ fabric kg
	Dues Type	Compliance with chemicals (ECHA European Chemical Agency)
B	CO_2 in the flax plantation	CO_2 (tonnes) the flax plant(ation) will absorb during growth
	Water in the flax plantation	Water used (%) to produce a bunch of flax
	Fertilisers	Fertilisers used (%) to produce a bunch of flax

12.4.3 New Linen Factory Sustainability Plan

Sustainability is increasingly recognised in the fashion and TI to counteract the current negative environmental impacts. Thus, many companies incorporate sustainable practices and environmental impacts into their annual reports. The Higg Index measures the TI's environmental sustainability. Material Sustainable Index, part of the Higg Index, measures five environmental impacts: global warming, eutrophication, water scarcity, chemicals, and fossil fuels. These indicators for a new linen company are shown in Table 12.4.

Applying these environmental indicators to a Linen Factory leads to the following results

1. Install a rainwater collection system to reduce water wastage for spinning processes. Assess and minimise water wastage in subsequent production stages.
2. Establish eco-locations throughout the factory for proper waste collection and recycling. In waste management, follow the three Rs (reduce, reuse, recycle) principle, emphasising quality and durable fibres.
3. To boost durability and conserve resources and energy, the Quality department mixes reused fibres with new ones for innovative reuse in spinning, weaving, and garment design.
4. Achieve CO_2 offsetting by producing flax linen (it absorbs large amounts of CO_2).
5. Monitor and reduce energy consumption, utilising solar cells to generate clean energy.
6. Prioritise non-toxic chemicals and strictly use substances approved by the European Chemicals Agency (ECHA) to reach the certified materials goal.

It is essential to assess the supplier's production type (B in Table 12.4) for a sustainable value chain, especially given the company's exclusive use of organic linen to reduce fertiliser use. Design plays a key role in sustainability by choosing quality materials and following principles like simple garment design, shape planning to minimise fibre waste, and efficient fabric cutting. Garments should include material information, replace traditional labels, and explore natural elements like wood buttons and natural dyes. Commercially, the manager and team must forecast activity, ensure proper production levels, and optimise resources like energy and fibre waste. Social responsibilities include providing good working conditions, promoting equality, maintaining a code of conduct, offering health insurance, and supporting professional development.

12.4.4 New Linen Factory Business Model Canvas (BMC)

The BMC is a strategic management tool that visually represents a company's key components and how they work together to create, deliver, and capture value. Businesses and entrepreneurs often use it to describe, design, challenge, and pivot their business models. The BMC typically consists of several building blocks or elements, each representing a fundamental aspect of a business. These elements include (i) Key partners, (ii) Key activities, (iii) Key resources, (iv) Value propositions, (v) Customer relationships, (vi) Channels, (vii) Customer segments, (viii) Cost structure, and (ix) Revenue streams. BMC is usually presented as a visual chart with each building block represented as a separate section, allowing for a comprehensive overview of the business model on a single page. This tool is top-rated in the startup and entrepreneurial communities for its simplicity and effectiveness in articulating and refining business models.

Figure 12.6 presents a detailed breakdown that offers a broad view of each field within the BMC for a Linen Factory, highlighting critical aspects, strategies, and contributions in each category.

12.5 CONCLUSION

The higher cost of natural fibres makes switching to them challenging for companies. Considering this, consumer education is key to making more sustainable choices; thus, manufacturers and fashion companies should promote natural fibre benefits. Fast fashion contributes significantly to environmental degradation and emphasises the need for consumers to choose sustainable clothing. Consumers can reduce the environmental impact of textile production by considering more sustainable fibre types and production locations. The Linen Factory's commitment to sustainable practices guided by CE principles minimises waste and encourages recycling. Strategic partnerships and a focus on resource efficiency emphasise the company's environmental commitment. Moreover, offering customisation options for textile manufacturers and fashion designers aligns with CE principles and promotes durable and sustainable fashion. Also, direct selling minimises the number of intermediaries and is an example of a lean and environmentally friendly business model. The emphasis on sustainability in research and development demonstrates a commitment to continuous innovation.

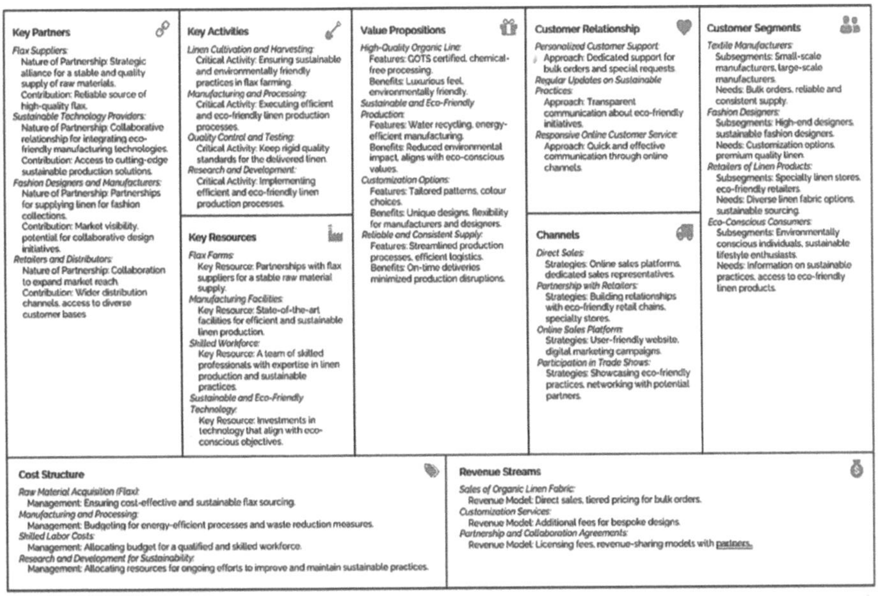

FIGURE 12.6 Canvas Model of a new linen company.

Diversified revenue streams, including organic linen fabrics and customised services, support circular consumption and a sustainable cost structure. Extending product life with high-quality, durable linen aligns with CE principles and fosters industry partnerships. However, challenges include high initial investment, funding barriers, supply chain uncertainties, regulatory compliance, and sustainable technology adoption. Maintaining competitiveness against unsustainable alternatives and navigating unpredictable markets require strategic planning. In regions lacking waste disposal infrastructures, compliance with circular principles can be difficult. Economic downturns and currency fluctuations also demand careful financial management. Despite these hurdles, Linen Factory's commitment to sustainability is key to its long-term success in a CE.

However, even if a linen company does not thrive in fabrics for fashion production in short run, after applying the Canvas model, it should consider that linen plays a significant role across various design sectors. Approximately 60% of its use is allocated to the fashion industry, 30% to lifestyle and interior design, and 10% to technical sectors. Beyond its traditional use in apparel, other industries have started incorporating flax into advanced applications like biocomposites (Váňa et al., 2024). Flax fibre is now being integrated into aerospace, automotive, and marine industries, and its potential is vast, especially in sectors where demand is growing and more industries are turning to flax as an alternative to materials like carbon fibreglass due to its exceptional qualities (Baley et al., 2021). Even in the fishing sector, there is a growing interest in flax and hemp fibres, which are considered sustainable alternatives to plastic-based fishing gear, such as nets and ropes (Deroiné et al., 2019).

According to the Alliance for European FlaxLinen and Hemp (2023), recent innovations have strengthened the flax and linen value chain, particularly with new production techniques that enable knitting with flax and expand its technical applications. These innovations include developments in linen knits, branded fibres, and innovative processes like creating washed or water-repellent linen. European companies, like Safilin (www.safilin.fr), are pioneering knit textiles made from linen. Traditionally, knit fabrics have been associated with cotton or synthetic materials, but linen now appears as a viable option.

The future of flax and linen is ready for further exploration and new product development. Numerous possibilities are just starting to emerge, combining environmental benefits and economic opportunities.

ACKNOWLEDGEMENTS

We would like to acknowledge the support of Center for Research in Architecture Urbanism and Design (CIAUD) and ISCTE-IUL and the contribution of the research carried out by Margarida Figueiredo, a master's student at ISCTE-IUL.

This work has employed ChatGPT, an AI language model, exclusively for language correction purposes. The model was used to improve grammar, spelling, and syntax in the manuscript. All content, ideas, analysis, and conclusions presented in this thesis are the sole work of the authors, and the use of the AI tool did not influence the development of the research, the interpretation of data, or any intellectual decisions. The authors assume full responsibility for the accuracy and integrity of the academic content

REFERENCES

Abdelmeguid, A., Afy-Shararah, M., & Salonitis, K. (2024). Towards circular fashion: Management strategies promoting circular behaviour along the value chain. *Sustainable Production and Consumption*, *48*, 143–156. https://doi.org/10.1016/j.spc.2024.05.010

Agyemang, M., Kusi-Sarpong, S., & Mani, V. (2019). Drivers and barriers to circular economy implementation: An explorative study in Pakistan's automobile industry. *ResearchGate*, *57*, 971–994. https://doi.org/10.1108/MD-11-2018-1178

Al-Issa, N., Khaki, A. R., Jreisat, A., Al-Mohamad, S., Fahl, D., & Limani, E. (2022). Impact of environmental, social, governance, and corporate social responsibility factors on firm's marketing expenses and firm value: A panel study of US companies. *Cogent Business & Management*, *9*(1). https://doi.org/10.1080/23311975.2022.2135214

Alliance for European FlaxLinen and Hemp. (2023). *Alliance FlaxLinen Hemp*. https://allianceflaxlinenhemp.eu/en

Allied Market Research. (2021). *Linen Fabric Market*. Allied Market Research. https://www.alliedmarketresearch.com/linen-fabric-market

Alves, L., Sá, M., Cruz, E. F., Alves, T., Alves, M., Oliveira, J., Santos, M., & Miguel, A. (2024). A traceability platform for monitoring environmental and social sustainability in the textile and clothing value chain: Towards a digital passport for textiles and clothing. *Sustainability*, *16*(1), 82. https://doi.org/10.3390/su16010082

Andersen, K. R. (2017). *Stabilizing Sustainability: In the Textile and Fashion Industry* (PhD thesis). Copenhagen Business School.

ATP- Associação Têxtil e Vestuário de Portugal. (2018). *Fashion from Portugal 4.0*. Diretório ATP. https://atp.pt/pt-pt/projetos/fashion-from-portugal-4-0/

ATP. (2019). Fashion from Portugal 4.0. Diret.rio 2019. ATP – Associa..o T.xtil e Vestu.rio de Portugal.

Baley, C., Bourmaud, A., & Davies, P. (2021). Eighty years of composites reinforced by flax fibres: A historical review. *Composites Part A: Applied Science and Manufacturing*, *144*, 106333. https://doi.org/10.1016/j.compositesa.2021.106333

Banco de Portugal. (2021). *Análise setorial da indústria dos têxteis e vestuário*. https://bpstat.bportugal.pt/conteudos/publicacoes/1292

BCG. (2020). *The Consumers behind Fashion's Growing Secondhand Market*. Boston Consulting Group. https://www.bcg.com/ja-jp/press/2november2020-the-consumers-behind-fashions-growing-secondhand-market

Berggren, N., & Bjørnskov, C. (2023). Does globalization suppress social trust? *Journal of Economic Behavior & Organization*, *214*, 443–458. https://doi.org/10.1016/j.jebo.2023.08.018

Bergkamp, C. (2024). *Textile Exchange Annual Report 2023*. Textile Exchange. https://textileexchange.org/app/uploads/2024/03/Textile-Exchange-Annual-Report-2023.pdf

BOF-Business of Fashion. (2021). *The Sustainability Gap – How Fashion Measures Up*. China Water Risk. https://chinawaterrisk.org/research-reports/the-sustainability-gap-how-fashion-measures-up/

Boston Consulting Group (BCG). (2020). *The CEO Agenda 2020: COVID-19: Accelerating a circular economy for fashion*. Global Fashion Agenda and Boston Consulting Group. https://globalfashionagenda.org/resource/fashion-ceo-agenda-2020-covid-19-edition/

BPstat. (2023). *Banco de Portugal, Eurosystem*. https://bpstat.bportugal.pt/

Canopy Planet. (2021). Annual report 2020–2021. https://canopyplanet.org/wp-content/uploads/2021/08/2020-2021-Canopy-Annual-Report.pdf

Chanda, T., Ahirwar, M., & Behera, B. K. (2020). Appraisal of bed linen performance with respect to sleep quality. *The Sleep Journal*, *3*(1), 19–29. https://doi.org/10.31881/TLR.2020.01

Chen, X., Memon, H. A., Wang, Y., Marriam, I., & Tebyetekerwa, M. (2021). Circular economy and sustainability of the clothing and textile industry. *Materials Circular Economy*, *3*(1), 1–12. https://doi.org/10.1007/s42824-021-00026-2

Costa, C., Viana, A., Silva, C., Marques, E. F., & Azoia, N. G. (2022). Recycling of textile wastes, by acid hydrolysis, into new cellulosic raw materials. *Waste Management*, *153*, 99–109. https://doi.org/10.1016/j.wasman.2022.08.019

Council of the EU. (2023). *European Green Deal*. Council of the European Union. https://www.consilium.europa.eu/en/policies/green-deal/

Dan, M. C., & Østergaard, T. (2021). Circular fashion: The new roles of designers in organisations transitioning to a circular economy. *The Design Journal*, *24*(6), 1001–1021. https://doi.org/10.1080/14606925.2021.1936748

Debnath, S. (2021). Flax Fibre Extraction to Textiles and Sustainability: A Holistic Approach. In: Gardetti, M.Á., Larios-Francia, R.P. (eds) Sustainable Fashion and Textiles in Latin America. Textile Science and Clothing Technology. Springer, Singapore. https://doi.org/10.1007/978-981-16-1850-5_4.

Deroiné, M., Pillin, I., Le Maguer, G., Chauvel, M., & Grohens, Y. (2019). Development of new generation fishing gear: A resistant and biodegradable monofilament. *Polymer Testing*, *74*, 163–169. https://doi.org/10.1016/j.polymertesting.2018.11.03

Dissanayake, G., & Perera, S. (2016). New approaches to sustainable fibres. In Muthu, S., & Gardetti, M. (Eds.), *Sustainable Fibres for Fashion Industry*. Springer. https://doi.org/10.1007/978-981-10-0566-4_1

European Environment Agency. (2019). *Textiles in Europe's Circular Economy*. https://www.eea.europa.eu/downloads/e27564dbf0f7462ea5c52e4c9aaf2775/1676462612/textiles-in-europe-s-circular-economy.pdf

Elkington, J. (2017). *Partnerships from Cannibals with Forks: The Triple Bottom Line of 21st-Century Business*. Capstone Publishers. https://doi.org/10.1002/tqem.3310080106

Ellen Macarthur Foundation. (2021). *Circular Business Models: Redefining Growth for a Thriving Fashion Industry*. https://emf.thirdlight.com/file/24/Om5sTEKOmm-fEeVOm7xNOmq6S2k/Circularbusinessmodels.pdf

EU. (2023). *Waste Framework Directive: A More Sustainable Use of Natural Resources*. European Parliament. https://www.europarl.europa.eu/RegData/etudes/BRIE/2023/757572/EPRS_BRI (2023)757572_EN.pdf

Euratex. (2020). *Facts & Key Figures of the European Textile and Clothing Industry*. Brussels: Euratex. https://euratex.eu/wp-content/uploads/EURATEX-Facts-Key-Figures-2020-LQ.pdf

European Parliament. (2023). *The Impact of Textile Production and Waste on the Environment (Infographics)*. https://www.europarl.europa.eu/news/en/headlines/society/20201208STO93327/the-impact-of-textile-production-and-waste-on-the-environment-infographics

Fashion Revolution. (2022). *Transparency Report 2022*. https://www.fashionrevolution.org/category/transparency/

Fashion Transparency Index. (2023). https://www.fashionrevolution.org/fashion-transparency-index/

FiGroup. (2022). *Recover and Resilience Plans (RRP)*. https://pt.fi-group.com/en/service/rrp-recovery-and-resilience-plan/

Fiore, F. (2005). *Write a Business Plan in No Time* (1st ed.). Indianapolis, IN: Que Publishing.

Gonzalez, V., Lou, X., & Chi, T. (2023). Evaluating environmental impact of natural and synthetic fibers: A life cycle assessment approach. *Sustainability*, *15*(9), 7670. https://doi.org/10.3390/su15097670

Hwang, C., Lee, Y. H., Diddi, S., & Karpova, E. E. (2016). Don't buy this jacket. *Journal of Fashion Marketing and Management*, *20*(4), 435–452. https://doi.org/10.1108/JFMM-12-2014-0087

International Labor Organization. (2019). *The Future of Work in Textiles, Clothing, Leather, and Footwear*. International Labour Organization. https://www.ilo.org/wcmsp5/groups/public/---ed_dialogue/---sector/documents/publication/wcms_669355.pdf

Lin, J., Li, J., Xu, Y., Xie, M., Zhao, S., & Ye, W. (2023). Editorial: Wastewater treatment & resource recovery technologies. *Frontiers in Bioengineering and Biotechnology*, *11*. https://doi.org/10.3389/fbioe.2023.1150044

Matušovičová, M. (2021). Sustainable fashion as a part of the circular economy concept. *Studia Commercialia Bratislavensia*, *13*(3), 29–37. https://doi.org/10.2478/stcb-2020-0009

Maximize Market Research. (2023). *Textile Industry Market Size and Share Analysis, 2029*. MMR Research. https://www.maximizemarketresearch.com/market-report/global-textile-market/29660/

Murray, A., & Skene, K. (2017). *The Circular Economy: An Interdisciplinary Exploration of the Concept and Application in a Global Context*. Journal of Business Ethics, 140(3), 369–380. https://doi.org/10.1007/s10551-015-2693-2

Niinimäki, K., Peters, G., Dahlbo, H., Perry, P., Rissanen, T., & Gwilt, A. (2020). The environmental price of fast fashion. *Nature Reviews Earth & Environment*, *1*(4), 189–200. https://doi.org/10.1038/s43017-020-0039-9

Peters, G., Li, M., & Lenzen, M. (2021). The need to decelerate fast fashion in a hot climate: A global sustainability perspective on the garment industry. *Journal of Cleaner Production, 295*, 126390. https://doi.org/10.1016/j.jclepro.2021.126390

Piippo, R., Niinimäki, K., & Aakko, M. (2022). Fit for the future: Garment quality and product lifetimes in a CE context. *Sustainability, 14*(2), 726. https://doi.org/10.3390/su14020726

PORDATA. (2023). *Estatísticas sobre Portugal e a Europa*. https://www.pordata.pt/pesquisa/textil

Porter, M. E. (1980). *Competitive Strategy: Techniques for Analyzing Industries and Competitors*. New York, NY: Free Press.

Provin, A. P., Dutra, A. R. D. A., De Sousa e Silva Gouveia, I. C. A., & Cubas, E. A. L. V. (2021). Circular economy for fashion industry: Use of waste from the food industry for the production of biotextiles. *Technological Forecasting and Social Change, 169*, 120858. https://doi.org/10.1016/j.techfore.2021.120858

Puig, F., Marques, H., & Pervez, N. G. (2008). Globalization and its impact on operational decisions: The role of industrial districts in the textile industry. *International Journal of Operations and Production Management, 29*(7), 693–697. https://doi.org/10.1108/01443570910971388

Rahla, K. M., Mateus, R., & Bragança, L. (2021). Implementing circular economy strategies in buildings: From theory to practice. *Applied System Innovation, 4*(2), 26. https://doi.org/10.3390/asi4020026

Resortecs. (2022). *Makes Clothes Recycling Easier with Dissolvable Thread*. https://resortecs.com/

Roos, S., Jönsson, C., Posner, S., Arvidsson, R., & Svanström, M. (2019). An inventory framework for inclusion of textile chemicals in life cycle assessment. *The International Journal of Life Cycle Assessment, 24*, 838–847. https://doi.org/10.1007/s11367-018-1541-y

Sandin, G. (2019). *Environmental Swedish Fashion Assessment*. ResearchGate (pp. 1–3). https://doi.org/10.13140/RG.2.2.30502.27205

Staber, U. (1997). Specialization in a declining industrial district. *Growth and Change, 28*, 475–495. https://doi.org/10.1111/1468-2257.00068

Stahel, R. W. (2019). *The Circular Economy*. Routledge.

Statista. (2024). *Retail E-Commerce Sales Compound Annual Growth Rate (CAGR) from 2024 to 2029, by Country [Graph]*. Statista. Retrieved October 21, 2024. https://www.statista.com/forecasts/220177/b2c-e-commerce-sales-cagr-forecast-for-selected-countries

Textile Exchange. (2022). *Preferred Fiber & Materials Market Report*. https://textileexchange.org/app/uploads/2022/10/Textile-Exchange_PFMR_2022.pdf

Tiwari, T., Kaur, G. A., Singh, P. K., Balayan, S., Mishra, A., & Tiwari, A. (2024). Emerging bio-capture strategies for greenhouse gas reduction: Navigating challenges towards carbon neutrality. *Science of the Total Environment, 929*, 172433. https://doi.org/10.1016/j.scitotenv.2024.172433

Váňa, M., Dvořáčková, Š., Knápek, T., & Kroisová, D. (2024). Elimination of delamination during the drilling of biocomposite materials with flax fibers. *Polymers, 16*(18), 2620. https://doi.org/10.3390/polym16182620

Werther, W. B., & Chandler, D. (2009). *Strategic Corporate Social Responsibility: Stakeholders in a Global Environment* (2nd ed.). SAGE Publications. https://books.google.pt/books?id=Ft_ndtEyJPkC&pg=PA5

World Bank Group. (2018). *What a Waste 2.0*. Retrieved September 15, 2021. https://datatopics.worldbank.org/what-a-waste

13 Sustainable and Innovative Mechatronic Prototype for Yarn Analysis in the Textile Industry

Filipe Pereira, José Machado, Filomena Soares, Rosa Vasconcelos, and Vítor Carvalho

This chapter is an extended version of the article presented in the Design Commit '24 conference proceedings.

13.1 INTRODUCTION

The textile industry is currently experiencing constant growth, with an emphasis on product quality as a market trend in its manufacturing (Pereira et al., 2021). The quality of textile products is dependent on the quality of the yarn, and it is essential to accurately determine the characteristics of the yarn (Pereira et al., 2017). Several systems are available on the market, such as the most recent USTER Tester 5 and 6 models (Uster Technologies, n.d.), which offer technologies to evaluate yarn quality; however, they are known for their high costs, raw material destructibility, and a limited ability to analyze only a few selected yarn quality parameters.

In recent years, there has been some ongoing research in an attempt to create an alternative method for collecting various yarn quality parameters using image processing techniques (Pereira et al., 2017; Sengupta et al., 2014). A correct and accurate analysis of yarns is essential for the textile industry, since the final quality of the fabric produced, which reaches the final consumer, is directly related to the quality of the yarns used (Pereira et al., 2023). Consequently, this research allowed the development of a physical prototype. This prototype aims to create a simple, economical, non-destructive, and portable system capable of automatically unwinding and winding a bobbin, while guiding the thread to pass in front of a high-resolution camera lens at a regulated speed (Pereira et al., 2021; Sengupta et al., 2014).

The main focus of this work is the design of a mechatronic system specifically adapted for the analysis of textile yarns. The objective is to provide an innovative

 DOI: 10.1201/9781032719290-16

solution to evaluate a wide range of yarn quality parameters, being a complement to the existing constraints of commercially available systems. This article focuses on the mechanical design of a yarn coil winding and unwinding system for application in a prototype for yarn quality analysis through computer vision and image processing. The primary objectives of this system include (Pereira et al., 2024):

- Guiding the yarn at a constant and precise speed in front of the lens;
- Minimizing yarn oscillations in front of the lens;
- Versatility to accommodate various bobbins and types of yarn;
- Designing a portable system with reduced mass;
- Creating a simple and cost-effective system;
- Develop a mechanical system for unwinding and winding yarn that emphasizes versatility, portability, simplicity, and non-destructive operation.

The structure of the paper is as follows: Chapter 2 presents the state of the art with the most significant works in context of design of mechatronics prototypes systems; Chapter 3 presents the methodology used in the project conceptual design; Chapter 4 details the design of the system plant, focusing on its subsystems, as well as the controller design, taking into account system's operating modes and the Supervisory Control and Data Acquisition (SCADA) system; Chapter 5 presents the integration of AI in the mechatronic prototype; Chapter 6 presents tests and results with the mechatronic prototype; and Chapter 7 enunciates the final remarks and future developments proposals.

13.2 STATE OF THE ART OF YARN ANALYSIS MECHATRONIC PROTOTYPES

The work developed by Carvalho et al. (2009) introduces a novel, cost-effective, and portable yarn evaluation system, the YSQ (Yarn System Quality), developed for thorough quality control of yarn properties in laboratory environments. Overall, the design of the YSQ system emphasizes modularity, precision, and signal processing techniques, aiming to revolutionize yarn evaluation by offering a cost-effective, portable, and high-precision solution for yarn quality control. Future plans involve enhancing the YSQ system into a more compact, user-friendly commercial prototype with additional features like a user interface and improved measurement efficiency (Carvalho et al., 2009; Pereira et al., 2024).

The work developed by Shi et al. (2022) presents the development of a small prototype for covered yarn, aiming to address the limitations of existing large-scale covered yarn machines, which are not suitable for small-scale laboratory research and production. The study analyses the principles underlying covered yarn formation, focusing on three systems: unwinding, covering, and winding. The study suggests several potential avenues for future work or research based on the findings and limitations highlighted.

However, this work has as its weaknesses the lack of innovation and sustainability in the following areas (Pereira et al., 2024):

- Innovation in the use of AI;
- Development of exclusive datasets;
- Focus on sustainability and efficiency;
- Integration of mechatronic prototyping.

The application of AI in mechatronic yarn analysis systems is important for improving the quality of the final product and the energy efficiency of the production process within the textile industry. More recent studies, such as that by Chauhan (2024), demonstrate that AI, integrated with image processing techniques and neural networks, allows an accurate analysis of yarn quality and at the same time reduces the consumption of energy. This allows a significant improvement in quality control, as also indicated by the work carried out by Semnani et al. (2006), El-Geiheini et al. (2020) and Chauhan (2024).

These AI systems not only allow the correct identification of defects in the yarn, but also an optimization of the energy resources used during textile production. The integration of AI into mechatronic systems thus represents a very significant advance in the evolution of the textile industry, making it more competitive and environmentally responsible according to the research work carried out by El-Geiheini et al. (2020).

According to the research work developed by Reddy (2022), AI allows the development of new innovative practices, such as "everyday artificial creativity", which allow us to transform the way we interact with computer systems. The application of these technologies in an industrial environment not only improves the accuracy of defect detection, but also reduces energy consumption, which will result in more sustainable and efficient manufacturing processes (Semnani et al., 2006).

These research studies allowed us to obtain a more comprehensive view of recent advances in mechatronic prototypes for wire analysis, with a focus on innovations in modular solutions, the application of signal/image processing techniques, and the use of machine learning systems. The results obtained through these studies allow for a new way of developing more accurate, efficient, and versatile systems for evaluating yarns. Future research work in this area indicates that it is necessary to further explore the integration of AI for the advanced identification and classification of yarn characteristics, as well as the development of practical and accessible solutions for a more widespread implementation within the textile industry (Pereira et al., 2024).

13.3 METHODOLOGY

13.3.1 Defined Requirements and Specifications for the Development of the Mechatronic Device

Adhering to the goal tree method, six primary objectives were identified for the project:

- Safety; Simplicity; Reliability; Portability; Versatility; and Cost Efficiency.

To achieve these objectives, specific parameters needed to be quantified and defined, aligning with performance specifications. Table 13.1 illustrates the outlined performance specifications for the project (Pereira et al., 2024).

The development of this project followed a structured approach, covering both the plant and controller aspects, and used the methodologies described in the VDI 2206 and VDI 2221 standards. Concept development began with an in-depth analysis of comparable yarn testing systems. Following the VDI 2206 V-model, the initial step

TABLE 13.1
Specifications for Project Performance (Pereira et al., 2024)

Objective	Specifications	Rationale
Analyze various types of yarn materials	Compatible with cotton and polyester; breaking force ≥ 1 N; compatible with variations in breaking tension ≤ |10|%.	Adaptability: The examination of different yarn materials, diameters, and bobbin dimensions enhances the adaptability of the prototype, accommodating a variety of commonly utilized materials and sizes within the textile sector. This adaptability allows for greater practical application and also compatibility with a wider range of yarn types, which meets production requirements.
Analyze various types of yarn diameters	Compatible with diameters ≤ 2 mm; compatible with variations in diameter ≤ |10|%.	
Analyze various bobbin dimensions	Compatible with bobbins with an outer maximum base diameter of ≤ 160 mm and height of ≤ 200 mm.	Accuracy and precision: the use of speed and position tolerances in the prototype allows for good accuracy and precision of the machine. These characteristics are important for accurate yarn analysis and consistency in textile production.
Adherence to speed tolerances	Speed ≤ 4.2 m/s with a maximum variation of 0.5%.	Decreasing position deviation improves safety and accuracy, reducing the risk of usage incidents.
Adherence to position tolerance	Deviation in position ≤ 1 mm.	Minimizing position deviation enhances operational safety and precision, reducing the risk of operational incidents.
Machine and user safety	Emergency machine stop within ≤ 1 second.	Safety: The fact that the prototype prioritizes the immediate stopping of the machine in the event of an emergency makes it safer and also reduces the risk of accidents during operation. Furthermore, the configuration and programming created is easy to use, as we have to apply these types of procedures.
Simple, intuitive, and quick setup	Setup time ≤ 2 minutes.	Durability and maintenance: When there is reduced wear on moving components, this allows for greater machine durability. At the same time, it minimizes the need for maintenance and ensures more sustainable performance over time.

(Continued)

TABLE 13.1 (*Continued*)
Specifications for Project Performance (Pereira et al., 2024)

Objective	Specifications	Rationale
Reduced wear of moving components	Optimal operating conditions ≥ 10 km.	Quality control: When yarn breakage is prevented, it allows the integrity of the raw material to be maintained, ensuring continuous analysis and good quality in textile production.
Avoid yarn breakage	Avoid ≤ 5 yarn breaks per 10 km.	Quality control: Avoiding yarn breaks sustains production integrity, enabling precise analysis without compromising quality.
Reduced mass	Mass ≤ 30 kg.	Portability and practicality: a low mass and smaller volume allow for high portability and efficiency in the space they occupy. In the case of textile industries with limited space, these characteristics allow easier transportation and installation in different types of industrial environments in the textile area.
Reduced volume	Volume ≤ 0.275 m^3.	Space efficiency: When there is a reduction in machine volume, it allows you to obtain more free space, which is beneficial in industrial environments where space is limited or where mobility is very important.
Lower cost	€1.300,00 (only development hardware and software).	Cost efficiency: The lower initial investment in this prototype, when compared to existing commercial alternatives, makes the prototype financially more viable for testing or development. These characteristics make it a much more economical solution compared to commercial alternatives that have higher prices, such as USTER.

of the project was to define objectives and requirements based on the information gathered, categorized into plant and controller domains (Buur, 1995; Elnashar, 2022; Zhu & Wu, 2014; Zhang et al., 2009; Pereira et al., 2024).

13.3.2 Conceptual Design

A Morphological Chart has been developed for the final prototype, as shown in Figure 13.1, in order to help decide the best configuration for the system to be developed (Pereira et al., 2024).

The solution chosen through the conceptual design is presented in Figure 13.2. The machine in the project, can be divided into three parts: the unwinding system (the entire part outside the box on the left side), the yarn guiding system (the entire part inside the box), and the winding system (the entire part outside the box on the right side). The unwinding system is responsible for holding the bobbin, releasing the yarn, and guiding it into the box (Pereira et al., 2024).

Functions	Possible solutions						
Yarn unwinding	Over-end withdrawal	Side withdrawal (vertical bobbin)	Side withdrawal (horizontal bobbin)	Bobbin placed in an eccentric plate	Frictional unwinding using a motor cylinder		
Yarn position control	Yarn guide roller (V profile)	Yarn positioning based on rollers	Additive tensioner and guiding system	Transparent tube	Ceramic yarn guide	Yarn positioning with an air blast	
Velocity control	Control of both bobbins	Single bobbin control with a custom intermediate PYF	One intermediate custom PYF	One commercially available PYF	Single bobbin control with two intermediates custom PYF	Single bobbin control without intermediate elements	Single bobbin control with two intermediates commercially available PYF
Yarn tension control	Angular position relationship of both bobbins	Position relationship of two PYF	Multiplicative tensioner	Additive tensioner	Active additive tensioner	Self-adjusting lever tensioner	Self-adjusting ball tensioner
Support of unwinding bobbin	Adjustable support with conical ends	Adjustable support with flat ends	Simple vertical support	Simple support with flexible fitting			
Support of winding bobbin	Adjustable support with conical ends	Adjustable support with flat ends	Simple vertical support	Simple support with flexible fitting			
Yarn winding	Guidance by servomotor and lever	Guidance by grooved drum	Grooved drum and yarn guide with transmission ratio (random winding)	Grooved drum and yarn guide with transmission ratio (precision winding)	Guidance by system of linear movement driven by servomotor	Guidance by counter rotating blades	

FIGURE 13.1 Morphological Chart for the final prototype Solutions selected for the final prototype are highlighted in green cells [20].

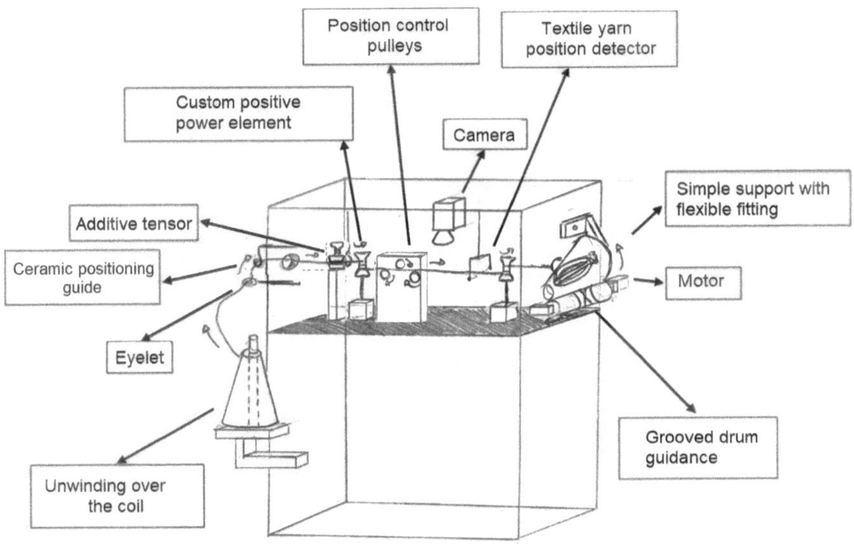

FIGURE 13.2 Conceptual design solution [20].

13.4 SYSTEM PLANT DESIGN

13.4.1 Mechanical Configuration

The final mechanical solution was carefully developed to match the project's requirements and goals. The prototype's final design, shown in Figure 13.3, is divided into three main components, along with the camera, lens, support, and other accessories. On the left side of the box is the unwinding subsystem, which includes parts numbered 2, 4, 6, and 49. Inside the box is the yarn guidance subsystem, made up of parts numbered 9, 11, 12, 13, 14, 20, 27, 28, 31, and 43. Conversely, on the right-hand side of the box, the winding system is situated, featuring parts labeled 33, 34, 38, and 40 (please note that some parts within all three subsystems are not visible or numbered in Figure 13.3) (Caldas et al., 2022; Pereira et al., 2024).

13.4.2 Unwinding Subsystem

This subsystem holds the bobbin, unwinds the yarn, and directs it into the box using the over-end withdrawal method. Initially, the bobbin is secured by part number 2, accommodating various shapes and sizes, up to 160 mm in diameter, allowing for multiple tests with the same yarn. The yarn is then unwound by the guide (part 49), supported by part 4. After passing through the yarn guide, it is directed through an opening on the left side of the box, leading to the yarn guidance subsystem (Caldas et al., 2022; Pereira et al., 2024).

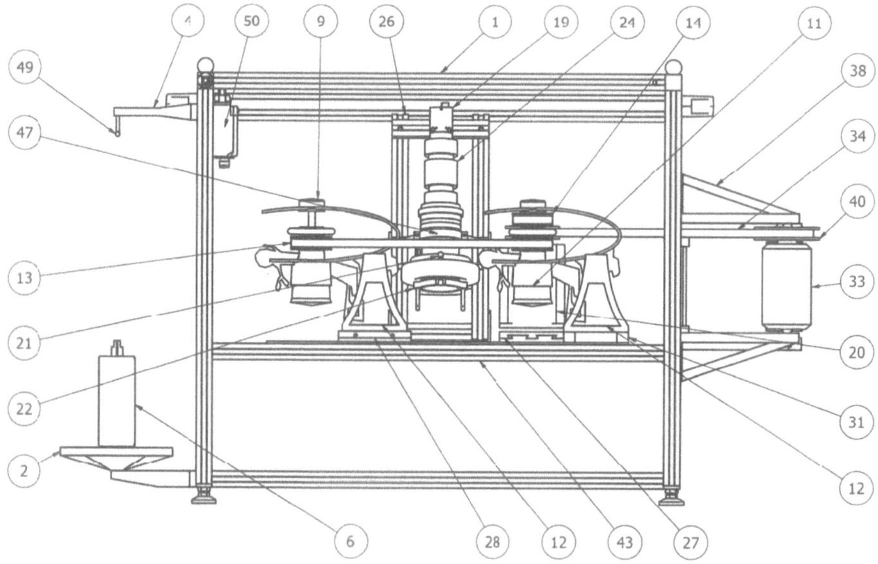

FIGURE 13.3 Front view diagram of the machine components. (Adapted from Caldas et al., 2022 [20].)

13.4.3 Yarn Guidance Subsystem

This subsystem plays a critical role in controlling the yarn's position, speed, and tension. Initially, the yarn is directed through the first Positive Yarn Feeder (PYF) located on the left (part 9), utilizing an included additive tensioner. The PYF insulates the yarn from tension fluctuations originating from the unwinding subsystem, maintaining a consistent feed rate regulated by a servomotor. The yarn then passes through components mounted on part 22, positioning it correctly in front of the lens. Subsequently, it passes through another PYF (part 11) to shield the yarn from tensions generated by the winding subsystem before exiting through a slot on the right wall of the box (Caldas et al., 2022; Pereira et al., 2024).

13.4.4 Winding Subsystem

This subsystem secures the bobbin and winds the yarn using a random winding method. A grooved drum guides the yarn and rotates the bobbin through surface friction. The yarn exits through a slot in the box wall, is routed around the grooves of the drum, and is then wound onto the bobbin positioned behind it. As the package diameter increases, the arm supporting the bobbin rotates to adjust, while springs maintain the proper pressure between the drum and the package. This setup is shown in Figure 13.4 (Caldas et al., 2022; Pereira et al., 2024).

FIGURE 13.4 Trajectory of the yarn during the winding process. (Adapted from Caldas et al., 2022 [20].)

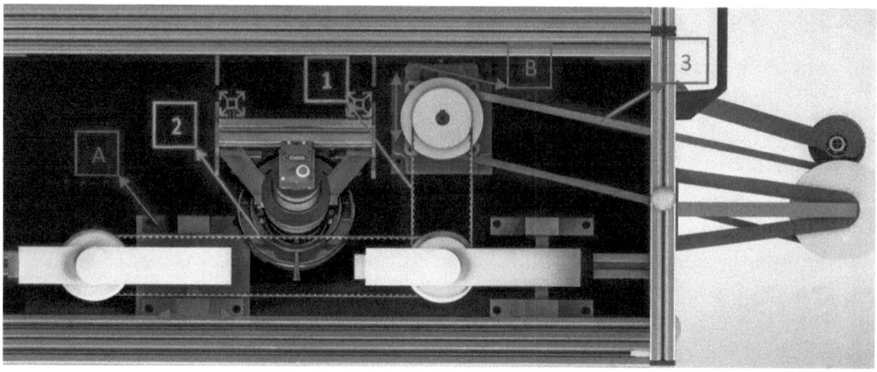

FIGURE 13.5 Path of the yarn during the winding process. (Adapted from Caldas et al., 2022 [20].)

13.4.5 Kinematic Chain

The entire system is driven by a single servomotor, which uses three separate sets of belts and pulleys to transfer movement. The first set (marked as 1 in Figure 13.5) consists of two toothed pulleys and a belt that transfers motion from the motor to the second PYF. The second set (labeled 2) also consists of toothed pulleys and a belt, connecting the first PYF to the second one, both sets having a 1:1 transmission ratio. The third set (labeled 3) uses a flat belt and pulleys to power the winding system, with a 3:4 transmission ratio to ensure a consistent linear speed between the guidance and winding subsystems. Two sliding supports are integrated into the servomotor (labeled B in Figure 13.5), and the first PYF (labeled A) accommodates belt switching and tensioning (Caldas et al., 2022; Pereira et al., 2024).

13.4.6 Controller Architecture

The controller setup, illustrated in Figure 13.6, integrates a Programmable Logic Controller (PLC) for managing a pulse command servomotor responsible for moving the yarn via a kinematic chain. It incorporates multiple yarn sensors in the PYF, associated with relays to detect yarn breakage across all subsystems. Additionally, there's a sensor for guidance subsystem cover detection, along with an emergency button for system automation (Caldas et al., 2022; Pereira et al., 2024).

This system incorporates a SCADA system to enable user interaction. It includes a local HMI and a LabVIEW program running on the local computer, as we can see in Figure 13.6. The controller specification encompasses various equipment behavior modes (Caldas et al., 2022; Pereira et al., 2024):

- Testing preparation mode: assists users in preparing yarn quality tests through semi-automatic system control;
- Testing mode: monitors and controls the system during yarn quality analysis, categorized by yarn quantity and velocity or test duration and speed;

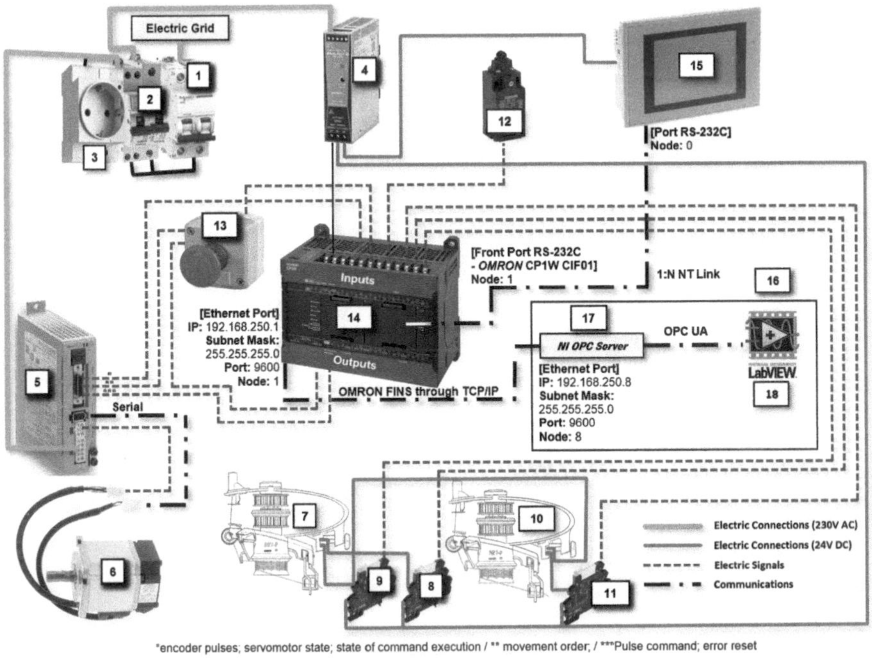

FIGURE 13.6 Overview of the automation system developed. (Adapted from Caldas et al., 2022)

- Completion mode for testing: helps users complete yarn quality tests using semi-automatic system control;
- Emergency stop mode: oversees system shutdown in emergencies or failures;
- Manual mode: enables users to configure, monitor, and test the system with authentication for authorized access.

13.5 INTEGRATION OF AI IN THE MECHATRONIC PROTOTYPE

This section presents how the innovative integration of AI, in this case, Deep Learning (DL), was developed in the mechatronic prototype developed specifically for the textile industry. The main focus of the developed system is on the sustainable and efficient analysis of yarn quality, and on the innovative integration of AI in the detection of defects such as hairiness in yarn, identifying protruding fibers and loop fibers (Figure 13.7) (Pereira et al., 2024).

13.5.1 Innovation in AI-Driven Sensing

The integration of AI, particularly through the use of advanced DL models like the YOLOv5s6 algorithm, enables significant innovation in textile engineering. The mechatronic prototype is equipped with a high-resolution image acquisition system

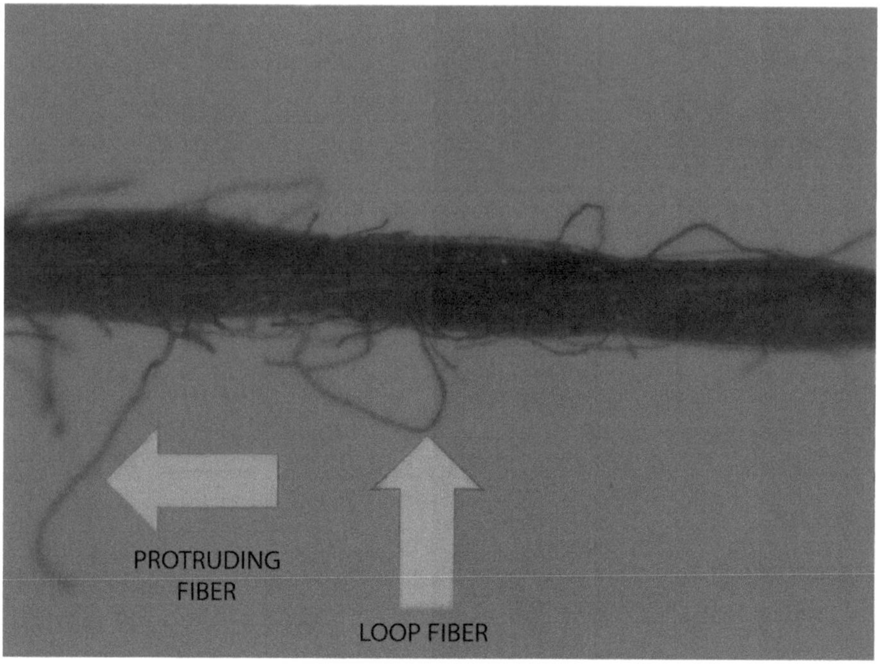

FIGURE 13.7 Example of an image with hairiness defects, in which loop-shaped fibers and fibers protruding from the thread are highlighted. (Adapted from Pereira et al., 2024.)

that captures detailed images of the moving yarn, allowing for accurate and timely detection and classification of defects such as hairiness, presented in Figure 13.7. Images are obtained, with a higher resolution, through a camera installed on the prototype while the wire is in motion. This can reach speeds of up to 4.2 m/s, and from that moment on, images with a resolution of up to 1.6 megapixels are generated. This new technique and characteristic allow immediate corrective actions, which improves product quality and minimizes yarn waste (Pereira et al., 2024).

13.5.2 Sustainable Production Practices and Innovative Design and Flexibility

Sustainability is the foundation of this AI-integrated mechatronic system. The AI system's ability to continuously monitor and optimize manufacturing parameters results in significant energy savings and is aligned with green production goals, making it a prototype that builds on sustainability. The design of the mechatronic prototype is innovative and adaptable, as it features modular components that facilitate upgrades and the integration of new technologies. This flexibility is very important in the textile industry in view of technological advances. Additionally, the prototype has an intuitive user interface, represented in Figure 13.8, that allows users to monitor performance and make necessary adjustments, improving operational efficiency and the effectiveness of AI integration (Pereira et al., 2024). Figure 13.8a shows the

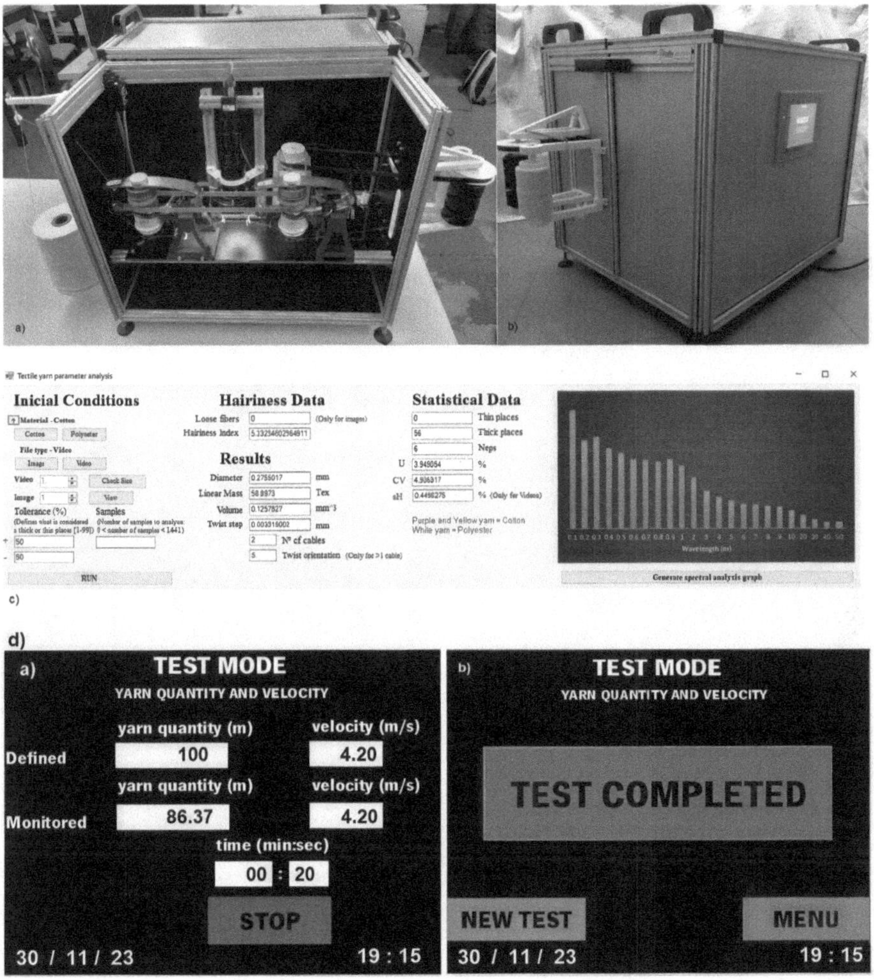

FIGURE 13.8 Mechanical system of the prototype with camera, HMI, interface for image and video acquisition, and user interface screens created for HMI (Pereira et al., 2024).

mechanical system of the prototype with the camera; Figure 13.8b displays the HMI in the prototype; Figure 13.8c illustrates the interface developed for image and video acquisition; and finally, Figure 13.8d presents the screens created for the HMI user interface, where the first screen shows the "yarn quantity and speed" test (a), and the second screen displays the "time and speed" test (b) (Pereira et al., 2024).

13.5.3 Data Set Creation and Algorithm Results

A dataset was created to support the AI system, with a focus on detecting hairiness of threads, including loop and protruding fibers. This dataset, consisting of 684 original images augmented to 1644 images with more than 11,000 annotations, allowed

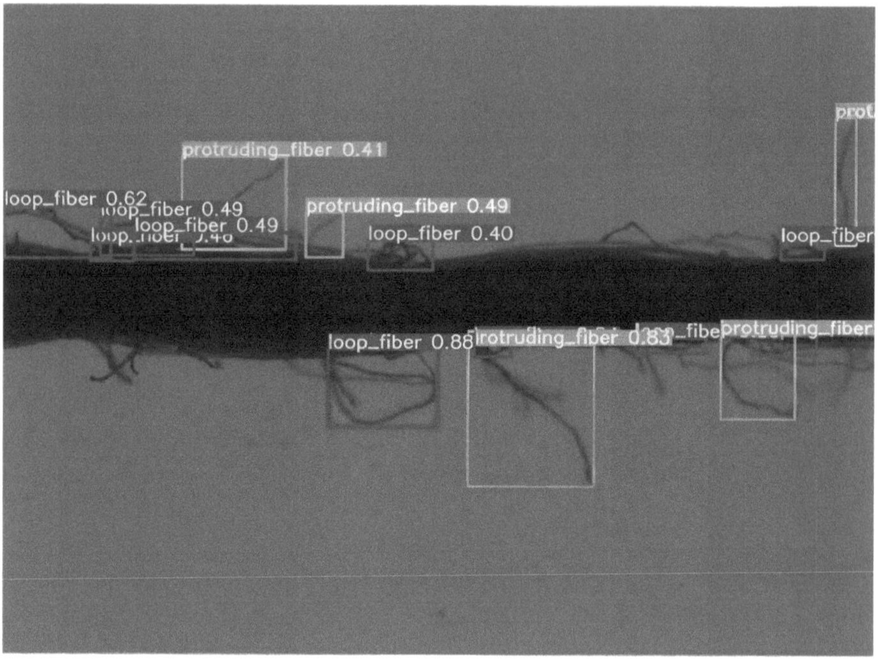

FIGURE 13.9 Validation of yarn hairiness defects with 0.3 of confidence with the implemented algorithm. (Adapted from Pereira et al., 2024.)

the training and validation of AI models. The results obtained from the YOLOv5s6 algorithm demonstrated accurate defect detection, and the model achieved significant improvements in key metrics such as F-score, recall, and precision compared to other existing models. These results enabled the effectiveness of the dataset and AI integration in improving the quality control process when analyzing defects such as hairiness in yarn. Figure 13.9 shows an image with hairiness detection with 0.3 of confidence with the implemented algorithm (Pereira et al., 2024).

13.5.4 Possible Impact on the Textile Industry

The AI-based mechatronic prototype allowed for the possible integration of new features into it, improving the quality of textile products but at the same time allowing for more sustainable practices. The system's ability to detect and correct defects reduces material waste and energy consumption, aligning with the industry's growing focus on environmental responsibility. As the prototype created is non-destructive equipment, this system will make it possible to change production processes in the textile industry and contribute to sustainability objectives in the sector (Pereira et al., 2024).

13.6 TESTS AND RESULTS

The development of a new physical system involves an almost infinite number of variables, many of which remain unknown despite extensive research on various components and subsystems. The goal of preliminary testing is to uncover some of

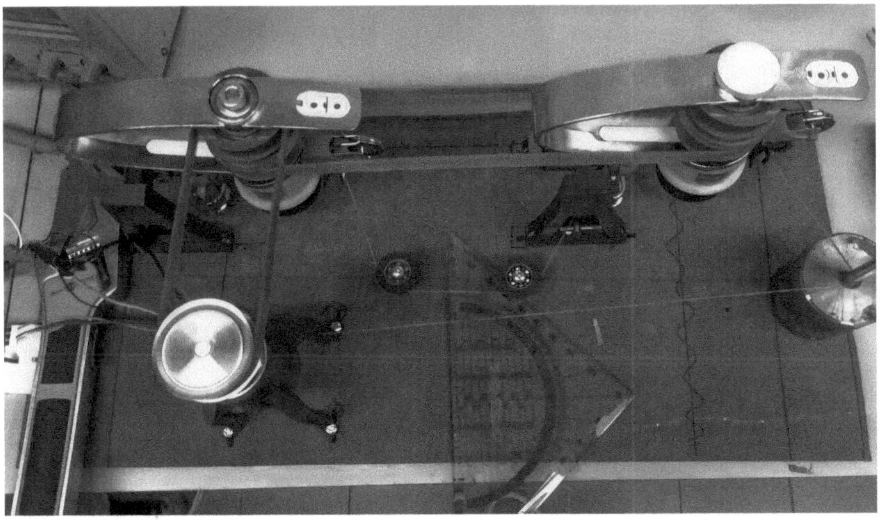

FIGURE 13.10 Test bench with looped yarn, deflected by pulleys [20].

these variables, the absence of which could potentially disrupt the intended operation of the machine upon completion, leading to increased development costs and time (Pereira et al., 2024).

In conducting preliminary tests, given that at that phase the entire machine was neither constructed nor operational by definition, it was essential to assemble a test bench. The sole purpose of this bench was to provide an environment similar to the desired final structure for testing the different subsystems intended for integration into the machine. The bench comprised a board with dimensions matching the inside of the box, the two acquired PYFs, and a temporary motor to activate them at the required speed. This setup facilitated subsequent load analysis during the trials. As the winding system was not yet in existence, the processed yarn traveled in a loop from the first to the second PYF and back to the first, as depicted in Figure 13.10. This loop arrangement allowed for continuous tests or was tethered to one of the PYFs for short unwinding trials (Pereira et al., 2024).

Table 13.2 contains detailed results analysis (Caldas et al., 2022; Pereira et al., 2024).

The camera positioning and support structure, the camera support adapter, the camera, the lens, the lens support arm, and the yarn illumination lamp are components belonging to the vision system. However, the adapter for guiding the yarn through the lens and the yarn guides for the adapter for guiding yarn into the lens belong to the yarn conduction. The guides, like the rest of the guides on the machine, will be ceramic guides. The shape chosen for these guides allows easy placement of the yarn; however, the diameter of the opening is slightly smaller than the rest of the machine guides to maximize the precision of positioning the yarn in front of the lens. The adapter for guiding the yarn through the lens is also a customized component and manufactured using additive manufacturing for the same reasons as the other supports, as we can see in Figure 13.11 (Pereira et al., 2024).

TABLE 13.2
Results Obtained from the Yarn Guiding Test

Test	Description	Yarn	Result (Horizontal osc.)	Result (Vertical osc.)
1.1.	Without guides or pulleys	19 Tex	≤1 mm	≤1.5 mm
		49 Tex	≤1 mm	1 mm
1.2.1	Guide distance 50 mm	19 Tex	≤1 mm (95%)	≤1 mm (95%) ≤2 mm (5%)
		49 Tex	≤1 mm (90%)	≤1 mm (90%) 1.5 mm (9%) 3 mm (1%)
1.2.2	Guide distance 86 mm	19 Tex	≤2 mm (95%)	≤2 mm (95%) ≤3 mm (5%)
		49 Tex	≤1 mm (95%)	≤1 mm (95%) 2 mm (5%)
1.3.1	With pulleys	19 Tex	≤1 mm	≤1 mm
		49 Tex	≤1 mm (90%)	≤1 mm (90%) ≤1.5 mm (10%) ≤1.5 mm (10%)
2.1.	With the PYF tensioner	49 Tex	Imperceptible	-------------
2.2.	Without the PYF tensioner	49 Tex	≤1 mm (at start)	-------------

The percentage of test time for each oscillation amplitude interval is indicated in parentheses (Pereira et al., 2024).

FIGURE 13.11 Final mechatronics prototype design solution with yarn [20].

TABLE 13.3
Estimated Cost of the Prototype in Mechanical and Design Terms (Pereira et al., 2024)

Component	Cost (€)
Prototype Box	150
Camera Support Structure	30
Servomotor	682
2 PYFs	80
Various 3D-printed parts	30
Belts	32
Ceramic guides	2
Grooved cylinder	8
Various screws and nuts	8
Bearings	3
Springs	3
Total	**1,028**

Following the implemented changes and the integration of all subsystems into the enclosure, along with the addition of the camera's support and accessories, the final solution is depicted in Figure 13.11. The cost of this mechatronic prototype is depicted in Table 13.3 (Pereira et al., 2024).

General tests refer to the evaluations conducted after the final prototype is completed. Unlike preliminary tests, these trials assess all subsystems simultaneously to evaluate the machine as a whole. The testing procedure involved the normal operational sequence of the machine, encompassing steps such as unwinding the coil, passing the yarn through various components, activating the control system, ensuring specified winding rotations, and, finally, closing the machine for the test run. These tests were conducted using various cotton yarns with different linear densities and a polyester yarn, utilizing coils of varying sizes and shapes (Pereira et al., 2024).

The outcomes indicated the machine's successful performance in unwinding, guiding, and winding the yarn, meeting speed requirements, oscillation standards, and coil uniformity without experiencing yarn breakages, equipment malfunctions, or safety failures. Furthermore, it was found that the portability of the machine presents characteristics that differentiate it, as it has a volume of less than $0.27\,m^3$ and a weight of less than 30 kg (Pereira et al., 2024).

13.7 FINAL REMARKS AND FUTURE DEVELOPMENTS

It can be concluded that the research work carried out allowed us to obtain an innovative mechatronic prototype for thread analysis, which uses AI to detect and analyze loops and protruding fibers. This technique used has a lot of potential for applications in the textile industry, offering the ability to analyze quickly and efficiently, guaranteeing the quality of the yarn. The use of AI in this system allows for the identification and characterization of defects in the yarn. In addition to the technology used, the system allows a wide range of yarn characteristics to be automatically obtained, providing a more detailed and rapid assessment compared to conventional methods and technologies used. The ability to identify imperfections such as loops and protruding fibers can be a significant added value for the textile industry, allowing early detection of problems in the yarn, contributing to improving the quality of the final product, and reducing waste. The mechatronic design of this prototype offers an integrated and versatile solution for wire analysis. The combination of mechanical and electronic components, along with created AI algorithms, creates a more robust and adaptable system for wire inspection in any type of industrial environment. The use of sensors, high-resolution cameras, and control systems allows a continuous and precise analysis process to obtain a wide range of characteristics and defects in yarn, while AI improves the capabilities of detecting and classifying hairiness in the yarn. Based on the design and development of this mechatronic prototype, it is important to highlight the main features it offers for wire analysis (Pereira et al., 2024):

- Low cost: The prototype was developed at an affordable cost, making it viable for application in industries with varying budgetary levels;
- Innovative design: It features an innovative design that efficiently integrates mechanical and electronic components, offering a unique solution for yarn analysis;
- Portability: Its portability allows for easy transport and use in different production locations, increasing the system's flexibility and adaptability;
- Simplicity: It was developed with a simple and user-friendly approach, aiming for practicality and ease of operation for users;
- Simple user interface: It has an intuitive and user-friendly interface, facilitating interaction between the operator and the system, making operation accessible even for less experienced users;
- Data analysis: AI analyses large volumes of data about yarns, identifying defects such as hairiness, identifying loops and protruding fibers;
- Performance prediction: Using AI Algorithms, the developed system can predict the state of the yarn based on historical data;
- Predictive maintenance: The use of AI in the prototype allows equipment failures to be anticipated, avoiding unexpected downtime and optimizing resources. In addition, the prototype allows remote maintenance and interventions;
- Resource optimization: The use of AI allows improving the use of materials and energy, promoting sustainability, as it is non-destructive equipment;
- Intelligent interface: The intuitive interface created for the user facilitates its use and is capable of automatically adapting to the operations required in its use;
- Contributions to the industry: The creation of specific datasets on hairiness defects in the hair allows future research and innovations in this area.

These characteristics converge to create a yarn analysis prototype that is affordable, innovative, easily transportable, simple to operate, and with a user-friendly interface, making it a valuable tool for yarn inspection in various industrial environments. Next, possible future work is presented both in terms of design and in terms of technology integration, namely (Pereira et al., 2024):

- Design optimization: Exploring further refinements in the design to enhance its ergonomics, ensuring easier handling and maintenance. Improving the overall aesthetics without compromising functionality could also be a focus;
- Feature enhancement: Adding additional functionalities or features to the prototype that might augment its capabilities, such as introducing wireless connectivity or enhancing its compatibility with various types of yarn materials;
- User interface refinement: Enhancing the user interface to make it more intuitive and adaptable, potentially incorporating customizable settings for different user preferences or roles;
- Intelligence augmentation: Developing more advanced algorithms for the AI component to improve its accuracy and efficiency in detecting and analyzing yarn imperfections;
- Manufacturing scalability: Possible scale production of the prototype: Investigate what methods could be used to speed up the manufacturing process of this prototype, making mass production possible without putting quality or low associated costs at risk.

The continuous evolution and possible improvements that could be implemented in this prototype will create a significant impact on the textile industry. The combination of a new design, innovative nature, portability, ease of use, and the use of cutting-edge technology means that the prototype created allows it to be a versatile and very useful tool for examining yarn in an industrial context. Its potential to detect small imperfections and obtain mass characteristics and the fact that it is very simple and adaptable to various industrial environments make it an essential system in the textile industry. The prototype not only allows for better quality control in textile production but also offers a very economical and accessible solution. Continuous development and improvements to be implemented in this prototype will allow it to be an innovative system in the way yarn analysis is carried out, offering efficiency, precision, and ease of use, ensuring above-average quality in the final product, and minimizing waste during manufacturing in the textile industries.

ACKNOWLEDGMENTS

The authors are grateful to FCT—*Fundação para a Ciência e Tecnologia* (Portugal), which partially financially supported this work through the RD Units Project Scope: UIDP/04077/2020, UIDB/04077/2020, UIDB/05549/2020, UIDP/05549/2020, and UIDB/00319/2020.

REFERENCES

Buur, J. (1995). The mechatronics design process. In *Springer eBooks* (pp. 27–32). https://doi.org/10.1007/978-94-011-0225-4_2

Caldas, P., Sousa, F., Pereira, F., Lopes, H., & Machado, J. (2022). Automatic system for yarn quality analysis by image processing. *Journal of the Brazilian Society of Mechanical Sciences and Engineering*, *44*(11). https://doi.org/10.1007/s40430-022-03875-3

Carvalho, V., Belsley, Vasconcelos, R., & Soares, F. (2009). Automatic yarn characterization system: Design of a prototype. *IEEE Sensors Journal*, *9*(8), 987–993. https://doi.org/10.1109/jsen.2009.2025810

Chauhan, A. (2024). Enhancing yarn quality in the cotton industry: AI- based NEP detection for improved manufacturing processes. In *Communications in Computer and Information Science* (pp. 88–103). https://doi.org/10.1007/978-3-031-68617-7_7

El-Geiheini, A., ElKateb, S., & Abd-Elhamied, M. R. (2020). Yarn tensile properties modeling using artificial intelligence. *Alexandria Engineering Journal*, *59*(6), 4435–4440. https://doi.org/10.1016/j.aej.2020.07.049

Elnashar, E. A. (2022). Applications of Mechatronics opportunities in textiles. *Trends in Textile Engineering & Fashion Technology*, *6*(3). https://doi.org/10.31031/tteft.2022.06.000640

Pereira, F., Carvalho, V., Soares, F., Vasconcelos, R., & Machado, J. (2017). Computer vision techniques for detecting yarn defects. In *Elsevier eBooks* (pp. 123–145). https://doi.org/10.1016/b978-0-08-101217-8.00006-3

Pereira, F., Macedo, A., Pinto, L., Soares, F., Vasconcelos, R., Machado, J., & Carvalho, V. (2023). Intelligent Computer Vision system for analysis and characterization of yarn quality. *Electronics*, *12*(1), 236. https://doi.org/10.3390/electronics12010236

Pereira, F., Machado, J., Soares, F., Vasconcelos, R., & Carvalho, V. (2024). Design of a yarn analysis mechatronic prototype for textile industry. In *DESIGNCOMMIT 2024*, Braga, Portugal.

Pereira, F., Oliveira, E. L., Ferreira, G. G., Sousa, F., & Caldas, P. (2021). Textile yarn winding and unwinding system. In *Lecture Notes in Mechanical Engineering* (pp. 347–358). https://doi.org/10.1007/978-3-030-79165-0_33

Pereira, F., Pinto, L., Soares, F., Vasconcelos, R., Machado, J., & Carvalho, V. (2024). Online yarn hairiness– Loop & protruding fibers dataset. *Data in Brief*, *54*, 110355. https://doi.org/10.1016/j.dib.2024.110355

Reddy, A. (2022). Artificial everyday creativity: Creative leaps with AI through critical making. *Digital Creativity*, *33*(4), 295–313. https://doi.org/10.1080/14626268.2022.2138452

Semnani, D., Latifi, M., Tehran, M. A., Pourdeyhimi, B., & Merati, A. A. (2006). Grading of yarn appearance using image analysis and an artificial intelligence technique. *Textile Research Journal*, *76*(3), 187–196. https://doi.org/10.1177/0040517506056868

Sengupta, A., Roy, S., & Sengupta, S. (2014). Development of a low cost yarn parameterisation unit by image processing. *Measurement*, *59*, 96–109. https://doi.org/10.1016/j.measurement.2014.09.028

Shi, Y., Yan, T., Cao, S., Zhuang, H., Lin, Y., Zhu, L., & Lu, D. (2022). Development of a small, covered yarn prototype. *Autex Research Journal*, *23*(2), 310–314. https://doi.org/10.2478/aut-2022-0007

Uster Technologies. (n.d.). Uster Tester 6 is the standard for yarn testing. https://www.uster.com/products/staple-yarn-testing/uster-tester/

Zhang, J., Wu, X., Zhu, L., Yu, C., Zhu, W., & Zhang, J. (2009). Embedded yarn tension control system based on the design of the sensor.In *2009 International Conference on Measuring Technology and Mechatronics Automation* (vol. 1, pp. 123–128). https://doi.org/10.1109/ICMTMA.2009.261

Zhu, L., & Wu, X. (2014). Design of yarn tension system and jacquard knitting machine. In *Advances in Intelligent Systems Research/Advances in Intelligent Systems Research.* https://doi.org/10.2991/mce-14.2014.40

14 Sustainable Futures
A Transdisciplinary Research Project for Dishwasher Wastewater Recycling in Domestic Vegetable Production

Annamaria Alabiso, Giorgio Buratti, Fiammetta Costa, Matteo Meraviglia, Luciana Migliore, and Attilio Nebuloni

14.1 INTRODUCTION

The problem of overuse and waste of water resources is a critical environmental issue. According to the 2030 Agenda's Sustainable Development Goal 6 (Clean water and sanitation), wastewater management through reuse and recovery is an urgent need (WWAP, 2017, WWAP, 2013). Despite this, wastewater from household appliances, such as dishwashers and washing machines, is rarely considered a recoverable resource. However, they fall into the category of grey water (Raschid-Sally & Jayakody, 2008), which is wastewater containing rapidly degradable organic matter with a lower bacterial load, as opposed to black water, which is characterised by a higher level of toxicity and contamination. It should also be noted that, assuming for the Italian user an average per capita consumption of 200L per day, the wastewater produced in a household consists of 60% grey water and the rest of black water (Conte, 2008). The two are usually mixed in the same sewer, wasting a potential alternative source that can meet a significant proportion of domestic water needs. In addition, the 2030 Agenda of the Sustainable Development Goals (Goal 2, Zero hunger) promotes sustainable food production systems and resilient agricultural practices capable of increasing productivity with low environmental impact. Zero Mile food production is gaining popularity worldwide due to the significantly reduced transport and the growing interest in fresh and healthy food. New innovative forms of cultivation are emerging in cities and urban contexts to complement the traditional community vegetable garden, giving people space to grow their own food. Vertical gardens are proliferating greenery that can be installed on facades and walls, usable indoors and outdoors, via structures on which plants root, fed by fertigation systems programmed according to the crop's needs.

DOI: 10.1201/9781032719290-17

Based on these assumptions, the Zero Mile research project aims to integrate the reuse of grey water from dishwashers with the cultivation of edible plants from a home vegetable garden (Costa et al., 2018, 2019; Alabiso et al., 2023) to create a closed, self-sustaining system. Through this integration, it aims to contribute to SDG 6 and 2 in a circular economy perspective, as proposed by Schroeder et al (2019). The scale and complexity of the project required the formation of a transdisciplinary research team involving the Department of Design of Politecnico di Milano and the Department of Biology of the University of Tor Vergata.

Private companies interested in the project's objectives joined the team later in the research and collaborated in the development of a joint patent. The research methodically approached the design problem from a set of initial hypotheses, developed in several successive iterations until a resolving convergence was reached. The integration of design solutions was articulated through cycles of prototyping – evaluation – and improvement, leading to the realisation of an interactive prototype, increasing the Technology Readiness Level of the system (TRL 3/4 – from experimental proof of concept to technology validated in lab) and anticipating the potential of environmentally and quality-of-life conscious living spaces.

14.2 PROJECT DEVELOPMENT

The overall aim of the research group is to reduce water consumption and wastewater, produce healthy plant food, and improve citizens' environmental awareness. The specific Zero Mile goal is to recycle domestic dishwasher wastewater in the indoor cultivation of edible plants. To achieve this, each research group has been assigned one or more objectives, developed according to their scientific field and coordinated by a transdisciplinary approach structured in five phases:

- Definition of the framework and state of the art: Department of Design, Politecnico di Milano and Department of Biology, Tor Vergata University (see Section 14.2.1);
- Development of the biological filter: Department of Biology, Tor Vergata University (see Section 14.2.2);
- Technical and functional design of the system: Department of Design, Politecnico di Milano (see Section 14.2.3);
- Edible plant monitoring and growth analysis: Department of Biology, Tor Vergata University (see Section 14.2.4);
- Investigating user needs and behaviours: Department of Design, Politecnico di Milano (see Section 14.2.5).

In addition, the heterogeneity of the fields involved has resulted in the need to define an appropriate research methodology capable of integrating such complexity of scientific approaches and disciplinary languages. The result, although provisional, is thus the outcome of a transdisciplinary research process involving companies and future users in the design practice. Different moments of the research are described below.

14.2.1 Definition of the Framework and State of the Art

In recent decades, biological treatment has been widely studied and applied to wastewater dealing with domestic, industrial, agricultural, and livestock (Goncalves et al., 2016; Fito & Alemu, 2019; Subashchandrabose et al., 2011). Cases of kitchen wastewater reuse for vegetable irrigation can already be found in the literature (Durán–Álvarez et al., 2014), showing that this kind of effluent is particularly suitable due to its high nutrient content from food waste and low levels of pathogens and chemical contaminants. Compared to black water, it is easier to recover grey water for irrigation purposes through appropriate treatment (Posadas et al., 2013; Costa et al., 2018; Congestri et al., 2020). Today's biological filtration and bioremediation strategies for water remediation are mainly based on microbial consortia, communities formed by photosynthetic microorganisms (cyanobacteria or microalgae) and heterotrophs (bacteria) in synergistic relationships. The combination of organisms with different metabolic activities, adapted to different environmental conditions, allows the development of a robust biological system that can operate in various conditions with different pollutants and nutrient loads.

In addition, microbial consortia reduce oxygenation costs and carbon dioxide emissions, contributing to increased system sustainability (Muñoz & Guieysse, 2006; Liu et al., 2017; Posadas et al., 2017). However, most of the currently available low-impact water treatment technologies are optimised for rural settlements or tertiary treatment and usually arise in response to a lack of connection to the collective sewerage system. Solutions that adopt phytoremediation as an alternative to sewerage septic tanks imply the use of extensive areas that are not always available in many locations, such as urban areas. The work we present adopts a different logic: Instead of using a single large area to collect all the effluent, we intervene upstream in the process with numerous filtering devices distributed per dwelling and located in the place of wastewater production. In addition to the water savings already mentioned, this solution allows a reduction in the volume and concentration of pollutants in the wastewater, making the purification process more effective.

Case studies on sustainable green solutions were also collected and critically analysed, focusing on the degree of interaction between a new technology and an existing artefact. Three main design strategies emerged: (i) applications by juxtaposition; (ii) applications by superimposing a new technology on present artefact or configurations, where each element retains its autonomy; and (iii) applications by integration, where all different components are merged into a single complex system. These design approaches were then used to define the relevant application strategies that were later applied in the research stages of the project (Buratti et al., 2019).

14.2.2 Development of the Biological Filter

At the heart of the Zero Miles system is a biofilter based on a custom-designed microbial consortium of different microorganisms selected for their ability to process and mineralise food waste. To achieve efficient wastewater treatment, both the selection of suitable microorganisms to form the consortium and detailed knowledge of

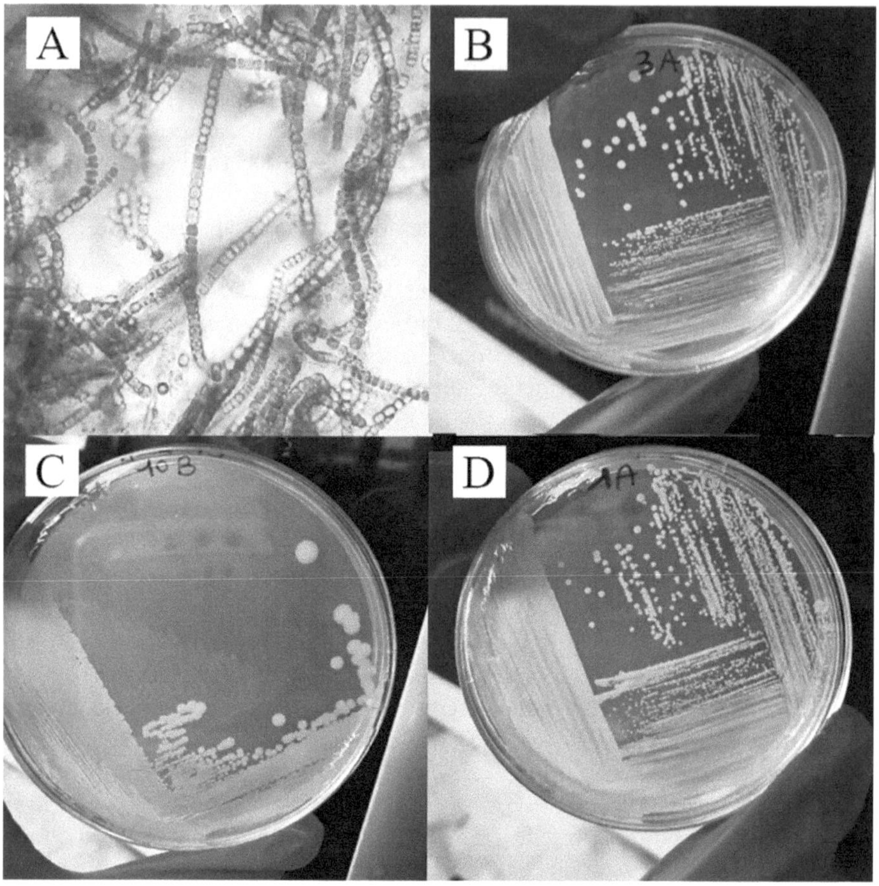

FIGURE 14.1 *Trichormus variabilis* (a) and the three aerobic heterotrophic bacterial strains isolated from dishwater wastewater: *Acinetobacter* sp. (b), *Aeromonas* sp. (c), and *Exiguobacterium* sp. (d). (Authors.)

their structure, behaviour, and interactions were essential. The consortium designed for our system consists of two different microbial partners (Figure 14.1). A photosynthetic, filamentous, nitrogen-fixing cyanobacterium called *Trichormus variabilis* (from the collection of the Algae Biology Laboratory of Tor Vergata. It was isolated from biofilms taken from the Cabras Lagoon in Sardinia, Italy, a rather stressful environment due to the contribution of fresh water from rivers and saltwater from the sea, resulting in the characteristic brackish water). In addition, three aerobic heterotrophic bacterial strains were isolated from dishwasher wastewater, and the association of these microorganisms was studied to determine the optimal ratio/conditions for their thriving in wastewater (Congestri et al., 2020; Alabiso et al., 2023, 2024).

The objectives of these experiments were:

- to evaluate the effect of abiotic conditions (irradiance, pH, and organic load) on the growth of the microbial consortium;

- to evaluate the consortium's ability to survive and thrive in different wastewater under batch (static system) or semi-batch operating conditions;
- to define the morphological characteristics of the biofilter container and to establish implementation requirements for the development of integrated technology and construction solutions.

The interaction between the research group that developed the biological system and the team responsible for the design was crucial for the industrial application of the designed system. Understanding the physiological characteristics of the consortium and optimising the culture conditions are crucial for biofilter design and topology, which must ensure the vital growth needs of the microorganisms for as long as possible. The results (Figure 14.2) of the tests carried out (Alabiso et al., 2023, 2024) highlighted several requirements.

The requirements for the development of the biofilter container are:

a. The importance of illumination (as both intensity and duration), which is essential to cyanobacterial growth and viability.
b. The use of a material resistant to the effluent alkaline pH. This is not harmful to the specifically selected microorganisms chosen for their resistance but may be destructive to amphoteric metals (e.g., aluminium).

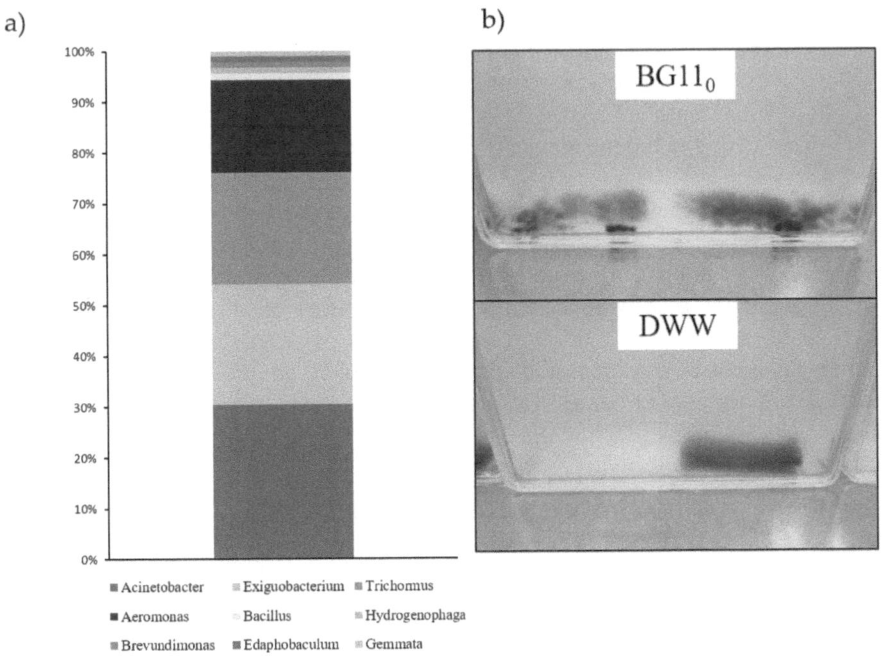

FIGURE 14.2 Bacterial genus distribution (as frequency) in the microbial consortium (left). Effect of different growth media on the consortium morphology (right). BG110 = Mineral growth medium, as control; DWW = Dishwasher wastewater. (Authors.)

c. The need to keep the microbial consortium stable. Microorganisms organise themselves three-dimensionally into an aggregate shaped according to the container, which floats by not adhering to surfaces. This structure is reversible: When subjected to strong shaking, it dissociates and reconstitutes itself within a few hours. Avoiding continuous stress and allowing its maximum purification efficiency is essential.
d. The consideration of the wastewater refill cycle. The capacity of the consortium to degrade a high organic load has already been tested over 4 days. Following the previous point, the effluent flow must replace the treated water on-site without damaging the consortium's three-dimensional structure.
e. The accessibility inside the filter. Since overgrowth of heterotrophic bacteria has been observed in some types of effluent, it is necessary that the filter, although watertight, can be easily opened to deal with any biomass excess.
f. The microbial consortium achieves maximum efficiency under operating conditions between 10°C and 35°C, a characteristic that makes it particularly suitable for indoor use but implies the need for insulation, heating, or cooling in the case of outdoor applications.

As a result of the interaction between the various (bio)technological components, the stated requirements direct the design of the space and building system in the following directions:

- use of transparent polymeric materials to allow the maximum use of natural light while maintaining the possibility of integration with artificial light;
- modularity and independence of the biofiltration, irrigation, and cultivation subsystems;
- flexibility of the plant cultivation subsystem in order to host different cultivation types;
- separation of building structure and plant support elements;
- easy access, maintenance, and movement of all elements and subsystems.

The development of demonstrators and prototypes, in addition to the consortium's experiments, has proved helpful in anticipating the various application scenarios in the home environment.

14.2.3 Technical and Functional Design of the System

The system was prototyped on a domestic dishwasher (energy class A), chosen as a reference appliance because of the higher organic load in the waste, compared to washing machine effluent. During the tests, the "eco" wash cycle, which produces about 12 L of discharged wastewater, was always used, supported by an EU Ecolabel-certified dishwasher tablet detergent, while rinse aid was excluded. The dishwasher was used daily to clean the bowls used by the staff of the School of

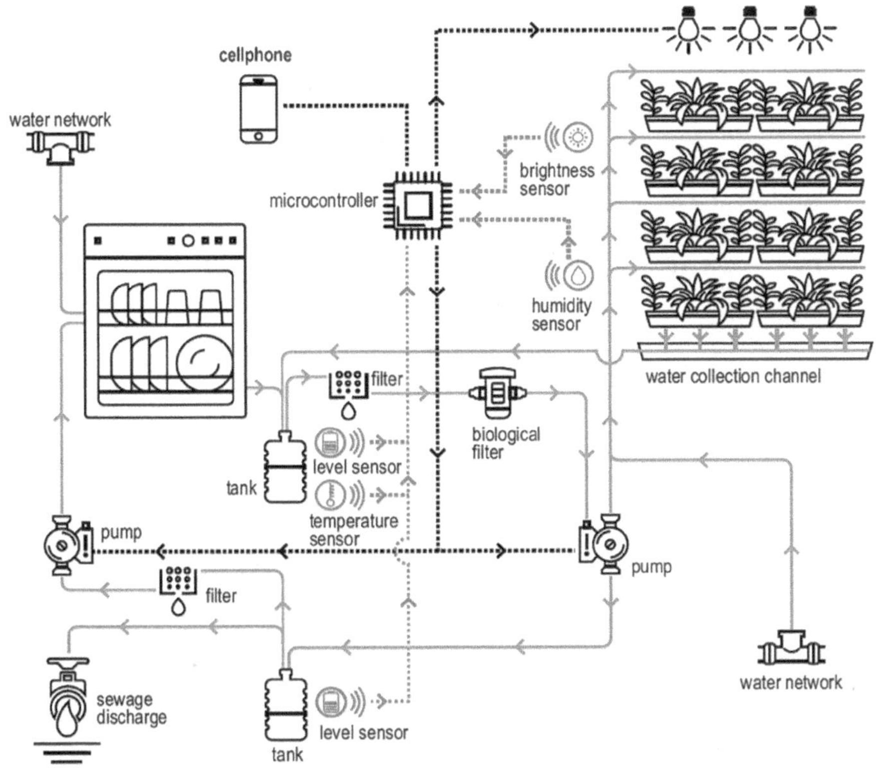

FIGURE 14.3 System functional diagram. (Authors.)

Design's several laboratories during their lunch break, in order to obtain as diverse a wastewater as possible. Figure 14.3 illustrates the integration of the dishwasher with an indoor growing apparatus that recycles dishwasher effluent with two parallel purposes: to make the organic nutrients in the wastewater available for growing vegetables in a domestic environment and to reuse the recycled water for subsequent washing cycles.

The prototype dimensioning considered:

- the average dishwasher use by a typical Italian family corresponds to approximately four washing cycles per week;
- the amount of wastewater discharged by each washing cycle;
- the daily water consumption of each vegetable grown;
- the number and type of vegetables/vegetation needed to phytodepurate the expected amount of water per cycle.

In addition to testing plant growth, the prototype was used to study how the different elements of the system, developed independently by different research groups, interacted. The research group developed the first prototype (Figure 14.4) in collaboration

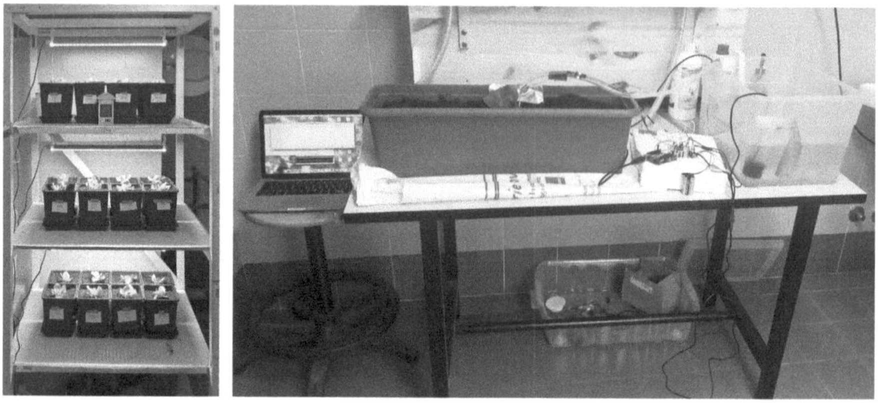

FIGURE 14.4 First maquette and implementation of the automatic fertigation system and management software. (Authors.)

with the Modelling Lab of the School of Design to conduct the first cultivation tests, verifying the lighting system and the first fertigation apparatus. In parallel, a dedicated information and communication technology system has been developed and implemented using MIT App Inventor and Arduino, which, using sensors to control environmental parameters (soil moisture, temperature, solar radiation, etc.), will ensure measurement of the optimised effluent over time, with the possibility of modification by the users. The implemented computer system plays an essential role in the possible industrial development of the product, helping the user to maintain the plant and ensure its correct growth.

At a later stage, the working group was extended to include several companies interested in the possible results of the project:

- D'Alesio & Santaro, a lighting design firm that works in the fields of architecture, product design, automotive, and art installations.
- Ka-We, a company that designs and produces valves and filters for water treatment, offering standard products and *ad hoc* solutions.
- Spinitalia, a company based on developing high-tech products in the electronics and information technology sectors.
- Verdeprofilo, a company specialised in creating and designing custom gardens, mainly vertical and vegetable gardens, in public and private areas.

Each of these partners offered its specific expertise to refine the system in its optimal configuration and also provided technological resources and materials to create a new functional prototype suitable for testing and refinement, with standard size and technical design solutions that could be replicated in series (Figure 14.5). The new prototype was then used for subsequent experimentation and user-centred development.

FIGURE 14.5 Advanced prototype developed with the latest components and technologies. The size and footprint achieved by the research team are currently the minimum required for optimal system operation. (Authors.)

14.2.4 Growth and Monitoring of Edible Plants

Before the setup of the microbial consortium, the possibility to use wastewater without any treatment was evaluated. Two wastewater samples (W1, W2) and one drinking water (C) sample were analysed for physicochemical parameters (pH, BOD_5, COD) and ionic composition. Analyses were performed using standard methods (APHA/AWA/WEF, 2012). The wastewater samples were collected under the same operating conditions as the dishwasher but with different food residue loads. 40 lettuce (*Lactuca sativa* L.) plants grown in 2.5L pots with peat-based substrate were divided into four batches: plants irrigated with wastewater (W1, W2) or clear water (C), and plants treated with an NPK mineral fertiliser (F) and irrigated with wastewater (W1/F, W2/F) or clear water (C/F). The experiments were conducted indoors under artificial lighting (TLED 26W Growing, Secret Jardin) with a photoperiod of L:D 16:8 h.

Throughout the growth cycle, the plants received regular watering at roughly 40 mm per plant daily. Following 48 days, the lettuce was harvested, weighed promptly (when wet), and then dehydrated in an oven at 105°C for 72 hours before being weighed again (when dry). Furthermore, analysis of plant functionality was done using biochemical parameters. In vivo, leaf chlorophyll levels were quantified utilising a CL-01 chlorophyll metre (Hansatech, King's Lynn, UK). In contrast, chlorophyll fluorescence measurements were conducted on dark-adapted leaves with a portable fluorometer (Handy PEA, Hansatech, King's Lynn, UK).

The salicylic acid method was used to measure the concentration of nitrate (Cataldo et al., 1975) and was calculated using a standard calibration curve of KNO_3. The analytical results indicate a significant divergence in the ionic composition of the drinking water sample and the two wastewater samples. Samples W1 and W2 are characterised by a higher pH (W1 = 9.3; W2 = 9.6; C = 6.9) and approximately 50 and 27 times higher sodium (Na+) ion concentrations, respectively, than that found in sample C (34 mg/L). The rise of Na+ results in an increase in salinity, evaluated indirectly by electrical conductivity. This is attributed to including Na_2CO_3 in the detergent and the ion exchange process in the water softener inside the dishwasher.

The organic content of W1 and W2 samples predominantly arises from food residues, thereby exhibiting substantial heterogeneity. Comparatively, the W2 sample contains significantly higher levels of organic content concerning COD (W1 = 730 mg/L; W2 = 2,600 mg/L) and BOD_5 (W1 = 210 mg/L; W2 = 1,500 mg/L). There were no significant differences in the production and quality of lettuce irrigated with potable water or wastewater regarding wet and dry weight or chlorophyll content. The dry weights indicate that wastewater irrigation allows for slightly higher growth compared to potable water and using wastewater (W) resulted in the highest accumulation of chlorophyll in leaves. Moreover, it was observed that plants irrigated with wastewater had higher leaf functionality and photochemical efficiency of PS II. The Fv/Fm ratio, which measures the quantum maximum efficiency of photosystem II, varied among different batches ranging from 0.79 to 0.84: The highest index value was found in W. At the same time, it was significantly lower in fertilised batches irrigated with potable water (C/F). Lastly, nitrate content was lower in plants irrigated with wastewater (W ≈ 2,600 mg/kg fresh weight [FW]) than with fertilisers (around 4,000 mg/kg FW), this last showing the highest value.

These relatively satisfactory results and, more importantly, the difficulty of maintaining the wastewater in a home environment for relatively long time without causing unpleasant effects (e.g., bad smell) or worse, health hazards, led to the design of the microbial biofilter, which avoids the negative side effects.

14.2.5 Investigating User Needs and Behaviours

- The research encompasses multiple levels and involves resolving several issues related to the bio-technological system. However, to enhance the likelihood of acceptance by the potential market, it is essential to analyse the expressed or unexpressed needs of the end-users. As per User-Centred Design guidelines, this includes engaging them in various steps of the design process (Figure 14.6), beginning with requirements analysis and culminating in final evaluation (Norman and Draper, 1986).

This inquiry is significant when intervening in a domestic setting, as individuals view this space as a personal territory that facilitates intimacy, well-being, safety, and self-expression (Pillan & Colombo, 2017). Consequently, exploring various factors linked to relevant social groups' attitudes, expectations, and behaviours was required.

FIGURE 14.6 Interviews, focus groups, and user involvement were used to identify the strengths and weaknesses of the system and improve the final design. (Authors.)

In particular, the inquiry concerned:

- factors related to users' interactions with the system and their surroundings, both spatially and temporally;
- factors related to users' sociocultural background, including motivations, expectations, values, psychological needs, and relationships;
- the ecosystem involving the organisational models and economic and transactional factors adopted by different households.

The research employed a multi-method approach (Seawright, 2016) – commonly referred to as the mixed method – which merged qualitative and quantitative methodologies. These comprised quantitative data analysis, focus groups involving prospective customer interviews, and stakeholder engagement. The sample group was selected to ensure maximal representation of diverse household types and comprised 20 participants who were unacquainted with each other and unfamiliar with the project. Furthermore, the prototype was presented alongside preliminary designs for the Zero Miles system, aiding in the comprehension of the group's viewpoint on the proposed project. The study analysed daily habits of reusing and recycling, dishwashing routines, and households' relationship with eco-sustainable behaviour to identify the strengths and weaknesses of the Zero Miles system.

Respondents displayed a heightened awareness of environmental protection. Recycling, reusing items and food, and conserving water and electricity were established habits among many. However, radical approaches were not widespread. Participants reported that adopting a completely eco-friendly lifestyle can be challenging, both objectively and subjectively.

Contrary to the research team's fears, no doubts or worries were expressed about the edible plants' quality irrigated with dishwasher effluent. Respondents do not report perplexity to edible use with adequate information and/or a guarantee from a reputable bureau or producer. However, worries about the trade-off between possible environmental sustainability improvements and cost have been raised.

The raised concerns relate to:

- the system's size and space requirements concern many participants, as they compete with other higher-priority uses;
- the complexity of managing the system and the limited range of plants that can be grown are drawbacks for some. The project would be more appealing to many if it could accommodate ornamental plants, which are generally less demanding;
- the system adaptability not only in the kitchen but also in other rooms of the house, such as the living room;
- the potential odours from the biofilter and the possibility of excessive plants attracting insects, although these worries were a minority aspect;
- the safety of children and pets mentioned by two participants.

Despite the potential for some observations to be misled by the unoptimised prototype, the research team considered many observations for the model's formal evolution (see Figure 14.7).

FIGURE 14.7 Preliminary design concepts based on developed technological requirements and on needs identified by users. (Authors.)

Overall, there was significant interest in the system. For most of the participants, whether they acquire the system depends on whether its technical-organisational aspects align with their household habits and spending power. It is worth noting that an ideological position that disregards cost as a central purchasing factor is held in a significant cluster. This group draws an analogy with purchasing an electric car, also characterised by higher costs. They are willing to pay more in the interest of sustainability, and for some, the proposed system would signify their commitment.

14.3 ACHIEVEMENTS AND FUTURE DEVELOPMENTS

Focusing on profit and production, water and food resources have been viewed as commodities to be maximised for cost-effectiveness. This approach has generated unsustainable patterns of both natural and energy resource use, which are not feasible given the continuous growth of the world's population. Over time, the debate has progressively broadened, involving researchers and politicians and increasing common population groups. The emergence of a conscious consumer market has resulted in a shift from hedonic purchasing to an awareness of the potential social and ecological impact. The emphasis is no longer solely on price, but also on the healthfulness of the food and the sustainability of production processes concerning social and environmental impact. It appears that the present moment is ripe for implementing technologically advanced biological structures similar to the one developed by our research team. These systems can purify dishwasher wastewater with an innovative biofilter and utilise it to cultivate nourishing vegetables with desirable organoleptic qualities.

The experimental phase carried out through the Zero Mile project produced several results.

a. An innovative biofilter was developed using a microbial consortium capable of thriving in dishwasher wastewater, with a basic pH due to the presence and composition of detergents. The consortium was designed to withstand variations in pH levels and organic loads typically found in an average family diet in terms of both quantity and composition. Under optimal conditions and if cyclically sprayed with wastewater, the consortium guarantees an excellent purification capacity.

b. The development of a technological-biological system, based on a biofilter integrated with the kitchen environment using vertical gardening solutions accommodating 40 plants. The system is equipped to treat the dishwasher wastewater produced by an average family. Electronic devices will monitor growth protocols by adjusting luminance with special LED lights, which have been modulated to optimise the life cycle of vegetables and control the necessary temperature and humidity levels regardless of actual conditions at any latitude. As a result, these vegetables will not be contaminated with potentially harmful chemicals or soils, ensuring fresh products, retaining their organoleptic properties.
c. The production of quality lettuce comparable to vegetables irrigated with potable and normally fertilised water. The experiment showed that wastewater irrigation resulted in even slightly greater growth than drinking water, and the highest chlorophyll accumulation in the leaves was observed using wastewater. Furthermore, it was observed that leaf functionality and photochemical efficiency were higher in plants irrigated with wastewater compared to those fertilised normally. Additionally, the nitrate content was lower in the former. These outcomes support that even small indoor crops can produce nutritious and healthy food.

So far, the project has obtained two patents (National Patent No. 102019000004583, 2019; International Patent 2020/194224 Al, 2020) and a significant number of publications.

The research team is still working to improve the system to be industrially produced and made available to the public. Based on the focus group's conclusions, the following obstacles must be overcome for this goal to be accomplished:

- increase the system's diversity by incorporating different vegetables and improving efficiency, leading to a smaller overall structure footprint and increased appeal to home environments;
- expand research to include non-edible flora to broaden the range of possible uses, including ornamental plants and different areas within the domestic environment.

When devising a domestic food system, it is crucial to regard food consumption as a means of assigning significance to everyday life. These influences create a consumer's identity (Blay-Palmer, 2010). The findings from the focus group indicate that the Zero Miles system must be as self-sufficient and efficient as possible, despite the customers' sensitivity. Although it appears as a conventional vegetable garden, the time spent growing it is not perceived as a hobby but as additional work.

The involvement of enterprises related to the project's target markets has expanded the research parameters. Indeed, the system's circular water economy creates opportunities for use in restricted and extreme environments, like refugee and emergency camps, polar bases, or space stations. The companies' experience with large-scale projects prompted an initial study (Buratti et al., 2019) on sizing the system for communal settings such as schools, care homes, flats, or eateries. Effectively, on a larger scale, the issue of space occupied by plants and technological devices would be reduced. The Zero Miles system has the potential to aid in the revitalisation of marginal urban spaces and environments, as it is highly adaptable and flexible. Its use in urban and architectural settings will allow experimentation with covering and façade construction systems, leading to new matrices for inhabited areas.

REFERENCES

Alabiso, A., Frasca, S., Bartolini, M., Congestri, R., D'Andrea, M.M., Costa, F., Nebuloni, A., Meraviglia, M., Buratti, G., & Migliore, L., Home sweet home: setting the best thriving conditions for the ad hoc engineered microbial consortium in the Zero Mile system. *Sustainability* 2024, 16, 2227. https://doi.org/10.3390/su16062227

Alabiso, A., Frasca, S., Cantelmo, V., D'Andrea, M.M., Braglia, R., Scuderi, F., Costa, F., Savio, S., Congestri, R., & Migliore, L., A dirty job: dishwasher wastewater reuse and upcycle through an ad hoc engineered microbial consortium. *npj Clean Water* 2023, 6, 66. https://doi.org/10.1038/s41545-023-00280-8

APHA/AWA/WEF, *Standard Methods for the Examination of Water and Wastewater*, American Public Health Association/American Water Works Association/Water Environment Federation, Washington, DC, 2012.

Blay-Palmer, A., *Imagining Sustainable Food Systems*, Ashgate Publishing, Aldershot, 2010.

Buratti, G., Nebuloni, A., & Meraviglia, M., Zero-kilometre plants production. An integrated design application, in Ambrosio, M., & Vezzoli, C. (eds), *Designing Sustainability for All. Proceedings of the 3rd LeNS World Distributed Conference, Milano, Mexico City, Beijing, Bangalore, Curitiba, Cape Town, 3–5 April 2019*. Milano: Edizioni POLI.design, pp. 319–324, 2019.

Cataldo, D. A., Maroon, M., Schrader, L. E., & Youngs, V. L., Rapid colorimetric determination of nitrate in plant tissue by nitration of salicylic acid. *Communications in Soil Science and Plant Analysis* 1975, 6(1), 71–80. https://doi.org/10.1080/00103627509366547

Congestri, R., Savio, S., Farrotti, S., Amati, A., Krasojevic, K., Perini, N., Costa, F., & Migliore, L., Developing a microbial consortium for removing nutrients in dishwasher wastewater: towards a biofilter for its up-cycling. *Water Science and Technology* 2020, 82 (6): 1142–1154.

Conte, G., *Nuvole e sciacquoni; come usare meglio l'acqua in casa e in città*, Edizioni Ambiente, Milano, 2008.

Costa, F., Aureggi, M., Migliore, L., Perego, P., Pillan, M., Standoli, C.E., & Vignati, G, Design for sustainability and ICT: a household prototype for wastewater recycling, in *(a cura di): Marcelo Ambrosio & Carlo Vezzoli, Designing sustainability for all*, Milano: Edizioni POLI.Design, pp. 869–872, 2019.

Costa, F., Amati, A., Antonelli, M., Cocetta, G., Mauro, M.D., Ferrante, A., Krasojevic, K., Mangiarotti, R., Meraviglia, M., Nebuloni, A., & Perego, P., Designing the future: an intelligent system for zero-mile food production by upcycling wastewater, in *MPDI, Proceedings 2018*, vol. 2, p. 1367, 2018. https://doi.org/10.3390/proceedings2221367.

Durán–Álvarez, J.C., & Jiménez–Cisneros, B., Beneficial and negative impacts on soil by the reuse of treated/untreated municipal wastewater for agricultural irrigation–A review of the current nowledge and future perspectives. *Environmental Risk Assessment of Soil Contamination* 2014, 24, 711–732.

Fito, J., & Alemu, K., Microalgae–bacteria consortium treatment technology for municipal wastewater management, *Nanotechnology for Environmental Engineering* 2019, 4(1), 1–9. https://doi.org/10.1007/s41204-018-0050-2

Goncalves, A.L., Pires Jose, C.M., & Simoes, M., A review on the use of microalgal consortia for wastewater treatment, *Algal Research* 2017, 24, 403–415.

Liu, H., Lu, Q., Wang, Q., Liu, W., Wei, Q., Ren, H., Ming, C., Min, M., Chen, P. & Ruan, R., Isolation of a bacterial strain, *Acinetobacter* sp. from centrate wastewater and study of its cooperation with algae in nutrients removal, *Bioresource Technology* 2017, 235, 59–69.

Muñoz, R., & Guieysse, B., Algal–bacterial processes for the treatment of hazardous contaminants: a review, *Water Research* 2006, 40(15), 2799–2815.

Norman, D.A., & Draper, S.W., *User-Centered System Design*, Lawrence Erlbaum Associates, Mahwah, NJ, 1986.

Pillan, M., & Colombo, S., Will smart homes improve our lives? A design perspective towards effective wellbeing at home, *The Design Journal* 2017, 20(sup1), S2580–S2591. https://doi.org/10.1080/14606925.2017.1352769

Posadas, E., Alcántara, C., García-Encina, P.A., Gouveia, L., Guieysse, B., Norvill, Z., Acién, F.G., Markou, G., Congestri, R., Koreiviene, J., & Muñoz, R., Microalgae cultivation in wastewaters, in Munoz, R., & Gonzalez, C. (eds.), *Microalgae-Based Biofuels and Bioproducts*, Woodhead Publishing, Sawston, Cambridge, pp. 67–91, 2017.

Raschid-Sally, L., & Jayakody, P., *Drivers and Characteristics of Wastewater Agriculture in Developing Countries: Results from a Global Assessment*, IWMI, vol. 127, International Water Management Institute, Colombo, Sri Lanka, 2009.

Schroeder, P., Anggraeni, K., & Weber, U., The relevance of circular economy practices to the sustainable development goals. *Journal of Industrial Ecology* 2019, 23(1), 77–95.

Seawright, J., *Multi-Method Social Science: Combining Qualitative and Quantitative Tools*, Cambridge University Press, Cambridge, 2016.

Subashchandrabose, S. R., Ramakrishnan, B., Megharaj, M., Venkateswarlu, K., & Naidu, R., Consortia of cyanobacteria/microalgae and bacteria: Biotechnological potential, *Biotechnology Advances* 2011, 29(6), 896–907. https://doi.org/10.1016/j.biotechadv.2011.07.009

WWAP – UNESCO World Water Assessment Programme, *The United Nations World Water Development Report 2023: Partnerships and Cooperation for Water*, Paris, 2013.

WWAP – United Nations World Water Assessment Programme, *UN World Water Development Report, Wastewater: The Untapped Resource*, Paris, 2017.

15 The Key Role of Life Cycle Design Strategies in the Eco-Design of a Kitchen Island Project

Carolina Polónio, J. Paulo Davim, and Fátima Pombo

This chapter is an extended version of the article presented in the Design Commit '24 conference proceedings.

15.1 INTRODUCTION

Planet Earth is the home of Humankind. This small island of life is what separates human beings from the vastness of the Universe and provides them with the resources to fulfil their needs. Consequently, the history of humanity is inextricably linked to the relationship that human beings establish with the natural environment in which they live (Zirnstein, 1996, as cited in Hofer, 2009). The industrial revolution created a context for reflection on the link between human beings and nature, considering this relationship as a foundation for sustainability (Edwards & Orr, 2005).

Unfortunately, 'the system of production and consumption, as a set of human actions in all its complexity, has determined and continues to determine an unsustainable situation of loading and unloading for the environment' (Manzini & Vezzoli, 2002, p. 325) meaning that planet Earth is today in an environmentally dramatic situation. Nevertheless, according to the same authors, it is not only possible to live in a society with sustainable development, but it is also the logical course to follow in order to manage available resources correctly and protect the planet's future. In addition, the eco-design of products and services is at the forefront of this process, since it is not enough to improve or repair existing products, but rather to rethink and redesign others with innovation (Sherwin et al., 1998). Therefore, the practice of eco-design has emerged as the imperative path to follow in order to find strategies and apply them to the creation of products/services that contribute to the environmental sustainability of the planet (Sherwin et al., 1998). Within this scope, we will discuss in this chapter the project of a kitchen island.

Today, companies and designers are getting more and more involved in the sustainable development of products, and, as a result, users are becoming more aware of them (Afonso et al., 2018). Regarding existing kitchen island projects, it

DOI: 10.1201/9781032719290-18

was recognised that the concept of sustainability can be expressed, for example, by encouraging habits that make the user more aware of the environmental consequences of their actions by incorporating a small growing space, composting or waste separation (Casais, 2011); by minimising resources and reusing them in the kitchen, such as using energy-efficient equipment and reusing water that has already been used (Yoneda, 2010); and by minimising the amount of virgin material resources used in production through the cyclical reintegration of discarded materials or those considered waste (IKEA Museum, 2021).

The existence of standards and regulations such as ISO 14001:2015 can be seen as a common starting point for the development of eco-design internationally (Charter, 1998). In effect, the creation of standard instruments and strategies for eco-design not only establishes the principles for implementing solutions, but also allows these principles to be disseminated around the world in order to improve the environmental performance of a product/service (DGEG, 2017).

In this context, in this chapter, an environmentally sustainable kitchen island that, although on a small scale of influence within the academic and commercial spheres, aims to contribute to reiterating the importance of product eco-design and to increasing the number of solutions available on the market is proposed. To this end, we have explored eco-design strategies and implemented their guidelines. With the same level of importance, this text emphasises the study of the environmental profile of materials through these eco-design tools so that the materials can be compared and the best ones selected to be incorporated into the project.

15.2 LCD THEORETICAL FRAMEWORK

The influence of sustainable development on design can be traced back to the beginnings of the modern environmental movement in the 1970s (Sherwin, 2000). Years later, the definition of sustainable development by the Brundtland Commission in 1987 encouraged designers and engineers to develop the first scientific articles focusing on environmental concerns (Vicente, 2012). Ulrich et al. (2020) state that the 1990s saw the emergence of tools that made it possible to quantitatively assess the overall environmental impact of products. Given this history, sustainable design became part of the sustainable development movement and is considered to be the design/engineering field's response to the interactions established between society, culture, and the environment over the years to the present day (Amland, 2004, cited in Vicente, 2012). This response is imperative and requires designers, manufacturers, and consumers to play a responsible role. It is up to designers to propose innovative product alternatives and up to consumers to practice these sustainable alternatives after recognising the positive effects of these conscious choices (Manzini & Vezzoli, 2002).

Sustainable product design can be defined as the balancing of economic, environmental, ethical, and social issues in the product development process (Jones et al., 2001), or, in other words, the eco-design of products aims to improve their energy and environmental performance throughout their life cycle (Tischner et al., 2000). Manzini and Vezzoli (2002) also embrace this concept, calling it Life Cycle Design (LCD). The authors state that LCD is the methodology, from a systemic view of the product, that analyses all the inputs and outputs of every stage of its life, with the aim of assessing the environmental, economic, and social consequences. From the same perspective, Ashby (2021) states that the general aim of designing the entire life cycle of the product is to reduce the environmental burden associated with all the activities required to produce, distribute, use, and dispose of that same product.

According to Manzini and Vezzoli (2002), the five LCD strategies that allow environmental aspects to be integrated into product development, that is, guide an eco-design project, are minimising resources, selecting low-impact resources and processes, optimising the product lifetime, extending the life of materials, and facilitating disassembly. The first strategy involves reducing the use of materials and energy required throughout the product's life cycle. The second strategy involves selecting materials, manufacturing processes, and energy sources with a low impact on the environment, human health, and the ecosystem. The third strategy proposes designing the product to increase its useful life and intensify its use. The fourth strategy concerns extending the life of the product's materials after it has fulfilled its function, grouping together guidelines orientated towards the final phase of the product's life cycle. Preferably the materials will be transformed several times into secondary raw materials and only then sent for incineration. Finally, the fifth strategy is a set of guidelines aimed at facilitating the separation of product components and materials that retrieve the optimisation of the products´ lifetime and the extension of the materials' lifespan strategies. According to the above-mentioned authors, the guidelines selected for an eco-design must be aligned with the objective and requirements of the project so that there are no situations of conflict between the strategies adopted and, above all, that they do not jeopardise other aspects of the product's performance, such as functionality, ergonomics, aesthetics, or safety.

In order to develop sustainable products, designers rely on a set of analysis and evaluation methods and tools that provide them with useful information for making decisions in the eco-design process (Fitzgerald et al., 2007). From the point of view of Tischner et al. (2000), eco-design tools can be used for four different purposes: analysing and evaluating the strong and weak environmental aspects of a product or concept; establishing and communicating priorities for improving the environmental aspects of a product; supporting the generation of product ideas with a low environmental impact; and coordinating with other important economic criteria. On the other hand, the tools can also be grouped into qualitative and quantitative (Vicente, 2012). The same author systematises the tools of the first group into four types: matrices, spiderweb diagrams, strategies lists, and checklists. It also highlights Life Cycle Assessment (LCA) as one of the quantitative tools.

Matrices are tables whose purpose is either to assess the environmental impacts of existing products, such as the MET matrix, or to help prioritise a project, such as the Eco-design Priority Matrix (Vicente, 2012). According to Tischner et al. (2000), spiderweb diagrams or polar diagrams are graphically distinguished by their spider web shape on which several axes are crossed, with the qualitative classification being conveyed by the area formed by these. The definition of the criterion for each axis differs depending on the purpose of using this tool, and there are several versions of this technique, for example, Brezet and Hemel's LiDS-Wheel and Dow Europe's Eco-Compass. The purpose of this technique is to provide an overview of the framework of eco-design objectives and to communicate eco-design priorities creatively and intuitively (Hemel, 1998, cited in Vicente, 2012). In turn, strategy lists are, as the name suggests, a grouping of strategies to be implemented in product eco-design. The simplicity of this qualitative tool lies in listing strategies that, in a more extensive or less focused way, allow the product development process to be guided. Checklists are tables that list a set of detailed criteria for design strategies or product life cycle phases. These criteria are then subjected to a qualitative assessment with the aim of achieving a goal or have a didactic vision of these same criteria (Vicente, 2012).

LCA is a methodology for analysing and evaluating the reciprocal relationships that occur between the product and the environment, made up of four phases: defining the objectives and scope of the evaluation to be carried out; gathering data; evaluating the impacts; and interpreting the results (Manzini & Vezzoli, 2002; Matthews et al., 2014). LCA is based on different databases relating to materials, manufacturing processes, and energy consumption, and there is software dedicated to this methodology. Despite the advantages of LCA, this quantitative tool requires large amounts of information, time, and experience to interpret the results (Vicente, 2012).

It is also worth mentioning in this section the EPD. The EPD is a technical and public document that provides information about the impact that a product has on the environment and human health. The EPD is usually provided by the product manufacturer, verified by an independent expert, is valid for five years (Bovea et al., 2019; Lapitec, 2020), and is based on the ISO 21930 standard and the scientific method of LCA. While an LCA calculates the environmental footprint of a product throughout its life cycle in an extensive and detailed way, the EPD is a simplified form of LCA, transparently displaying the assessment values for a set of parameters related to environmental impacts, the consumption and valuation of resources, and the types of waste generated (EPD International AB, n.d.). Table 15.1 details the set of environmental parameters present in an EPD to assess materials. Thus, the EPD can be seen as a practical tool to support the work of designers seeking to know the environmental profile assessment of a particular product or material to be included in an eco-design project. It can even be used to make comparisons between similar products and make the most ecological design choices.

15.3 METHOD TO COMPARE THE KITCHEN ISLAND MATERIALS

The aim of the project to be explained in this text was to create an environmentally sustainable kitchen island. After the theoretical framework presented previously, it was decided that the eco-design of the kitchen island would be done by implementing LCD strategies 1, 2, and 3, namely minimising resources, selecting low-impact resources and processes, and optimising the life of products, respectively. All three strategies were explored with creativity into the design process, although the second strategy required additional attention due to the nature of its guidelines. That is, for this eco-design project, the selection of low-impact resources and processes meant researching and choosing the most environmentally sustainable materials.

While in the design of a kitchen island the choice between numerous possible colours and material finishes is fundamental to forming the aesthetic identity of the

TABLE 15.1
Environmental Parameters Used in the Comparison of Materials

Environmental Parameter	Abbreviation	Units
Global warming potential Measure that compares the capacity of different greenhouse gases to retain heat in the atmosphere.	GWP	Kg CO_2 eq.
Ozone depletion potential Measurement of atmospheric emissions that contribute to the depletion of the stratospheric ozone layer responsible for absorbing ultraviolet radiation from the Sun.	ODP	Kg CFC-11 eq.
Acidification potential Measure that compares the ability of different chemical substances, such as sulphur oxides and nitrogen oxides, to contribute to the acidification of the environment.	AP	mol H^+ eq.
Eutrophication potential Measure that comprises the impacts on terrestrial and aquatic environments due to excessively high levels of nutrients such as nitrogen and phosphorus.	EP	Kg PO_4^{-3}eq.
Photochemical ozone creation potential Measurement of atmospheric emissions such as volatile organic compounds that, in the presence of solar radiation and nitrogen oxides, contribute to the formation of smog.	POCP	Kg NMVOC eq.
Mineral abiotic depletion potential Measure related to the extraction of virgin abiotic materials, compromising their availability in the future, considering the reserves that exist.	ADPm	Kg Sb eq.
Fossil abiotic depletion potential Measure related to the extraction of fossil fuels that are essential as a source of energy and as raw materials; however, they are finite and generate polluting emissions.	ADPf	MJ

Source: Adapted from Steven (2023).

product, consideration of the environmental profile of the materials is a priority for selecting the most environmentally sustainable ones. The present section is therefore dedicated to presenting the method used to compare the possible materials to be applied on the cupboard fronts and to compare the possible materials to be applied on the kitchen island worktop. The aim of comparing the environmental profile of the materials was to find the two materials, one for each type of application, with the lowest environmental impact so that they could be integrated into the kitchen island eco-design project to fulfil the second LCD strategy.

The method for selecting the most environmentally sustainable materials was carried out by assessing the impacts that promising materials have on the environment throughout their life cycle. Figure 15.1 illustrates the steps of the method used to compare the environmentally promising materials, which is now explained.

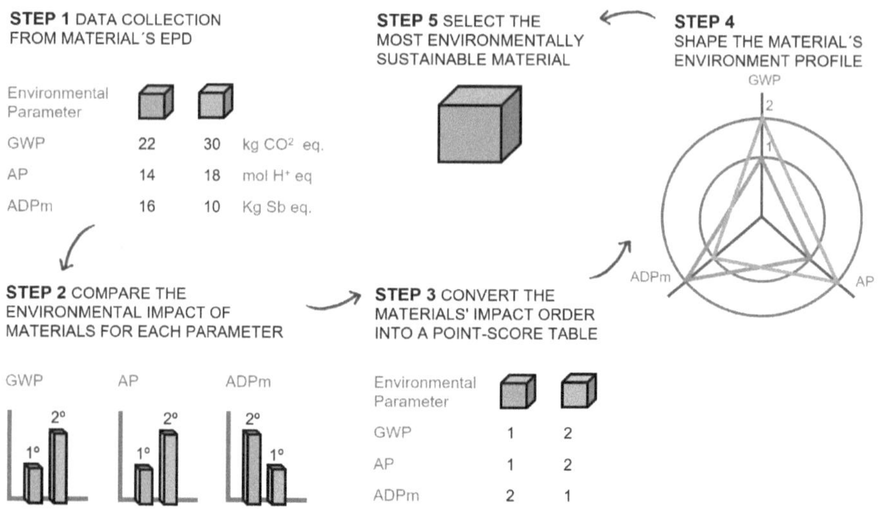

FIGURE 15.1 Example of the method used to compare the materials. (Picture by Carolina Polónio.)

The environmental profile of each material was analysed on the basis of its EPD, which was used to extract the values for each environmental parameter assessed and declared by the manufacturer (step 1). The data collected were then analysed using tables and charts to compare the environmental impact of the various materials and thus conclude which were the most environmentally sustainable for the cupboard fronts and for the kitchen island worktop (step 2).

It should be noted that as the environmental parameters have different units of measurement, a point-score equivalent system was used to make it possible to compare the different materials. In other words, this step consisted of making a bar chart for each environmental parameter in order to assign points to the materials in terms of the environmental impact they cause. In this way, the material that causes the most damage to the environment in a specific parameter takes last place and receives the highest possible point-score. Converting the units of environmental parameters into an equivalent point-score scale makes it possible to quickly conclude, by adding up the points in a table, which materials generate the most and least environmental impact (step 3). However, to find the most environmentally sustainable material, the score table must be depicted in a spiderweb diagram, especially when materials end with the same total points. This type of diagram is composed of a set of concentric levels that correspond to the possible equivalent scores and each axis corresponds to an environmental parameter. In this diagram, the points scored to each material become a perimeter of linked vertices that shape the environmental profile of that same material (step 4). In short, the most environmentally sustainable material is the one with the most points concentrated on the inner levels of the diagram, while materials with a perimeter extending to the outer levels of the diagram have an environmental profile with room for improvement (step 5). The method described for comparing materials was inspired by the 'matrix method' (p. 53) presented by Ashby (2021) and the already mentioned tools: EPD and spiderweb diagram.

15.4 APPLICATION OF LCD STRATEGIES

This section presents the practical results of the kitchen island eco-design and how the strategies and guidelines investigated were applied to design this sustainable product. Figure 15.2 summarises its main features: a water reuse system (to implement the first LCD strategy), the sustainable materials for the worktop and cupboard fronts (to implement the second LCD strategy), and an aromatic herbs nook and a sliding surface (to implement the third LCD strategy), now explained in more detail.

As for the first LCD strategy – minimising resources – the aim of the eco-design of the kitchen island was to minimise the consumption of resources during the use of the product, specifically the consumption of water needed to care for the kitchen island's aromatic herbs nook. To achieve this goal, a water reuse system was incorporated into the product. In this way, the plants are watered using the water collected in the wet channel when the dishes are drained or the food is washed.

With regard to the second LCD strategy – selecting low-impact resources and processes – the results of the method for comparing environmentally promising materials are now presented, firstly for cupboard fronts and secondly for the kitchen island worktop. The environmentally promising materials to be incorporated into kitchen cupboard fronts were Solid Panel from Moso® (bamboo plywood), painted material depletion factor (MDF) from Ballingslöv®, melamine Particleboard (PB) from Sonae Arauco®, and thermolaminated PB from Polyrey®. We began by consulting the EPD of each material in order to extract the value of its impact on the various environmental parameters. The compilation of those data is shown in Table 15.2.

As the environmental parameters had different units of measurement and the order of scale of the values is not the same, a bar chart was drawn up for each environmental parameter assessed in order to compare the impact caused by each material. Figure 15.3, for example, illustrates how much the materials contribute to global warming potential. After analysing them, it was possible to conclude that painted MDF was the material with the greatest impact on this environmental parameter

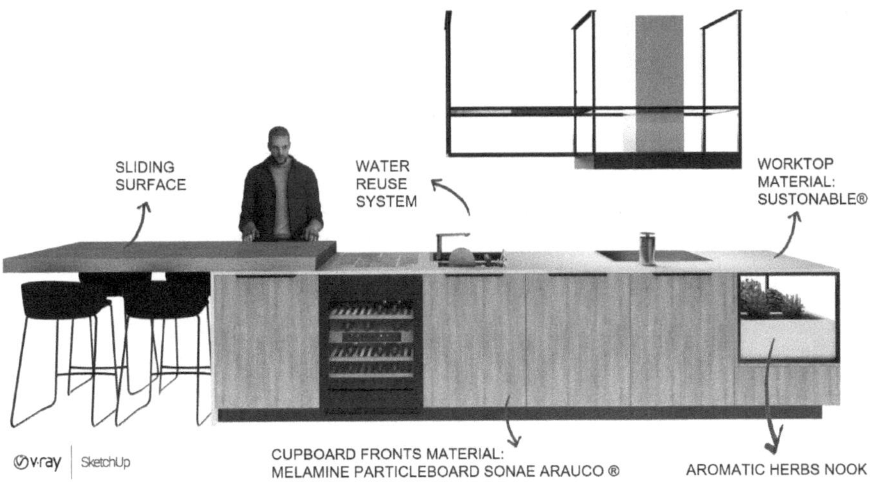

FIGURE 15.2 The designed kitchen island and its main features. (Render by Carolina Polónio.)

TABLE 15.2
Data Collection of the Environmentally Promising Materials for Cupboard Fronts

Environmental Parameter	Units	Solid Panel	Painted MDF	Melamine PB	Thermolaminated PB
GWP	Kg CO_2 eq.	9,53E-02	1,26E+00	1,84E-01	5,47E-01
ODP	Kg CFC-11 eq.	9,27E-09	[−]	−7,18E+00	3,29E-08
AP	Mol H^+ eq.	4,16E-03	1,23E-02	3,79E-03	1,27E-02
EP	Kg PO_4^{-3}eq.	1,00E-05	6,25E-04	4,65E-04	1,80E-03
POCP	Kg NMVOC eq.	3,86E-03	5,45E-03	3,83E-04	9,86E-04
ADPm	Kg Sb eq.	5,63E-06	1,36E-07	2,98E-06	2,96E-06
ADPf	MJ	−2,40E+00	2,25E+01	−3,26E-02	1,26E+01

Source: Table by Carolina Polónio.

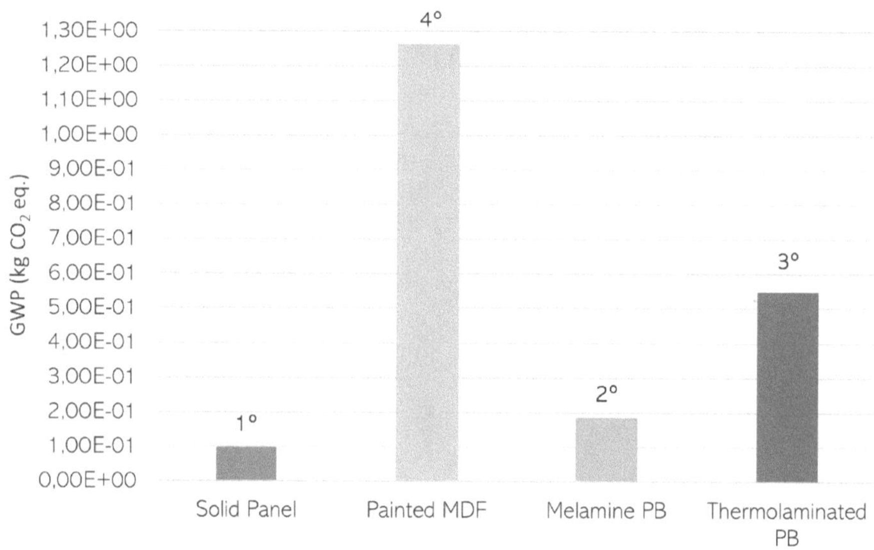

FIGURE 15.3 Environmental impact of the different materials for cupboard fronts relative to the GWP parameter. (Bar chart by Carolina Polónio.)

and Solid Panel was the material with the least impact. In this way, it was possible to order the materials by the scale of their impact, which was equivalent to a numeric point-score.

The same procedure was applied to the other environmental parameters, so it was possible to group all the scores generated by the different graphs in Table 15.3. It should be noted that the ozone depletion potential (ODP) parameter was not included in the material´s point scoring because the assessment of this environmental parameter was not declared in the EPD for Ballingslöv's painted MDF.

TABLE 15.3
Point-Score Table for the Cupboard Fronts Materials

Environmental Parameter	Solid Panel	Painted MDF	Melamine PB	Thermolaminated PB
GWP	1	4	2	3
AP	2	3	1	4
EP	1	3	2	4
POCP	3	4	1	2
ADPm	4	1	3	2
ADPf	1	4	2	3
Total	12	19	**11**	18

Source: Table by Carolina Polónio.
The bold value in Table 15.3 highlights the material (Melamine PB) with the lowest total environmental impact score among the options evaluated. It indicates the most environmentally favourable cupboard front material according to the criteria used.

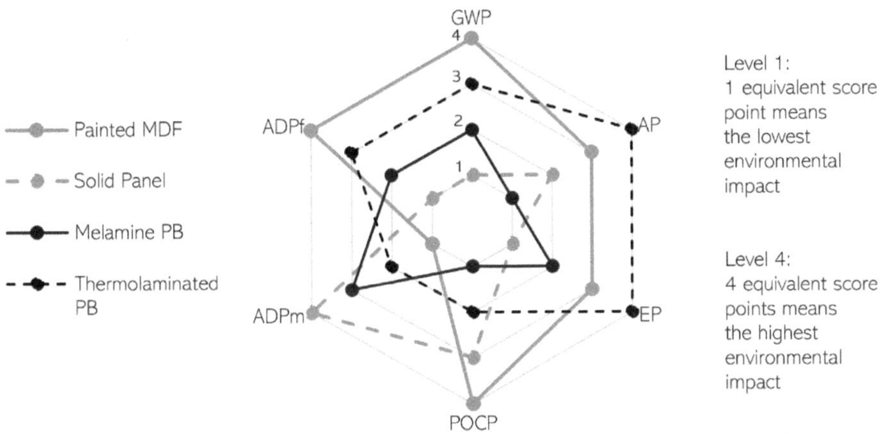

FIGURE 15.4 Environmental profiles of the cupboard fronts materials. (Diagram by Carolina Polónio.)

A horizontal reading of Table 15.3 showed that the materials had equal scores in the parameters: global warming potential and fossil resource depletion potential. In this general overview, the ascending order of the materials according to their total score was melamine PB, Solid Panel, thermolaminated PB, and painted MDF. By analysing Table 15.3 and its graphical representation (Figure 15.4), it was concluded that melamine PB and Solid Panel were the materials that contributed the least to environmental problems and that, on the other hand, painted MDF and thermolaminated PB were the materials that contributed the most.

Solid Panel was the material with the least environmental impact on the ADPf, global warming potential (GWP), and eutrophication potential (EP) parameters, while at the same time, it was one of the materials with the greatest impact on

ADPm and photochemical ozone creation potential (POCP). Solid Panel's disparate point-score was reflected in its broader profile (greater environmental impact) than that of melamine PB. The PB melamine's environmental profile was drawn closer to the inner levels of the graph, its vertices were not as dispersed, and, as a result, melamine PB was the most environmentally sustainable material for kitchen island cupboard fronts.

On the other hand, the set of environmentally promising materials to be incorporated into the kitchen island worktop were stainless steel, Richilite® (phenolic composite made from recycled paper), Durat® (100% recyclable polyester-based solid surface), Sustonable® (100% recyclable material made from recycled photochemical ozone creation potential [PET] plastic), Silestone® (quartz surface), Dekton® (carbon-neutral ultracompact surface), and Neolith® (ultracompact surface with 100% natural raw materials based on granite and glass minerals). As was the case to comparing the environmental profile for the cupboard fronts materials, the same method was used to select the most environmentally sustainable material for the kitchen island countertop. The equivalent scores given to the set of promising materials are compiled in Table 15.4.

A horizontal reading of Table 15.4 showed that the materials had very similar scores in the GWP and ODP parameter pair and in the acidification potential (AP) and EP pair. In this general overview, the ascending order of the materials according to their total score was Sustonable and Dekton, Silestone, Neolith, stainless steel, Richlite, and Durat. Looking at Table 15.4 and its graphical representation (Figure 15.5), it was concluded that Sustonable and Dekton were the materials that contribute the least to environmental problems and that, on the contrary, Durat was the material that contributes the most. The environmental parameters that contradict this generalised conclusion were AP, EP, and ADPf, where Dekton, Neolith, and

TABLE 15.4
Point-Score Table for the Kitchen Island Worktop Materials

Environmental Parameter	Stainless Steel	Richlite	Durat	Sustonable	Silestone	Dekton	Neolith
GWP	4	6	7	1	3	2	5
ODP	4	5	7	1	3	2	6
AP	7	5	6	3	2	1	4
EP	7	6	5	3	4	2	1
POCP	4	3	7	1	6	2	5
ADPm	6	4	7	1	3	2	5
ADPf	1	6	7	3	4	2	5
Total	33	35	46	**13**	25	**13**	31

Source: Table by Carolina Polónio.

The bold values in Table 15.4 indicate the materials with the lowest total environmental impact scores across all parameters. In this case, Sustonable and Dekton, each with a total score of 13, were identified as the most environmentally favourable options.

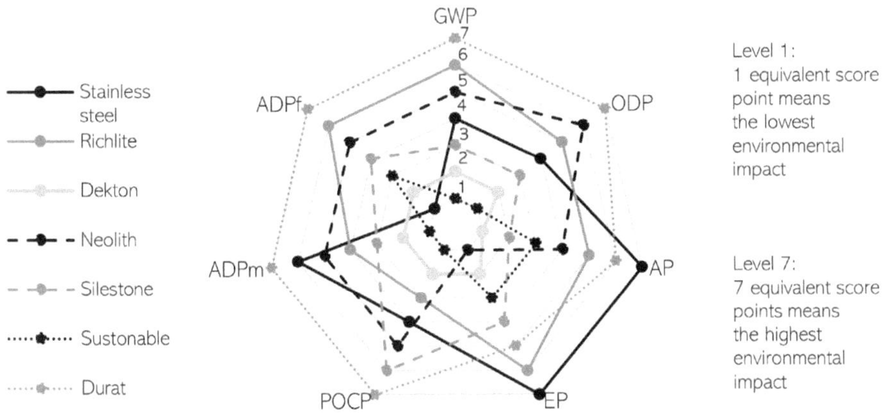

FIGURE 15.5 Environmental profiles of the kitchen island worktop materials. (Diagram by Carolina Polónio.)

stainless steel were the materials with the least environmental impact, respectively. Similarly, Durat was not always the material with the greatest environmental impact, as the highest score in the AP and EP parameters was attributed to stainless steel.

The spiderweb diagram of environmentally promising kitchen worktop materials has seven concentric levels corresponding to the equivalent score range of 1 (least impact) to 7 (highest impact). The comparison of the environmental profile (the figure shaped by the perimeter) of each material allowed us to conclude which material is the most environmentally sustainable, even when Sustonable and Dekton obtained the lowest and equal total scores compared to the other materials. Dekton material has the same classification in practically all environmental parameters evaluated (with the exception of Acidification Potential). Graphically, the environmental profile of this material was limited to the second score level, which means that Dekton is the second material with the lowest impact on environmental parameters. The Sustonable material occupied the first position in most of the environmental parameters evaluated; that is, it was the material that caused the least environmental impact in global warming, in the depletion of stratospheric ozone, in the formation of photochemical ozone, and in the depletion of mineral resources. In the remaining environmental parameters evaluated (ADPf, AP, and EP), Sustonable had a greater impact than Dekton; however, the environmental profile stretched up to level 3 in these aspects should not be overvalued because other materials overlap (including Dekton). For example, stainless steel and Neolith occupied the lowest environmental impact position in the ADPf and EP parameters, respectively. Therefore, as Sustonable concentrated most of its vertices (points) at the lowest level of environmental impact and Dekton's profile was practically drawn on level 2 of the graph, it can be concluded that Sustonable was the most environmentally sustainable material for the worktop of the kitchen island.

Several guidelines were taken into consideration when implementing the second LCD strategy, such as avoiding toxic and environmentally dangerous materials achieved through Sustonable being the composite stone with the lowest carbon footprint (Sustonable Data EPD, 2022); using recycled materials separately or together with other virgin materials achieved through Sustonable being the first material made with recycled PET plastic bottles (100 bottles for each m^2 of material); using renewable materials achieved through melamine PB being a material made with wood; and using materials from production process waste achieved through melamine PB being a material made from by-products of the flax, hemp, straw, and bagasse industries.

And lastly, the third LCD strategy – Product Lifetime Optimisation – implemented in the eco-design of the kitchen island took into consideration the following guidelines: designing reconfigurable and/or multifunctional products and facilitating their adaptability, manifested by the incorporation of a sliding surface. Thinking about the different needs that the user has in their daily life, with this component, the kitchen island presents a multifunctional dimension that allows it to adapt to different context scenarios such as cooking, eating, socialising, and working. Furthermore, as the sliding surface can easily hide the kitchen island top, thanks to the user-friendly concealed telescopic slides system 'Breakfast Top' (Emuca, n.d.), the kitchen island, for example, when not cooking, flows naturally from the kitchen space to the living room.

The creation of an empathetic connection between the user and the product means that, when well maintained, the product's lifetime is optimised (greater durability) and fewer resources are used to make new products. Therefore, creating a horticultural nook in the kitchen island calls for the user to take care of the aromatic herbs, which can awaken a sense of responsibility and assign it a meaning with a value associated with the product. Consequently, the incorporation of an aromatic herbs nook in the kitchen island also responds to the third strategy of LCD since the maintenance relationship that the user has with the product will contribute to the preservation of the kitchen island for as long as possible, that is, the optimisation of the product's lifetime.

15.5 CONCLUSION

In accordance with current guidelines of environmental sustainability, it is concluded that if the practice of eco-design is imperative and irrevocable to minimise the planet's environmental crisis, it is also important that the methodologies and tools of eco-design are widely disseminated among professionals in this area so that they are increasingly used and explored with creativity by them.

This eco-design project is the result of applying strategies and practices aimed at developing sustainable products, expressed in this case by incorporating a water reuse system to minimise resource consumption (first LCD strategy), environmentally sustainable materials for the kitchen island to select low-impact resources and processes (second LCD strategy), and a nook for aromatic herbs and a multifunctional sliding surface to optimise the product's lifetime (third LCD strategy).

This collaborative research with Design and Engineering showcases a very privileged position, serving consumers and working with the industry, intending to contribute to a more sustainable future for the planet. In this sense, we hope that the project of this kitchen island highlights the need for society's transition towards sustainability and the urgent dissemination of the aforementioned concepts, intentions, and aspirations.

ACKNOWLEDGEMENTS

Carolina Polónio is the author of the kitchen island project.

This work was financed by national funds through the FCT – Fundação para a Ciência e a Tecnologia, I.P., under the scope of the project UIDB/04057/2020.

SketchUp 3D modelling software was used under the licence of the SketchUp Studio product purchased with the first author´s subscription for academic purposes only.

All the product representation simulations were carried out under a licence for the software V-Ray Solo software licence purchased with the same subscription and customer number 7F78WRWDR4FC.

This chapter was supported by the projects UIDB/00481/2020 and UIDP/00481/2020- Fundação para a Ciência e a Tecnologia, DOI 10.54499/UIDB/00481/2020 (https://doi.org/10.54499/UIDB/00481/2020) and DOI 10.54499/UIDP/00481/2020 (https://doi.org/10.54499/UIDP/00481/2020).

REFERENCES

Afonso, C., Gavilan, D., García-Madariaga, J., & Gonçalves, H. M. (2018). Green consumer segmentation: Managerial and environmental implications from the perspective of business strategies and practices. In A. Leal-Millan, M. Peris-Ortiz, & A. L. Leal-Rodríguez (Eds.), *Sustainability in Innovation and Entrepreneurship* (pp. 137–151). Springer International Publishing. https://doi.org/10.1007/978-3-319-57318-2_9

Ashby, M. F. (2021). *Materials and the Environment Eco-Informed Material Choice* (3rd ed.). Oxfor & Cambridge, MA: Butterworth-Heinemann.

Bovea, M., Alberola, M. M. C., Izquierdo, A. G., & Mendoza, F. J. (2019). *Declaración ambiental según ISO 14025: una oportunidad para informar de los aspectos ambientales de los servicios de gestión de residuos. Gestión de Residuos*, 20.26. https://core.ac.uk/download/pdf/61392106.pdf

Casais, M. (2011). *Design for the "New Consumer": The Green Kitchen as a Catalyst for Sustainable Living* [Master's thesis, Universidade Técnica de Lisboa, Repositório da Universidade de Lisboa]. https://hdl.handle.net/10400.5/3939

Charter, M. (1998). Welcome to the seventh issue of The Journal of Sustainable Product Design. *The Journal of Sustainable Product Design*, *7*, 5–6. https://proyectaryproducir.com.ar/public_html/Seminarios_Posgrado/Material_de_referencia/ABB%20-%20Ecoeffective%20Product%20Design.pdf

DGEG. (2017). *Ecodesign e Etiqueta Energética*. https://www.dgeg.gov.pt/pt/areas-setoriais/energia/eficiencia-energetica/ecodesign-e-etiqueta-energetica/

Edwards, A. R. & Orr, D. W. (2005). *The Sustainability Revolution: Portrait of a Paradigm Shift*. Gabriola Island, BC: New Society Publishers.

Emuca. (n.d.). *Corrediças para mesa deslizante Twice Top*. https://www.emuca.pt/cozinha/mesas-extensiveis-cozinha/mesas-deslizantes-cozinha/mesa-deslizante-twice-top/emuca-mesa-deslizante-twice-top-para-cozinha-aluminio-anodizado-mate

EPD International AB. (n.d.). *Environmental Product Declarations*. https://www.environdec.com/all-about-epds/the-epd

Fitzgerald, D., Herrmann, J., Sandborn, P., Schmidt, L., & Gogoll, T. (2007). Design for environment (DfE): Strategies, practices, guidelines, methods, and tools. In *Environmentally Conscious Mechanical Design* (pp. 1–24). https://doi.org/10.1002/9780470168202.ch1

Hofer, R. (2009). History of the sustainability concept - renaissance of renewable resources. *RSC Green Chemistry*, *4*, 1–11. https://www.researchgate.net/publication/286374614_History_of_the_sustainability_concept_-_Renaissance_of_renewable_resources

IKEA Museum. (2021). *Kungsbacka*. https://ikeamuseum.com/en/explore/product-stories/kungsbacka-kitchen-fronts-2017/

Jones, E., Harrison, D., & McLaren, J. (2001). Managing Creative Eco-Innovation – Structuring outputs from Eco-innovation projects. *The Journal of Sustainable Product Design, 1*(1). https://doi.org/10.1023/A:1014494005565

Lapitec. (2020). Sustainability Summary Lapitec. https://www.lapitec.com/ContentsFiles/Lapitec-SustainabilitySummary(4).pdf

Manzini, E., & Vezzoli, C. (2002). *Desenvolvimento de produtos sustentáveis* (1st ed.). São Paulo: Editora da Universidade de São Paulo (Edusp).

Matthews, H. S., Hendrickson, C., & Matthews, D. H. (2014). *Life Cycle Assessment: Quantitative Approaches for Decisions that Matter*. https://lcatextbook.com/

Sherwin, C. (2000). *Innovative Ecodesign: An Exploratory and Descriptive Study of Industrial Design Practice* [Doctoral dissertation, Cranfield University, Cranfield CERES]. https://hdl.handle.net/1826/3543

Sherwin, C., Bhamra, T., & Evans, S. (1998). The 'eco-kitchen' project – using eco-design to innovate. *The Journal of Sustainable Product Design, 7*, 16–25. https://pure.royalholloway.ac.uk/en/publications/the-eco-kitchen-project-using-ecodesign-to-innovate

Steven. (2023). *Impact Assessment Categories (CML, TRACI and PEF)*. One Click LCA. https://oneclicklca.zendesk.com/hc/en-us/articles/360015036980-Impact-Assessment-Categories-CML-TRACI-and-PEF-

Sustonable Data EPD. (2022). *Environmental Product Declaration*. https://www.sustonable.com/wp-content/uploads/2025/04/EPD-IES-0022352-Sustonable-rGlassrPETFeldspar-Circular-Design-Surface_220425_compressed.pdf

Tischner, U., Masselter, S., & Hirschl, B. (2000). *How to Do EcoDesign?: A Guide for Environmentally and Economically Sound Design*. German Federal Environment Agency. https://archive.org/details/howtodoecodesign0000unse/page/n5/mode/2up

Ulrich, K. T., Eppinger, S. D., & Yang, M. C. (2020). *Product Design and Development* (7th ed.). New York, NY: McGraw-Hill Education.

Vicente, J. (2012). *Contributos para uma metodologia de design sustentável aplicada à indústria do mobiliário. O caso português*. [Doctoral dissertation, Universidade de Lisboa, Repositório da Universidade de Lisboa]. https://hdl.handle.net/10400.5/5645

Yoneda, Y. (2010). *Whirlpool's Green Eco System Kitchen Debuts in Milan*. Inhabitat. https://inhabitat.com/whirlpools-green-eco-system-kitchen-debuts-in-milan/

Zirnstein, G. (1996). *Ökologie und Umwelt in der Geschichte* (Reihe Geschichte, Vol. 080). Marburg: Metropolis Verlag.

16 Neo-Rural Bioconstruction
Bridging Past and Present in Asturias

Llara Fuente Corripio

16.1 INTRODUCTION

The rural exodus to urban centers, spurred by industrialization and intensified, particularly from the mid-20th century onward, was one of the triggers for significant transformations in the rural environment. Industrialization does not only impact the economic aspect in terms of opportunities, but in the imagery of this rural exodus, the countryside is perceived as synonymous with precarity. The trauma of the past gives way to a more appealing perspective. The demographic loss was linked to a shift in traditional agricultural functions, which also underwent industrialization, diminishing its prominence compared to the secondary sector. Population loss has further consequences, such as the gradual decline of services and the lack of infrastructure. From the 1970s, in an inverse manner, there has been a gradual return to the countryside motivated by a dual crisis documented by various authors, the economic crisis and a crisis of values associated with the deficiencies and excesses of urban life. Although this repopulation occurs on a much smaller scale than the rural exodus, it will have a significant impact on the ecosystem of rurality.

The population exchange between the city and the countryside is not the only one occurring, and spaces become permeable. Various authors document how rural areas take on attributes of urban settings, incorporating functions required by cities. Notably, the tertiarization of space linked to tourism stands out, but also other forms of leisure associated with cultural and natural heritage. This includes the establishment of industrial zones with factories and large shopping centers, especially in the rural–urban interface areas. The analysis of this exchange is prominently featured in the work of Roseman et al. (2013).

To address the study of these changes, concepts and theories will be formulated and developed, some focused on the economic aspect, such as the Multifunctionality of Agriculture, and others like the New Rurality,[1] which, along with the change in functions, focus on social changes and the responses and relationships produced by the local and the foreign. It is through the concept of New Rurality that issues such as new ways of inhabiting the territory and the meanings of the rural are considered.

DOI: 10.1201/9781032719290-19

Within this context, the focus of this study is directed toward the investigation of neo-rural bioconstruction and its philosophy, analyzing the impact of this trend. While necessarily related to many of the concepts mentioned earlier regarding the general causes and consequences of these initiatives, the study focuses on points connected with cultural and natural heritage. Direct and indirect repercussions are considered: the integration of the constructions into the environment, the dialogue between new and pre-existing constructions, materials, techniques, and typologies, the use of elements and knowledge, transmission methods, and the revival of cultural practices.

The study area is centered on those regions of Asturias where neo-ruralism has a stronger presence—Cabranes, Villaviciosa, and Piloña. However, the similarity to occurrences in other territories in the national and European panorama allows many of the discussed aspects to be extrapolated to various geographical contexts. The case of neo-ruralism in Asturias significantly reflects the implications of the changes in territorial dynamics. As one of the regions in the peninsula with the largest rural area, the highest demographic decline, and the greatest aging population, in recent years, it has experienced a new influx of population linked to the neo-rural phenomenon. This trend is particularly noticeable in the aforementioned territories. The causes are varied: the attraction of those already settled in the area, increased administrative facilities, the appeal of the region as it combines a purely rural environment with notable accessibility compared to other areas, and cultural dynamics that encourage a reconnection with a more sustainable and authentic way of life. This demographic resurgence not only reconfigures economic and social activities, traditionally based on agriculture and livestock (García, 2008, 2011), but also has a direct impact on the landscape (Hernández, 2009; Gómez 2013) as the revitalization of villages, the restoration of traditional buildings, and the use of specific construction techniques are transforming the rural area. Among the various cases examined, in addition to housing, neo-rural bioconstruction projects connected to other practices will also be addressed. A notable example includes those related to the museological and tourism sectors, two areas that are often interconnected. Tourism, in particular, stands out as one of the primary drivers of economic growth in the region, bringing both benefits and new challenges. In this context, the analysis will also focus on emerging tourism practices that align more closely with sustainability, benefiting both the community and the environment (Grande, 2006; Herrera et al. 2005). These alternative models offer a more balanced approach to economic development while preserving the integrity of the local territory.

16.2 METHODOLOGY

To address the study theme, a dual approach research has been conducted: documentary and fieldwork. A review of existing literature from various disciplines has been carried out, focusing on key aspects: issues and phenomena associated with rural environments, neo-ruralism, bioconstruction, traditional housing, related crafts, and rural landscapes. Scientific articles from various repositories, doctoral theses, and books have been used. We have attempted to use recent bibliography to ensure a contemporary perspective, although some of the documents used also conduct reviews of past literature. Likewise, we have sought to focus on Spanish publications. Some of the consulted books are widely used manuals by neo-rurals, such as Bueno (1998) or Khan (2008), making their consultation relevant to understanding their theoretical foundations. To update some demographic data, regional and national statistical databases have been consulted. Furthermore, fieldwork has been carried out in the region of Asturias. The interview format was in-person and semi-structured. Three study groups were focused on architects and educational institutions, neo-rural inhabitants and/or builders, and the local population. The selection criteria for informants aimed to encompass both the perspectives of long-time residents and newcomers, as well as perspectives from the native residents.[2] Throughout all stages of the study, there has been permeability, comparing the obtained data for subsequent analysis, allowing us to draw conclusions that take into account the diversity of factors and agents involved.

16.3 THE NEW RURALITY

'The onset of the neo-rural mode has been associated with the 1960s, the counterculture and the hippie movement in the United States, as well as the so-called May '68 revolution in France' (Roseman et al., 2013, p. 17). Over time, the neo-rural movement will incorporate new ethical aspects that, while related to social and environmental issues at its genesis, also bring novelties. It is important to consider that this is not a static philosophy, despite common points throughout its development: rejection of the capitalist economic system, its rhythms, and the frenetic and linear consumption, affinity with the environmental movement, or simply a connection with nature. Similarly, there are periods in which neo-rurality has a greater presence, often coinciding with specific historical moments. In this sequential development, crises mark key moments. The clearest example is found in economic crises, precisely linked to issues within the system that are often identified with urban problems, from which there is a desire to escape or from which there is no choice but to escape. In the words of Rodríguez & Trabada, 'Escape, from what? From where?... From the city and everything it symbolizes as excess or deficiency' (1991, p. 74). In the case of Spain, it is from the economic crisis of 1975 onwards that the neo-rural movement gains momentum, and the countryside experiences an increase in its population. Another example, very recent in this case, is the revaluation of the rural environment that occurred after the health crisis of the COVID pandemic in 2019.[3] Another factor is the idealized imagery surrounding the countryside, whether through text or image. In his study on Asturian housing, Rodríguez Pérez (2021) highlights how, from early times and various areas, there has been an idealization of the rural. This is achieved through texts and paintings associated with regionalist movements (Litvak, 1911), or from architecture, with the dissemination of picturesque housing types. Short (2006) and Bell (2006) also refer to the recent role of mass media, whether through magazines, series, movies, or even music. In addition, we must consider the predominant and decisive role of social media today. Through all these channels, the rural is presented as bucolic, a refuge from the difficulties of modern life, overlooking the inherent problems and precarization of the rural environment. Finally, in the return to rural life, we must not forget the importance of improvements in transportation or communication technologies. However, not all returns to the countryside are related to what is generally classified as neo-rural, but they do form part of the dynamics of the New Rurality: retirees who decide to settle in the village after having spent their working lives in the city; or occasional returns, such as tertiary sector workers who settle during the summer season, as well as urban workers with a second residence (Roseman et al., 2013; Rodríguez Eguizabal & Trabada Crende, 1991).

When describing neo-rurals, it is essential to consider various aspects. First, the majority belong to a young or middle-aged social group, have medium or higher education, and an economic position, at least, and are stable when opting for a lifestyle change. However, there are exceptions. For instance, establishing in rural areas temporarily is a response to economic difficulties. Also, we can find differences in the forms of settlement in the territory: those who settle in communes, closely linked to the early stages of the movement; those who opt for a communal life, sharing some elements in contrast to the prevailing individualism in the city; and those who retire with their partner or family unit (Nogué i Font, 1988), with individual retirement being an option that has become more popular in recent decades. Another difference lies in the type of economic activity performed, some of which were documented by Rodríguez & Trabada as early as 1991 and are still observed today, as evidenced by the fieldwork conducted. On one hand, there are those who maintain their jobs in the city but reside in the countryside, constituting the minority. On the other hand, there are digital nomads who, like the former group, have a home in the rural environment but typically do not integrate into the local community. Lastly, there are those with occupations related to the environment. These include the revival of agriculture, horticulture, beekeeping, or livestock farming—with emphasis on sheep and goat farming—those linked to traditional gastronomy—such as breadmaking, pastry, cheese production, etc.—craftsmanship—leather, ceramics, wood, textile, jewelry, stonework, cosmetics, etc.—and workshops focusing on various areas such as craftsmanship or bioconstruction. A special mention goes to neo-rurals engaged in tourism-related activities, often framed in new forms such as rural ecotourism. Additionally, completely new activities have emerged, such as those associated with new age culture, like yoga centers. Perhaps, as the authors support and as has been confirmed through interviews, this revival of traditional practices is in line with the need to restore balance with nature, committing to its care, after exploitation from the city, and in pursuit of maximizing economic benefits. Regarding the various forms of returning to the rural that we have outlined, it is illustrative to consider the habitat scheme for the Asturian case proposed by García Martínez (2008). Using several concentric circles, the anthropologist delimits and distributes various rural areas, with constructions and housing at the center. Moving outward, we find the areas closest to the homes, including threshing floors and orchards. Further away, there are 'erías' (uncultivated areas), followed by meadows, and, finally, mountains. Based on this scheme and the research conducted, we can observe how the repopulation of rural areas is not always done comprehensively. Often, the return to the countryside focuses solely on the first circle, corresponding to housing and, at most, the nearest constructions that, sometimes, when they serve economic or social functions from the past, end up being forgotten. The rural area transitions from being a continuum with interconnected elements, with a global use and care of the territory, to focusing on a single component, residential, detached from the environment. The key actors for the study are those who are more closely connected to the territory, as a potential agent of change that reverses heritage loss and contributes to its recovery. However, this recovery is not mimetic; it involves reinterpretations and the incorporation of new concepts, such as sustainability, bioconstruction, design, or ecotourism. Yet, even in these aspects, we can find agreements. In the words of Fernández Ans & Molina Huelva, 'Vernacular architecture already implicitly carries passive strategies in response to the environment' (2013, p. 102).

Despite the aforementioned, the return to rural areas should be approached with caution, avoiding exaggeration. Despite media coverage, as a phenomenon repeatedly highlighted by the media, the population balance continues to experience a noticeable imbalance. In Asturias, despite a considerable increase with a total increment of 1114 inhabitants in 2023 compared to 2022, the growth remains concentrated in urban areas, while the rural sector continues to lose population. Additionally, there is a second imbalance among the rural areas themselves, with a higher population in those close to the cities in the central area (IDEPA, 2023; INE, 2023; Datos Macro, 2023).

16.4 BIOCONSTRUCTION, TRADITIONAL CONSTRUCTION, AND CRAFTSMANSHIP

Despite the population increase, it has not occurred on a large scale. The difficulties in carrying out repopulation are numerous, but we will focus on the issue directly affecting the bioconstruction movement: the complexity of access to housing in rural areas. Initially, most neo-rurals used to opt for leasing or illegal occupation. However, due to apprehension about instability, interest in property acquisition has been progressively growing. In Asturias, we encounter a challenging scenario, both for renting and buying: numerous houses used as second homes or rural accommodations; many complete abandonments with property problems due to inheritances; and houses available but in very bad condition, requiring a significant economic investment for restoration and rehabilitation. In this situation, bioconstruction emerges as an alternative for middle-class neo-rurals looking to restore or build a new eco-friendly home and also for young individuals opting for self-construction due to economic reasons. Many times, the reason for opting for new construction, despite the perspective of sustainability, where the most ethical choice would be to rehabilitate existing structures, is the cost that can sometimes exceed that of new construction. Therefore, rehabilitation or restoration through bioconstruction linked to self-construction emerges as an interesting alternative.[4] This approach helps lower construction costs and, in the long run, reduces the home's maintenance expenses. It's essential to clarify here that, contrary to the widespread perception that bioconstruction, such as straw bale construction, is more economical, it's important to note that while significant reductions in material, transportation, and energy-related costs are achieved, labor costs increase. A comparative analysis from the administration of Lausanne, Switzerland, is noteworthy. Using a graph, the costs of a wall constructed with bioconstruction techniques are compared with a concrete wall. With the same budget, in the former, 98% is allocated to the artisan performing the work, while only 2% is reserved for tools, transportation, materials, etc. On the other hand, in reinforced concrete, labor costs would be around 45%. This, coupled with the shortage of professional builders, often associates self-construction with bioconstruction. However, the research indicates an interest from a percentage of groups in hiring professionals for these tasks. This aspect is particularly relevant in our study, as it would lead to a redistribution of economic resources, highlighting the significance of artisan work.

In bioconstruction, there isn't a predetermined typology or material, with notable differences and formulas,[5] more or less adapted to the territory. However, we can provide a clear definition of the concept. The term originated in Germany in the 1970s and reached Spain in the 1990s through pioneering architects like Petra Jebels, publications, and associations. Gradually, with the integration of the internet into daily life, the digital realm became a channel for interaction and dissemination. While various terms exist, bioconstruction encompasses the meanings of all these terms (Neila González, 2004). It's crucial to make a particular differentiation from the contemporary term 'passive house', widely used in current architecture. A passive house focuses on the energy efficiency of the building and its energy consumption, while bioconstruction takes a process-oriented approach. It considers the overall impact, from material extraction and use, which should involve minimal transformation, to transportation, preferring local materials, to construction and its costs—energy, water, etc.—to the use of the dwelling during its lifespan, and potential dismantling without waste production. Additionally, bioconstruction also considers finishes and interiors, taking care with paints, varnishes, etc., avoiding or reducing toxins. With the contemporary change in lifestyle, the number of hours spent inside buildings has increased significantly. The interior serves as a place for rest, leisure, and work (Lage et al., 2015), noticeably reducing our outdoor activities. The World Health Organization defines Sick Building Syndrome as a condition where a building may harm our health. According to the organization, 30% of new or renovated buildings suffer from this pathology. Finally, bioconstruction must adapt to the terrain and the inhabitants' needs, with conscious designs and space distribution. One could draw a connection between the passive house concept and high-tech, with high technology as a tool to seek solutions, as opposed to bioconstruction, generally more linked to low-tech, where existing knowledge is leveraged but with a look to the past, implementing elements that can be built, maintained, and repaired (Mandala, personal communication, November 19).[6] Traditional construction, often perceived in the collective imagination as inherently precarious, could serve as an inspiration, revaluing construction systems adapted to a specific space and time. Bioconstruction doesn't aim to alter the autochthonous but contributes to its improvement (De Carvalho, 2015).

The greatest concordance regarding works carried out in bioconstruction is in the use of materials, with a predominance of wood, straw, clay, and occasionally stone. Among these, straw is the most commonly used material, although it is not always employed with the same technique, highlighting straw infill and construction with straw bales using the Nebraska or Greb system.[7] While traditional houses made of straw are not found in the region beyond vegetal coverings, the use of straw is documented in other regions. Actually, the straws usually come from Castilla y León. The rise of straw bioconstruction has also led to a certain increase in the price of the material. However, to replace it with more locally sourced materials, experiments are being conducted with dried ferns, algae, or wood shavings to utilize carpentry waste, among other solutions for infill straw. From the neo-rural perspective, there is also an interest in engaging with formal education and academia. A clear example is the tests that will be conducted with the Department of Materials at the University.

This technique involves a mixture of plant fiber with a clay slip, placing two layers, one interior and one exterior, within the light substructure with wooden studs, which can be removed or act as a permanent formwork, with structures about 15 cm thick. Concerning traditional housing, examples of vegetal elements like maize have been found in infill. Additionally, it is a bioclimatically interesting technique, as clay, due to its hygroscopic capacity, absorbs a large amount of moisture from the environment, achieving great thermal comfort in indoor spaces. It also serves as a thermal accumulator, absorbing heat during hours of solar radiation to release it gradually afterward. In any case, large overhanging eaves are required to protect the facades, especially considering the climatic conditions of Asturias with abundant rain and wind (Figures 16.1 and 16.2).

Although self-construction, as mentioned, is often a prominent choice among self-promoters, we can classify different types of practices: self-construction—whether carried out autonomously, with self-construction guidance, or through some form of community collaboration or volunteering—and construction by professionals, often with a lack of architecture studies and specialized builders. Many times, we find solutions that, while seeking materials considered sustainable, are not always local, crafted, or linked to traditional methods, and they are more related to the passive house concept. Here, we include constructions with straw using industrialized and often prefabricated wooden structures, made from imported wood species such

FIGURE 16.1 Construction process of a bioconstruction house using the straw-bale infill technique. Macario Iglesias, Jorge Rojas and Maria Arce.

FIGURE 16.2 Traditional constructions that can be linked to self-construction including maize fillings, and plaster finishes. Macario Iglesias, Jorge Rojas and Maria Arce.

as pine or spruce treated or contemporary solutions such as oriented strand board boards. In terms of appearance, these houses are often given a more contemporary look, with straight and geometric lines. On a different note, in self-construction, it is essential to delve into a particular dynamic: the houses that are based on more or less standardized plans, often taken from the internet by self-builders, which a priori incorporate elements that clash with the traditional. This could be related to other types of constructions that have also proliferated in the landscape, such as tipis or domes. Among all the models, we can highlight the Earthship by Michael Reynolds, a typology born in the desert of New Mexico and widely disseminated by the media. Despite this, we also find typologies that may be linked to pre-existing constructions in the area, also in more organic forms, such as the circular-plan thatched roof houses known as 'teitos'. Generally, many of these constructions are perceived with a pejorative view, although we can find examples of similar organicist architectures considered part of high culture. While in the collective imagination, bioconstruction was associated with these concepts, the perception of it is changing, and it is beginning to be used in numerous public buildings in Europe. The most interesting approach regarding heritage recovery is the one that links bioconstruction with traditional construction. This can be carried out by self-builders with a special sensitivity, as well as by some architects and builders.

While it is not the object of this study, if we want to make a comparison between bioconstruction and traditional construction, it is necessary to try to define, despite its obvious peculiarities, vernacular architecture. As the terminology itself indicates, it is intimately linked to the place where it is located. Therefore, it is complex to speak of a single typology. Geographical, historical, material, and economic constraints result in various typologies collected by various authors (Rivas Andina, 2004). In general, the most commonly used material is stone, with sandstone, limestone, and slate predominating, with mud mortar, not always plastered on the interior or exterior, especially in the early stages. Still, when it is done, it is done in lime, something highly praised from a hygienic point of view. As for its internal division, this will undergo numerous variations depending on the historical period and economic capacity, ranging from a single room to two, with one reserved for livestock, on the lower floor if there is one, to several rooms in more advanced times or houses of economic standing. Internal divisions were initially made with 'cebatu', hazelwood rods braided together, which is sometimes filled with clay, or wooden planks. The roof was never flat and usually covered with tiles in the central and eastern areas and with slate in the western areas, although some vegetative coverings were also made. Inside, there are finished floors with packed earth. However, with historical changes brought about by industrialization, new materials such as glass, wooden floors, tiles, brick, and cement are incorporated (Rodríguez, 2021; Martínez Torner et al., 2005) used both structurally and in coatings. Over time, prefabricated elements, insulation, plastic-derived materials, and, with this, new construction techniques such as reinforced concrete and typologies will be incorporated. That is, there is a transition from more natural materials to industrialized ones.

In material terms, bioconstruction establishes many connections with pre-industrial vernacular architecture. Although it incorporates foreign elements like straw, wood plays a crucial role. Traditionally, wood has had a notable presence in the region (Merino Cristóbal, 2004; Abella, 1985). It is an abundant material, deeply rooted in local craftsmanship, and has been used for iconic architectural elements such as galleries, turned balustrades, carpentry, roofing, and other structures like 'hórreos' and 'paneras', as well as in numerous artisanal pieces. Although to a lesser extent, historical evidence shows its presence in dwellings in working-class housing or various popular solutions (Fernanda Fernández, 2023, November 17–18). In bioconstruction, wood is used both structurally and for interiors and decorations with artisanal pieces. In relation to the recovery of crafts, it is relevant how often artisanal pieces find better marketability when they are integrated into construction. This can happen because of larger budget allocations, in which the cost of the piece, which is usually higher than an industrialized one, is dispersed; due to their connection with a practical element; because of the sense of exclusivity and value they bring to the house; or because of the importance of having well-crafted and custom-made elements. Rarely do we find houses entirely constructed with wood, and regulations often restrict its use if left exposed externally without cladding. Its use is prominent in interior divisions using 'cebatu' and solutions like straw bale construction. The preferred wood in this case is often oak or chestnut—even when using industrialized woods like laminates—due to their performance and resistance against pests. Frequently, this wood is reclaimed from old structures or obtained through bartering with neighbors. In bioconstruction, efforts are made to minimize logging, and when necessary, it is done consciously and following traditional practices. Forestry work is typically carried out in December or January, or in some cases, November or February, when the tree is dormant and sap circulation is minimal. This results in better performance against moisture and woodworm and reduces the likelihood of wood splitting. Historically, the wood acquisition process was a circular one, where obtaining wood was both preceded and followed by planting. This practice has been lost in most cases, and although there are efforts to recover it, there are currently no similar neo-rural initiatives. New solutions aligned with the circular economy,[8] resource utilization, and reuse are also employed. Alongside the recovery of wood, traditionally considered lower-value or waste materials, such as chestnut shavings or wood used in the production of oak barrels,[9] are utilized. To a lesser extent, oak, larch, boxwood, or beech are used, sometimes resorting to bartering or collection. If treatment is necessary, it is done with borax salt. For finishes, beeswax, walnut stain, olive oil, and linseed oil are preferred, avoiding varnishes and other synthetic enamels that not only produce harmful elements but also prevent the material from breathing, making it more susceptible to pathologies. Regarding stone as a material, it is striking that there is almost no active master in the field, even though it traditionally was the predominant material. The neo-rural interest in bioconstruction is evident, but there is a perceived lack of training and a perception of complexity. Clay, on the other hand, is reserved as a secondary material, complementing various techniques such as formwork, 'cebatu', or as wall plaster and floor finishing.[10] Another point of convergence between bioconstruction and traditional construction is the use of interior and exterior finishes. External plaster is usually done with lime, and interior plaster with clay. If clay is used externally, it is always mixed with lime to protect against moisture and prevent disintegration while maintaining the breathability of the construction. These layers also serve to protect against rodents and insects. Considering the environmental impact of lime production, efforts are made to make it as sustainable as possible. Modern bioconstruction techniques include External Thermal Insulation Systems with various materials. Lime is also used indoors, creating bright environments with whitish tones, in line with contemporary minimalist aesthetics. Floors are mostly made of rammed earth. All these elements are documented in historical texts referring to traditional housing, as seen in medical topographies collected by Rodríguez Pérez (2021) (Martínez et al., 2005). A notable fact regarding materials is the prevalence of ceramic tile construction over green roofs. Although the former is more widespread by current regulations in the region, self-construction tends to prefer green roofs for bioclimatic reasons. We should not forget the historical presence of these types of roofs, especially in western homes, and one of the most iconic heritage elements: 'teitos' (Fernández Ans & Molina Huelva, 2013; Menéndez Martínez, 2008) (Figures 16.3–16.5).

While traditional houses had basic foundation solutions, bioconstruction pays special attention to this aspect: first, because it does not use stone in walls due to material constraints like straw, which is weak against humidity; second, addressing one of the main concerns of bioconstruction: capillary moisture. Systems like caviti, reused materials of various types, including tires, modern bioconstruction materials like lightweight ceramic block or thermoclay with a cork interior to break the thermal bridge, etc., are used. Some designs are inspired by traditional granaries ('hórreos') and barns, with a base of wooden posts on stone slabs. New components like arlite are added to traditional solutions like rammed earth floors to improve insulation. Additionally, bioconstruction employs solutions for old materials, such as bricks, to make foundations. There is a need to incorporate the material peculiarities of construction into regulations. Many bioconstructors use old materials from existing structures that cannot be reused for a new construction, such as stone and wood. However, this practice should be carried out following an intermediate process:

FIGURE 16.3 Artisanal pieces linked to self-construction. On the left, a lamp made with natural materials, and on the right, a stool with reclaimed wood, similar to the floor it sits on. Guillermo Rozalén (Artesanía la Luz del Bosque).

FIGURE 16.4 Wooden elements recovered from ceilings with beams and revalued woods by bioconstruction, such as chestnut wattle. Guillermo Rozalén (Artesanía la Luz del Bosque) and El Mandala Centro de Permacultura.

debris removal and treatment of materials for subsequent use and acquisition. This process is oriented toward the treatment of industrialized materials. Bioconstruction also incorporates insulation materials such as cork in sheets or bulk treated with borax, wool, recycled paper cellulose, straw, clay, or even reused materials like textile waste. From a contemporary perspective, and beyond the materials, we must take into account the bioclimatic use of certain traditional elements, such as galleries (Lage Cal et al., 2015).

FIGURE 16.5 The use of the "cebatu" technique in the restoration of the Benéfica, carried out collaboratively with artisans. Macario Iglesias, Jorge Rojas and Maria Arce.

In addition, self-construction is also seen as a way to recover intangible cultural heritage. In traditional practice, only professionals—acquired through oral learning, workshops, or other informal training—were sought for certain complex constructions (Cantero Fernández, 2016). Elements such as 'andechas'—community members perform a favor for another member, expecting it to be reciprocated when needed—and 'sestaferias'—work that benefits the entire community—contribute to a sense of belonging, where people collaborate on a task; these practices were predominant. Construction, whether of traditional housing or bioconstruction, is seen as a process of exchange and participation, involving cooperation throughout all project phases. Additionally, barter systems are implemented during construction, exchanging materials, handicrafts, or knowledge. Figures like the 'curiosu', as mentioned by Cantero Fernández in his study on dry stone techniques, refer to 'a person with special skills and proven experience in a discipline [...] who did not reach the level of a professional and did not demand economic compensation for their services' (2016, p. 16). In straw bale self-construction, despite being a relatively simple technique, this figure is important for more complex elements such as the foundation, corners, and contact with carpentry. Generally, in bioconstruction projects, the presence of other neo-rurals who have already carried out constructions and guide new builders throughout the process is common, providing support for self-construction. However, this figure can also be that of a professional who, while not part of the project, offers voluntary advice. Regarding volunteer work, another common practice in bioconstruction with initiatives such as participatory construction or construction schools, traditional construction is the collaboration in exchange for accommodation and meals (2016, p. 16). Likewise, non-formal education through workshops with centers like Mandala in Asturias is employed. Digital tools are also integral to the knowledge of self-construction in contemporary settings.[11]

Despite these connections between rural and traditional practices, we must consider that not everything traditional is bioclimatic. A clear example in Asturias is the presence of poorly oriented houses that do not face south or those attached to hillsides, leading to moisture-related issues. Similarly, there are problems with insulation, ventilation, or sunlight due to small windows, an issue prevalent until the advent of glass (Rodríguez Pérez, 2021, pp. 133–134) or the study by Fernández Ans and Molina Huelva (2013) on the thermal behavior of a stone and straw-roofed house and another more contemporary traditional house in Asturias which revealed that neither met current energy efficiency standards.

To conclude, we must address that in the last few decades, there has been a proliferation of various types of constructions with a reductionist and superficial interpretation of natural, rural, or bioclimatic, with elements that are not but are perceived as such. This association can take two forms: either based on superficial aesthetics with contemporary construction methods and materials, such as highly treated imported woods, artificial stone veneers, PVC windows with carpentry finishes, etc., or in the collective imagination of the rural and natural in a general sense, with constructions of Nordic style, tree houses, and structures inspired by cinematic representations. It is necessary to highlight the role of numerous rural tourism establishments, which, as mentioned, have exponentially increased their presence due to initiatives promoted by the administration and trends such as healthy lifestyles, emphasized on social media, and in the wake of events like the Covid-19 pandemic or the climate crisis. In contrast to these tourist initiatives, architects trained in bioconstruction, such as Macario Iglesias, advocate for experiential tourism with contextual architecture. An example would be the use of thatching workshops that are complemented with overnight stays in a 'brañeiru', combining material and immaterial recovery with tourism, having an impact on the community. Similarly, neo-rural centers like the Asturcón Ecomuseum propose combining overnight stays in traditional 'hórreos', historically documented (Rodríguez, 2021), but facing challenges related to the changing functions of buildings from their traditional agrarian connection. Regarding these cultural practices, we conclude the section with initiatives in bioconstruction or traditional construction carried out in museums, such as the Asturian People's Museum with pieces created by Macario Iglesias: a traditional hazel wood partition 'cebatu' for a basketry exhibition and a false ceiling in the peasant house recreation. Other examples include interventions in the Teitos Museum, experimenting with local clay to find the most water-resistant and enhancing it with lime, a project managed by Heritage. Future projects include the restoration of the circular structure at the Campa Torres Archaeological Museum, using woven walls and thatching. Finally, a participatory example is the rehabilitation of the Benéfica de Piloña, closely linked to neo-ruralism and new ways of recovering traditional cultural heritage through contemporary means as a social cultural center. In this project, various artisans and volunteers, including the carpenter Rafa Martín, intervened, creating structures with recovered wood, among other tasks. Following this line of recovery of traditional construction for the community, the Anima Mundi initiative stands out in Cabranes, which, along with the rehabilitation of buildings using traditional stone techniques and including a Cultural Interest Site (BIC), aims to bring the area closer to community service through the creation of a social center for people with special needs. This center would also serve as a catalyst for the revitalization of craftsmanship through the sale of products and workshops. Also noteworthy is the rehabilitation in Fresneu of a former cider mill and dance hall, which will unite material and intangible heritage through music and dance. These approaches can also serve as a resource in the rehabilitation of vernacular housing, often in a precarious state and subject to aggressive interventions due to the lack of protection (Cantero Fernández, 2013) and insufficient training. Examples include widespread renovations using cement mortar—which also affect other constructions like washhouses,—'hórreos', etc.—or those carried out on thatched roofs with fiber cement or metallic plates (Fernández Ans & Molina Huelva, 2013), leading to the emergence of new pathologies in the constructions. Bioconstruction is also considered an alternative for buildings with new museological approaches such as the ecomuseum, sharing elements with such concepts. This is exemplified by the Asturcón Ecomuseum, whose architectural design is based on the Asturian seafaring house schema, both in volumetric aspects and distribution. Thus, it draws from a reference inherent to the rural landscape of the coastal area, including elements typical of bioconstruction.

16.5 CONCLUSIONS

In the recovery of rural areas, most activities have been oriented toward the industrialization of agriculture and the tertiary sector of the economy. Through this study, the possibility of enriching the territory through additional contributions related to neo-rural bioconstruction has been considered: revitalizing traditional construction, crafts, and other productive activities. It is essential to address territorial challenges from a holistic perspective, paying close attention to all the possibilities offered by rural environments. An approach that considers multiple aspects enables a more effective allocation of efforts and results.

The architecture and construction techniques used in their development are among the most significant imprints left by humans in shaping landscapes. Therefore, it is crucial to carefully choose which tools are employed and for what purpose. Bioconstruction is intertwined with material and inmaterial culture and heritage from its conception, with a strong connection to the hosting territory, renewing and experimenting with traditional techniques, some of which are resumed without modifications. In this way, bioconstruction acts as a catalyst and a connecting link with other cultural manifestations.

In the case of crafts, it is necessary to delve into this connection between construction and craftsmanship, considering it as a pathway for recovery. Crafts are closely linked to alternative modes of production and consumption that are more sustainable and respectful of the environment. Their recovery and revitalization in the context of the current crisis is particularly compelling, as they offer valuable opportunities to promote ethical practices, strengthen local economies, and foster a deeper connection with cultural heritage.

These activities can, in turn, contribute to slowing down depopulation and the neglect of traditional housing constructions, with conscious rehabilitations and restorations in line with new bioclimatic approaches. Although self-construction may not always be viable without professional guidance, depending on the technique used, it is important to remember that hiring skilled artisans, with a more balanced distribution of construction costs between materials and human labor, is a crucial consideration. This approach ensures both quality craftsmanship and a fairer allocation of resources. In this regard, it would also be worthwhile to reflect on the need to invest in training for these areas of knowledge, and, if possible, offer such education within rural environments. This approach could foster local expertise, preserve traditional techniques, and contribute to the sustainable development of rural communities. Similarly, it is worth highlighting once again the connection that these types of architecture have with the investment in furniture and decorative objects that are also of an artisanal nature. This synergy enhances the coherence of the overall design and further promotes sustainable and locally rooted production practices.

Finally, it is proposed as a measure to transform and enrich existing tourism, helping conservation and heritage restoration. Considering that tourism is a reality we must anticipate and manage, it is valuable to support local and respectful initiatives, which, as seen in some examples, are closely connected to the principles behind bioconstruction. These initiatives not only promote sustainable tourism but also align with environmentally friendly practices and the preservation of cultural heritage.

This approach aligns with issues of utmost relevance today, such as the commitments of nations to the Sustainable Development Goals. It addresses global problems like the climate crisis through local initiatives. Design, in all its forms and scales, can be a tool for change in our production and consumption system. Its importance is undeniable, especially when considering design linked to construction, one of the most polluting industries today. Bioconstruction emerges as a proposal for the future.

However, it is important to note that within bioconstruction, there are diverse approaches, more or less respectful of the concept itself, the environment, heritage, and the community. There are also divergences and associated issues, making it pertinent to conduct more detailed studies, including specific case studies and open-access manuals and guidelines for best practices. While architectural permeability has always existed, and the landscape has been shaped by the succession of historical cultural strata, of which bioconstruction is part, contributions must be made with sensitivity and criteria. Furthermore, ongoing research on traditional heritage is necessary, improving its protection, and ensuring the respect and conservation of representative examples and environments. Lines of study should also be established linking these forms of construction with urban planning laws and land treatment in rural areas. Lastly, the fragile balance between neo-rural, traditional rural, and administration must be considered. These new movements add a new reading to the rural, generating new identities. Resignification takes place through a new perspective, dynamic, youthful, and incorporating contemporary concepts. It is essential to continue establishing connections between the past and the present, not only for their recovery in contemporaneity but also to establish dialogues based on connections. The metamorphosis of the rural demands not only a reconsideration from a theoretical perspective with studies and methodologies but also from practice, with actions on the political territory of European rural development at the regional level.

NOTES

1 To delve deeper into the abundant existing literature, the studies by Ginés Sánchez and Querol Vicente (2019) stand out.
2 In this complex rural fabric, with numerous participants and factors, reactions and responses vary. Although cultural clashes between the dynamism of neo-rural culture, more youthful and contemporary, and the entrenched rural immobility of traditional inhabitants are frequently mentioned, disagreement is not the only plausible response. During our fieldwork, we have determined that while situations of conflict or mutual ignorance exist, in many cases, particularly in areas with lower population density, the reception of the neo-rural phenomenon is positively embraced by traditional inhabitants as a way to counter depopulation. In other cases, there may be an initial clash, but both groups tend to assimilate over time, discovering numerous points of convergence between the traditional and innovative approaches introduced by neo-rural culture.
3 Many interviewed neo-rurals also refer to the importance of the pandemic, not only in terms of crises but also in terms of the time it provided, necessary for rethinking lifestyles, but also for training and starting in craftsmanship and bioconstruction.
4 In Spain, self-construction is not regulated by legislation beyond what is stipulated in *Real Decreto 1627/1997 sobre disposiciones mínimas de seguridad y salud en las obras de construcción*, the architect's project that must be approved, compliance with the technical code, and oversight by the municipal architect.
5 Although documented and well-known techniques are employed, in many analyzed cases, experimentation through trial and error is an inherent part of the practice of bioconstruction.
6 In Spain, 20% of energy is allocated to the needs of the home, according to the Ministry for the Ecological Transition. Meanwhile, the International Energy Agency estimates that, on a global scale, construction is responsible for 36% of worldwide greenhouse gas emissions. It's essential to consider that emissions are not only produced during the building's lifespan but also the extraction of materials, with 40% of these materials in Spain going to this sector, as well as during the construction phases and subsequent demolition.
7 The techniques of straw construction are detailed in numerous manuals, such as the one by Martínez Roca and Nitzkin and Termens (2016). However, in the documentary research and the analyzed cases, various small variations are identified compared to the original models of each technique.
8 The circular economy is a sustainable approach that aims to maximize resource utilization, minimize waste, and encourage a continuous cycle of production and consumption (Cerdá & Khalilov, 2016).
9 The latter is a very hard wood, both due to the nature of oak and the method of cutting without sectioning the veins to make it impermeable for barrel production.
10 Clay has been widely used in bioclimatic and traditional architecture in other regions (Castilla Pascual, 2004). A very close example of its use in structures can be found in León.
11 Indeed, the numerous initiatives being carried out in Latin America to establish community ties are noteworthy (Huerta & Esteban, 2013). These efforts are not only seen in depopulated areas due to economic changes but also in post-conflict zones.

REFERENCES

Abella, I. (1985). *El hombre y la madera*. Barcelona: Egedsa.
Bell, D. (2006). 'Variations on the rural idyll'. In: Cloke, P., Marsden, T. & Mooney P. (Eds.), *Handbook of rural studies* (pp. 149–160). London: Sage Publishing. https://doi.org/10.22325/fes/res.2021.39

Bueno, M. (1998). *El gran libro de la casa sana*. Barcelona: Ediciones Martínez Roca.

Cantero Fernández, C. (2013). El patrimonio etnográfico asturiano en los Catálogos Urbanísticos. In: D. Ruiz de la Peña (Coord.), *Los catálogos urbanísticos en el Principado de Asturias: Una perspectiva pluridisciplinar* (pp. 125–155). Oviedo: Universidad de Oviedo.

Cantero Fernández, C. (2016). Técnica constructiva de la piedra en seco en Asturias. Oviedo: Gobierno del Principado de Asturias, Consejería de Educación y Cultura, Dirección General de Patrimonio Cultural.

Castilla Pascual, F. J. (2004). *Estabilizacion de morteros de barro para la proteccion de muros de tierra* [PhD]. Madrid: Universidad Politecnica de Madrid.

Cerdá, E., & Khalilov, A. (2016). Economía circular. *Economía Industrial, 401*(3), 11–20.

Datos Macro. (2023). Asturias: Población en diciembre 2023 con variación respecto a diciembre 2022. Datos Macro. Retrieved from https://datosmacro.expansion.com/demografia/poblacion/espana-comunidades-autonomas/asturias?anio=2023

De Carvalho, M. (2015). *Herramienta de Certificación para la Bioconstrucción* [PhD]. Madrid: Universidad Politécnica de Madrid.

Fernanda Fernández, M. (2023, November 17–18). *Hacer de la necesidad virtud: la madera en su paso de la construcción tradicional al contexto industrial en Asturias* [conference]. Jornadas de Japón en la Ribera. Ribera de Arriba: Un diálogo a través de la madera, Ayuntamiento de Ribera de Arriba y Universidad de Oviedo.

Fernández Ans, P., & Molina Huelva, M. (2013). Evolución del comportamiento térmico en viviendas tradicionales de piedra y cubierta de paja. Puesta en valor de un modelo sostenible en el noroeste de España. *Revista de la construcción*, 12(2), 102–115. https://dx.doi.org/10.4067/S0718-915X2013000200008.

García Martínez, A. (2008). *Antropología de Asturias. I. La cultura tradicional, patrimonio de futuro*. Oviedo: KRK.

García Martínez, A. (2011). *Antropología de Asturias II: El cambio: la imagen invertida del otro*. Oviedo: KRK Ediciones.

Ginés Sánchez, X., & Querol Vicente, V. A. (2019). Construcción social de lo rural y Nueva Ruralidad. Una aproximación al marco de interpretación de lo rural de agentes políticos y sociales. *Economía Agraria y Recursos Naturales*, *19*, 37–57. https://doi.org/10.7201/earn.2019.01.03

Gómez, J. (2013). Del patrimonio paisaje a los paisajes patrimonio. *Documents d'Anàlisi Geogràfica*, *59*(1), 5–20.

Grande, J. (2006). 'La evolución del turismo rural en España y las nuevas oportunidades del turismo de naturaleza'. In: *Estudios Turísticos* (pp. 169–170, 85–102). Madrid: Secretaría de Estado de Turismo y Comercio, Instituto de Estudios Turísticos.

Hernández, M. (2009). El paisaje como seña de identidad territorial: Valorización social y factor de desarrollo, ¿utopía o realidad? *Boletín de la Asociación de Geógrafos Españoles*, *49*, 169–183.

Herrera, L., Blanco, A., & Cànoves, G. (2005). Turismo rural en España: Un análisis de la evolución en el contexto europeo. *Cuadernos de Geografía*, *77*, 41–58.

Huerta, J. E. T., & Esteban, J. (2013). *Bioconstrucciones, una alternativa a la ocupación y el desarrollo de comunidades rurales* [conference]. Instituto Tecnológico Superior de Pátzcuaro.

Instituto de Desarrollo Económico del Principado de Asturias (IDEPA). (2023). Demografía de Asturias 2023. Retrieved from https://www.idepa.es/conocimiento/asturias-en-cifras/demografia

Instituto Nacional de Estadística (INE). (2023). Revisión del Padrón municipal a 1 de enero de 2023: Serie histórica por municipios — Principado de Asturias. Madrid: INE. Retrieved from https://www.ine.es

Khan, Ll. (2008). *Builders of the Pacific Coast*. California: Shelters Publications.

Lage Cal, S., Folgueras Díaz, M. B., Alonso Hidalgo, M., & Luengo García, J. C. (2015). 'Identificación de los elementos singulares de una casa tradicional asturiana y análisis de su influencia en el comportamiento térmico de la misma'. In: *IX Congreso Nacional de Ingeniería Termodinámica: libro de actas* (pp. 1032–1040). Cartagena: Universidad Politécnica de Cartagena, Servicio de Documentación.

Litvak, L. (1911). *El tiempo de los trenes. El paisaje español en el arte y la literatura del realismo (1849–1918).* Barcelona: Ediciones del Serbal.

Martínez Torner, F., López, Álvarez, J., & Ortiz García, C. (prol.) (2005). *Dos estudios geográficos y etnográficos sobre Asturias.* Quirós: Fundación Municipal de Cultura, Educación y Universidad Popular, Muséu del Pueblu d'Asturies and Museo Etnográfico de Quirós.

Menéndez Martínez, C. O. (2008). *Teitos. Cubiertas vegetales de Europa Occidental: de Asturias a Islandia.* Oviedo: Colegio Oficial de Aparejadores y Arquitectos Técnicos de Asturias.

Merino Cristóbal, L. (2004). *Usos Tradicionales de los Árboles en el Occidente de Asturias.* Madrid: Consejo Superior de Investigaciones Científicas.

Neila González, F. J. (2004). *Arquitectura bioclimática en un entorno sostenible.* Madrid: Munillaleria.

Nitzkin, R., & Termens, M. (2016). *Casas de paja. Una guía para autoconstructores* (2th ed.). Olba, Teruel, Spain: EcoHabitar Visiones Sostenibles S.L.

Nogué i Font, J. (1988). El fenómeno neorural. *Agricultura y Sociedad, 47*, 145–175.

Rivas Andina, J. A. (2004). *El hórreo y la arquitectura popular en Asturias.* Oviedo: Picu Urriellu.

Rodríguez Eguizabal, A. B., & Trabada Crende, X. E. 1991. De la ciudad al campo: el fenómeno social neorruralista en España. *Política y Sociedad 9*, 73. https://doi.org/10.30827/Digibug.44291

Rodríguez Pérez, S. (2021). Notas sobre la vivienda rural en Asturias entre los siglos XIX y XX a través de las topografías médicas. *Res Mobilis, 10*(12), 127–153. https://doi.org/10.17811/rm.10.12.2021.127-153

Roseman, S. R., Prado Conde, S., & Pereiro Pérez, X. (2013). Antropología y nuevas ruralidades. *Gazeta de Antropología, 29*(2). https://digibug.ugr.es/handle/10481/28509

Sánchez, J., & Querol, C. (2019). *Nova Ruralidade: mudança social, paisagem e inovação na agricultura.* In I. Silva & A. Baptista (Eds.), *Territórios, ruralidades e paisagens: contributos para o ordenamento do território em Portugal* (pp. 123–138). Lisboa: Colibri.

Short, B. (2006). 'Idyllic ruralities'. In: Cloke, P., Marsden, T. & Mooney, P. (Eds.). *Handbook of rural studies* (pp. 133–148). London: Sage Publishing. https://sk.sagepub.com/reference/hdbk_rural

17 The Use of Natural Materials in Architecture

Exploiting Cellulose for the Additive Manufacture of Innovative Sustainable Construction Systems

Tatiana Campos, Paulo Cruz, and Bruno Figueiredo

17.1 CONTEXT

The concept of sustainability, according to Franklin and Till (2018), "has previously focused on reducing consumption, but the growing acknowledgement that the hunger for 'new' is far from satiated is causing a shift in the way that designers conceive high-volume, low-cost items." With synthetic materials, particularly plastics, having dominated our planet for an extended period, their purported benefits of greater longevity and quality are now viewed less favourably in light of current environmental concerns (Siegel, 2018). Transitioning from synthetic to natural materials and incorporating organic matter, while seemingly straightforward, introduces additional ecological considerations (Treggiden, 2020). It is crucial to seek solutions that are consciously sustainable, natural, and biodegradable, ensuring a lower negative impact on the environment. As a result, it is necessary to re-evaluate the natural resources available and look for alternative raw materials in order to promote more sustainable and innovative practices. Groves and Murakami, inspired by maritime vessels, have developed a line of furniture made from plastics found on the planet. The Nienke Hoogyliet studio has produced a sustainable wire by cultivating seaweed, which is not only economical and abundant, but also detoxifying and the main source of oxygen for the planet. Chilean designers Sisa Collection has developed an experimental skin from Cochayuyo seaweed for garment production, one of the world's most polluting sectors (Franklin & Till, 2018). Dunia, a student at TU Darmstadt, has developed an experimental study that consists of producing three-dimensional paper moulds using cellulose pulp as the base material (Rosendahl et al., 2024). The study essentially focuses on optimising cellulose pulp in order to improve the shrinkage and deformation of the material after drying, taking into account the pulp's extrusion capacity. On the other hand, Christopher Bierach at Laboratory for Additive

DOI: 10.1201/9781032719290-20

Manufacturing in Architecture (LAMA) developed a study that consisted of idealising a biodegradable pulp by combining lignin and cellulose (Rosendahl et al., 2024). By changing the different proportions, he was able to develop a paste with the constructive capacity to produce window frames, taking advantage of AM. Gabriela Rossi, from CITA, developed a wall 3 m long and 1 m high, made up of 44 components with different configurations (Rosendahl & Gaudillière-Jami, 2023). Each component was produced according to a wavy pattern, in which the integration of a fastening system was explored, taking advantage of the geometry to produce male and female fittings.

Taking advantage of natural local raw materials, namely the use of cellulose – a sector with a high prevalence in the north of Portugal – and common food waste, a set of new natural sustainable materials has been developed (Campos et al., 2021a). Each material was designed with the aim of being applied in AM, so it was necessary to carry out a set of preliminary tests that resulted in the study of the material's fluidity, workability, and resistance. Subsequently, a set of new contemporary architectural construction systems was developed – naturally sustainable and biodegradable – with the premise of sustainable construction. BioTILES promotes the development of natural façade coverings through the use of shrimp shells, resulting in highly mouldable, customisable, and textured bioplastics (Campos et al., 2021b). The Kusudama (Campos et al., 2019) and Cellusound prototypes are based on the use of cellulose to build highly customisable blocks – Cellusound in particular by studying the propagation and absorption of sound – capable of adapting to different contexts and functions. The use of AM technologies increases the versatility and applicability of these natural materials, as well as making it possible to develop more sustainable and biodegradable construction methods by reducing CO_2 emissions, mostly caused by the transformation and recycling of raw materials, which are highly recurrent in this sector, such as cement, steel, and concrete.

17.2 BIOTILES

The fishing industry produces tonnes of shellfish waste every year, resulting in huge losses due to poor management of marine resources. Shellfish waste – shells, fish scales, squid skin, and fungal cells – is rich in chitin, the second most abundant biopolymer on the planet (Azevedo et al., 2007). Chitin is a monomer similar to cellulose and is obtained through a series of chemical processes. The search for more sustainable, biodegradable materials is encouraging the use of chitin in various fields of research, particularly in biotechnology, due to its antibacterial and anti-allergenic properties. As a way of proving the versatility of the material under study, an architectural cladding system was produced, consisting of a set of individual customisable modules.

17.2.1 Chitosan

Chitin first appeared in France with the analytical study of fungi by the chemist Braconnot, who obtained an insoluble residue after treatment with alkali, which he called "fongine" – a biopolymer composed of nitrogen and insoluble with water

(Crini, 2022). Later, Odier found the same substance in the wings of insects, equally insoluble except for the presence of the chemical element nitrogen (N), which he called "chitin." Odier argued that chitin contained cellulose, not nitrogen, and behaved similarly to plant cell walls. This controversy of ideas went on for some time until Lassaigne, during his analytical study of the exoskeleton of the silkworm butterfly, proved that chitin was not a type of cellulose, but a new material containing nitrogen. Later, Rouget discovered that treating chitin with a concentrated solution of caustic potassium transformed this new material into a "new modified chitin," and chitosan – the main derivative of chitin – was discovered.

17.2.2 Mixture

The material is prepared by combining 80%–90% water (v/v), 4%–8% chitosan (w/v), 4%–8% starch (w/v), 4% glycerine (v/v), and, depending on the final properties required, 0%–12% cellulose (w/v). After homogenising the components, the mixture is heated until it forms a viscous pulp. As the temperature decreases, 2%–4% acetic acid (v/v) is added to the material (Figure 17.1).

17.2.3 Properties

Changing the proportions of the different components of the mixture under study gives it different mechanical and physical properties, such as elasticity, rigidity, strength, flexibility, shape recovery, transparency, density, and colour. Cellulose controls the level of mechanical strength of the final object produced. Starch aggregates the virgin cellulose fibres. Chitosan – antibacterial and anti-allergenic – controls the final appearance of the object produced, such as transparency and colour (Figure 17.2, left side). Acetic acid controls viscosity. Glycerine gives the material flexibility (Figure 17.2, right side).

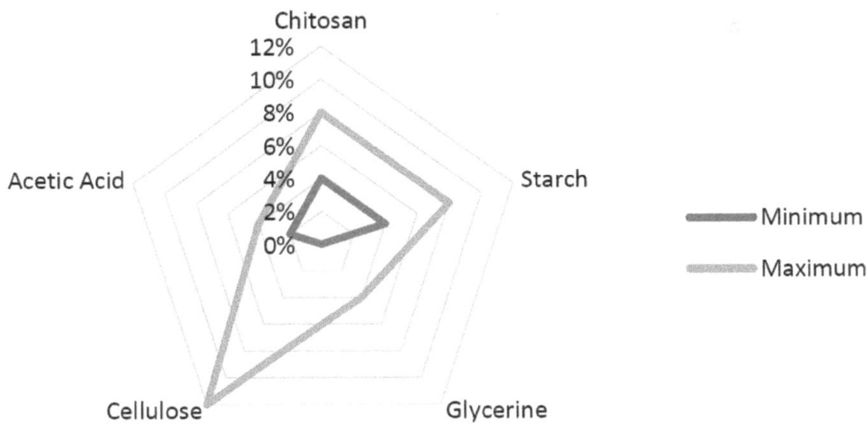

FIGURE 17.1 Graph showing the proportions of materials added to the mixture to be used for the production of BioTILES.

FIGURE 17.2 Final finish of samples with varying proportions of chitosan, left figure. Final behaviour of samples with varying proportions of glycerine, right figure. (Photograph by the author.)

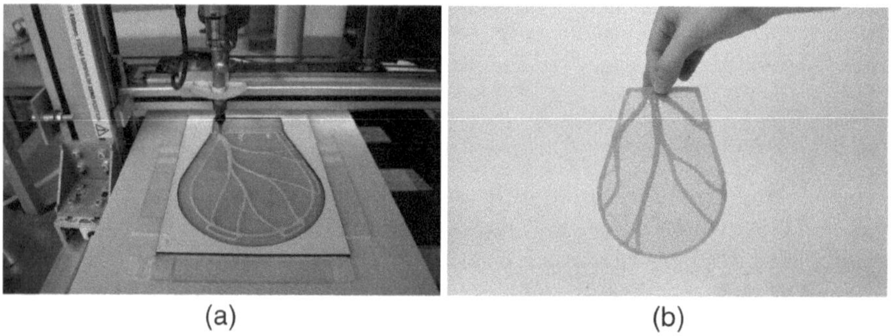

FIGURE 17.3 Printing of a BioTILES prototype module, left figure. Final finishing of a module, right figure. (Photograph by the author.)

17.2.4 Prototype

Taking advantage of the constructive properties and limitations of the material, a constructive cladding system was proposed, the BioTILES, a reinterpretation of the old soletos of historic Portuguese buildings (Figure 17.4) (Rosendahl & Gaudillière-Jami, 2023). Soletos, also known as fish scales, are small tiles made from slate rocks, with thin thicknesses, superimposed on each other and fixed to a wooden batten (Silva, 2018). A vertical sample was therefore developed comprising 35 individual modules (Figure 17.3, left side). Each module was designed taking into account the shape and fixing system. To create greater mechanical resistance, a variable pattern inspired by the silhouettes of leaf veins was developed (Figure 17.3, right side).

To produce the different modules, a levelled substrate was first placed on a mould with a high degree of flexibility and then, using the Lutum Mini 2.0®, the PEM (Paste Extrusion Modelling) AM method was used to obtain the variable standard. The building system was generated using three-dimensional modelling tools, Rhinoceros and Grasshopper. The different modules were dried at a temperature of 22% and 36% humidity. This drying phase spanned a full five days to ensure thorough and complete drying (Figure 17.4).

FIGURE 17.4 Final finishing of the prototype developed. Final composition made up of 35 individual modules with an organic internal pattern inspired by leaf veins. (Photograph by the author.)

17.3 KUSUDAMA

We are facing a material revolution that will allow us to build our relationship with the environment and consequently change our society (Solanki, 2018). Today, we transform raw materials into products, which are then distributed around the world, satisfying our needs until they are disposed of. However, this Make-Take-Discard cycle is depleting raw materials that are fundamental to our survival and generating huge amounts of waste (Treggiden, 2020). The unsustainable rate at which we are consuming our planet's finite resources is evident. We need to change this paradigm and use more sustainable solutions, generating an infinite cycle in which everything we produce can be returned to the earth and then reused as new raw materials, such as cellulose, a natural material. As a way of proving the viability of using sustainable and biodegradable raw materials, a self-supporting wall construction system was developed, consisting of a set of individual modules, using cellulose as the construction material (natural material), sawdust (organic waste), and clay (Campos et al., 2019).

17.3.1 Cellulose

Cellulose is a compound and structural component found in the cell walls of plants, algae, and oomycetes – filamentous, single-celled organisms similar to fungi – and, in turn, the most abundant organic polymer on earth. It was discovered by Anselme Payen, a French chemist, in the mid-19th century, after isolating cellulose from the other components during a scientific study. It comes from the eucalyptus globulus, "an important plantation in subtropical regions, including southern Europe (Spain, Portugal)" and selected for its quality for paper production, since 50% of its fibres are cellulose. "The tree major species used for pulp production are E. grandis, E. globulus, and E. camaldulensis (E. rostrata)" (Sisko & Pfaffli, 1995). All the constituent elements in eucalyptus are utilised, from the leaves for the production of essential oils, cellulose for the production of paper, and lignin, a macromolecule, for the production of thermal insulation foams. After its discovery, new materials derived from it began to emerge, from celluloid – the first thermoplastic derived from cellulose – to viscose – a semi-synthetic fibre based on cellulose – or cellophane – a relatively thin, transparent sheet made from regenerated cellulose. The discovery of this biopolymer was revolutionary, and even today, its use stands out, particularly in the paper industry.

17.3.2 Mixture

Prepare the mixture of natural materials by combining 10% corn starch (w/v) (VRW, Radnor, PA) with 76% water (v/v). After homogenising the materials, heat them with vigorous stirring until a highly viscous hydrogel is formed. After the temperature of the hydrogel has dropped, gradually add 14% micronised cellulose (w/v) in small amounts until a completely homogeneous mixture is formed. If desired, add a percentage of additives – sawdust – to give the mixture more resistance to drying. Finally, an extremely viscous natural mixture is obtained. Each natural material was properly selected with the aim of developing a natural, biodegradable, and sustainable mixture capable of being returned to the environment. The use of local materials was also taken into consideration, thus reducing possible environmental impacts associated with transport and later production.

17.3.3 Properties

Changing the percentage of cellulose introduced into the mix strongly influences the material's behaviour in AM. By looking at Table 17.1, we can compare the printing factors based on changing the proportions of added cellulose.

The clay mixture – GRES 130 MP – has a moisture content of between 19%–21%, requiring the subsequent addition of water until it reaches a moisture content of between 32%–33%. When cellulose is added, it will absorb the water present in the mixture, so it is necessary to carry out the Pfefferkorn test to calculate the percentage to be added.

17.3.4 Prototype

As a proof of concept, Kusudama, a self-supporting wall construction system made up of a set of individual blocks based on natural materials, was designed. This prototype is made up of geometrically variable modules produced using three-dimensional modelling programmes, Rhinoceros and Grasshopper (Campos et al., 2021c). Generated from a regular hexagonal mesh, a True and False pattern was introduced to divide the components into two lists, the lower triangular blocks and

TABLE 17.1
Comparative Analysis of the Properties of a Cellulose-Based Mixture by Changing the Amounts of Cellulose

	>10%	10% < 20%	<20%
Fluidity	High water content	Ideal	Low water content
Material Extrusion	High material spread, difficult to print an object	Possibility of printing an object	Low water caused by a high proportion of cellulose, difficulty of extrusion
Retraction	>10%, objects retract in height	10%, objects retract in height	0% < 10%, objects retract in height
Delamination	High between layers	Additives decrease delamination	Low between layers
Deformation	Deformation is controlled in the act of code. Continuous printing = low deformation. The presence of stitches in the same place generates greater deformation		
Cracking	As these are mixtures based on natural materials, the rupture between connections occurs mostly at seam points between layers		

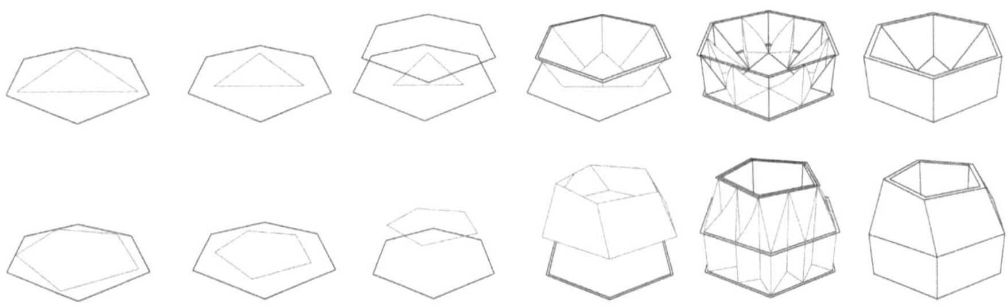

FIGURE 17.5 Generation of the individual Kusudama modules. (Produced by the author.)

the upper pentagonal ones. The lower triangular surfaces consist of a hexagonal base with a central triangular opening, of variable dimensions, defined by a mathematical algorithm. First, the surfaces were generated, and then, the internal walls were developed (Figure 17.5) to connect the surfaces generated previously.

The upper pentagonal surfaces consist of a hexagonal base with a central pentagonal opening and variable dimensions. First, the surfaces were generated, and then, the internal walls were developed to connect the surfaces generated previously (Figures 17.5 and 17.6).

After the parametric computational model was generated, a manufacturing sample was selected, a wall section, for the manufacture of the components that make it up. This section was divided into three horizontal zones (Figures 17.7 and 17.8). After making all the pieces that make up the sample, they were assembled, defining a 10 mm thick medium-density fibreboard support system and internal connections to support the components.

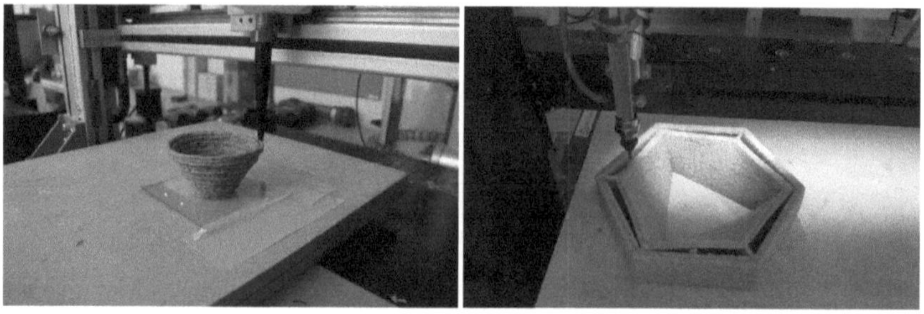

(a) (b)

FIGURE 17.6 Mechanical tests with cellulose – testing the maximum degree of curvature supported by the mixture, left figure. Printing a block of the Kusudama construction system, right figure. (Photograph by the author.)

(a) (b)

FIGURE 17.7 Final finish of samples with varying proportions of chitosan, left figure. Final behaviour of samples with varying proportions of glycerine, right figure. (Photograph by the author.)

FIGURE 17.8 Final prototype assembled. Tonal variation, texture between the different pieces that make up the sample. (Photograph by the author.)

17.4 CELLUSOUND

Cellusound is an evolving wall prototype designed to take advantage of the use of natural materials, such as cellulose, to develop a system for sound absorption. Cellulose is a natural biopolymer, widely abundant in nature and with extraordinary mechanical properties, especially when combined with organic substances. Is cellulose a good material for sound absorption?

17.4.1 Concept

When a sound wave propagates through a given environment and collides with a surface, it can be dissipated, reflected, or transmitted. These reactions are mainly due to the type of material that makes up the different surfaces in the environment, so we can classify materials as reflectors, diffusers, or absorbers (Figure 17.9).

Reflective materials have smooth, hard surfaces, reflecting a large part of the incident energy. They also increase the internal reverberation of the room and consequently amplify the sound, making the environment livelier. Diffusing materials reflect the sound incident in different directions, promoting greater sound dispersion in the environment, which directly guarantees greater sound quality and uniformity, thus sending out the echo. Absorbing materials reduce the reflection and transmission of sound by promoting the absorption of sound energy. Fibrous and porous, the air inside makes it difficult for sound to pass through, reducing internal reverberation time. The greater the surface area and thickness of the material, the greater the sound absorption.

17.4.2 Material

For the development of the Cellusound study, the mixture previously used in the production of the Kusudama prototype was optimised with the aim of improving the shrinkage and deformation behaviour after drying. The optimisation of the formulation involved two main modifications:

- Restructuring the basic composition by changing the proportions of materials, in particular reducing the water content. This change resulted in a decrease in the fluidity of the paste and, consequently, an increase in its constructive capacity.

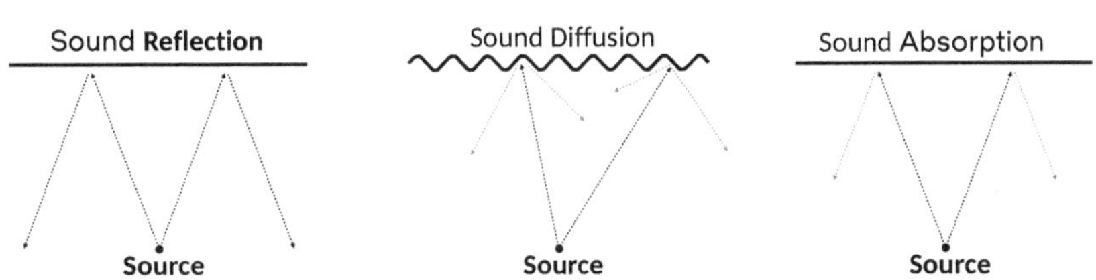

FIGURE 17.9 Behaviour of the propagation of a sound wave after collision on a given surface, based on the type of material. (Produced by the author.)

- The addition of a new component, a natural and biodegradable glue, helped to reduce delamination between layers, minimise possible deformation, and reduce the shrinkage index.

After developing the base mix, additives of natural origin were introduced throughout the material study, with the aim of evaluating the impact on mechanical and physical properties. These additives were subjected to preliminary tests, which involved analysing possible deformations, shrinkage, and the structural integrity of the material. The test results showed a significant improvement in the material's behaviour, especially in terms of reduced deformation. Based on the data obtained, black cork agglomerate was selected as the main additive. This material, widely available in Portugal and recognised for its exceptional acoustic properties, was chosen for its lightness, biodegradability, and ability to improve sound absorption, as well as giving the material greater resistance to deformation. The addition of cork to the base mix not only improved acoustic performance but also contributed to the development of a more sustainable material in line with ecological guidelines.

17.4.3 Design

Cellusound is a modular acoustic wall system made up of individual blocks produced from cellulose-based materials. Each block is made up of a customisable external wavy geometry – a rigid surface – responsible for reflecting, dissipating, or absorbing sound waves.

Wavy geometry is made up of a set of inner and outer circles, connected to each other by tangents, which we call waves. The outer waves don't intersect, so they create a set of envelopes that can easily be manipulated according to the needs of the space. The wavy geometry can act as a dissipation surface (Figure 17.10), through the reflection of the incident energy after the first contact, or as an absorption surface

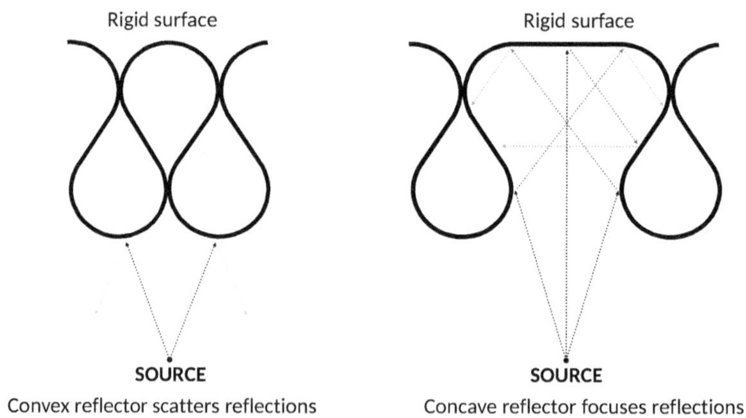

FIGURE 17.10 Diagrams representing the propagation of sound waves, analysing the dissipation and absorption of incident energy in two different wave situations. (Produced by the author.)

(Figure 17.10), through the presence of openings between waves, allowing the incident energy to propagate within each individual comb. The behaviour was different when analysing two possible options.

A parameterised corrugated geometry was developed, varying a set of different parameters: (1) radius of the circles; (2) number of double openings; (3) openings between waves; and (4) wall thickness. With the multiplicity of possible situations, by altering the parameters outlined, a study focused essentially on optimising the shape using the Galapagos evolution solver, in order to find the best possible solution – with the greatest acoustic absorption capacity – and in turn, the worst solution. The radius of the circle made it possible to control the size of the gap between waves: the larger the radius of the circle, the smaller the gap between waves, the greater the dissipation, and the lower the absorption. The smaller the radius of the circle, the wider the gap between waves, and the greater the sound absorption. For greater sound absorption, the tangents connecting the different circles must not be parallel to each other. The thickness of the wall is a crucial point for greater or lesser acoustic absorption, since the greater the thickness, the greater the depth of the enclosures, the greater the sound contact surface. However, the greater the number of double openings, the wider the enclosures, the lower the sound absorption.

17.4.4 Optimisation

To determine the best and worst solution for absorbing the incident energy, an evolutionary solver, Galapagos, a Grasshopper tool, was used to optimise the shape, varying the parameters mentioned above – radius of the circles, number of double openings, openings between waves, and wall thickness. The tests performed were developed using the three-dimensional modelling software Rhinoceros 3D – a Computer Aided Design (CAD) program – and the Galapagos tool on Grasshopper (Figure 17.11).

The Galapagos component evaluated the optimisation of the shape for a set of samples after the first collision with the wavy surface, and the number of waves projected into the environment (Table 17.2).

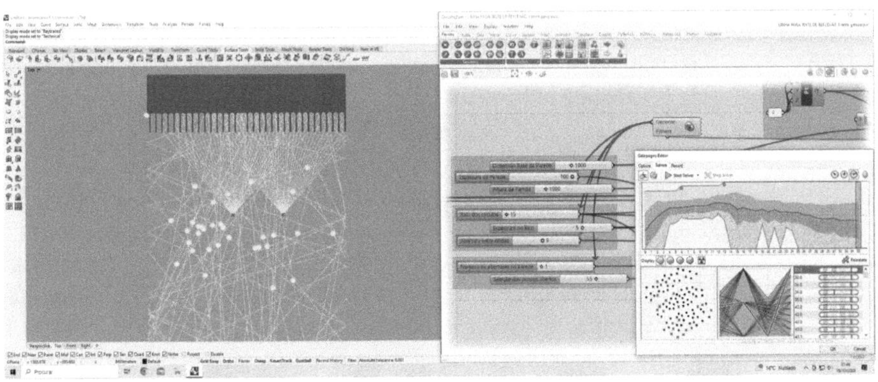

FIGURE 17.11 Optimisation of undulated geometry using an evolutionary solver, GALAPAGO. The best solution found. (Produced by the author.)

TABLE 17.2
Comparative Analysis between Solutions with (Best, Middle, and Worst) and without Walls

	Best Solution	Middle Solution	Worth Solution	Without Wall
Radius of circles	15 mm	20 mm	18 mm	–
Number of double openings	1	1	1	–
Openings between waves	9 mm	9 mm	12 mm	–
Wall thickness	100 mm	55 mm	50 mm	–
Number of sound waves projected after the first contact	32	48	64	70

FIGURE 17.12 Individual workspace and collective workspace. (Produced by the author.)

17.4.5 Architectural System

Taking advantage of the optimised surface for sound reflection, dispersion, and absorption, a modular construction system was developed, made up of customisable wall sections connected to each other by double or triple vertical connections. Each section of the wall can be customised, taking into account the sound intensity present in the space, the direction of sound incidence, and the amount of energy to be absorbed, by varying the size of the radius of the circles, the gap between waves, and the thickness of the wall. So, a system was designed with four possible spaces with different functions and different sound levels: individual work spaces, leisure spaces, exhibition spaces, and collective work spaces (Figures 17.12 and 17.13).

FIGURE 17.13 Representative axonometry of option 2. Individual workspace and collective workspace. (Produced by the author.)

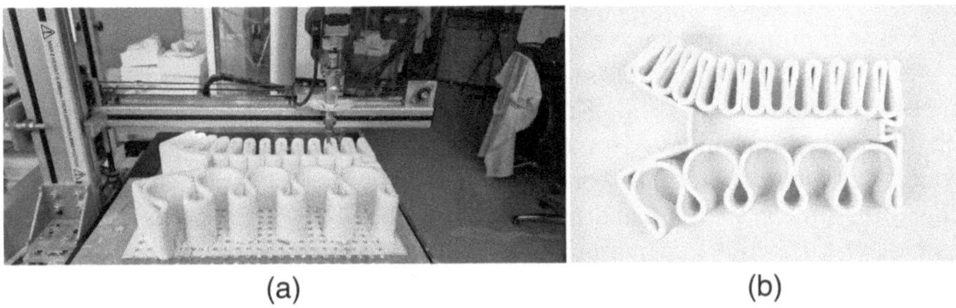

FIGURE 17.14 Printing of a module of the Cellusound prototype, left figure. Variation of the constituent walls of each individual module, taking into account the space to be applied, right figure. (Photograph by the author.)

In conclusion, the aim of this construction system is based on the possibility of grouping together different spaces with different functions and sound intensities, using mixtures based on natural materials such as cellulose as a building material. The possibility of combining spaces with different functions will determine different modules, varying the type of wall to be applied in each space (Figure 17.14).

17.5 DISCUSSION

The use of materials of natural origin in the development of construction systems is still below what we want to see. It is necessary to change society and the way we think and produce things, introducing the concept of circularity, in which everything we produce can be returned to the planet to feed new cultures and promote the growth of new natural materials. The promotion of a material should always take into account the impacts associated with its use, always seeking to promote local materials such as cellulose and low-cost materials such as shrimp shells that produce chitosan. Despite the fact that these materials are so similar in that they are natural, biodegradable, and recyclable, when used in acoustic foam, they behave differently. Cellulose has great three-dimensional construction capacity, versatile applicability, and a wide range of combinations with natural additives that give it greater resistance and distinctive finishes. However, when the mixture is exposed to variations in humidity, it tends to grow mould due to the presence of starch in the mixture. As a way of avoiding this factor, we intend to introduce a biocide into the mixture, eliminating the future possibility of fungi breeding and spreading on the piece when exposed to changes in humidity. As a way of extending the useful life of the Cellusound prototype, we are developing a new formula together with RAIZ – Institute Research of Forest and Paper. Unlike cellulose, chitosan has limited two-dimensional constructive capacity and is limited to planimetric moulds. In order to circumvent this factor, we have developed a manufacturing technique that consists of printing a geometry on a gelatinous base. The final finish of the pieces is essentially due to the proportion of chitosan added. The flexibility is due to the proportion of glycerine added.

17.6 CONCLUSION

This research promotes the use of natural materials and their viability for the development of sustainable architectural construction systems, as is the case with the prototypes presented – BioTILES, Kusudama, and Cellusound. These were produced with different compositions of mixtures resulting from the same natural base, taking advantage of the use of local raw materials and the possibility of customising systems using three-dimensional modelling tools. Developed from the same perspective, they behave differently depending on the material used. BioTILES are smaller, flexible, translucent, textured components produced using moulds and low construction capacity. Kusudama and Cellusound are made up of larger, more robust, rigid components with high construction capacity, both for the production of components and formwork. In short, chitosan acts as a coating – a skin; cellulose acts as a structure.

ACKNOWLEDGEMENTS

This work was financed by the Project Lab2PT – Landscapes, Heritage and Territory laboratory – UIDB/04509/2020 through FCT – Foundation for Science and Technology and the FCT Doctoral Grant with the reference SFRH/BD/144794/2019. We thank RAIZ – Institute Research of Forest and Paper for their support and partnership in this research, namely through the supply of cellulose. We are grateful to the Institute of Design of Guimarães for hosting and supporting the Advanced Ceramics R&D Lab on the use of their facilities and equipment.

REFERENCES

Azevedo, V., Chaves, S., Bezerra, D., Lia Fook, M., & Costa, A. (2007). Quitina e Quitosano: aplicações como biomateriais. *Revista eletrónica de materiais e processos*, Vol. 2–3, pp. 27–34. Retrieved from https://www2.ibb.unesp.br/Museu_Escola/4_diversidade/alimentacao/Textos/Quitina_Quitosana.pdf

Campos, T., Cruz, P. J. S., & Figueiredo, B. (2019). *Exploração da utilização de pasta de papel na fabricação aditiva em arquitetura* (Dissertação de mestrado). Universidade do Minho, Guimarães.

Campos, T., Cruz, P. J. S., & Figueiredo, B. (2021a). BioTILES: a sustainable interior wall panels. In *BE-AM Symposium and Exhibition: Proceedings of the of BE-AM 2022 – Symposium for Additive Manufacturing in the Built Environment, Frankfurt, Germany, November 15–18, 2022* (Vol 1, pp. 25–32).

Campos, T., Cruz, P. J. S., & Figueiredo, B. (2021b). Experimentation of natural materials: the use of chitin in additive manufacturing. In Hvejsel, M., & Cruz, P. J. S. (Eds), *Proceedings of the fifth International Conference on Structures and Architecture (ICSA 2022), Aalborg, Denmark, 6–8 July 2022* (pp. 7–8).

Campos, T., Cruz, P. J. S., & Figueiredo, B. (2021c). Exploration of natural materials in additive manufacturing in architecture: use of cellulose-based pulps. In Hvejsel, M., & Cruz, P. J. S. (Eds), *Proceedings of the Fifth International Conference on Structures and Architecture (ICSA 2022), Aalborg, Denmark, 6–8 July 2022* (pp. 87–88).

Crini, G. (2022). *Chitin and Chitosan: Discoveries and Applications for Sustainability.* London: Academic Press.

Franklin, K., & Till, C. (2018). *Radical Matter: Rethinking Materials for a Sustainable Future.* London: Thames & Hudson.

Rosendahl, P., & Gaudillière-Jami, N. (2023). *Be-AM Building Environment Additive Manufacturing. Be-AM Symposium.* Darmstadt, Germany.

Rosendahl, P. L., Figueiredo, B., Turrin, M., Knaack, U., & Cruz, P. J. S. (2024). *AM Perspectives | Research in Additive Manufacturing for Architecture and Construction.* BookRxiv.

Siegel, L. (2018). *Turning the Tide on Plastic: How Humanity (And You) Can Make Our Globe Clean Again.* London: Trapeze.

Silva, C. (2018). *Identidade Cultural: A casa do centro histórico de Guimarães.* Covilhã: Universidade da Beira Interior. Retrived from https://ubibliorum.ubi.pt/bitstream/10400.6/8470/1/6420_13493.pdf

Sisko, M., & Pfaffli, I. (1995). *Fiber Atlas: Identification of Papermaking Fibers.* Berlin & Heidelberg: Springer Science & Business Media.

Solanki, S. (2018). *Why Materials Matter: Responsible Design for a Better World.* New York: Prestel Publishing.

Treggiden, K. (2020). *Wasted: When Trash Becomes Treasure.* Brussels: Ludion Publishers.

18 Harnessing Biomimicry in Product Design for Higher Education

Lessons from Workshop Biomimicry 101 for a Sustainable Future

Jeremy Aston, Ana Duque, and Luciana Barbosa

18.1 INTRODUCTION

This work is a reflection on the author's participation in the workshop "Biomimicry 101—Designing for a Sustainable Future" at ESADA Granada, a course facilitated by Chris Gauthier, Theresa Millard, and Matthew Neiman in Granada (ES), accredited professionals as Manager of Environmental Sustainability, Creative Director, and Sustainability Analyst. The workshop was organised into five distinct and sequential modules, with a progressive approach to the content: from the presentation of the main concepts of biomimicry to the potential of its biological models, to observation in fieldwork, to the implementation of design thinking in practical work. The presentation of various case studies and real projects demonstrated how learning with nature can be achieved and how the *praxis* of biomimetics presupposes a methodology of innovation. The fieldwork, *in situ* observation, aimed to verify some aspects of the survival of certain organisms and how they have found efficient and innovative solutions to survive. At the end of the course, we have realised a practical approach to design thinking, making a prototype, and extrapolating the lessons learned.

This chapter is a proposal for reflection and discussion based on this experience and is organised into five parts:

1. *The meaning of nature* provides a contextualisation of the meaning of nature, the artificial-natural dichotomy (natural world and built world), and the scope of biomimicry in terms of its philosophical and practical essence. In the era of the Anthropocene, the depletion of the natural world and the consequent imbalance of its systems have led humanity to become aware of the implications of the built environment in its most varied expressions. We consider the importance of biomimetic practices in activating a new awareness of nature and sustainability.

DOI: 10.1201/9781032719290-21

2. *Biomimicry:* an inspiring science that explores the intersection of biology and technology through biomimicry, positioning it as a key driver of 21st-century innovation in design, especially in creating sustainable and efficient products. It highlights the potential of biomaterials and bio-inspired solutions, such as mycelium and lab-grown plant materials, while also addressing the ethical concerns surrounding the manipulation of natural systems for human benefit, urging caution in balancing innovation with ecological integrity.
3. *The Role of Design in Integrated Biomimicry* highlights the critical role of design in integrating biomimicry, emphasising that biomimetic practices require designers to draw inspiration from natural systems to create innovative, sustainable solutions. It underscores the importance of interdisciplinary collaboration between designers, biologists, and engineers in academia and industry to overcome challenges, enrich design processes, and effectively implement biomimicry in education and product development for sustainable innovation.
4. *Biomimicry tools for designers* include several biomimicry practices using methods like the Biomimicry Design Spiral, which includes steps such as defining challenges, biologising functions, and emulating nature's strategies to create sustainable solutions. By utilising nature's patterns, taxonomy, and evolutionary lessons, even those without specialised biological training can develop innovative, bio-inspired products that enhance user experience and promote environmental sustainability.
5. *Biomimicry for Product Design* is the integration of biomimicry into product design that can occur at different stages of the design process, such as leading with biomimicry research, inspiring through expert collaboration, or embedding experts within the design team. By studying natural systems and principles, designers can create more sustainable, efficient, and innovative products, as exemplified by projects like bio-inspired orthotic devices, educational tools, and eco-friendly packaging.

18.2 THE MEANING OF NATURE

Humans' relationship with nature is multifaceted and historically deep. Beyond survival and dependence on natural resources, human action in the natural world has always been intervening and transforming, representing the construction of the artificial world in its multiple expressions.

It is not our intention to delve into the concept of nature to its full extent, as there is a vast multiplicity of approaches from various fields of science, revealing the complexity of the concept and its implications. We approach the subject from the point of view of biomimicry, as this is the starting point for all its theory and practice, and because planet Earth is our ultimate homeland, in the words of Edgar Morin. There is an urgent need to raise awareness of the issue of climate change, which is the consequence of imbalances in the various ecological systems. Environmental organisations, supported by the scientific community, have issued severe warnings, and, faced with this escalating situation, it is possible that a certain feeling of helplessness and pessimism prevails, leading to denial and consequent non-action. We highlight the importance of biomimetic practices because they advocate a proactive, effective, and real plan, leading to a new awareness of ecosystem maintenance. In this way, we believe that the methods and results patented in biomimetic projects and products convey a sense of real action in building a possible future for the planet and all the beings that inhabit it.

In addition to this change in the understanding of the natural world, we believe that biomimicry, according to Beynus, presupposes an approach based not only on biological principles but also on philosophical ones, understanding nature as *physis*,[1] that is, an inherent and self-generating force that governs the natural order of the world. The word nature, according to its etymology, comes from the Latin *natura* and has a direct relationship with the act of being born, of creation. The Greek word for nature is physis, which includes human aspects.

18.2.1 Between Natural and Artificial

The history of the impact of human intervention on the environment is as remote as the evolution of humanity. Homo sapiens developed in southern Africa around 260,000–350,000 years ago, living in dryland areas and migrating to different ecosystems. The need for food and shelter is thought to have led humans to intervene in some way in the multiple environments where they have lived and settled throughout history, especially since sedentarisation (Schlebusch et al., 2017 in Keune et al., 2022).

Throughout the modern era, the Western world has emphasised the separation between culture, humans, and nature. The nature/culture divide reveals a dichotomy that places nature as a separate entity whose knowledge and conquest are for the benefit of humanity (Keune et al., 2022). Historian Donald Worster states that, during the Industrial Revolution, the biological classification of species and scientific exploration took place, thus the scientific rationalisation of nature (Worster, 1977: 2 in Keune et al., 2022). This rational way of perceiving nature is contemporary with other trends, namely Romanticism, which questions civilisation's dominance over nature, idealising it and glorifying its purity and splendour of beauty.

It is worth noting that the history of industrialisation and consequent technological advances has led to a distancing of humanity from nature and a denial of human dependence on non-human nature (Mathews, 2011). In the 20th century, anthropocentric and biocentric environmentalist theories were defended, valuing humanity to the detriment of nature or highlighting nature as its untouchable system, equating the value of the human with the natural. Kuperus and Oele (2017) state that, in postmodernity, the notion of biotic[2] nature and abiotic[3] non-human nature is perhaps not yet clear enough. In the age of environmentalism, questions are being raised about concepts of nature, including the value of the human and the value of the natural, in short, a reflection on the existence of human life in the non-human biotic and abiotic world.

About the concept of sustainability, it can be noted that the notion of reconnection aims to integrate socio-economic dynamics with ecological systems. It is important to establish reconnection strategies for all human societies that depend on the ecosystem services[4] provided by the biotic and abiotic environment (Kuperus & Oele, 2017). Relevantly, it could be said that the very notion of sustainability indicates that we should model and incorporate the entire built environment with nature, i.e., reorganise all productive and industrial sectors and systems.

Mathews (2011) refers to another definition of nature where the human being has a new place: She uses the concept of inclusivity as a notion of a larger life system, contemplating the differentiation and co-relation of human and non-human aspects. In short, an environmental and moral ethic that places humans and non-humans on the same level. The author adds that it is important to rethink humanity's place in the natural dimension also because it includes the artificial, which is an expression that enhances the natural. "Nature will no longer be understood as something untouched by human hands, but rather as something deeper—something that can be expressed through our craftsmanship, just as it is in the work of the spider or the bee" (Mathews, 2011).

18.2.2 Biomimicry

In this context, the concept of biomimicry emerged with the biologist Janine Beynus (1997), considering that this new area would represent a new era, not in a relationship of extraction from nature, but of deep learning, mimicking the principles of the life of systems and their modes of adaptation. The *biomimicry* movement emerged around the publication by Benyus (1997) *Biomimicry: Innovation Inspired by Nature.* This work popularised the term, defining the principles for a possible method. Over the last few decades, Benyus' work has stimulated a considerable amount of biomimicry research projects, as well as the production of literature on the subject.

> In a biomimetic world, we would manufacture as animals and plants do, using the sun and simple compounds to produce fibres, ceramics, plastics, and biodegradable chemicals. Our farms, inspired by the prairies, would be self-fertilising and pest-resistant. To find new medicines or crops, we would consult animals and insects that have been using plants for millions of years to stay healthy and nourished. Even computing would be inspired by nature, with software that "evolves" solutions and hardware that uses the lock and key paradigm to compute by touch.
>
> *(Benyus, 1997, 10)*

Following in Benyus' footsteps, economists and designers such as Amory and Hunter Lovins and William MacDounough[7] think that the human species will necessarily have to rethink its place on the planet and in the universe. In the words of Benyus, "one vote in a parliament of 30 million (maybe even 100 million), one species among species" (1997, 16). On an ethical and moral level, biomimicry should be at the service of all beings who have their meanings and purposes.

The intersection between biomimicry and product design could mean a paradigm shift in terms of creativity applied to design. Biomimicry is a practice that proposes a new technical culture underpinned by ecological principles, following an iterative thought process. It applies a conceptual and technological methodology to discover and imitate the design of organisms. In the words of Fisch (2017), biomimicry promotes inspiration, escaping epistemological analysis, and its modality is ontological. Biomimicry is considered ontological because it relates to the fundamental nature of being or existence. By emulating processes found in nature, biomimicry reflects an understanding of the inherent principles and structures that underpin the functioning of living organisms. It goes beyond imitation, seeking to understand and apply the essence of natural solutions to various problems. In this way, biomimicry delves into the ontology of nature to find inspiration and solutions to technological and design problems. Thus, the object of biomimicry is to emulate the construction of an organism, its ecosystem strategies, and its chemical processes. To implement the design process, the Biomimicry Institute (AskNature, 2008) has developed a classification system, a taxonomy organised according to eight operating processes and various subcategories.

Mathews (2011) states that to practise biomimicry in a deeper sense is to understand the very will and interests of nature itself, assuming a change in what humanity wants for itself and this relationship with nature.

18.3 BIOMIMICRY: AN INSPIRING SCIENCE

"I think the biggest innovations of the twenty-first century will be the intersection of biology and technology. A new *era is beginning, just like the digital one was when I was his age*" (Isaacson, 2011).

The fusion between biology and technology, highlighted by Walter Isaacson, is shaping up to be the epicentre of 21st-century innovations. Biomimicry, going beyond aesthetics, is emerging as an inspiring science for design, particularly in products and industrial design.

Since biomimicry is the empathetic and interconnected understanding of how life works, it is a practice that learns and imitates the strategies used by species living today, offering practical and scientific inspiration for design, engineering, and innovation (Biomimicry Institute, 2019).

Benyus (1997) proposes a comprehensive approach, encouraging an in-depth understanding of the efficient adaptive strategies developed by nature over millions of years of evolution.

The integration of biomimetic principles into industrial design promises to revolutionise innovation, enhancing not only aesthetics but also the functionality, efficiency, sustainability, and resilience of products. This approach represents a new paradigm for nature-centred design, driving an era of solutions inspired by biological evolution.

18.3.1 Designing for Sustainability: Design, Biomaterials, and Biomimetics

Design is a crucial process that translates ideas and materials into functional objects, structures, or systems with a clear purpose. In the context of sustainability, it plays an essential role in reducing environmental impact while offering innovative solutions. By integrating biomaterials and biomimetic principles, designers can move away from traditional processes, which are demanding in terms of resource use and wear, and adopt more circular and ecological methods. With the growing awareness of environmental problems, the scarcity of resources, and the urgent need for sustainable solutions, industries are undergoing a global transformation. In this context, we believe that design, biomimicry, and biomaterials can converge in a viable approach to sustainability.

Biomaterials, derived from renewable biological resources, provide ecological alternatives to fossil fuel-based materials. On the other hand, biomimicry, which emulates nature's strategies, offers a model for creating resilient and efficient systems. In an ideal scenario, these disciplines allow designers to reimagine products, structures, and systems that not only meet human needs but also integrate harmoniously with natural ecosystems.

In this quest for sustainable innovation, the design process connects scientific discoveries with practical applications. Some authors argue that the production of alternative materials has given rise to a new design practice at the intersection of design, materials science, biology, and the arts, which can significantly change the role of the designer from a mere receiver to an active creator of materials (Karana et al., 2018). Neri Oxman's vision of "Material Ecology" reimagines the future of design, linking it closely to ecological principles and materials science. Her work suggests that sustainable innovation can be achieved by treating materials as dynamic and intelligent entities capable of responding to their environment. By drawing inspiration from nature and harnessing technological advances, Oxman proposes a future in which design becomes more integrated with living systems, reducing environmental impact and promoting a more sustainable interaction between humans and their environment (Doordan, 2021).

18.3.2 When Design Meets Science

Antonelli (Kune, 2021) points out that the multiple forms of expression offered by bio-design, combined with the ability to re-evaluate and reinvent the paradigms of design and production, favour an environment conducive to innovation. Bio-design, therefore, represents a field where designers have a responsibility to conceive and develop sustainable products and habitats, involving themselves directly with living systems. This intersection between biology, art, architecture, and design is motivated by the opportunities that arise when working with living organisms, creating solutions that meet human needs and integrate natural ecosystems.

The idea persists that design finds inspiration in nature to create forms. One example is the work of Jane Scott, which includes the development of fabrics that can change shape in response to environmental stimuli, taking inspiration from natural processes such as the movement of pinecones. Leading the Living Textiles Research Group at Newcastle University, Jane Scott studies how textiles can serve as a sustainable and responsive building material (Kune, 2021). Overall, Jane Scott's work pushes the boundaries of design by integrating biological principles but also invites critical reflection on the implications and effectiveness of biomimetic approaches in the quest for true sustainability.

In an interview (BIOTOPIA, n.d.), Carole Collet, founder and researcher at the Design & Living Systems Lab, a pioneering laboratory that explores the interface between biological sciences and design, says that the future of design and manufacturing lies in biology and that there is an urgent need to radically rethink production systems so that they become more ecological, thus eliminating any kind of chemical processes. Carole Collet has imagined synthetic and programmable scenarios in which the planet faces limited resources and plants, genetically modified, face not only food needs but also textile production. Her BioLace project proposes four genetically modified plants: tomato, basil, strawberry, and spinach, whose DNAs have been adapted to produce food from their aerial parts and lace fabrics from their roots (Dezeen, 2013).

The pernicious action of plastics has been a relevant motivation for scientific research, and as the problems related to plastics become increasingly evident, the demand for biodegradable products has grown. It is important to remember that the production of plastics for the packaging industry has a significant impact, accounting for approximately 40% of plastic production (Pohan et al., 2023). Polystyrene, or Styrofoam, is one of the most widely used products in packaging and is also one of the most environmentally damaging waste products. Despite being an effective solution for protecting food, electronics, or other fragile items, studies indicate that it contains substances that are harmful to human health. Prolonged exposure can cause neurological effects and increase the risk of diseases such as leukaemia (Pohan et al., 2023).

Some alternatives to these scenarios are mentioned; the use of mycelium, a component of fungi which offers a biodegradable alternative to polystyrene, stands out. This material is formed through a network of microscopic white fibres, is light, is mouldable, and has potential applications in packaging, furniture, and organic plastics. Studies indicate that mycelium-based materials are a viable solution for small and medium-sized enterprises (SMEs) dealing with plastic waste (Pohan et al., 2023).

In the field of biomaterials, we mention *Mylo*, developed by Bolt Threads, as one of the innovative applications of biomaterials as a sustainable alternative to leather. *Mylo*'s production process includes the cultivation of mycelium in a vertical farming facility powered by 100% renewable energy. This material is biodegradable and requires fewer resources, making it an alternative for industries such as fashion, furniture, and car interiors. Brands such as Adidas and Stella McCartney have incorporated *Mylo* into their products, demonstrating the versatility of this material. Bolt Threads also produces *MicroSilk*, based on the study of silk proteins produced in spider webs, determining their properties by studying DNA. Proteins inspired by these natural silks are developed using bioengineering to insert genes into yeasts. Spider webs contain silk fibres with relevant properties, such as high tensile strength, elasticity, durability, and tactile softness (Bolt Threads, 2024). It is worth noting that, with today's technology, it is possible to create these proteins without using the spiders themselves.

MIT researchers have developed an innovative technique to grow wood-like plant material in the laboratory, offering a sustainable alternative to deforestation. This material can be chemically adjusted to have specific properties, such as density and rigidity, and be moulded into precise shapes via 3D bio-printing. Using plant cells such as *Zinnia elegans*, the process is much faster than growing natural trees and generates less waste. The research could, in the future, enable the creation of customised construction materials without the need to cut down trees (Beckwith et al., 2022; MIT, 2022).

VTT and ISKU have developed the first bio-composite chair using foaming technology, initially created for paper production. Made from cellulose fibres, wood pulp, and polypropylene, the chair is a sustainable design object made from natural materials from Finnish forests. The bio-composite maintains good mechanical properties and can be moulded in 3D and worked with conventional tools. According to the company, as well as being light and resistant, the material is recyclable and versatile, with the potential to be used in furniture, packaging, and construction. According to the company, the future biotechnological revolution will transform industrial processes in the chemical, energy, food, pharmaceutical, and cellulose areas. At VTT, living cell factories are developed using microorganisms such as bacteria, yeasts, and moulds, as well as algae, plants, and plant cells, for the sustainable production of biomass, various ingredients, high-value chemicals, and proteins. VTT intends to expand these innovations to the market (VTT, 2024).

We believe that the proposals of biomaterials and design, or the integration of life into design practices, are not yet a simplified solution and could generate ethical and moral controversies. In addition to the use of living constructions and systems, bio-design also encompasses synthetic biology, which can destabilise natural ecosystems. Bio-design raises the ethical question of to what extent humans should attempt to redesign natural systems, especially when the long-term effects are unknown. William Meyers believes that the potential benefits and the need to adapt current practices to biological systems outweigh these risks. Thus, according to the author, integrating nature into design is an expected and promising approach for the future (Meyers, 2012). However, the ethical debate prevails, which revolves around respecting the intrinsic value of all living beings and balancing innovation with ethical limits.

The desire to emulate and manipulate nature has deep historical roots in human civilisation, embodying our desire to understand, control, and benefit from the natural world. From ancient agriculture to modern biomimicry and genetic engineering, we have sought to harness nature's processes. However, this raises critical questions about whether we have really evolved from this mindset or whether we are just continuing to shape nature in ever more sophisticated ways.

Is there a real sense of evolution in the mindset, or are we just refining our tools of intervention? The concern is whether this emulation can turn into exploitation, where natural systems are manipulated for human gain, potentially to the detriment of ecological balance.

18.4 THE ROLE OF DESIGN IN INTEGRATED BIOMIMICRY

The synergy between biomimicry and design, especially in the context of products, constitutes an innovative approach that transcends the simple manifestation of creativity, demanding true innovation. The proposal aims not only to imitate natural morphology but, more intrinsically, to capture the essence of nature to achieve a balanced fusion between aesthetic and functional aspects, emphasising resilience as a central component. Within biomimetic design, creativity plays a crucial role, keeping the focus on the centrality of the human experience in the face of rapid technological transformations and global challenges.

The practice of biomimicry can be categorised into three distinct levels. At the first level, it seeks to emulate the natural "form," as proposed by Benyus. At the second level, biomimicry explores biological "processes" and their application in design and engineering. At the third level, the biomimetic approach reaches the "systemic" level, focusing on principles, patterns, and strategies present in ecosystems (Benyus, 2002). This classification highlights the complexity and scope of biomimicry as an innovative tool in contemporary design.

In this way, we can affirm that design plays a fundamental role as a mediator and scenario builder; its presence is vital not only to incorporating biomimetic elements but also to conceiving scenarios and behavioural prototypes (Myers & Antonelli, 2018). This capacity goes beyond aesthetics, assuming a leading position in defining human interactions with products, especially in times of rapid environmental, political, and technological transformations.

The design doesn't just follow nature; it moulds human experiences through its unique mediation.

Furthermore, design is recognised as a solver of challenges; although adaptation and imitation in biomimicry are fundamental, they present complex challenges for designers. Meticulously analysing biological solutions requires not only technical knowledge but also a creative mind capable of extracting biomimetic principles and applying them innovatively and sustainably (Benyus, 2009). In this context, design emerges as the solver of these challenges, transforming biological complexity into practical functionality.

18.4.1 Importance of (Inter)Multidisciplinary Teams and Collaborative Process in Biomimicry

Design is usually a collaborative process, encompassing multidisciplinary teams, involving at least marketing professionals, engineers, and clients. On the other hand, biomimicry, by transcending the mere imitation of natural patterns, incorporates the collaborative logic of ecosystems in solving human challenges. This process requires a multidisciplinary approach, reflecting the efficient collaboration found in nature, where different species play specific roles for the collective benefit. Biomimetic design, on the other hand, also includes biologists, increases complexity, goes beyond simply viewing the design challenge from different perspectives (multidisciplinary), and also involves the integration of disciplinary knowledge (interdisciplinary) (Cohen & Reich, 2016).

Effective biomimetic design must, therefore, reflect the diversity of nature, involving biologists, designers, engineers, and other specialists. Effective collaboration between disciplines creates synergies of knowledge, overcoming disciplinary barriers and enriching the understanding of natural systems.

The convergence of knowledge in academia is not just a strategy but an imperative necessity to drive biomimetic innovation (Vincent & Mann, 2002). Biomimicry, fuelled by different academic perspectives, transcends imitation, achieving sustainable and ethical innovations through interdisciplinary collaboration.

The multidisciplinary team is crucial to the success of biomimetic design, playing a central role in the holistic understanding of natural systems and the effective application of biomimetic principles. Product design inspired by biomimicry is intrinsically collaborative and interactive. Close collaboration between designers and biomimetic experts is crucial to deepening the understanding of natural systems, driving an iterative process of adaptation and innovation (Vincent & Mann, 2002). The ability of design to integrate feedback and continually adjust is essential to the success of biomimicry in practice.

The collaborative process in both academia and industry catalyses the practical translation of biomimetic principles. The continuous interaction between designers, biologists, engineers, and other specialists results in more efficient and ethical solutions, driving innovation and enabling the agile application of biomimetic concepts.

18.4.2 Integrating Biomimicry into Design: Challenges and Opportunities

The contemporary design landscape is constantly evolving, entering fields previously considered outside its traditional scopes, such as scientific visualisation, interfaces, sociological theories, and nanotechnology. This expansion requires essential collaboration between designers and experts from different fields, resulting in a symbiosis between creativity and scientific knowledge (Myers & Antonelli, 2018).

For the successful integration of biomimicry into education, an in-depth understanding of natural systems is imperative, from a specialised education to gaining insights into biological and ecological principles. However, the effective introduction of biomimicry encounters obstacles in traditional educational curricula, requiring the revision and adaptation of educational programmes to include biomimetic components (Vincent & Mann, 2002). The lack of specialised courses can result in a shortage of specialists, requiring the implementation of dedicated programmes and specialised teachers (Benyus, 1997). Biomimicry also requires access to specific resources, such as laboratories, making it crucial to ensure their availability (Benyus, 2009).

In academia, interdisciplinary integration involving biology, design, and engineering professors is advantageous. Practical, collaborative projects involving students in real biomimetic challenges provide a concrete application of the principles learned. Collaborations between academic institutions and the biomimetic industry enrich the students' experience (Benyus, 2009).

The lack of knowledge about biomimetics is a significant barrier, requiring efforts to overcome this gap. The complexity of biological processes requires in-depth research and the involvement of multiple specialists. Resistance to innovative approaches, both cultural and organisational, is an additional barrier (Benyus, 1997; Vincent & Mann, 2002).

Inappropriate application of biomimetic principles can result in ineffective or unsustainable solutions. High initial costs and investments in research can increase expenses, but promise long-term efficiency and sustainability (Benyus, 2009; Vincent & Mann, 2002).

Biomimicry offers opportunities to create innovative, efficient, and sustainable products in line with the principles of nature. Companies that adopt biomimicry increase their competitiveness by offering unique and environmentally friendly solutions. Biomimetic research contributes to the development of advanced materials, while educational programmes centred on biomimetics transform the mindset of designers, preparing a new generation for innovative and sustainable approaches (Benyus, 1997; Vincent & Mann, 2002).

Thus, the successful integration of biomimicry into design requires a holistic approach, overcoming educational challenges, and practical barriers, and managing risks. The opportunities generated by biomimicry have the potential to transform design, driving sustainable innovation and aligning industrial design practices with the fundamental principles of nature.

18.4.3 Workshop Experience

Participating in the "Designing a sustainable future BIOMIMICRY 101" workshop in November 2023 in Granada, over three intensive days, proved to be an immersive experience in innovation inspired by nature, where there was the opportunity to absorb not only the fundamental standards, methods, and tools on biomimicry but also to be guided by a multidisciplinary team led by experts.

Focused observation and targeted research into the patterns and forms of survival refined over millions of years in nature proved to be extremely relevant to our professional career as designers, researchers, and lecturers in design. The effectiveness of the workshop's theoretical–practical format stands out from this experience, as does the ability to immerse oneself, learn, and practically apply biomimetic projects.

The workshop not only laid a solid foundation but also projected a future vision, where we can see a deeper integration of biomimetic approaches into the teaching component and anticipate a more innovative practice aligned with the fundamental principles of nature. This experience reflects not only a successful immersion but also signals a promising and sustainable trajectory in design and higher education.

18.5 BIOMIMICRY TOOLS FOR DESIGNERS

This research prompts the fundamental inquiry: "In what manner can designers effectively incorporate biomimicry into their professional practice?" This question gains significance because the typical educated designer often lacks specialised training in biological sciences. The Biomimicry 101 workshop conducted in Granada unveiled compelling tools and perspectives that hold potential for adaptation, transfer, or integration into design practices. Particularly, these insights could be instrumental in developing novel products imbued with bio-inspired attributes, aimed at enhancing user experience, contributing to societal welfare, and fostering environmental sustainability.

Biomimicry tools and perspectives: essential elements; evolutionary timeline; case studies; nature's unifying patterns; biomimicry taxonomy; biomimicry design spiral.

18.5.1 Essential Elements

In the process of translating nature's strategies into design, the scientific practice entails three crucial elements: Emulate, Ethos, and (Re)Connect (Biomimicry.net, n.d.). These three foundational components are intrinsic to every dimension of biomimicry, embodying its fundamental principles and fostering collaboration among individuals sharing common objectives.

Ethos: This embodies the philosophy centred on comprehending life's workings and crafting designs that consistently uphold and cultivate conditions conducive to life;

(Re)Connect: This concept emphasises our inherent connection with nature, acknowledging that we are an intrinsic part of Earth's interconnected systems. Practising (Re)Connect encourages us to actively engage with and spend time in nature, gaining insights into life's mechanisms. This understanding cultivates a stronger ethos, enabling us to emulate biological strategies effectively within our designs;

Emulate: This involves the scientific and research-driven approach of studying and replicating nature's forms, processes, and ecosystems to foster the creation of more regenerative designs.

18.5.2 Evolutionary Timeline

As per scientific estimations, considering the creation of Earth approximately 4.54 billion years ago (National Geographic, n.d.), if this vast timeline were condensed into a single year, the existence of modern Homo sapiens would equate to merely one day, spanning around 195 thousand years (Britannica, n.d.). Hence, in comparison to other life forms, such as fish, which have thrived for approximately a month and a half, encompassing about 530 million years (University of Birmingham, n.d.), our presence on this planet is relatively nascent. This evolutionary context spans the development of Earth and the estimated count of 4 billion living species (Biology-StackExchange, n.d.), representing a substantial reservoir of case studies encompassing both thriving and extinct organisms, offering valuable insights for learning and exploration.

18.5.3 Case Studies

For novice designers exploring biomimicry design methodologies, numerous compelling case studies exist that elucidate their capacity for fostering innovative products with resource efficiency, thereby enhancing societal well-being. Among the renowned examples, the bullet train in Japan and Velcro stand out as notable illustrations of this approach.

One of the most renowned instances of biomimicry involves the redesign of bullet trains. Initially, the original bullet trains in Japan generated disruptive sonic booms upon exiting tunnels, causing inconvenience to nearby residents while also lacking optimal efficiency. Drawing inspiration from kingfisher birds, renowned for their sleek, sharp beaks that enable them to effortlessly dive into water without disturbing its surface, engineers embarked on a transformative redesign. They reconfigured the bullet trains to mimic the elongated, pointed front characteristic of a kingfisher's beak. The resulting new design significantly mitigated noise levels and enhanced energy efficiency, marking a notable success in biomimetic engineering.

A transformative moment during an outdoor excursion served as the catalyst for the inventor of the hook-and-loop fastener, commercially known as Velcro.

George De Mistrials' curiosity was piqued during a walk with his dog as both returned home covered in plant burs that adhered to their clothing and the dog's fur. Intrigued by these persistent seeds, he conducted a microscopic examination, revealing their surface adorned with minuscule hooks. This discovery inspired him to develop the iconic adhesive fabric featuring one side equipped with hooks interlocking with tiny loops on the other, resulting in the creation of Velcro (environment.co/biomimicry-examples).

18.5.4 Nature's Unifying Patterns

Nature's Unifying Patterns represents an endeavour aimed at discerning the ten pivotal lessons derived from observations in the natural world that warrant incorporation within a design framework (Biomimicry Institute, 2023). These patterns have been meticulously documented through the observation of diverse life forms inhabiting our planet. They serve as compelling sources of inspiration to inaugurate design endeavours and function as invaluable evaluation tools throughout the design process: Nature uses only the energy it needs and relies on freely available energy; nature recycles all materials; nature is resilient to disturbances; nature tends to optimise rather than maximise; nature rewards cooperation; nature runs on information; nature uses chemistry and materials that are safe for living beings; nature builds using abundant resources, incorporating rare resources only sparingly; nature is locally attuned and responsive; nature uses shape to determine functionality.

This manifestation of nature's intricate patterns is notably discernible across multiple case studies, wherein specific flora, insects, and fauna adeptly harness water amidst exceedingly arid environments, devise sophisticated mechanisms for self-preservation, and exhibit remarkable strategies to incentivise and enhance reproductive processes.

18.5.5 Biomimicry Taxonomy

The process of identifying, analysing, and assessing natural patterns for design purposes is inherently intricate, requiring specialised skills for deciphering and categorising the diverse traits and functionalities observed in living organisms. As a response to this complexity, the Biomimicry Institute has pioneered the development of a comprehensive classification system (see Biomimicry Taxonomy system. Source: biomimicry.org) to assist designers in defining functions from nature's patterns. This system serves as a visual representation of the taxonomy of functions, effectively organising strategies and innovations derived from nature. Furthermore, it systematically categorises the various methods through which organisms and natural systems address functional challenges, grouping them based on their interrelated functions (Biomimicry Institute, n.d.).

18.5.6 Biomimicry Design Spiral

The Biomimicry Design Spiral serves as a valuable tool for comprehending the essential steps crucial for successful biomimetic design (Figure 18.1). It proves beneficial

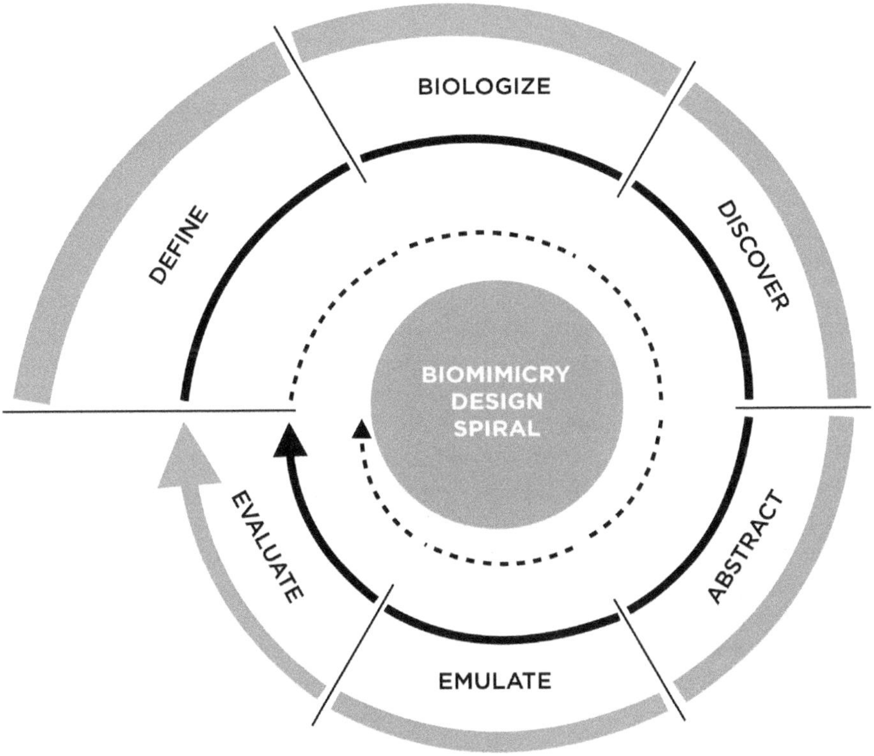

FIGURE 18.1 Biomimicry Design Spiral. (Authors' work.)

when addressing a specific problem (referred to as a "challenge") or identifying a design opportunity where drawing inspiration from biological models is sought. This tool shares similarities with established methodologies in design, such as the Double Diamond and Design Thinking, which designers, through formal training, are more familiar with.

The Double Diamond, introduced in 2004, offers a framework for innovation. Its two diamonds symbolise a process involving key principles and design methods. It begins with an exploration of an issue at a broader or deeper level (divergent thinking) and then progresses towards focused action (convergent thinking). The outcome is a clear, comprehensive, and visually descriptive representation of the design process (Design Council, n.d.).

Design thinking, on the other hand, is a non-linear, iterative process employed by teams to comprehend users, challenge assumptions, redefine problems, and innovate solutions for prototyping and testing. It encompasses five phases—Empathise, Define, Ideate, Prototype, and Test—making it particularly effective for addressing vague or unknown problems (Interaction-Design.org, n.d.).

The Biomimicry Design Spiral offers a concise delineation of fundamental components within a design process that draws upon nature's principles as a framework for devising solutions. It outlines six pivotal steps that a design team ought to undertake while exploring biomimetic resolutions for a design challenge. These steps are presented sequentially, serving as an initial guideline. Nonetheless, design teams frequently navigate fluidly between these steps or revisit them iteratively. This practice is commendable since each step often unveils fresh insights that can either validate or challenge assumptions formulated in prior stages (Biomimicry.net, n.d.):

Define: Precisely articulate the intended impact your design aims to achieve in the world (i.e., the challenge you seek to resolve). Define the specific criteria and constraints pivotal in determining the success of your design;

Biologize: Analyse the fundamental functions and context that your design solution must address. Rephrase these aspects using biological terminology, enabling an inquiry into nature for guidance and insights;

Discover: Explore natural models encompassing organisms and ecosystems dealing with analogous functions and contexts as your design solution. Identify the strategies employed by these natural models that support their survival and prosperity;

Abstract: Diligently examine the core features or mechanisms responsible for the success of these biological strategies. Translate these findings into non-biological terms, formulating them as actionable "design strategies;"

Emulate: Identify recurring patterns and interconnections among the strategies discovered, focusing on the crucial lessons that should shape your solution. Develop design concepts grounded in these identified elements;

Evaluate: Assess the design concept(s) against the criteria and constraints delineated for the design challenge. Consider the alignment with Earth's systems, as well as technical and business model feasibility. Continuously refine and revisit preceding steps as necessary to generate a viable solution.

18.6 BIOMIMICRY FOR PRODUCT DESIGN

Hence, arises the fundamental query of "when, where, and how within the design process can biomimicry find application?" As deliberated earlier, higher education in design often familiarises students with methodologies like Double Diamond and Design Thinking. However, these established methods do not explicitly delineate the specific placement of biomimicry within their frameworks. In contrast, while the Biomimicry Design Spiral exhibits a shared structure akin to the aforementioned methodologies, it also showcases an independent and distinct process (Figure 18.2).

In an endeavour to conceptualise the integration of biomimicry and design methodologies into a unified process, we explored various concepts aimed at augmenting project activities within the framework of the Double Diamond:

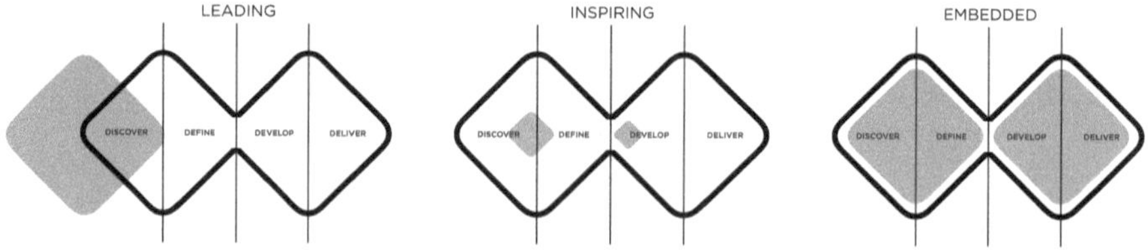

FIGURE 18.2 Double Diamond with biomimicry. (Authors' work.)

Leading: Envisaging a multidisciplinary project where Biomimicry research serves as a precursor to the design process;

Inspiring: Considering a transdisciplinary project model that involves inviting biomimicry experts at pivotal stages throughout the design process to offer insights and guidance;

Embedded: Deliberating on an interdisciplinary project approach where biomimicry experts collaborate closely with designers, actively participating in the design process.

These assumptions represent our observations, each posing inquiries regarding their implementation into higher education for design. The concepts of Leading and Influencing lean towards fostering collaboration and partnership with scientific institutions or industries. Notably, these approaches do not necessarily disrupt existing project teaching models; rather, they involve sourcing partners with shared objectives and coordinated management.

Conversely, the Embedded concept implies a fundamentally new course of action. This specialised approach in bio-inspired product development suggests the necessity for additional specialised facilities and human resources that typically extend beyond the scope of resources found within traditional art and design departments. Implementation of this concept would entail substantial structural modifications within educational frameworks.

18.6.1 Seventeen Sustainable Development Goals (SDGs)

It is common teaching practice to introduce students (at all levels of education) to the Seventeen SDGs, an agenda for 2030 adopted by all United Nations Member States in 2015. This agenda provides a shared blueprint for peace and prosperity for both people and the planet, now and in the future (United Nations, 2015). For product designers, this entails a more holistic approach that addresses society's needs—such as ending poverty and other deprivations—while aligning with strategies to improve health and education, reduce inequality, and foster economic growth. All of this must be achieved while simultaneously tackling climate change and working to preserve our oceans and forests.

Through higher design education, the role of a product designer is to meet user and customer needs by focusing on improved functionality, aesthetics, cost efficiency, and product life-cycle considerations. Beyond these core aspects, designers are encouraged to take a broader perspective on their potential influence, recognising their role in proposing products that not only enhance people's lives but also contribute positively to the environment. By integrating sustainable practices into their design processes, product designers can drive innovation that addresses societal challenges while minimising environmental impact, thus aligning with the larger goals of sustainable development.

18.6.2 Masters Project Examples

As a result, the expected teaching and learning outcomes for product design courses are increasingly orienting topics and themes, particularly at the master's level, toward sustainable design thinking and biomimicry inspirations. This shift encourages students to explore how nature's principles can inform innovative and responsible design solutions. The following examples from ESAD Escola Superior de Arte e Design and IPCA Instituto Politécnico do Cávado e do Ave showcase final projects from master's students that illustrate the advantages of integrating bio-influences into the design process. These projects demonstrate how studying natural structures, mechanisms, and materials can lead to more efficient, sustainable, and innovative product designs that reflect a deeper understanding of ecological systems.

The first exemplary project, "Design as a Preventive Strategy for Work-Related Musculoskeletal Disorders," was completed by José Almeida at IPCA in 2023. His master's research aimed to create a product solution to reduce musculoskeletal disorders (MSDs), with a particular focus on Carpal Tunnel Syndrome (CTS), which affects individuals who spend long hours working on computers, often adopting improper postures and performing repetitive movements (Almeida, 2023). Almeida's design research led to the development of a personalised orthotic device that utilises additive manufacturing technologies (ADM), commonly known as 3D printing. This approach enabled the creation of complex, intricate structures that would be difficult or impossible to achieve through conventional mass production methods.

As the product evolved, Almeida explored the use of lattice structures commonly found in nature (e.g., beehives, plants, and fungi) to design a lightweight yet durable product with a soft tactile quality. This bio-inspired lattice design not only enhanced the product's functionality—by providing ergonomic support and reducing strain on the user—but also introduced a visually dynamic and aesthetically appealing form. As a result, the final product transcended its purely medical function, becoming an object of desire that combined practicality with beauty, thanks to its nature-inspired design (Figure 18.3).

The second example, "Fixa, a Hybrid Pedagogical Product to Assist Creative Learning Activities," was completed by Maria Melo at ESAD in 2021. This project tackled a significant challenge in modern primary education: the overwhelming amount of visual and cognitive stimuli that children in the digital age are exposed to,

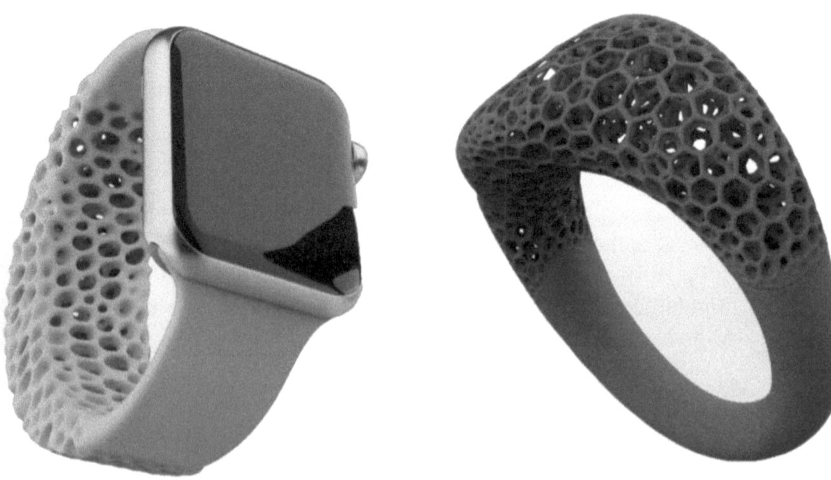

FIGURE 18.3 MSD/CTS bracelet. (Almeida 2023.)

which conventional teaching methods often struggle to compete with (Melo, M. J. C. F., 2021). After conducting extensive research, Melo proposed a hybrid pedagogical game designed for classroom use. The game combined a rewarding physical experience with digital interaction, promoting student collaboration while simultaneously providing teachers with tools for individual assessment.

During the exploratory phases, Melo sought mechanical solutions that would engage children more creatively than traditional toys, such as simple puzzles or vertical stacking blocks. Her research led her to study biological systems, particularly the ball-and-socket joints found in the human body, as well as in animals, birds, and insects. These natural joints offered a model for designing a system that allowed flexible and dynamic movement between connected pieces. Using 3D printing, Melo prototyped the concept to test the manoeuvrability of the components, experimenting with various materials to develop durable snap-fittings that could withstand repeated manipulation by young children.

The final product was an interactive problem-solving game that rewarded children for building unique structures using universally connecting pieces inspired by the mechanics of human anatomy. By adapting the ball-and-socket joint for a new educational purpose, Melo's design allowed for endless creative possibilities while maintaining durability and ease of use, making it both an effective teaching tool and an engaging learning experience (Figure 18.4).

The third example, "Tailor-Made Packaging for Azores Food Produce," was completed by Patricia Santos in 2020 as her final master's thesis at ESAD. Combining her passion for sustainable product design with her background in graphic communication, Patricia sought a design challenge that would make a visual impact while incorporating material innovation through "smart packaging." With personal ties to the Azores islands, she connected with the materials engineering department at Innovation Green Azores (IGA) and collaborated with local food producers to develop her proposal. Her objective was to design sustainable food packaging made from *Conteira*, an invasive plant that grows abundantly in the Azores, and to use it to create containers for locally produced gourmet food products (Santos, 2020).

During her experimentation with *Conteira*, Patricia worked alongside research scientists at IGA who developed a process for breaking down the plant and adding liquid additives to create a mouldable pulp. This process, similar to how egg cartons are press-moulded from paper pulp, was inspired by nature—particularly birds like swifts and swallows, which build nests from mud and straw pulp (Rushwood, n.d.), and hornets, which construct their nests using a mix of saliva and wood pulp (Redit.com). Just as these creatures rely on the precise combination of ingredients to build their homes, the success of the final product depends on achieving the right balance of materials, proportions, and consistency of the *Conteira* pulp. With the support of IGA's scientists and engineers, Patricia was able to explore the mouldable limits of the material and unlock its design potential.

Although achieving an industrial-scale solution without access to advanced machinery was challenging, Patricia used manual prototyping methods and 3D computer renderings to present a tangible representation of the packaging concept. The final result showcased an innovative, sustainable packaging solution that local gourmet food producers could use, utilising a regionally sourced material and promoting a circular economy for the Azores (Figure 18.5).

FIGURE 18.4 Fixa Hybrid Pedagogical Product. (Melo 2021.)

FIGURE 18.5 Smart packaging for the Azores. (Santos 2020.)

18.7 CONCLUSIONS

The application of biomimicry methods in situ has generated a differentiated awareness of this vast global archive of organisms, revealing an enormous field of possibilities. In this way, we believe that the methods and results patented in biomimetic projects and products convey a sense of real action in building a possible future for the planet and all the beings that inhabit it. Biomimicry offers designers the chance to tap into nature's 3.8 billion years of evolutionary wisdom, which, as demonstrated through case studies like the redesign of Japan's bullet train and the invention of Velcro, can result in innovative solutions that are both efficient and sustainable.

The Biomimicry 101 workshop held at ESADA Granada concluded after an intensive three-day session filled with immersive experiences. It introduced innovative teaching and learning methodologies, fostering an increased awareness of the immense potential this mindset holds for future design endeavours. Our primary goal as student participants in this event was to acquire knowledge and evaluate this experience. Furthermore, in our roles as educators, we aimed to propose and explore the potential implementation of these mechanisms within the realm of art and design higher education. The workshop emphasised the need for collaboration between designers and biological experts, a concept reinforced by the Biomimicry Design Spiral, which provides a structured pathway to integrate nature's strategies into product development.

As PhD holders and Specialists in Design, we swiftly realised that biomimicry demanded a deeper understanding of biology beyond the foundational knowledge typically imparted in basic education. This experience provided us with a foundation for reflection and the deepened need to analyse biomimicry's role in design. As mentioned, students must acquire these tools and principles early in their education, not just in theory but also through project practices. Examples such as José Almeida's bio-inspired orthotic device, Maria Melo's hybrid pedagogical game, and Patricia Santos' sustainable packaging show how biomimicry can be seamlessly integrated into design education. These projects demonstrate that understanding natural patterns and processes can lead to more efficient, innovative, and sustainable product designs.

Young designers should master their art and design craft but understand that biomimicry design thinking will require collaboration with other specialised scientific disciplines. The integration of biomimetic principles into industrial design can revolutionise innovation, representing a new paradigm for nature-centred design. This approach drives an era of solutions inspired by biological evolution, enhancing aesthetics, functionality, and the efficiency, sustainability, and resilience of products—ultimately contributing to a more sustainable and harmonious relationship between human design and the natural world.

ACKNOWLEDGEMENTS

This work is the result of the participation of ESAD (PT) lecturers and researchers Jeremy Aston[5], Luciana Barbosa[6], and Ana Lucia Duque[7] in the Biomimicry 101 2023 workshop, which took place at ESADA in Granada (ES) between 7 and 9 November

2023. The facilitators were Theresa Millard, Chris Ghautier, and Matthew Neiman, who presented the theoretical assumptions of biomimicry and emblematic case studies that represent it. The sessions were highly collaborative and interactive, with *in situ* nature observation work being carried out and participants proactively involved in solving practical problems, with the results being presented afterwards. With clear and objective communication, the trainers and other participants showed their passion for the subject and their awareness of and deep respect for nature conservation. This experience was fundamental in effectively approaching the subject and understanding the assumptions of biomimicry practices as a territory of possibilities for both science and project culture. It should be noted that this markedly positive experience was only possible with the indispensable and dedicated support of the Erasmus office at ESAD (PT) and esad-idea (ESAD's research centre).

NOTES

1 Is a concept rooted in ancient Greek philosophy, especially in the works of thinkers such as Heraclitus and Aristotle, who explored the underlying principles that shape the natural world.
2 Biotic nature refers to the living elements that are part of an ecosystem, namely animals, plants, fungi, bacteria and others. These living beings interact dynamically with each other and with the physical environment, forming complex networks of ecological relationships. Biotic dynamics are crucial for maintaining ecological balance in different ecosystems on the planet.
3 The concept of abiotic nature refers to the non-living components of an ecosystem, such as physical and chemical elements: water, rain, soil, temperature, sunlight and minerals. These elements play vital roles in characterising an environment and directly influence the life of the organisms that inhabit these spaces.
4 Understanding that valuing ecosystem services is fundamental to sustainable resource management and environmental conservation. Ecosystem services are benefits that humans obtain from ecosystems, e.g., products such as food, water, wood and fibres; regulating services, about the control of natural processes, including climate regulation, pollination, disease regulation and water purification.
5 Jeremy Hugh Aston (Specialist Title), Coordinator at ESAD (PT).
6 Luciana Moreira Barbosa (Specialist Title), Lecturer at ESAD (PT).
7 Ana Lúcia Pinto Duque (PhD), Lecturer at ESAD (PT) and affiliated researcher at ESAD.idea (PT).

REFERENCES

Benyus, J. M. (1997). *Biomimicry:* Innovation Inspired by Nature. Harper Perennial. Available at: https://www.academia.edu/38300413/.

Almeida, J. C. F. (2023). *Design as a Preventive Strategy for Work-Related Musculoskeletal Disorders*. IPCA Instituto Politécnio do Cávodo e do Ave. Available at: https://ciencipca.ipca.pt/jspui/handle/11110/2925

AskNature. (2008). *Biomimicry Taxonomy*. AksNature. Available at: https://asknature.org/resource/biomimicry-taxonomy/.

Beckwith, A. L., Borenstein, J. T., & Velásquez-García, L. F. (2022). Physical, mechanical, and microstructural characterisation of a novel, 3D-printed, tunable, lab-grown plant materials generated from *Zinnia elegans* cell cultures. *Materials Today*, 54, 27–41. https://doi.org/10.1016/j.mattod.2022.02.012

Benyus, J. M. (2002). *Biomimicry: Innovation Inspired by Nature*. New York: Harper Perennial. Available at: https://elmoukrie.com/wp-content/uploads/2021/11/janine-m-benyus-biomimicry_-innovation-inspired-by-nature-harper-perennial-2002-2.pdf

Benyus, J. M. (2009). *A Biomimicry Primer* (2nd ed.). Helena, MT: Biomimicry Institute.

Biology-StackExchange. (n.d.). *How Many Different Species Have Lived on Earth?* Biology-StackExchange. Available at: https://biology.stackexchange.com/questions/85634/how-many-different-species-have-existed-on-earth.

Biomimicry Institute. (2019). *What Is Biomimicry*? Biomimicry Institute. Available at: https://biomimicry.org/what-is-biomimicry/

Biomimicry Institute. (2023). *Nature's Unifying Patterns*. Biomimicry Toolbox. Available at: https://toolbox.biomimicry.org/core-concepts/natures-unifying-patterns/

Biomimicry Institute. (n.d.). *AskNature – Innovation Inspired by Nature*. Available at: https://asknature.org/

Biomimicry.net. (n.d.). *A Biomimicry Primer*. Biomimicry.net. Available at: https://biomimicry.net/

BIOTOPIA. (n.d.). *Lab behind the Scenes: Interview with Biodesigner Carole Collet*. BIOTOPIA. Available at: https://biotopia.net/en/biotopia-lab/popupexhibition/9-english/310-carole-collet-interview-en

Bolt Threads. (2024). *Mylo™*. Bolt Threads. Available at: https://boltthreads.com/technology/mylo/

Britannica (n.d.). *Just How Old are Homosapiens?* Britannica. Available at: https://www.britannica.com/story/just-how-old-is-homo-sapiens

Cohen, Y. H., & Reich, Y. (2016). *Biomimetic Design Method for Innovation and Sustainability*. Springer. Available at: https://doi.org/10.1007/978-3-319-33997-9

Design Council. (n.d.). *Framework for Innovation*. DesignCouncil. Available at: https://www.designcouncil.org.uk/our-resources/framework-for-innovation/

Dezeen. (2013). *Genetically Engineered Crops could Grow Lace Amongst Their Roots*. Dezeen. Available at: https://www.dezeen.com/2013/11/30/genetically-engineered-plants-that-produce-edible-textiles-by-carole-collet/

Doordan, D. (2021). Neri Oxman: Material ecology. *Design Issues*, 37(1), 83–84. Available at: https://doi.org/10.1162/desi_r_00626

Interaction-Design.org. (n.d.). *Design Thinking*. Interaction-Design.org, Available at: https://www.interaction-design.org/literature/topics/design-thinking#:~:text=Design%20thinking%20is%20a%20problem,solutions%20catering%20to%20their%20needs.

Isaacson, W. (2011). *Steve Jobs*. New York: Simon & Schuster.

Karana, E., Blauwhoff, D., Hultink, E.-J., & Camere, S. (2018). When the material grows: a case study on designing (with) mycelium-based materials. *International Journal of Design*, 12(2), 119–136. Available at: https://www.ijdesign.org/index.php/IJDesign/article/view/2918

Keune, H., Immovilli, M., Keller, R., Maynard, S., McElwee, P., Molnár, Z., … & Van Reeth, W. (2022). *Defining Nature*. Cambridge University. Available at: https://doi.org/10.1017/9781108856348.003

Kune, R. (2021). Designing ecologies: The role of systemic thinking in regenerative practice. In M. Chapman & C. Gant (Eds.), *Design and Nature: A Partnership* (pp. 87–104). London: Routledge.

Kuperus, G., & Oele, M. (Eds.). (2017). *Ontologies of Nature – Continental Perspectives and Environmental Reorientations*. Springer. Available at: https://doi.org/10.1007/978-3-319-66236-7

Mathews, F. (2011). Towards a deeper philosophy of biomimicry. *Organization & Environment*. Available at: https://doi.org/10.1177/1086026611425689

Melo, M. J. C. F. (2021). *Fixa: Um produto pedagógico híbrido papra apoiar atividades de aprendizagem criativa*. Available at: https://comum.rcaap.pt/handle/10400.26/36905

Meyers, W. (2012). *Bio Design: Nature + Science + Creativity*. London: Thames & Hudson.

MIT. (2022). *Toward Customisable Timber Grown in a Lab*. MIT. Available at: https://news.mit.edu/2022/lab-timber-wood-0525

Myers, W. & Antonelli, P. (2018). *Bio Design: Nature + Science + Creativity*. New York: The Museum of Modern Art.

National Geographic. (n.d.). *Age of the Earth collection*. National Geographic. Available at: https://education.nationalgeographic.org/resource/resource-library-age-earth/

Pohan, J., Kusumawati, Y., & Radhitanti, A. (2023). Mushroom mycelium-based biodegradable packaging material: A promising sustainable solution for the food industry. *E3S Web of Conferences*, 426, 02128. Available at: https://doi.org/10.1051/e3sconf/202342602128

Rushwood, D. (n.d.). *The Mud Builders*. Available at: https://www.krugerpark.co.za/krugerpark-times-3-13-mud-builders-23287.html

Santos, P. (2020). *Tailor made Packaging for Azores Food Produce*. ESAD Escola Superior de Artes e Design. Available at: https://comum.rcaap.pt/handle/10400.26/33789

Schlebusch, C. M., Malmström, H., Günther, T., Sjödin, P., Coutinho, A., Edlund, H., … Jakobsson, M. (2017). Southern African ancient genomes estimate modern human divergence to 350,000–260,000 years ago. *Science*, 358(6363), 652–655. https://doi.org/10.1126/science.aao6266

United Nations. (2015). *The Seventeen Sustainable Development Goals*. United Nations, Department of Economic and Social Affairs. Available at: https://sdgs.un.org/goals

University of Birmingham. (n.d.). *Vertebrate Collection (Fish)*. University of Birmingham. Available at: https://www.birmingham.ac.uk/facilities/lapworthmuseum/collections/palaeontology/vertebrate.aspx

Vincent, J. F., & Mann, D. L. (2002). Systematic technology transfer from biology to engineering. *Philosophical Transactions of the Royal Society of London. Series A: Mathematical, Physical and Engineering Sciences*. Available at: https://doi.org/10.1098/rsta.2001.0923

VTT. (2024). *Making Furniture with Paper Machines*. VTT. Available at: https://www.vttresearch.com/en/news-and-ideas/making-furniture-paper-machines-vtt-and-isku-have-created-novel-biocomposite-chair

Worster, D. (1977). *Nature's Economy: A History of Ecological Ideas*. San Francisco: Sierra Club Books.

19 Design Awareness
A Collaborative Initiative for Climate Change Engagement in Porto

Susana Barreto and Juliana Pires

19.1 INTRODUCTION

This chapter is an extended version of the article presented in the Design Commit '24 conference proceedings.

Beginning at the intersection of design, sustainability, and societal responsibility, this chapter delves into the imperative for designers to contribute to a future marked by both ecological resilience and social equity.

> The thoughtful designer of the 21 century will design with integrity, sensitivity, and compassion. He/she will design products (…) they serve human needs without depleting natural and manmade resources, without damage to the carrying capacity of ecosystems and without restricting the options available for present and future generations.
>
> *(Fuad-Luke, 2009, p. 15)*

Drawing on this optimistic vision, a project emerges from a collaboration between the Master's in Graphic Design and Editorial Project at the Faculty of Fine Arts in Porto and the Porto Climate Pact, an initiative by the Municipality of Porto, against the backdrop of an imminent climate crisis, urging us to act decisively in the current decade.

This chapter presents the initiative, detailing the collaboration and the resulting graphic design project. The tangible result of this collaboration materialises in an upcoming communication campaign, which includes eight dynamic posters for digital billboards throughout the city (Figure 19.1), an informational webpage, and alternative graphic elements to encourage sustainable behaviours in public spaces. Essentially, this collective project addresses the urgent need to enhance public awareness of climate change, recognising the crucial role that individual actions play in shaping a sustainable future.

In this context of a climate crisis, where the immediacy of action is paramount, communication design emerges as a tool for social change. It can wield visual communication as a force for behavioural change, instigating a shift towards healthier and more positive individual behaviours. As noted by Rawsthorn and Antonelli

DOI: 10.1201/9781032719290-22

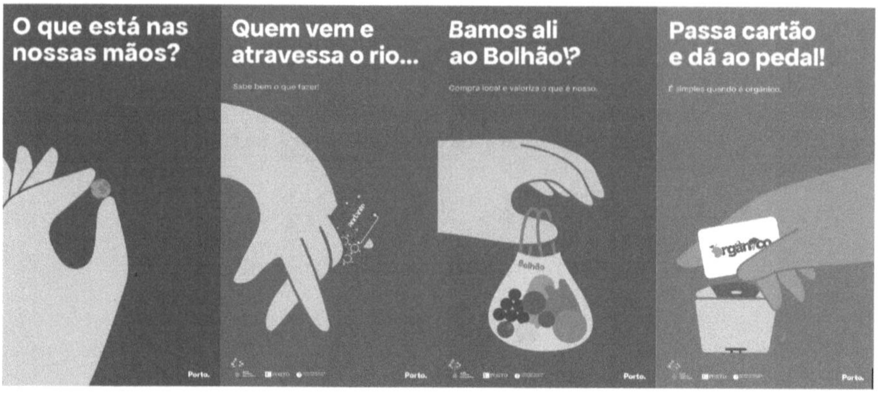

FIGURE 19.1 Four posters designed for the Porto Climate Pact campaign. (Juliana Pires design 2023.)

(2022, p. 12), "Design is not a miracle cure, but it is a powerful tool that can help us to address these issues [*the urgent challenges of our era*] if it is applied sensitively and responsibly".

This project aims to harness the potential of communication design as a powerful tool for enhancing public awareness of climate change in Porto, with the overall goal of instigating positive behavioural changes and fostering the adoption of sustainable habits among citizens.

To achieve this, the project sets out several specific objectives:

- Investigate the capacity of communication design to serve as a catalyst for citizen action by examining its effectiveness in conveying the urgency of climate change and motivating behavioural shifts.
- Explore the psychological and sociological aspects involved in the process of behaviour change, aiming to discern key factors and triggers that influence individuals to adopt more sustainable practices.
- Identify and articulate strategies within communication design that have the potential to inspire and sustain positive behavioural change among citizens.
- Consolidate the research findings to offer insights and recommendations for future design practices and communication campaigns.

This chapter serves as both an academic exploration and a narrative of this collaborative effort, highlighting the responsibilities of designers and the urgent need to elevate awareness of climate change.

19.1.1 Porto Climate Pact

The Porto Climate Pact, initiated by the Porto City Council, is a strategic project aimed at expediting the journey towards carbon neutrality. This initiative strives to spark people and organisations to take action on climate change by fostering a sense of community among citizens, businesses, and institutions and promoting a collective commitment to learning, sharing, and mutual support.

As Santos (2021, p. 96) noted:

> Cities can make a decisive contribution to solving the problem of climate change. City councils have the capacity to promote mitigation policies in the municipality through regulations, instruments, and policies, in town planning, energy efficiency in buildings, smart electricity grids (...) On an individual level, it is important to consume less, reuse what is possible, repair rather than discard, and recycle.

Recognising the pivotal role cities can play in combating climate change, mainly through municipal policies and regulations, the initiative aligns with the broader European mission for carbon neutrality (Pacto, 2023). Porto, chosen as one of the 100 Mission Cities by the European Commission, assumes a leadership role in national and European climate action, with a target of achieving carbon neutrality by 2030.

The emergence of the Porto Climate Pact within the framework of this Master's project signifies a unique opportunity to bridge academic endeavours with tangible, real-world challenges. The project, initiated with the goal of influencing behaviour and fostering environmental awareness, found resonance with the objectives of the Porto Climate Pact, evolving to encourage the incorporation of small sustainable practices into the daily lives of Porto's inhabitants. The emphasis now lies in motivating individual actions – clear, straightforward, and appealing behaviours that, when collectively adopted, contribute to the greater good. The collaborative engagement between the project and the Pact represents an earnest effort to translate design concepts into impactful societal outcomes.

Upon initiating contact and proposing a collaboration, the response from Eng. Daniel Freitas and Susana Alves of Porto Ambiente, the initiative's coordinators, was met with enthusiasm. This informal conversation marked the beginning of a partnership culminating in the results presented herein.

19.1.2 Research Methods

This project employs a mixed methodology integrating qualitative and quantitative methods across its study and development phases. The entire process drew significant influences from the Disruptive Design Methodology, a methodology developed by designer and social scientist Leyla Acaroglu, which aims to guide creative practice in the exploration of complex social and environmental problems and have a positive impact on societies. This methodology is based on the premise that all of us, as individuals, can make a positive impact in the world. As designers, this ability is multiplied thanks to the tools we have at our disposal (Acaroglu, 2017).

The methodological process has an iterative approach with three stages described as Mining, Landscaping and Building (Table 19.1).

1. In the beginning, the main objective is **Mining** into a problem set, studying it, digging deeper, and turning it into an opportunity. To this end, a research plan was organised with the various topics of interest, including the main

TABLE 19.1
Stages of the Methodological Process

1. Mining	2. Landscaping	3. Building
Scoping The responsibility of design and designers in the face of the climate crisis and the challenge of influencing sustainable behaviour	**Plan** Awareness and correlation of the concepts covered so far. Development and exploration of strategies for the project – Fogg's Behavioural Model as a theoretical basis	**Ideation** Conception of creative ideas and concepts
Study • Reflection on the role of the discipline of Design in relation to social and environmental issues; Papanek (1972), Fuller (1969), Margolin (2014), Norman (2023) • Study of the mechanisms essential to human behaviour, and theories for changing behaviour. Wendel (2014), Hitier (2021), Ajzen (1991), Lockton (2009), Ludden & Hekkert (2014), Fogg (2020)	**Collaboration** Start of collaboration with the Porto Climate Pact	**Design** Graphic design development, image exploration, and prototyping
		Refinement Work evaluation, analysis, reflection, and redesign

Source: Juliana Pires (2023). Adapted from Acaroglu (2017) and Applied to the Project.

authors and concepts to be explored. The responsibility of design(ers) towards social and environmental problems was chosen as a topic to study and reflect on, based on authors such as Papanek (1972), Margolin (2014), Rawsthorn and Antonelli (2022), and Norman (2023), among others, who think in different ways and times about the moral and social role this discipline and its professionals can play in society and the world. Another theme to explore was the challenge of influencing human behaviour, where Wendel (2014) and Hitier (2021) were essential authors to understand the mechanisms of human behaviour, paving the way for the study of four models of behaviour change by Ajzen (1991), Lockton's (2009), Ludden and Hekkert (2014), and Fogg (2020). However, following Leyla Acaroglu's maxim that research is curiosity in action, new questions and other authors of interest emerged as the research progressed.

2. In the next stage, the aim is to **Landscape**, or in other words, to plan. Based on all recognised learning and with a panoramic view of the problem, the

goal is to identify possible paths to follow, points of intervention, and to map out a creative solution to influence citizens of Porto to adopt more sustainable behaviours. At this point in the process, we started our communications to collaborate with the Porto Climate Pact, which was very important to understand the context in which the project would be placed. The proposal was well received, and we ended up deciding to go ahead with a communication campaign to promote citizen action.

Thus, after studying strategies to influence behaviour in the Mining phase, we decided to select and apply a single model for behavioural change, which would be further explored to contribute as a theoretical basis for the campaign. Fogg is the author chosen due to his complete model that explains the mechanics of behaviour – Fogg's Behaviour Model – and his simple and effective methodology for putting into practice – Tiny Habits (Fogg, 2020).

3. Finally, the last and bigger step, **Building** it. We started with a phase of ideation, where the project began to be written and structured. Several ideas emerged; different concepts, slogans, and idealisations were worked out and discussed. We used a shared spreadsheet containing all the proposed moments of the campaign, as well as the different visual and text options for all the planned graphic elements. This made it easy for every part involved to make comments and suggestions that kept the project moving forward.

Then, with most of the content thought out and approved, we moved on to design the campaign. There was a period of graphic experimentation around colours, typography, layout, and even the illustrated narratives and copies, which were rethought and redesigned a few times. All of this led to the maturation of the project.

As this was a long and complex process, it naturally became an iterative journey. We went back several times to make the most of new research, strategies, and forms – it turned out to be a flexible and ever-changing methodology.

19.2 CAMPAIGN

After understanding and studying the issues of interest to the project, we presented a proposal to collaborate with the Porto Climate Pact, with the aim of working together on a project to raise public awareness of climate change. The focus would be on answering the questions that immediately arise after realising that we "must do something" about the climate crisis: What? and How?

The idea was to communicate useful information to people, simple and practical solutions to change small behaviours in their routine. Thus, with a communication aimed at ordinary citizens, we filled a gap in the Porto Climate Pact, which until now had a communication only aimed at companies and institutions, and encouraged the community to join this municipal initiative.

We began by discussing various proposals for action and opted to develop a communication campaign that could be displayed on the city's digital billboards, an easily accessible medium with few associated costs. However, the exploration of creative solutions beyond these limitations was strongly encouraged, and the possibility of extending the project and maximising the campaign's impact to reach a wider and more diverse audience was highlighted.

Even before starting work, seven themes related to the goal of carbon neutrality in the city of Porto were selected, which the campaign should focus on: using public transport, installing solar panels, saving water, saving energy, buying local, separating waste, and reducing food waste.

19.2.1 Concept

The question "What is in our hands?" served as the opening query, setting the tone for this communication campaign.

Hands, a part of our body associated with the concept of action, symbolically represent our human capability to perform tasks, engage with the world, and convey intentions through gestures or movements. They serve as the tools enabling us to shape our surroundings, create, build, cook, write, and accomplish various other tasks. Given their fundamental role and pervasive presence in our lives, there exists a rich array of expressions attributing diverse meanings to them – hands-on, hands-off, butter hands, to have one's hands full, and first-hand, among others. This is akin to the highlighted question "What is in our hands?" which, beyond its literal interpretation, can also be understood as what is within our reach, under our domain or control, and, consequently, what actions we can take.

In this context, hands symbolise the capacity, capability, and responsibility to adopt more environmentally conscious behaviours. They serve as the central element of the entire campaign, aiming to emphasise the ability of each individual to contribute with their own hands to a sustainable future.

The primary image of the campaign, featuring the question "What is in our hands?" showcases a small planet Earth, the size of a marble, being cradled in the fingers of a hand (see Figure 19.1). This hyperbolic representation underscores the metaphorical significance of our hands in relation to the planet. By amplifying the true scale of things, the goal is to underscore the power and impact of our behaviours – actions that, despite appearing simple and innocuous, can have significant repercussions.

This theme of implied exaggeration persists throughout the campaign, where an illustrated hand appears to represent simple behaviours that can be embraced in support of sustainability and the pursuit of the city's carbon neutrality.

The concept is to communicate in a positive way, avoiding the institutional or even pedagogical language associated with this complex subject. The aim is to convey the message in a light and fun tone, with strong colours, dynamic illustrations, using humour as an ally, and making some references to the city's popular culture so that people can relate to it.

19.2.2 Aim

There are several objectives proposed for this campaign:

- Promote the adoption of small daily gestures for the city's sustainability;
- Motivate individual action, making simple, clear, and attractive behaviours that can be adopted for the benefit of the collective;

- Address the seven practices selected and proposed by the Porto Climate Pact: use of public transport, installation of solar panels, water savings, energy savings, buying local, waste separation, and reduction of food waste;
- Make the Porto Climate Pact known;
- To be another stage for the topic of climate change: raise awareness, influence, inform, and cause a positive impact, encouraging sustainable practices in our city.

19.2.3 Target Audience

As mentioned above, this campaign, developed in collaboration with the Porto Climate Pact, aims to stimulate individual action and encourage citizens to adopt more sustainable behaviours. Therefore, the target audience could only be all the citizens of the city of Porto, people of all ages present in public spaces.

The aim is to reach as many people as possible. To this end, the campaign was designed to be flexible and changeable, and so that it can grow and penetrate through various means within the city, thus targeting more particular audiences – in schools, libraries, swimming pools, and other public buildings, for example, as we'll see later.

19.2.4 Strategy

In a campaign primarily aimed at influencing behaviour, it was essential to study the mechanisms and gears of the human brain that resist changing our behaviour, such as our two modes of thinking, the emotional and the rational (Wendel, 2014), and the cognitive biases that can distort our decisions (Cherry, 2022). On this basis, mechanisms to promote behavioural change were also explored and, as mentioned above, it was decided to select Fogg's approach as a theoretical basis to contribute to the strategic construction of the project. According to his theory of design for behaviour change, there are three fundamental factors for behaviour change to occur: Motivation, Skill, and Stimulus (Fogg, 2020). Therefore, the strategy of this communication campaign is to develop graphic elements that respond to these three factors at different times.

The table below shows the different moments planned, the objectives of each phase, and the ways to achieve them. These moments involve different communication channels and tactics and are part of an overall process that has been strategically developed to ensure that the message is effective and has the widest possible reach (Table 19.2).

So, in addition to the moving posters (Figure 19.2) that raise awareness and suggest behaviours to empower and motivate people, the campaign also includes a website (Figure 19.3) to inform and provide tips and tools for more sustainable practices. And it extends into the physical realm of the city, with alternative graphic elements to encourage behaviour in public spaces, such as messages in public toilets to encourage water saving, in public canteens to highlight food waste, or even on public transport to thank passengers for their contribution to a carbon neutral city (Figure 19.4). In this way, the campaign covers all the factors that Fogg advocates, increasing the possibility of influencing citizens to change their behaviour.

TABLE 19.2
Campaign Strategy Outlined: Three Stages, Its Tangible Designs, Aims, and the Framework of Fogg's Theory

Awareness	Information	Nudge
→ Eight animated posters for the city's digital billboards	→ Online page for the campaign on the official website of the Porto Climate Pact	→ Alternative graphic elements, such as stickers or installations
Present the slogan "What's in our minds?"; propose seven behaviours deconstructed through the logic of Tiny Habits; invite people to sign the Pact	Provide information, data, tips, and curiosities on the seven themes covered by the posters	To provoke reactions, be they behaviours or reflections, through interventions in strategic places in the public space
MOTIVATION + SKILL	SKILL	PROMPT

Source: Juliana Pires (2023).

FIGURE 19.2 Mockups of the animated posters in the streets of Porto. (Juliana Pires design 2023.)

19.2.5 Development

The programme presented so far is the result of a whole development process. From start to finish, it has been a rigorous, albeit volatile, process in which we have sought to achieve the proposed objectives.

It is important to note that this entire project began as a result of the designer's will, without any prior briefing or request. It was therefore essential to adopt a broad and inclusive approach, which proved to be a great opportunity for creative freedom. All conceptual and strategic choices were guided by the goals and primary objective of the Master's project, as well as the formal options and graphic solutions. Throughout this subchapter, we highlight the key decisions and challenges faced in order to provide a comprehensive overview of the work carried out.

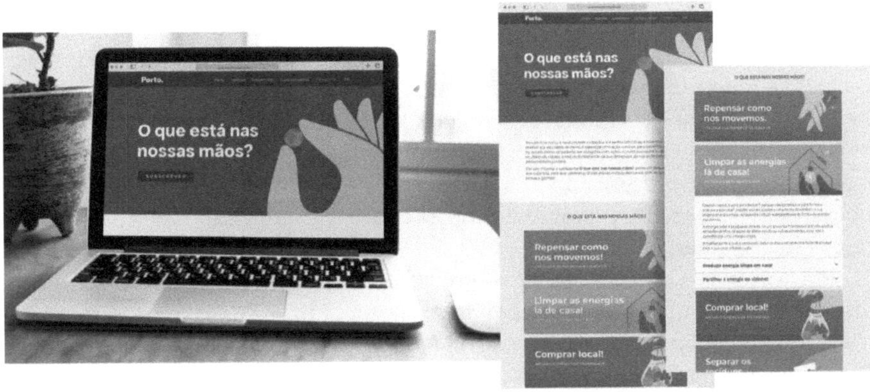

FIGURE 19.3 Campaign webpage on the Porto Climate Pact website. (Juliana Pires design 2023.)

FIGURE 19.4 Some of the stickers to hand out in public places and nudge. (Juliana Pires design 2023.)

After initial contacts with the partner, the first conceptual idea to emerge was inspired by an old Porto Ambiente campaign entitled "Porto is our home". In this way, we would give continuity to a slogan that had previously been disseminated throughout the city, presenting a set of behaviours to be adopted for a better home, in other words, a more sustainable city. The idea was well received, but we quickly realised that this project needed a new look, freshness, and creativity. However, we realised that the concept of "small gestures at home" could be a good starting point to get the desired message across. So, we came up with the idea of creating an illustrated mascot that practices various sustainable behaviours within the formal confines of the home.

With this in mind, we focused on balancing two tasks: the strategic planning of the campaign and the maturing of the creative concept. We sought to develop and structure the ideas within the seven themes and plan the copy content to find the ideal tone to convey the desired message. Collecting, curating, and maturing the ideas was a long and complex process and was probably the most challenging part of the whole process, perhaps because it moved a little away from graphic design skills and closer to areas such as marketing and copywriting.

But it was also a moment of great creativity, where all ideas were welcome, no matter how impractical or distant they might seem at first.

To accomplish this task, we created an online shared document containing all the information about the project, from objectives, target audience, concept, copywriting suggestions, and even visual narratives. This document served as a database for all ideas and acted as a strong organisational support. It was always accessible to everyone involved, making communication easy and active.

At the same time, the graphic design decisions were made: the colour palette, the choice and treatment of the typography, the grid and composition of the visual elements, and the drawing of the illustrations that would bring the project to life. This process of graphic development was marked by extensive experimentation and various transformations, determined by conceptual choices.

During this process, the phrase "What's in our hands?" emerged and became the campaign's motto. At this stage, when the written content and visual elements were idealised, it was natural to adopt a new approach. Acknowledging the significance of the gesture in conveying the message with efficacy, we opted to streamline the design, foregoing the concept of a mascot within a house and instead focusing on a hand executing sustainable gestures. This approach serves to emphasise the action being undertaken.

19.3 CONCLUSION

In conclusion, this collaborative design initiative has navigated the intersection of academia and real-world impact, forging a meaningful partnership between the student, the Porto Climate Pact, and the University of Porto. With this opportunity, the student was able to apply the knowledge she had acquired in an academic context to real circumstances, with the added challenge of coordinating the project, managing the opinions, contributions, and expectations of the client, and dealing with the constraints of creative practice. The Faculty of Fine Arts and the University of Porto have also benefited from this collaborative project, creating synergies and cooperative relationships that could open doors to future projects.

Porto Ambiente and specifically the Porto Climate Pact are enriching their communication with this new campaign that aims to reach ordinary citizens and inspire sustainable behaviour. As a result, the city of Porto also benefits from this collaborative project, both environmentally and socially, through the expected impacts of raising awareness, motivating and empowering the population on the subject of environmental sustainability and the goal of carbon neutrality.

The focus on enhancing public awareness of climate change and fostering individual engagement underscores the vital role of design in addressing pressing environmental concerns. We highlight the enormous value that collaborative work can bring to common causes. This discipline can play a truly significant role in finding solutions to raise awareness and change people's behaviour, but it is unlikely to do so alone. As we have shown in this project, you need partners with the will and the means to get involved in social and sustainable projects, as well as designers and other creatives with the flexibility to go beyond their creative skills and play a collaborative role.

As we move towards the public exhibition of the posters in September 2024, it becomes increasingly clear that this project has the potential to grow even further. The design proposals extend into the physical realm of the city by nudging people, as we see in some of the examples given in the previous chapter. Incorporating similar subtle yet impactful interventions in public spaces could further reinforce the message of sustainability. Street-level interventions can serve as constant reminders of each individual's environmental responsibility. Moreover, these types of elements can be easily inserted and developed in specific contexts, such as schools, for example, or even in public libraries, municipal swimming pools, and gardens, among other places. Considering the education dimension, the tray liner (Figure 19.5) messages in school canteens indicate the potential for impactful communication in educational settings. Expanding this concept to include more educational materials, workshops, or even a curriculum module could embed sustainable practices into the learning experience of students. Furthermore, the hand, as an expressive object for narrating and externalising an idea of action, can be an infinitely explored concept in various media and contexts. That's why we hope to see a potential expansion of this

FIGURE 19.5 Canteen tray liner and trolley to raise awareness of food waste. (Juliana Pires design 2023.)

concept – which could also include the design of educational materials, the development of interactive installations, or even workshops for the integration of sustainable practices.

Ultimately, if successful, this initiative will not only reside in the exhibition but also in its lasting impact on the community. By incorporating design proposals that reach into the streets and educational institutions, we can aim at a sustained dialogue on climate change awareness and active engagement beyond the initial campaign.

ACKNOWLEDGEMENTS

We extend our gratitude to the Climate Porto Pact, represented by Eng. Daniel Freitas, En. Susana Alves and Eng. Ana Rita Barros for their invaluable support in realising this project. Additionally, we would like to express our appreciation to Professor Pedro Amado, Director of the Master's in Graphic Design and Editorial Project at the Faculty of Fine Arts in Porto, for his support throughout this course, and to Professor Antonio Modesto for his contribution to the graphic and strategic development of the project. Their contributions have significantly enriched the quality and depth of this work.

REFERENCES

Acaroglu, L. (2017). *Disruptive Design Method Handbook*. Disruptive Design LLC.

Ajzen, I. (1991). The theory of planned behavior. *Organizational Behavior and Human Decision Processes*, *50*(2), 179–211.

Cherry, K. (2022). *Designing for Behavior Change: Applying Psychology and Behavioral Economics*. O'Reilly Media.

Fuad-Luke, A. (2009). *Design Activism: Beautiful Strangeness for a Sustainable World*. Earthscan.

Fogg, B. J. (2020). *Tyni Habits: The Small Changes That Change Everything*. Houghton Mifflin Harcourt.

Hitier, R. (2021). *O Clima: A Ambiguidade do Cérebro U*. F. à. l. P. ARTE France.

Lockton, D. (2009). *Design for Sustainable Behaviour: Influencing Users to Improve Efficiency of Product Use* ('My PhD' series).

Ludden, G., & Hekkert, P. (2014). *Design for Healthy Behavior: Design Interventions and Stages of Change*.

Margolin, V. (2014). *Design e Risco de Mudança*. ESAD, Verso da História.

Norman, D. (2023). *Design for a Better World – Meaningful, Sustainable, Humanity Centered*. MIT PRESS LTD.

Pacto. (2023). *Pacto do Porto para o Clima*. Porto Digital. https://pactoparaoclima.portodigital.pt/

Papanek, V. (1971). *Design for the Real World*. Bantam Books.

Rawsthorn, A., & Antonelli, P. (2022). *Design Emergency*. Phaidon Press Limited.

Santos, F. D. (2021). *Alterações Climáticas*. Fundação Francisco Manuel dos Santos.

Wendel, S. (2014). *Designing for Behavior Change: Applying Psychology and Behavioral Economics*. O'Reilly Media, Inc.

20 Teachers' Perspectives on Metaverse Integration in Education

Yannara Negre, Rui Patrício, and Élmano Ricarte

20.1 INTRODUCTION

What happens when virtuality, gaming, and learning intersect, as observed in the metaverse? Take this question as a starting point for this chapter, which is an extended version of the article presented in the Design Commit '24 conference proceedings.

In our digital age, students are constantly exposed to various high-tech devices, such as personal computers, digital streaming systems, and smartphones, all vying for their attention (Barry et al., 2015). This makes it essential to design compelling learning scenarios that capture students' interests and facilitate comprehension through innovative active learning strategies, as exemplified by the metaverse, a concept born from gaming. In this expanding technological landscape, the disruption driven by the metaverse may consolidate immersive digital worlds using game mechanics to enhance skills, knowledge, and competencies (Camilleri, 2023; Wang et al., 2022).

Mark Zuckerberg, owner of Facebook – now rebranded as Meta (Kelly, 2021, October 29) – has justified his significant investment in this sector, describing it as the "next chapter of the internet" (Mark Zuckerberg, as cited in Marr, 2022, August 15). Although immersive worlds' educational and knowledge acquisition effects remain preliminary, they should not be overlooked. Some educators are eager to employ metaverse applications to leverage their three-dimensional capacities (Camilleri, 2023).

Despite its roots in digital gaming, the metaverse diverges from these by introducing an almost spiritual component (Jiang, 2021). Online social games share similarities with the metaverse but are limited by specific rules, objectives, and boundaries that define their gameplay. Conversely, the metaverse does not require defined gameplay, allowing users to participate in various activities, such as live events, concerts, exhibitions, meetings, and classes. According to Yang et al. (2022), while the fundamental characteristics of the metaverse can be explained and demonstrated, no game has fully achieved an ideal metaverse. This underlines the open-ended and limitless nature of the metaverse compared to traditional games.

Radoff (2021) fervently argues that the metaverse is "the true gamification" due to its emotive, interactive, immersive, and absorbing potential. The Metaverse enables multiple users to synchronize on platforms, and even in groups, they experience an

DOI: 10.1201/9781032719290-23

individual sense of presence (Beerda, 2022). Telepresence, a defining feature distinguishing metaverses from traditional digital media, allows users to coexist and interact with other avatars in a shared world (Schlemmer et al., 2008), creating a "virtual digital life." Cavalheiro (2010) found that the etymology of "avatar" implies an esoteric dimension – a "god who materializes." da Silva (2010) encapsulates the Western definition of the term, viewing an avatar as the player's graphical representation in electronic games.

Barry et al. (2015) depict the metaverse as a three-dimensional virtual world where avatars act on our behalf. Digital avatars and technology foster an immersive interaction that challenges the distinction between the virtual and real worlds. The Metaverse is not merely an extension of the physical realm but a domain of possibilities that transcend conventional reality.

The notion that media plays a crucial role in guiding and shaping human subjectivities aligns with Sodré's (2009) concept of a "media bios," where human biology gains a new layer as our presence becomes mediated. Mediation occurs as technologies and communication practices influence societal interactions (Hepp, 2014). In contrast, mediatization describes how social or cultural areas, such as politics, religion, and education, become dependent on media (Hjarvard, 2012).

The telepresence and immersion provided by the metaverse represent a revolution in human interaction with digital environments, blurring the line between virtual and real worlds. Game scenarios are contingency models to explore unpredictable outcomes and skill development opportunities. However, "games alone do not facilitate learning" (Hanghøj, 2013, p. 18), nor does the metaverse alone. Educators play a crucial role in translating these environments into gamified, immersive learning spaces, ensuring an equitable and intimate interaction among participants.

When students are not encouraged to apply and enhance their cognitive and learning skills, the traditional education model heavily depends on the educator (Sugeng & Suryani, 2018). Findings by Arcagök (2021) indicate that game-based teaching practices motivate students to develop socio-emotional and problem-solving skills within a safe, enjoyable environment, positively impacting academic performance. However, during these activities, educators must assume the role of facilitators or take on a role akin to a coach, guiding students (Barry et al., 2015).

The passive knowledge acquisition model, typical of traditional classroom structures, does little to prepare individuals for society's evolving demands (Duarte, 2018). In the traditional model, knowledge is generally transferred from top-down, with experts imparting information to passive "receivers" (Patricio et al., 2022). In contrast, gamification promotes a collaborative, bottom-up team approach, where knowledge is created through collective effort and diverse perspectives.

This paper posits that traditional education models' challenges in fostering effective learning may be mitigated through games, mainly by applying gaming elements and mechanics within the metaverse. This research begins with the assumption that the metaverse has the potential to revolutionize future teaching processes. Although Tayal et al. (2022) highlight a gap in studies on gamification within the metaverse, Reyes (2020) counters that the metaverse enhances educators' ability to construct stimulating, immersive, and disruptive learning spaces.

This research envisions future trends for using and implementing an Edu-Metaverse, offering a fresh perspective on media elements and gamification as tools to enhance interaction and knowledge exchange. Concepts such as "game-based learning," "telepresence," "mediation," and "mediatization" form a philosophical research paradigm and are described and interlinked within a diachronic and synchronic empirical analysis.

A holistic approach is crucial for this study, as it aims to draw connections between seemingly disparate concepts and identify patterns and trends for innovative applications in educational contexts. This investigation strives to align theoretical insights with field research findings to reveal how games, particularly those designed within the metaverse, could be adopted as a transformative educational tool for future learners and educators.

20.2 LITERATURE REVIEW

20.2.1 The Relationship between the Metaverse and Gamification

Over time, several technological advances and initiatives have been carried out in the context of metaverses (Tayal et al., 2022). The starting point and reference platform of the metaverse are, without a doubt, video games, and many scholars point directly to the game Second Life® as the milestone of its creation (Jiang, 2021). The "MUVE" (Multi-User Digital Environments) present in the metaverse has a structure based on "MMORPG" (Massively Multiplayer Online Role-Playing Games), but, unlike the latter, they do not have a pre-defined objective, as is the case with video games (Díaz et al., 2020). Simply seeing the metaverse as an expansion or derivation of video games is a superficial and mistaken view. Fundamentally, the metaverse was born from video games, but, in turn, it transformed games themselves.

Games in the metaverse are and will continue to be radically altered in every aspect, from form, function, method, and purpose, becoming increasingly serious, intellectual, and spiritual (Jiang, 2021). This evolution represents a significant and profound change, redefining the nature and scope of the gaming experience and gamification itself. After all, "gamification is considered one of the most intuitive ways for users to engage with the metaverse" (Dwivedi et al., 2022, p. 39).

The gaming experience is related to gamification, which, as explained in the previous chapter, incorporates classic elements such as points, badges, levels, and leaderboards into a non-game context for better user engagement and motivation (Deterding et al., 2011). According to Tayal et al. (2022), the gamified elements in the metaverse are avatars, badges, digital goods, augmented reality, machine learning, and artificial intelligence (AI). However, this does not explain why people like games so much.

The problem identified by Radoff (2011) in his work "Game On: Energize Your Business with Social Media Games" breaks with the logic of Skinner Boxes, which would indicate operant conditioning about games through rewards in point systems and merits for specific behaviors/actions. B.F. Skinner is the best-known figure in the behaviorist school of psychology for inventing the operant conditioning box, which is used to study the behavior of laboratory animals and is famous in the world of games (Scimeca, 2011 August 10). For Radoff (2021, September 1), the fascination of the metaverse lies in its ability to generate genuine game approaches by seamlessly integrating game elements to improve various facets of daily life.

This is what is interesting about the metaverse. It offers true gamification. The metaverse is not just the next generation of the Internet, but a generation shaped by several decades of video game development. In contrast to the unfortunately named gamification fad of a few years ago, the metaverse is bringing more challenging aspects of gaming experiences to enhance use cases such as education, shopping, live music, fitness, and many others (Radoff, 2021, September 1).

In the logic of motivation, what makes games and gamified systems so appealing does not involve graphic and structural elements but incorporates the real sense of gaming approaches, aspects such as emotion, interactivity, immersion, and absorption, which are present in the metaverse (Radoff, 2021). Or even Steuer's own "telepresence" (1992 apud Chen & Yao, 2022), as the feeling of "being there" or even "teleportation," described by Green and Brock (2000 apud idem), as the experience of being so consumed by the virtual world that the connection with the real world can be lost. In contacts mediated by interfaces, users chat, play, post photos and videos, give "likes," and shape the image each one will have of the other. Together, but apart. The need for physical presence gives way to mediated presence. Cowan et al., 2021; Jang et al., 2019; Tsai et al., 2020; and apud Chen & Yao, 2021 state that, although telepresence can be induced by all forms of media content, immersive virtual reality screens create vivid telepresence through perceptual inputs, such as 3D vision, spatial audio, and control. The fact that they contain elements very similar to those of the real world, or experiences that can be as vivid and captivating as real ones, increases the sense of presence brought by virtual reality, augmented reality, or mixed reality (MR), which the metaverse uses as a basis. Telepresence is often associated with the ability to teleport and quickly move to other locations, which provides a sense of pleasure through discovery and the enjoyable experience of interacting with the avatar in an exciting and fun way (Schlemmer et al., 2008). Beerda (2022) agrees with Radoff (2021). For them, a metaverse is a collection of physical and digital experiences rendered in real time, which overlap and integrate seamlessly to create a simulated and shared 3D world. There is a sense of continuity that mimics real life, such as persistent objects and identities, shared stories, rights/properties, and the ability to make payments. Any number of people can synchronously enter the metaverse and still experience an individual sense of presence (Beerda, 2022). Gamified elements in the metaverse create an immersive and engaging atmosphere, positively impacting users' mindsets in their social interactions, engagement, and user immersion (Tayal et al., 2022).

According to the ISFE (Interactive Software Federation of Europe, 2022), video games can aid in acquiring skills needed to thrive and function in a digitalized society. While they allow players to immerse themselves in rich, imaginative worlds and collaborate or compete with friends or other players worldwide, they also present significant intellectual challenges, specifically information processing, problem-solving, creating strategies and plans, and interpreting verbal and visual information. Learning new skills, concepts, and logical and critical thinking skills can be aided by video games (ISFE, 2022).

The Council of the European Union (2022) now recognizes the importance of strengthening the cultural and creative dimension in the European video games sector through skills and training. Among the advantages of this training with video games, ISFE (2022) pointed out the unlocking of essential skills to thrive in a digital society, improve the performance of educators, reduce gender gaps, promote social inclusion, and combat early school leaving. The number of school dropouts decreased with the use of gamified methods in Isabelle's research (2020).

Some online social games are, in fact, very similar to the metaverse. However, despite the similarities, some aspects separate the two. Games conventionally have rules, objectives, and limits that allow them to shape specific gameplay. In contrast, the metaverse does not require any particular gameplay; its users can even participate in live events such as concerts, exhibitions, meetings, and classes (Yang et al., 2022). The essential characteristics of the metaverse can be perfectly explained and displayed by approaching digital games; however, no game has fully achieved an "ideal metaverse" (Yang et al., 2022).

Gaming approaches have been widely used in education as an effective way to engage students, promote active learning, and improve information retention. By studying these approaches, a solid theoretical basis can be established for applying game design principles in creating educational experiences in the metaverse. As seen in the literature review, there are parallels between the two scenarios. There is even the inclusion of board games within the metaverse that replicate those existing in the physical world (Russell, 2022, May 30; Bower, 2022, October 30) in addition to the already used online multiplayer virtual reality games. Many principles that make games engaging and practical can also be applied to designing educational experiences in the metaverse. Still, one aspect that should not be ignored is that they are different media. The Metaverse emerges as a technology in constant development, but with a new and innovative form of communication (Cruz Franco et al., 2022). Analyzing how human relationships behave and change when mediated by the metaverse can raise profound questions about its various uses, including educational ones.

20.2.2 The Metaverse as a Mediatic Environment

Over time, human understanding of communication technologies and their effect on life has undergone significant transformations (Hudson-Smith & Shakeri, 2022). Transitions arising from the COVID-19 pandemic have caused much of the world to accelerate its digitalization processes. Tayal et al. (2022) point out that time spent on the internet increased significantly after the pandemic, as humans were left disconnected from physical contact to maintain safety protocols. The Metaverse took shape due to this period, amalgamating the internet, 5G, and virtual convergence technologies to reflect an evolutionary trajectory that mirrors the phenomenon of global expansion (Kye et al., 2021).

Scholars have analyzed not only the context in which these technologies are used and their predecessors, but also the environment created by them, which introduces new habits of perception, forms of understanding, and even the monopoly of knowledge (Hudson-Smith & Shakeri, 2022). The Metaverse connects users globally, offers virtual environments for social interactions, and enables the creation of immersive experiences that transcend the limitations of the real world, with the internet as a prerequisite, capturing different types of user behavior based on different situations, emotions, personal preferences, and modes of communication in a virtually real world (Tayal et al., 2022).

In this context, it is pertinent to verify the influence of the interpellation between changes in media communication and the sociocultural transformation in mediated communication practices. Today's society is permeated by the media to such an extent that it can no longer be seen as something distinct from cultural and social institutions (Hjarvard, 2012). According to (Hjarvard, 2015, p. 53 *apud* Ricarte, 2019), "by mediatization of culture and society, we refer to the processes through which culture and society become increasingly dependent on the media and their *modus operandi* [original emphasis], or media logic," in other words, the patterns of social behavior change when using contact mediated by devices.

The media increasingly mediates institutions, resulting in the virtualization of social practices (Hjarvard, 2013). Although such institutions still maintain their central physical locations, such as homes, schools, and parliament, physical spaces are increasingly connected to virtual spaces, allowing social practices to occur outside physical locations (Hjarvard, 2014). To understand the metaverse as a medium, it is essential to consider that the shaping forces of the media only materialize in human action (Hepp, 2014). From a sociological perspective, the distinction between mediated and unmediated interactions loses meaning, making it irrelevant to differentiate the state of reality between them (Hjarvard, 2012). The Metaverse will allow simple online interactions to transform into interactions where emotions and culture are shared, especially for Generation Z, due to their familiarity with information technologies (Park et al., 2021).

Currently, the metaverse enables virtual interactions through personalized avatars, overcoming real-world limitations of distance and time (Tayal et al., 2022). "Your avatar can look however you want it to, even beyond the limitations of your equipment. If you are ugly, you can make your avatar beautiful … You can look like a gorilla, a dragon or a giant talking penis in the Metaverse" (Mirzoeff, 1999, p. 103). This feature allows users to participate in conferences, meetings, and events without the need to be physically present (Tayal et al., 2022), such as being telepresent.

The transformation into avatars impacts online and offline connections, evidencing the normalization of the creation of digital representations in the current era (Cavalheiro, 2010). Self-representation and mediated communities create new forms of interaction in the digital world, working together to project expressions of shared values and identities (Ricarte, 2019). The Metaverse is and will be the repository of all the legacy we want and must digitize to achieve universal accessibility (Cruz Franco et al., 2022).

Digital Twin's characteristics are virtual and real symbiosis, high degrees of simulation, real-time interaction, and intimate understanding (Ahmad et al., 2022). For Cruz Franco et al. (2022), Digital Twins are a way to preserve our present. However, compared to the industrial sector, the application of Digital Twins in education is still in its infancy (Ahmad et al., 2022).

Although virtual learning is not new, the growing use of the metaverse brings new opportunities and perspectives in the hybrid educational scenario, which mixes the offline world with the virtual and online. It is also good to think about this from a concrete point of view: Is it possible that, with the expansion of the metaverse, these means will gain more engagement as a pedagogical practice? Beyond academic environments, can gamification in the metaverse be used for learning in companies or autonomous lifelong learning proposals? Although this trend is already a reality, it is still in the early stages of where it can go. Therefore, it arouses interest from investors, researchers, the public, and teaching institutions. In this sense, gamification will offer user engagement, and the metaverse will provide a genuine human connection in the virtual world (Park & Kim, 2022).

20.2.3 Metaverse inside Education Contexts

The metaverse for education has been employed in many contexts, including primary and secondary education, higher education, and corporate training. It has the fundamental advantage that improving courses and curricula not only optimizes the transmission of knowledge but also improves the proper use of the learning network when well executed (Wang et al., 2022). The metaverse may expand learning possibilities and reshape the student ecosystem. Therefore, it is promising to develop an exciting digital universe and improve the physical world through exploring the metaverse (Yang et al., 2022).

Creating "metaverses" is an affordable alternative for companies and educational institutions, saving logistical and material expenses and allowing prior training practices (Alpala et al., 2022). To further address the limits of online learning, several education sectors began using metaverse, and as a result, its use skyrocketed in 2021 (Chua & Yu, 2023). The metaverse is and will be the repository of all the legacy we want and need to digitize to achieve universal accessibility (Cruz Franco et al., 2022).

The introduction of this new approach arouses students' interest. Still, it is in their integration into the classroom that the role of the teacher becomes even more vital, as in addition to involving students in activities that stimulate concentration, this method also might inspire teachers to new ideas and pedagogical approaches (Estudante & Dietrich, 2020). They need to be aware of and reflect on the materials selected as tools to innovate in the classroom and develop appropriate resources (Franco-Mariscal et al., 2021). Furthermore, it is important to highlight once again that the transposition to the Educational-mediated ecosystem requires the crucial role of teachers and instructors, ensuring that it is a platform for egalitarian and intimate interactions between people (Gee, 2020).

Despite the growing interest in using the metaverse in the educational scenario, it is necessary to remember that using something considered new does not mean an innovation in the context of education. For innovation to be possible, teachers/researchers must appropriate this technology to understand it in the context of its specific nature, which requires new methodologies, new practices, and pedagogical mediation processes for the potential that it offers (Schlemmer et al., 2008). For example, VictoryXR employs visual immersion technologies to recreate Digital Twins of high schools and universities. Students can choose to participate in learning on a conventional campus or in the metaverse (Wang et al., 2022). A Digital Twin is essentially a digital replica of a real-world entity. It maintains a dynamic bidirectional mapping between the physical object and its digital version, including connected data and metadata (Holopainen et al., 2022). This allows real-time data transfer between the physical and virtual realms, enabling them to coexist simultaneously (ibid).

Digital Twins are virtual and real symbiosis, high degrees of simulation, real-time interaction, and intimate understanding (Ahmad et al., 2022). According to Cruz Franco et al. (2022), Digital Twins are a way to preserve our present. However, compared to the industrial sector, which is booming, the application of Digital Twins in education is still in its early stages (Ahmad et al., 2022). Of the few cases mentioned in the literature review, Vaquero's Hermitage case and MR technology bring a set of substantial advantages, especially in the educational context. Through this approach, Digital Twins have emerged as a tool that provides practical and dynamic experiences that significantly enhance the understanding and application of knowledge (Cruz Franco et al., 2022).

To facilitate the implementation of the metaverse in education, educational institutions must improve their technical infrastructure and provide adequate training to teachers to use these tools (Dwivedi et al., 2022). They must also promote training inside and outside the classroom in synchronous and asynchronous modes and provide students with a dynamic, engaging, and collaborative virtual platform (Tlili et al., 2022).

The implementation process must be meticulous so that the impression of teachers and instructors does not become negative due to the mandatory use of the new technology (Hwang et al., 2016). Hjarvard (2014) says that the media relieve social actors – such as teachers – of the multiple organizational responsibilities that arise from operating within large-scale institutions – such as universities. In contrast, Kenski (2014) warns that work overload is also a risk that needs to be addressed. According to the author, old-fashioned recruitment models do not foresee the additional work imposed by mediation technologies since the preparation, production, and presentation of mediated educational activities require substantial effort and a shortage of time.

Suppose the metaverse is, for example, easy to access and useful for teaching. In that case, teachers are more likely to use the technology, as platforms influence perceived usefulness (PU) and ease of use among their users (Chua & Yu, 2023). In other words, the educational transformation that the metaverse offers must pass through the hands of teachers, who have the power to shape the future of virtual learning in a meaningful and impactful way.

Teachers are outlining future agendas for implementing the metaverse in their learning practices, driven by the countless possibilities offered by the real-like virtual space to enhance the social aspect of teaching (Tlili et al., 2022). However, there is a clear need for adequate training for teachers and students, which is essential to maximize the benefits of technology in the educational environment (Díaz, 2020). After all, simulating social interactions, gestures, and body language in the virtual environment also presents significant challenges (Dwivedi et al., 2022). According to Dwivedi et al. (2022), even though teachers and students can create their avatars in the metaverse, it is essential to enable students to communicate information felt in virtual worlds, such as facial expressions, and for teachers to learn students' body language.

Integrating the metaverse into education might enable teachers and instructors to simulate the physical classroom experience more effectively than traditional online teaching tools (Barry et al., 2015). This approach could enhance pedagogical practices, requiring adaptations in teaching methods and course programs to optimize these enriched experiences.

Therefore, this research is focused on two research questions:

RQ1: Explore the challenges and opportunities related to how teachers and instructors deal with the technological infrastructure necessary for the metaverse and the development of pedagogical approaches to do so. Study the current adverse effects of the immersive environment and how to improve the usability of platforms, as well as explore possibilities of use for academic institutions that wish to incorporate the metaverse into their educational programs. RQ1 (**What are the future trends in using the metaverse in education from the perspective of teachers and instructors?**) aims to address this research gap.

RQ2: Although the published literature mentions the importance of teachers and instructors in implementing the metaverse, it addresses very few specific pedagogical strategies for the metaverse and aspects such as training and professional development on the part of this group. In addition to the advantages, practical, and ethical challenges of implementing the metaverse in education, there is the possibility of exploring in empirical studies how professional teachers and instructors perceive their own experience of teaching in the metaverse. RQ2 (**What is the role of teachers and instructors in transitioning to an educational model mediated by the metaverse?**) aims to address this research gap.

20.3 MATERIALS AND METHODS

Bearing that the delimited design does not limit itself to reaching conclusions, it aims to understand a more significant phenomenon in concepts, uses, developments, and future trends. Agreeing with Sá et al. (2021, p. 99), we believe that "by analyzing trends, the information obtained can be used to predict future events, or even to explain occurrences in the past."

This study applies a qualitative research approach where the emphasis is placed on the qualities of entities, processes, and meanings that are not examined or measured experimentally in quantity, value, intensity, or frequency (Denzin & Lincoln, 2005). Adopting a qualitative perspective was a way of understanding how social experience is created and gains meaning (Denzin & Lincoln, 2005). Furthermore, a reflective process supports each step of this study to ensure that the researcher's assumptions and opinions are visible, facilitating the viewer to understand better the investigation's overall truth (Chenail, 2010).

This way, a netnography session was held to study virtual worlds such as the metaverse (Kozinets, 2022). Four teachers and instructors were invited to participate in an immersive netnography from the perspective of simulating a virtual class in a 3D environment.

According to Kozinets (2022), netnography is a qualitative research method that adapts ethnographic techniques to study behaviors, interactions, and cultures online. Focused on virtual practices and communities, this approach enables the analysis of meanings and social dynamics in digital environments. Netnography involves collecting and interpreting data from social networks, forums, and other online spaces, like the metaverse.

The other participants acted as students during the exhibition, watched the class, and then interacted with questions. Within the Spatial platform, a kind of simulated outdoor auditorium was created with a traditional stage and chairs but with trees and a lake around it (see Figures 20.1 and 20.2); after all, in immersive netnography, the particular elements of the environment of a scenario are experienced more directly than the elements would be in social media netnography (Kozinets, 2022), hence the adoption of a scenario with possibilities for exploration.

During the experiment, which included technical help, other virtual objects – tables, books, papers, and even a virtual cat – were added to the environment to show participants some possibilities for using the space. Taking advantage of this potential

FIGURE 20.1 A mosaic simulating the platform's internal view, created via screenshots and edited by the authors.

FIGURE 20.2 Participants' avatars simulate a class in the metaverse. (Photo by the authors.)

for immersion, qualitative research was used here to illustrate phenomena vividly (Anderson, 2010; Merriam, 2009), such as class simulations, and provide conditions for understanding the meanings of participants' actions (Denzin, 1989).

The choice of the netnography technique is complemented by an ethnographic analysis, which relates both the physical part of observing the use of equipment during the group interview in an offline mode and the online observation of participants within the virtual auditorium. The Oculus Meta Quest 2, which consists of a virtual reality headset/glasses and manual controls, is a type of device already foreseen in Kozinets (2022) for the chosen approach.

20.3.1 Participant Observation

Two researchers were present in the classroom during the experiment. They were instructed to use an immersion diary to reflect on and intellectually record their subjective experiences in digitally mediated meetings and interactions (Kozinets, 2022). Records were made immediately in a notebook and, in addition, after leaving the social situation (Mónico et al., 2017).

Researchers have been representing themselves digitally, shaping how others perceive them instead of following a single fixed role in an online community (Caliandro, 2016), time observers without interaction, 1-hour students, and 1-hour technical assistants. The interactions would initially have a less active role on the part of the researchers. However, the participants constantly requested active participation within and outside the metaverse (for help with equipment setup).

In this technique, we observe the phenomenon of interaction between teachers and instructors and physical and technological resources (headset and controls), the relationships between them and their avatars and with each other, as well as the virtual tools present in the metaverse (video, images, emojis, scenery) to evaluate the usability and immersion of participants, both with the physical equipment and the virtual elements.

With participant observation, researchers could share the same conditions as those being investigated and access the same experiences, living situations, and problems (Lessard-Hébert et al., 1990). However, as these data were insufficient, the methodology was complemented with another approach to deepen the understanding of the problem under analysis through different perspectives (Mónico et al., 2017), the focus group.

20.3.2 Focus Group

The focus group had a single round session to generate insights promptly into the topic studied (Masadeh, 2012). The nature of participation was voluntary, with data previously approved for the scope of the investigation. The duration was approximately 45 minutes since "opinions on the ideal length of sessions fluctuate between half an hour to two-and-a-half hours per focus group" (Stewart & Shamdasani, 1990, p. 65 apud Masadeh, 2012). Time was optimized so that the focus on the topics raised did not deviate and to obtain as much information as possible in the minimum amount of time (Masadeh, 2012).

The script structure followed categories, which were rearranged or suppressed by moderation. As the research is a changeable process in response to the field, the possibility of creating new questions by the moderator was admitted and encouraged.

To extract an understanding from participants of the acceptance and use of metaverse technology, minimize resistance from potential users, and maximize its practical use, some questions followed the Technology Acceptance Model (TAM). TAM seeks to explain the behavior of using information systems, focusing mainly on two dimensions: PU, that is, the degree to which a person believes that using a given system will improve their performance, and the perception of ease of use (PFU), relating to the degree to which a person believes that using a specific system will not require efforts (Parreira et al., 2018). Table 20.1 presents the profile of the focus group sample.

20.3.3 Data Analysis

Thematic analysis was used to understand the objectives and, foremost, teachers' and instructors' expectations concerning using metaverse in the classroom based on data collected in individual interviews and focus groups. This coding procedure is commonly used in qualitative research to identify patterns and themes in data, providing valuable insights into complex phenomena (Braun & Clark, 2006; Gioia et al., 2013; Strauss & Corbin, 1998). The data were transcribed, labeled, and organized in the MAXQDA software, whose procedure followed six steps by Braun and Clark (2006).

TABLE 20.1
Focus Group Sample

Interviewee	Experience as an Educator	Received Any Pedagogical Training	Scientific Area of Teaching	Number of Students	Previous Experience with the Metaverse for Education
M.1	4 Years	No	Design	75	No
M.2	6 Years	No	Computer Science	100	No
M.3	5 Years	Yes	Digital communication; Innovation and Marketing	90	No
M.4	3 Years	Yes	Digital communication	70	No

Following this procedure, as pointed out by the authors, obtaining a more in-depth understanding of the data and finding more significant themes that could influence the development of theory and future research was possible. Therefore, the first-order codes were reduced to second-order themes. Then, aggregated dimensions were generated. The aggregated dimensions refer to higher-order themes or categories, representing an interpretation of trends throughout data collection and a more holistic understanding of the phenomenon (Braun & Clark, 2006; Gioia et al., 2013; Strauss & Corbin, 1998). Gioia et al. (2013) argue that aggregated dimensions should be data-driven and reflect emerging themes or patterns. Similarly, Strauss and Corbin (1998) emphasize the importance of linking second-order themes to the data to ensure that they are grounded in the empirical material. Overall, aggregate dimensions provide a way to synthesize the results of qualitative analysis and identify overarching patterns or themes relevant to the research question.

20.4 RESULTS

The observations recorded the experimental session both offline (observation of participants in the presential room) and within the metaverse through netnography (teachers and instructors within the platform, observer wearing glasses and capturing the computer screen), in which they explored the virtual environment and conducted simulated classes.

Participants engaged with the metaverse on their mobile devices in the session, embracing the platform's playful nature and game-like aspects. They explored the virtual environment enthusiastically, using game-related terms and expressing excitement about their avatars. The gamification aspect was evident in their interactions, with participants discussing their experiences in previous avatar creation for games and testing platform tools like chat, emojis, and avatar movements.

However, the participants faced challenges related to equipment setup, discomfort due to the headsets, and difficulties navigating through the physical space while using virtual reality devices simultaneously. Technical support played a crucial role in assisting them. Once online, they needed to get used to the available tools, such as handwriting and visual resources, to enhance their teaching. They also encountered issues like scale perception, which initially surprised some participants, and the need for assistance with tasks like a slide presentation and object manipulation in the virtual environment.

The participants showed enthusiasm while exploring the metaverse, but relied on support to handle basic tasks. They engaged in social interactions within the virtual environment, replicating classroom behaviors and attempting social connections by sending "heart-shaped" emojis and supportive gestures. However, their relief upon removing the headset indicated the physical discomfort experienced during the session. Participants recognized the platform's potential for younger generations more familiar with gaming and spatial navigation, suggesting its potential for future educational applications.

The focus group section, which involved four teachers and instructors, discussed the results using the three aggregated dimensions, which provided insights into their experiences and perceptions.

20.4.1 Immersive Context

The participants highlighted the close relationship between the metaverse and gaming elements, suggesting that prior familiarity with electronic games was advantageous for practical exploration. Additionally, they emphasized the potential for storytelling in the metaverse to enhance educational experiences. The participants found the immersive nature of the metaverse beneficial, especially in scenarios involving visual and narrative elements, such as music analysis or the exploration of complex objects. They discussed the importance of avatars and their need to align with the educational contexts, although they recognized flexibility in avatar representations to promote inclusivity and eliminate discrimination. Furthermore, the participants discussed the implications of technology in modern educational settings, focusing on communication techniques and adapting strategies to engage students more effectively within the broad field of view offered by the metaverse.

The teachers and instructors also compared the metaverse to traditional classrooms and considered whether it should mimic these physical settings or be viewed as a unique space. They emphasized the significance of incorporating the virtual environment into the educational process rather than replicating traditional classroom dynamics. Despite recognizing the immersive nature of the metaverse, participants discussed challenges related to non-verbal communication limitations, including restricted movements and natural gestures when using virtual reality devices. However, they also expressed optimism about future improvements in non-verbal communication within the metaverse through technological advancements. The participants acknowledged that these limitations could be addressed by developing features such as gloves or cameras capable of accurately capturing and reproducing physical movements. Overall, they recognized the metaverse's immersive, interactive, and emotionally engaging aspects despite some initial challenges (Figure 20.3).

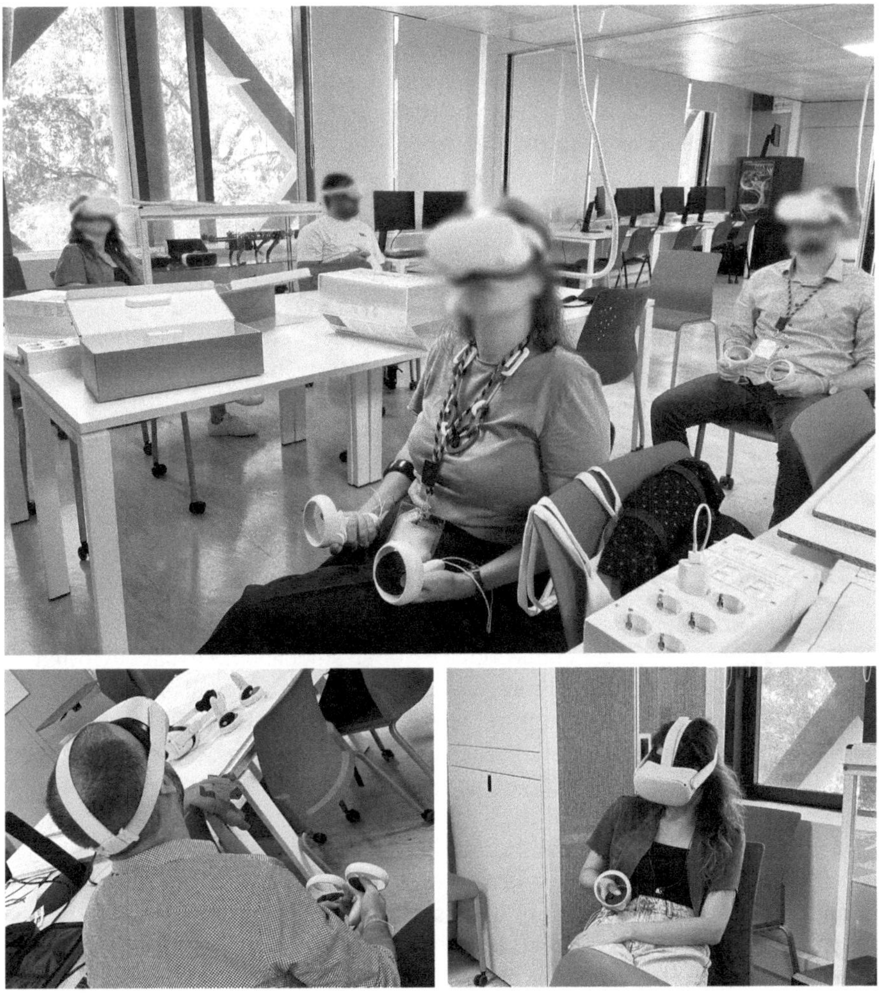

FIGURE 20.3 Mosaic of photos showing participants using the equipment. (Photo and edit by the authors.)

20.4.2 Challenges and Opportunities

Following their experience in the metaverse, teachers and instructors discussed the potential future applications of this technology in education. They highlighted the ability of the metaverse to create a sense of proximity and engagement between educators and students, addressing some of the limitations of traditional distance learning. They emphasized the importance of visual and interactive elements in enhancing the learning experience, with the metaverse allowing students to explore subjects in an immersive and engaging manner. The expansion of the virtual classroom beyond the confines of traditional screens was also noticed, offering new forms of interaction and learning. As pointed out by the participants, this technology's real potential lies in its simulation capabilities. Additionally, the metaverse can eliminate the need for physical travel, making distant educational environments more accessible and fostering a sense of community among students, especially in international educational contexts.

However, several challenges were identified in the use of the metaverse for education. Not all subjects are seamlessly adaptable to this virtual environment, resulting in difficulties for certain activities. Concerns regarding optimizing virtual teaching tools were raised, with considerations for the effectiveness and usability of 3D formats. Space limitations and visibility issues were also highlighted in the case of larger classes. Addressing those challenges in a structured and collaborative manner is crucial for successful integration. Teachers' and instructors' training and clear guidelines for educational use are essential, and it is vital to consider the potential issues of user fatigue and the additional planning required when transitioning to teaching in the metaverse. Having technology-savvy support can also be valuable in assisting teachers and instructors with technical issues on the platform.

20.4.3 Teachers' Role

Teachers and instructors envision a promising future for the metaverse in education, as it can create stronger connections between students in virtual environments, enhance the sense of presence, and reduce absenteeism. They also see its value in facilitating international classes through real-time translation and improving communication during tutorial meetings across different locations. Nevertheless, teachers and instructors stress the importance of establishing clear rules and guidelines within the metaverse to govern interactions, set boundaries between teachers and instructors and students, and uphold codes of conduct.

Furthermore, teachers and instructors recognize the pressing need for training and preparation to effectively harness this new educational tool. They advocate for teachers and instructors to play an active role in shaping the metaverse for educational purposes, emphasizing that they should not be passive users but active co-creators. Collaborative involvement from the education community is essential for the platform to ensure teachers' and instructors' specific needs and expectations. This approach is envisioned as pivotal to developing more effective platforms aligned with educational objectives, enhancing the overall virtual teaching experience.

20.5 DISCUSSION

20.5.1 RQ1: What are the Future Trends in Using the Metaverse in Education from the Perspective of Teachers and Instructors?

This research reveals a series of participants' perceptions and considerations regarding the potential use of the metaverse in education. After practical experience with this technology, participants expressed various opinions and views on how it can be applied in the educational context.

Participants highlighted the ability of this technology to provide a sense of closeness and involvement between teachers, instructors, and students, as seen in Yang et al. (2022), overcoming some of the limitations associated with conventional distance learning. For participants, the potential to represent a shared space was identified as especially relevant for distance learning, creating a sense of community and belonging among students.

Participants emphasized the importance of visual and interactive elements to create a richer and more dynamic learning experience. Furthermore, they discussed the possibility of immersing study objects, enriching the learning process, and making it more engaging and practical. According to participants, the immersion characteristic of the metaverse can capture students' interest, resulting in greater participation in learning activities.

Although the experimental case is modeled after a real-world classroom setting, Ahmad et al. (2022) state that using Digital Twins in education is still in its early stages. In the case of medicine, participants value improving their vision of items or even human organs more than duplicating a classroom situation. Participants also highlighted the usefulness of the metaverse in architecture and interior design classes, where students can explore three-dimensional spaces and evaluate aspects such as size and lighting, providing unique opportunities for practical learning, and with the advantage of not incurring travel expenses. This would make access to distant educational environments more accessible for students with mobility difficulties. The conceptual harmony between the metaverse and Digital Twins creates exclusive educational prospects within three-dimensional virtual object settings.

Interestingly, all perspectives of use by participants referred to the synchronous nature of metaverse activities but did not point out the asynchronous potential or *online* and offline connections, which follows the claim of Dwivedi et al. (2022) that the most traditional use of the metaverse is to replicate the general in-person classroom experience in a more immersive way than previous technologies.

However, participants observed that not all themes are ideally suited to this virtual environment, as evidenced by the difficulty of handwriting or using static slides within the metaverse or the simple fact that the computer screen sometimes meets their demands. The group's perception of the current and short-term use (PU) of technology is for the simulation and modeling of virtual objects in 3D, very much in the logic of Digital Twins.

Regarding the long-term perception of the possibilities of using technology (PU) – a substantial difference to be highlighted, according to Chau (1996 apud Parreira et al., 2018) – is the use of simultaneous translation within the metaverse to facilitate communication in international educational contexts or even the use of AI to integrate groups of students. This reveals an openness on the part of the participants toward a broader conceptualization of the metaverse with the combination of other technologies, following Chua and Yu (2023).

Another important aspect highlighted in the results is the expansion of the limits of the traditional screen, with mentions by M.2, who commented on being able to write "infinitely," unlike in conventional classes where they must erase the board. The participant seems intrigued by the concept of writing in a limitless manner. Seeing it as an advantageous aspect of the metaverse, it suggests an approach to writing with no limited length, allowing the writer to explore ideas or narratives without pre-defined boundaries encountered in the traditional classroom whiteboard. This approach encourages creativity and the free flow of thoughts without conforming to traditional constraints or limitations often associated with more structured writing forms, and was not found in the literature review.

On the other hand, the results also point to concerns. They highlighted issues related to optimizing the tools used and space limitations, especially in large classes where visibility can be challenging. Another major challenge was that participants needed technical assistance to perform simple tasks, such as passing slides and positioning screens in a way that allowed students to view the content appropriately presented. Participants would need the help of a "learning architect" during classes to manage the platform and minimize failures. The difficulty with connectivity, brought up by Tlili et al. (2022), wasn't brought to the fore in the discussion, giving way to challenges with a more direct pedagogical focus. Finally, the biggest complaint presented by the participants was physical discomfort with dizziness and nausea.

These challenges highlight the importance of carefully evaluating the effectiveness and usability of virtual teaching tools, considering the needs of teachers, instructors, and students. Chua and Yu (2023) say that the use of metaverse must be perceived as accessible and useful for teaching, which will influence the adoption of the TAM method described by Parreira et al. (2018) by teachers and instructors, the PFU. These difficulties reported by participants may impact teachers' and instructors' adherence to the tools if a correct transition between educational models is not made.

After overcoming these initial challenges and becoming familiar with the technology, participants explored the tools available in the metaverse, including notes and visual resources to enrich their classes. However, interactivity revealed that there was still a learning curve to master the use of those tools entirely.

Teachers' and instructors' perspectives regarding the metaverse in education were generally positive, recognizing its potential to improve teaching and learning dynamics. However, they also emphasized the importance of defining clear rules and guidelines for using this technology and the need for training and preparation to deal with this new educational tool effectively. "Students should be sitting on the chairs" or "teachers on the stage and students in the audience" were examples of norms of conduct in the classroom that should be replicated in the metaverse, as well as the use of avatars that represent the seriousness of the educational space.

Although the experiment used only one educational model of the metaverse, the participants' speech showed interest in other models.

Other topics covered in the literature review that the participants did not highlight during the session and group interview are the increase in the volume of data and the consequent change in evaluation metrics (Dwivedi et al., 2022) or possible mediation of potential risks such as identity theft, data breaches, financial fraud (Tlili et al., 2022), and students' addiction to games (Chen, 2022).

In summary, the results of this academic research indicate that the metaverse has significant potential to transform education, providing new forms of engagement and learning. However, they also highlight the importance of addressing the challenges and concerns associated with this technology to ensure successful integration and a more efficient and practical teaching experience. There is room for teachers and instructors to continue creating virtual environments to improve their students' motivations, skills, and learning outcomes. Immersion, exploration of study objects, and the ability to customize avatars are essential aspects in this context.

Linking participants' perspectives on the metaverse's educational potential, the focus group also highlighted how integrating game elements, like avatars and storytelling, can enrich this immersive environment, enhancing engagement and learning experiences.

The focus group participants recognized the importance of the metaverse's relationship with game elements, highlighting the need to incorporate narrative and visual elements of games to enrich the educational experience. The integration of gamified elements into the metaverse has proven to be an effective strategy (Park et al., 2021). The relationship between the metaverse and the gamification process was evident during the experiment in that the participants naturally associated the virtual environment with game elements. From the beginning, while the equipment was being configured, the participants began to navigate the platform through their mobile devices, and, in a short time, a playful tone emerged organically among them with comments such as "I'm going to play the game." The language and attitude of the participants were intrinsically linked to the active exploration of the virtual scenario and the use of game-related terms. Participants even invited each other by saying "Everyone is already playing."

This relationship appears in Díaz et al. (2020), who state that MUVEs in the metaverse are based on MMORPGs, but that they differ from each other because the metaverse does not have pre-defined objectives like traditional video games. The metaverse incorporates elements of game-based approaches, such as avatars and storytelling, to engage users more deeply. This goes beyond simple rewards in the form of points and merits, as suggested by Radoff (2011).

Immersion is not exclusive to the digital environment, as Calleja (2022) highlighted, as even board games can offer immersive properties. Vasalou et al. (2017) demonstrated that students' behaviour during interaction with games depends not only on the format (digital or analog), but also on the characteristics of the game and the opportunities for interaction in the environment. Games have evolved, from board games to virtual experiences. (Huizinga, 1949) notably pointed out in his Homo Ludens that humanity has been related to games as a cultural factor since ancient times. One of the participants in the immersive netnography stated that it was easier to use the equipment due to personal familiarity with electronic games.

Regarding the subdivisions between types of media, as previously discussed, studies comparing analog and digital games in middle school students concluded that students had more fun interacting with digital games. Unlike the authors, the individual interviews in this study referred to higher education students. Some interviewees pointed out a possible greater involvement of younger students in dealing with electronic games, including the participants in the focus group in the metaverse session, who stated that younger students would also have an easier time relating to games.

Radoff (2021, September 1) suggests that "The Metaverse is true gamification" due to elements such as emotion, interaction, interactivity, immersion, and absorption, which are fundamental in motivating users. Telepresence, the feeling of "being there," regardless of time and space, is the most significant differentiator for the metaverse compared to conventional two-dimensional learning platforms.

The focus group also pointed to the psychological effect of "telepresence" as a unique experience enabled by the metaverse, beyond mere interaction in a virtual classroom setting. This concept refers to the sense of being physically present in a digital space, transcending time and distance limitations and significantly enhancing engagement and immersion. According to Park et al. (2021), the depth of this experience sets the metaverse apart from traditional *online* learning platforms, creating an environment that fosters emotional engagement and a feeling of togetherness among users.

Ultimately, the study suggests that the metaverse has the potential to reshape the educational landscape, offering educators and students alike an engaging, immersive, and interactive environment. However, participants stressed that careful planning, clear guidelines, and robust technical support are crucial to address the initial difficulties. Embracing these considerations will be essential for teachers, instructors, and institutions to fully harness this new technological tool's benefits and optimize its integration into current and future educational practices.

To fully leverage the metaverse's potential in education, participants also emphasized the need for dedicated support roles, like a "learning architect," to aid with technical management during classes. This role could ensure smooth navigation, avoid interruptions, and minimize technical issues, particularly in managing visual aids and interactive elements. These support structures were deemed essential for educators who might find adapting to the metaverse's technological demands challenging without assistance.

Despite the enthusiasm surrounding the platform, participants expressed concerns about physical discomfort, especially dizziness and nausea, which were experienced by some during immersive interactions. These reactions highlight the need for ergonomic improvements and breaks during sessions to mitigate discomfort, ensuring that extended use remains practical and beneficial for all users. The impact of such physical issues on prolonged metaverse engagement was an unexpected challenge that diverged from previous studies on traditional digital or 2D platforms, where these discomforts were less prominent.

Another key finding relates to the integration of gaming elements within the metaverse. Participants naturally associated the virtual setting with game mechanics, like avatars, storytelling, and dynamic narratives, enhancing the overall experience. These elements contributed significantly to engagement, fostering a playful atmosphere that made learning more interactive. Park et al. (2021) suggest that gamification in the metaverse context provides intrinsic motivation, promoting active participation and creative exploration. This aligns with feedback from educators, who reported positive impacts on students' motivation, collaboration, and problem-solving skills when game-based approaches were utilized in metaverse simulations.

In conclusion, the study indicates that while the metaverse presents considerable educational potential through immersive and gamified learning experiences, careful attention to logistical and ergonomic considerations will be crucial. Thoughtful integration of support roles, ergonomic adaptations, and game-like elements can bridge gaps between traditional and metaverse-based education. This approach not only makes the platform more accessible but also aligns with the evolving needs of both educators and students, setting a foundation for future advancements in virtual learning environments.

20.5.2 RQ2: What is the Role of Teachers and Instructors in the Transition to an Educational Model Mediated by the Metaverse?

Participants emphasized the critical role of educators in integrating the metaverse within educational contexts, arguing that teachers and instructors should actively participate in shaping these platforms rather than merely using them. This viewpoint aligns with Franco-Mariscal et al. (2021), who emphasize that teachers and instructors need to consider resources for classroom innovations and thoughtfully create appropriate materials. Unlike other studies, this research points out the value of involving educators in platform development to meet instructional needs.

For participant M.2, the popularity of the metaverse calls for a significant restructuring of pedagogical training programs by higher education institutions. They suggest that training on metaverse use should be mandatory, in line with Díaz (2020), who believes that adequate training for teachers, instructors, and students maximizes the educational benefits of the metaverse. Dwivedi et al. (2022) further argue that institutions must strengthen technical infrastructure and offer comprehensive training, highlighting the need to teach tool usage and improve support structures.

Participants noted that, as metaverse technology is new, educators must acquire skills to navigate and teach effectively in this virtual environment. This includes learning to operate within the metaverse and adapting course content to align with educational goals, which may require detailed planning and adjustments. Kenski (2014) warns of potential work overload with new technologies, a risk also voiced by participants.

Caution should be exercised in implementing the metaverse, as mandatory use due to external pressures can negatively impact educators' perceptions (Hwang et al., 2016; Parreira et al., 2018). While motivation was highlighted as a benefit of the metaverse, some participants, like M.1, noted that non-verbal communication limitations hindered their engagement, revealing areas for improvement.

Netnography participants further stressed the need to establish clear guidelines for educational use of the metaverse, including conduct and ethics standards within virtual environments. Schlemmer et al. (2008) similarly argue that educators and researchers should develop a deep understanding of metaverse technology and adopt new methodologies aligned with its capabilities.

The metaverse stands out as an innovative media environment that redefines social interactions, digital culture, and educational methodologies. This study highlights the potential of the metaverse to support educational advancements while offering an immersive experience that challenges traditional concepts of identity, classroom interaction, and pedagogical structure.

20.6 CONCLUSIONS

Literature shows that education is digitalizing to embrace innovations such as the metaverse, where the relationship between the metaverse and gaming approaches is evident from the perspective of teachers and instructors. They emphasize the importance of gaming literacy in enriching your educational experience in the metaverse. The results of the immersive netnography field research indicate that the relationship between the metaverse and gaming approaches is evident from the perspective of teachers and instructors, with language and attitude intrinsically linked to the active exploration of the virtual environment and the use of gaming-related terms. Educators highlight the importance of having gaming literacy to enrich their educational experience in the metaverse. Furthermore, the metaverse represents a significant shift from conventional educational technologies, highlighting the importance of immersion, storytelling, and avatar customization. The communication dynamics in the metaverse are perceived as different, with changes in teachers' and instructors' communication techniques. However, there are limitations in non-verbal communication due to virtual reality devices.

In contemporary society, the mediatization of educational processes is occurring, reflecting the virtualization of social practices. The Metaverse, a result of the acceleration of digitalization during the COVID-19 pandemic, is a mediatic environment that offers new social interactions and immersive experiences and creates new habits of perception. Virtual identities and the relationship between the "virtual self" and the "real-world self" are complex in the metaverse, affecting self-perception and creating an additional layer of mediatized existence. The study also examines the intricate relationship between virtual and non-virtual identity, discussing how this new type of media challenges conventional conceptions of identity, presence, and users' self-perception, and how this can affect learning.

Teachers and instructors see the metaverse as a synchronous teaching tool. However, they also recognize its asynchronous potential and highlight the primary uses of the metaverse in general for observing three-dimensional objects, simulations, and immersive narrative scenarios, which finds similarity in the concept of Digital Twins. Despite the challenges, teachers and instructors have a favorable view of the metaverse's potential in education, highlighting its ability to promote proximity and engagement between teachers and students, overcoming the limitations of traditional distance learning and even the classroom structure like the whiteboard. The research highlights the need for a careful and strategic approach to dealing with challenges such as technological infrastructure and privacy issues. It highlights the importance of active collaboration between teachers and instructors in developing the metaverse and restructuring pedagogical training programs. Teachers and instructors believe that the metaverse has the potential to transform education, although there are challenges to be overcome. This study is a theoretical and practical starting point for exploring this constantly evolving area. Trying to clarify all these points (benefits and non-benefits) related to the utilization of the metaverse inside pedagogical contexts, study findings help understand how the metaverse can contribute to education and define the teachers and instructors' more active role in this new digital immersive ambiance, e.g., on the study, participants anticipated the metaverse's positive impact in education, foreseeing stronger student connections and reduced absenteeism. They highlight its value in facilitating international classes but stress the need for clear rules and advocate for them to be active co-creators rather than passive users. This aligns with existing literature emphasizing educators' instrumental role in introducing innovations and reshaping pedagogical training programs. There's a growing call for mandatory education about the metaverse to equip educators and students adequately, ensuring effective use, but to active the full potential of these new technologies is essential to acknowledging potential challenges such as work overload and reinforcing the need for clear rules and ethical guidelines in the metaverse's educational use that can and should be brought by the teachers and instructors as educators.

20.6.1 Contributions to Theory and Practice

The study offers significant theoretical and practical contributions. Theoretically, it advances the understanding of the metaverse as an educational space, exploring its ability to reconfigure the learning environment and the changes in human relationships mediated by this technology. Furthermore, it highlights the relationship between virtual and non-virtual identity. It also advances theoretical frameworks surrounding "gamification," "immersion," and "mediatization" within the context of the metaverse, examining its potential integration into educational environments to facilitate a more dynamic and participatory learning approach. Gamification is posited as a core element, extending beyond traditional game components to encompass a "true gamification" that fosters a more profound, immersive, and transformative experience. Telepresence is reconceptualized in this context, as the depth of immersion provided by the metaverse is argued to exceed that of conventional online games, offering continuity with real-life experiences through persistent identities, objects, and opportunities for essential digital skill development. Furthermore, the study positions the metaverse as a "mediated" educational environment, where mediatization redefines learning as an interactive and collaborative process. By leveraging avatars and simulations, the metaverse challenges conventional pedagogical models, fostering opportunities for socio-emotional skill development, critical thinking, and problem-solving, thereby expanding the scope of mediation within educational contexts.

Practically, the study lays the groundwork for exploring the metaverse's impact on education, emphasizing teachers' and instructors' pivotal role in shaping its use, ranging from technological training to creating guidelines, best practices, and codes of ethics for effective implementation. Participants anticipate positive effects, advocating for clear rules and mandatory metaverse education to maximize benefits while addressing potential challenges. It identifies technical tools and challenges, offers suggestions for sustainable business models, and highlights the importance of active educator collaboration in developing metaverse platforms aligned with specific educational objectives.

The practical components of the study provided in this paper primarily examine the interaction between the metaverse and classroom instruction in the context of the higher education level.

Despite this, the research presented has several limitations, including a geographic sample restricted to teachers and instructors from a single academic institution, logistical limitations in the availability of equipment, a homogeneous sample of participants in immersive netnography, and the need for prior experience with Edu-Metaverse. Furthermore, most of the teachers and instructors studied were linked to academic institutions with innovative teaching approaches, which may restrict the generalization of the results. For future research, several directions are suggested, including investigating factors that influence the adoption of the metaverse technologies, how institutions are incorporating the metaverse into learning programs, comparative studies between the metaverse and traditional teaching methods, exploration of the metaverse in specific areas of teaching, the development of skills in the context of the metaverse, and studies in other asynchronous educational modalities in the metaverse, such as lifelogging.

ACKNOWLEDGMENTS

This research was supported by UNIDCOM/IADE-UE (Faculdade de Design, Tecnologia e Comunicação, da Universidade Europeia) – Design and Communication Research Unit, Lisbon, Portugal, under a grant from the Fundação de Ciência e Tecnologia (FCT).

It draws on an exploratory study for a master's degree in Multimedia and new social spaces of human interaction.

REFERENCES

Alpala, L. O., Quiroga-Parra, D. J., Torres, J. C., & Peluffo-Ordóñez, D. H. (2022). Smart factory using virtual reality and online multi-user: towards a metaverse for experimental frameworks. *Applied Sciences*, 12(12), 6258. https://doi.org/10.3390/app12126258

Anderson, C. (2010). Presenting and evaluating qualitative research. *American Journal of Pharmaceutical Education*, 74, 141. https://doi.org/10.5688/aj7408141

Ahmad, I., Sharma, S., Singh, R., Gehlot, A., Priyadarshi, N., & Twala, B. (2022). MOOC 5.0: a roadmap to the future of learning. *Sustainability*, 14(18), 11199. https://doi.org/10.3390/su141811199

Arcagök, S. (2021). Designing augmented reality experiences for cultural heritage education: a constructivist approach. *Education and Information Technologies*, 26(6), 7521–7543. https://doi.org/10.1007/s10639-021-10583-5

Barry, D. M., Ogawa, N., Dharmawansa, A., Kanematsu, H., Fukumura, Y., Shirai, T., … & Kobayashi, T. (2015). Evaluation for students' learning manner using eye blinking system in metaverse. *Procedia Computer Science*, 60, 1195–1204.

Beerda, T. (2022). *Designing Human-Centric Digital Futures: Ethics, Systems and Innovation*. Amsterdam: BIS Publishers.

Bower, M. (2022). Physical play in virtual spaces: Board games and learning design. XR Learning Weekly.

Braun, V., & Clarke, V. (2006). Using thematic analysis in psychology. *Qualitative Research in Psychology*, 3(2), 77–101. https://doi.org/10.1191/1478088706qp063oa

Caliandro, A. (2016). Ethnography in digital spaces: ethnography of virtual worlds, netnography, & digital ethnography. In R. Denny & P. Sunderland (Eds.), *Handbook of Anthropology in Business* (pp. 658–679). New York, NY: Routledge.

Calleja, G. (2022). *Board game immersion*. In T. Kikkawa, W. C. Kriz, & J. Sugiura (Eds.), *Gaming as a Cultural Commons: Risks, Challenges, and Opportunities* (pp. 43–55). Singapore: Springer.

Camilleri, A. R. (2023). *Designing the Metaverse: Metadesign, Ethics and Futures*. London: Routledge.

Cavalheiro, G. M. (2010). *Interações e interfaces do design com os jogos digitais: uma abordagem centrada no usuário*. São Paulo: Estação das Letras e Cores.

Chen, X., & Yao, Y. (2022). The impact of gamified learning environments on student engagement: A meta-analysis. *Computers & Education*, 190, 104607. https://doi.org/10.1016/j.compedu.2022.104607

Chenail, R. J. (2010). Introduction to qualitative research design [Workshop presentation]. In: *23rd Annual Primary Care Research Methods & Statistics Conference, San Antonio, Texas, USA*. Retrieved from https://nsuworks.nova.edu/cahss_facpres/1899/

Chua, H. W., & Yu, Z. (2023). A systematic literature review of the acceptability of the use of Metaverse in education over 16 years. *Journal of Computers in Education*, 11, 1–51.

Council of the European Union. (2022). *Council Conclusions on Building a European Strategy for the Cultural and Creative Industries Ecosystem*. Brussels: Council of the EU. Retrieved from https://www.consilium.europa.eu/en/press/press-releases/2022/05/18/council-conclusions-on-building-a-european-strategy-for-the-cultural-and-creative-industries-ecosystem/

Cowan, B. R., Branigan, H. P., Obregón, M., & Clark, L. (2021). Voice interfaces in everyday life: Implications for trust, privacy and social interactions. *Proceedings of the ACM on Human-Computer Interaction*, 5(CSCW1), 1–25. https://doi.org/10.1145/3449104

Cruz Franco, P. A., Rueda Márquez de la Plata, A., & Gómez Bernal, E. (2022). Protocols for the graphic and constructive diffusion of digital twins of the architectural heritage that guarantee universal accessibility through AR and VR. *Applied Sciences*, 12, 8785. https://doi.org/10.3390/app12178785

da Silva, D. T. (2010). *Design de Jogos: uma abordagem centrada no jogador*. São Paulo: Blucher.

Denzin, N. K., & Lincoln, Y. S. (2005). Introduction: the discipline and practice of qualitative research. In: Denzin, N. K., & Lincoln, Y. S. (eds), *The Sage Handbook of Qualitative Research* (3rd edn). Thousand Oaks, CA: Sage.

Denzin, N. K. (1989). *Interpretive Interactionism* (Vol. 16). Newbury Park CA: Sage.

Deterding, S., Dixon, D., Khaled, R., & Nacke, L. (2011). From game design elements to gamefulness: Defining "Gamification". In *Proceedings of the 15th International Academic MindTrek Conference: Envisioning Future Media Environments* (pp. 9–15). ACM. https://doi.org/10.1145/2181037.2181040

Díaz, J. E. M. (2020). Virtual world as a complement to hybrid and mobile learning. *International Journal of Emerging Technologies in Learning (iJET)*, 15(22), pp. 267–274. https://doi.org/10.3991/ijet.v15i22.14393

Díaz, L., Martín, E., & García, A. (2020). Gamification and augmented reality in education: A systematic review. Education Sciences, 10(1), 52. https://doi.org/10.3390/educsci10010052

Duarte, A. (2018). *Design Thinking e Educação: A Prática do Design Centrado no Ser Humano*. São Paulo: Editora Contexto.

Dwivedi, Y. K., Hughes, L., Baabdullah, A. M., Ribeiro-Navarrete, S., Giannakis, M., Al-Debei, M. M., … & Wamba, S. F. (2022). Metaverse beyond the hype: multidisciplinary perspectives on emerging challenges, opportunities, and agenda for research, practice and policy. *International Journal of Information Management*, 66, 102542.

Estudante, A., & Dietrich, N. (2020). Using augmented reality to stimulate students and diffuse escape game activities to larger audiences. *Journal of Chemical Education*, 97(5), 1368–1374.

Franco-Mariscal, A. J., Cebrián-Robles, D., & Rodríguez-Losada, N. (2021). Impact of a training programme on the e-rubric evaluation of gamification resources with pre-service secondary school science teachers. *Tech Know Learn* 28, 769–802 (2023). https://doi.org/10.1007/s10758-021-09588-1

Gee, J. P. (2020). *Games Change Learning: Game Literacy, Critical Thinking, and the Future of Education* (Sun, J., translator). Shanghai: East China Normal University Press

Gioia, D. A., Corley, K. G., & Hamilton, A. L. (2013). Seeking qualitative rigor in inductive research: notes on the gioia methodology. *Organizational Research Methods*, 16(1), 15–31. https://doi.org/10.1177/1094428112452151

Green, M. C., & Brock, T. C. (2000). The role of transportation in the persuasiveness of public narratives. *Journal of Personality and Social Psychology*, 79(5), 701–721. https://doi.org/10.1037/0022-3514.79.5.701

Hepp, A. (2014). *Mediatization: A Concept, a Theory, a Field*. Cambridge: Polity Press.

Hjarvard, S. (2012). *The Mediatization of Culture and Society*. London: Routledge.

Hjarvard, S. (2013). The mediatization of culture and society: Theoretical and empirical perspectives. In K. Lundby (Ed.), *Mediatization of Communication* (pp. 199–226). Berlin: De Gruyter Mouton.

Hjarvard, S. (2014). Midiatização: conceituando a mudança social e cultural. *Matrizes*, 8(1), 21–44. https://doi.org/10.11606/issn.1982-8160.v8i1p21-44

Holopainen, M., Saunila, M., Rantala, T., & Ukko, J. (2022). Digital Twins' Implications for Innovation. *Technology Analysis & Strategic Management*. https://doi.org/10.1080/09537325.2022.2115881

Hudson-Smith, A., & Shakeri, M. (2022). *The Future's Not What It Used to Be: Urban Wormholes, Simulation, Participation, and Planning in the Metaverse. Urban Planning*, 7(2), 214–217. https://doi.org/10.17645/up.v7i2.5893

Huizinga, J. (1949). Homo Ludens. Routledge & K. Paul. https://books.google.pt/books?id=ALeXRMGU1CsC

Hwang, Y., Al-Arabiat, M., & Shin, D.-H. (2016). *Understanding technology acceptance in a mandatory environment: A literature review. Information Development*, 32(4), 1266–1283. https://doi.org/10.1177/0266666915593621

Interactive Software Federation of Europe. (2022). *Video Games Industry Insights Report 2022*. Brussels: ISFE. Retrieved from https://www.isfe.eu

Jang, Y., Vitale, J. M., Jyung, R. W., & Black, J. B. (2019). Direct manipulation is better than passive viewing for learning anatomy in a three-dimensional virtual reality environment. *Computers & Education*, 106, 150–165. https://doi.org/10.1016/j.compedu.2016.12.009

Jiang, F. (2021). *Designing User Interfaces in the Age of AI: Principles and Practical Techniques*. Beijing: Tsinghua University Press.

Kelly, E. (2021). *The Future of Design: Global Product Innovation for a Complex World*. New York: Bloomsbury Visual Arts.

Kenski, V. M. (2014). *Tecnologias e tempo docente*. Campinas, SP: Papirus Editora.

Kozinets, R. V. (2022). Immersive netnography: a novel method for service experience research in virtual reality, augmented reality and metaverse contexts. *Journal of Service Management*, 34(1), 100–125.

Kye, B., Han, N., Kim, E., Park, Y., & Jo, S. (2021). Educational applications of metaverse: Possibilities and limitations. *Journal of Educational Evaluation for Health Professions*, 18, 32. https://doi.org/10.3352/jeehp.2021.18.32

Lessard-Hébert, M., Goyette, G., & Boutin, G. (1990). *Investigação qualitativa: Fundamentos e práticas*. Lisboa: Instituto Piaget, 63–68.

Marr, B. (2022). *Future Skills: The 20 Skills and Competencies Everyone Needs to Succeed in a Digital World*. Chichester, UK: Wiley.

Masadeh, M. A. (2012). Focus group: reviews and practices. *International Journal of Applied Science and Technology*, 2(10), 63–68.

Merriam, S. B. (2009). *Qualitative Research: A Guide to Design and Implementation*. San Francisco, CA: Jossey-Bass.

Mirzoeff, N. (1999). *An Introduction to Visual Culture*. London: Routledge. ISBN: 9780367235345.

Mónico, L., Alferes, V., Parreira, P., & Castro, P. A. (2017). A Observação Participante enquanto metodologia de investigação qualitativa. *Investigação qualitativa em ciências sociais*, 3(1), 972–978.

Park, Y., Kim, H., & Cho, H. (2021). Smart learning in the metaverse: the case of South Korea. *Journal of Educational Evaluation for Health Professions*, 18, 10. https://doi.org/10.3352/jeehp.2021.18.10

Park, S., & Kim, S. (2022). Identifying world types to deliver gameful experiences for sustainable learning in the metaverse. *Sustainability*, 14(3), 1361. https://doi.org/10.3390/su14031361

Parreira, P., Proença, S., Sousa, L., & Mónico, L. (2018). Technology Acceptance Model (TAM): Modelos percursores e modelos evolutivos [Technology acceptance model (TAM): pathway models and evolutionary models]. In: *Competências empreendedoras no Ensino Superior Politécnico: Motivos, influências, serviços de apoio e educação* (pp. 143–166), Guarda, Portugal: Instituto Politécnico da Guarda.

Patrício, R., Negre, Y., & Ricarte, É. (2022). Gamification in immersive learning environments: An experience in virtual reality design education. In A. Reis & J. Amorim (Eds.), *Digital Learning and Teaching in Art and Design Education*. Hershey, PA: IGI Global. https://doi.org/10.4018/978-1-7998-8816-1.ch010

Radoff, J. (2021). The Metaverse Value-Chain. Available at: https://medium.com/building-the-metaverse/the-metaverse-value-chain-afcf9e09e3a7

Reyes, M. (2020). *Designing for Virtual Reality: User Experience Design and 3D Environments*. New York: Apress.

Russell, D. (2022). How board games are entering the metaverse. *The Metaverse Journal*.

Schlemmer, E., Trein, D., & Oliveira, C. (2008). Metaverso: a telepresença em Mundos Digitais Virtuais 3D por meio do uso de avatares. In: *Brazilian Symposium on Computers in Education (Simpósio Brasileiro de Informática na Educação-SBIE)* (Vol. 1, No. 1, pp. 441–450).

Sá, P., Costa, A. P., & Moreira, A. (Org.). (2021) *Reflexões em torno de Metodologias de Investigação: recolha de dados* (Vol. 2). UA Editora. https://doi.org/10.34624/ka02-fq42

Scimeca, D. (2011). *Defining Gamification: From Buzzword to Game-Changer*. Gamasutra. Available at: https://www.gamedeveloper.com/design/defining-gamification-from-buzzword-to-game-changer

Sodré, M. (2009). *Antropotécnica: para uma estética da comunicação*. Rio de Janeiro: Vozes.

Strauss, A., & Corbin, J. (1998). *Basics of Qualitative Research Techniques: Techniques and procedures for developing grounded theory* (2nd ed.). Sage Publications, Inc.

Sugeng, B., & Suryani, N. (2018). The development of augmented reality-based learning media to improve students' critical thinking skills. *International Journal of Emerging Technologies in Learning (iJET)*, 13(10), 231–238. https://doi.org/10.3991/ijet.v13i10.8470

Tayal, D., Gupta, R., & Singh, R. (2022). Immersive learning technologies: A framework for future education. *Education and Information Technologies*, 27, 4091–4110. https://doi.org/10.1007/s10639-021-10713-z

Tlili, A., Huang, R., Shehata, B., Liu, D., Zhao, J., Metwally, A. H. S., ... & Burgos, D. (2022). Is Metaverse in education a blessing or a curse: a combined content and bibliometric analysis. *Smart Learning Environments*, 9(1), 1–31.

Tsai, C.-W., Fan, Y.-T., & Huang, W.-H. (2020). A research framework of design thinking-based teaching in STEM education. *Thinking Skills and Creativity*, 37, 100681. https://doi.org/10.1016/j.tsc.2020.100681

Vasalou, A., Khaled, R., Holmes, W., & Gooch, D. (2017). Digital games-based learning for children with dyslexia: A social constructivist perspective on engagement and learning during group game-play. Computers & Education, 114, 175–192. https://doi.org/https://doi.org/10.1016/j.compedu.2017.06.009

Wang, M., Yu, H., Bell, Z., & Chu, X. (2022). Constructing an Edu-Metaverse ecosystem: a new and innovative framework. *IEEE Transactions on Learning Technologies*, 15(6), 685–696.

Yang, Q., Zhang, Y., & Lin, Y. (2022). Study on the influence mechanism of virtual simulation game learning experience on student engagement and entrepreneurial skill development. *Frontiers in Psychology* 12, 772157. https://doi.org/10.3389/fpsyg.2021.772157

Part 4

Thematic – SPECULATIVE

Subthemes: Futures – Critical Thinking – Design Research

21 Genre Differentiation in Design Futures

An Empirical Study on Audience Perception of Practice Cases

Tiantian Li and Zhiyong Fu

21.1 INTRODUCTION

Over the past two decades, the future and design have developed into a more intimate collaborative relationship. On the one hand, designers are becoming more future-oriented, integrating long-term thinking, scenarios, and other foresight tools into their work. On the other hand, an increasing number of futurists are not just studying various possibilities from a safe analytical distance but are participating in prototype design and setting a background for future experiences while using design (Candy & Dunagan, 2017). Design practitioners and researchers have extensively explored mixed methods that combine design and foresight processes, such as creating provocative fictional prototypes (Dunne & Raby, 2013), making future scenarios experiential and immersive (Candy & Dunagan, 2017), and generating video prototypes to enrich storytelling (Sterling, 2009). Hartman (van der Duin, n.d.) argues that design works – visualisations, videos, and prototypes – can concretise and translate foresight outcomes into vivid and engaging forms, which can be appreciated by a broad audience that is not necessarily well versed in foresight to a wide audience. Against this backdrop, more and more future-oriented design typologies are being conceived. In short, DF is the combination of design and futurology, with tremendous prospects for development.

Another noteworthy issue is that DF is a cross-cutting field that combines a large number of paradigms and approaches from different fields, driven by individuals or organisations from different perspectives (Bhatti et al., 2014; Bleeker, 2009; Candy, 2010; Dunne & Raby, 2013). DF has been described as a confluence of many tributaries (Angheloiu et al., 2020). Admittedly, this rich diversity reflects the field's burgeoning popularity, but it also brings with it a great deal of complexity. DF is a general category, and typologies are hugely problematic. On the one hand, this rapidly evolving field is constantly producing new genres or reshaping established ones through crossover. On the other hand, there are overlapping concepts and ill-defined terms within the field, making it difficult to develop a clear consensus. This is simply

DOI: 10.1201/9781032719290-25

a fatal problem for the development of the discipline, as no one has been able to unify so many concepts. How to contribute to the establishment of a more scientific typology of practice cases became the research motivation of this paper.

To achieve these aims, this paper takes an alternative path where we study audience perception to provide further reference and insights for DF subtype divisions. We hope to go beyond relying solely on literature research by obtaining an understanding of the audience's cognitive structures regarding these types. The research findings will then be applied to further clarify the differences between various subtypes and assist in revising the preliminary typological divisions proposed.

The above assertion raises two research questions for this chapter:

RQ1: What are the semantic perception dimensions associated with DF practice cases, and how can these dimensions be extracted?

RQ2: How do audiences perceive the differences between subtypes, and can these perceptions be visualised (through perceptual mapping) to provide insights for revising the typology?

This study is structured into the following sections: Introduction, Related Works, Methodology, Results, Discussion, and Conclusion. Initially, the study introduces five subtypes of DF practice cases (DF subtypes). Second, the research methodology involves a literature review and expert questionnaire survey (involving 20 experts) to construct and refine semantic perception dimensions for DF practice cases (RQ1). Third, an evaluation experiment was conducted with 85 professional participants to assess 15 key cases using the extracted perception dimensions. Fourth, the experimental data were thoroughly analysed and visualised. By creating perception maps, we were able to visually represent the relationships between 15 typical cases and the semantic terms obtained (RQ2). Following this, we utilised the perception maps to uncover the audiences' cognitive structures regarding subtypes, examined the differences between types, and provided insights and support for the refinement of the typology. The findings of this study are instrumental in establishing a more scientific typology of practice cases, thereby contributing to the resolution of a critical issue in the discipline's development.

21.2 RELATED WORKS

21.2.1 A Preliminary Typology of DF Practice Cases

A clear typology of DF practice cases is needed for further research. However, the current classification of cases faces several issues, primarily in terms of its ambiguity in both extension and inclusion. (i) Extension refers to the fact that DF is a rapidly evolving field that is continuously giving rise to new types or redefining existing types through crossover. For instance, design innovation driven by forward-looking technologies with a strong interdisciplinary nature has become a significant area that cannot be overlooked. Within this domain, participatory design, concepts and methods of futures studies, and human–computer interaction or other fields biased towards engineering technologies have merged (Kelliher & Byrne, 2015; Mankoff et al., 2013; Pierce, 2021). Therefore, these types must be considered when establishing a typology. (ii) Inclusion refers to the various perspectives from which to classify, leading to overlaps between categories. For example, within the design field, Dunne and Raby (2013) list the following predecessors for their method: speculative design, critical design, design fiction, DF, anti-design, radical design, interrogative design, design for debate, adversarial design, etc. In contrast, Galdon takes a broader view, delineating six representative samples of practice in the field of DF, which includes types related to strategic foresight, such as scenario planning, etc. We believe that the root of the problem lies in the fact that DF, as an interdisciplinary category, encompasses not only emerging design types but also the background of futures studies. Given that this paper examines practice cases, we have carefully examined scenario planning, which is more oriented towards strategic foresight, and replaced it with a typology that is more oriented towards the combination of strategy and design: visionary design. Based on these considerations, this paper proposes a typology consisting of five categories: speculative design, design fictions, visionary design, transition design, and technological innovation as follows.

1. Speculative design: Speculative design materialises critical thinking, enabling people to ponder possible futures and employ these to comprehend the present and discuss the future they aspire for. Its emphasis is not on solving problems, but on posing "well-designed questions". Speculative design practices utilise fictional design objects to explore potential scenarios for the future, stimulating imaginations about possible technologies and their significance in contextualised situations, and engaging in debates about the social, cultural, and ethical impacts of emerging technologies.
2. Design fiction: Design fictions combine elements of science fiction, science fact, and design to frame the imaginative creation of diegetic prototypes – that is, prototypes that exist within story worlds. Design fiction proceeds by imagining and prototyping ideas or artefacts in fictional or story worlds, providing insights, feelings, and ideas about possible or probable futures. It originated from the creative practice of combining the traditions of writing and storytelling with the material production of objects, and is widely present in printed materials, images, and physical artefacts.
3. Visionary design: Visionary design is based on extensive or focused research on factors or key areas such as business, education, futures, and social politics. This process aims to formulate a guiding content of tasks, goals, and visions for the future of countries and industries. Visionary design presents cultural values and missions to government agencies, enterprises, institutions, or cooperative organisations, offering a blueprint for their shared pursuit. Visionary design assesses probable or plausible futures to inform and build capacities for strategic risk reduction. By emphasising future well-being and striving to plan for a more preferable future, visionary design seeks to create a collective sense of direction and purpose. Visionary design conveys the important characteristics that future studies have developed in conjunction with design: experiential and participatory.

4. Transition design: Transition design is based on long-term visioning and the recognition of the need for solutions rooted in new, more sustainable socioeconomic and political paradigms. It originates from long-term thinking, lifestyle-oriented, location-based solutions, and always acknowledges the natural world as the larger context for all design solutions. Transition Design not only emphasises more compelling future-oriented visions to inform and inspire current projects but also requires the development of these visions through design tools and methods. The transition vision provides greater leverage for service and social innovation projects, forming more effective transition steps towards an ideal future.
5. Technological innovation: Technological innovation is not only driven by forward-looking technology, but also integrates art, science, design, and engineering, possessing strong interdisciplinary attributes. Technological innovation focuses on creating and commercialising transformative future technologies, emphasising experimentation in the space of interaction between people and technology. In the practice of technological innovation, prototype(s) of one (or more) features of approximate products, services, or systems are developed to explore and develop new concepts, or to conduct user research and testing. At the same time, participatory design practices, technology futures perspectives, and human–computer interaction or other cross-cutting fields of engineering and design have converged and are becoming popular methods and emerging traditions in the practice of technological innovation.

21.2.2 User Perception Studies

The basic idea of perceptual mapping is to construct a multi-attribute perceptual space, which is widely applied in design and marketing (Holak, 1994). Perceptual maps are based on attribute rating or similarity judgements and use perceptual mapping algorithms to create low-dimensional maps, where each point on the map corresponds to a product alternative, and the distance between any two points corresponds as closely as possible to the perceived differences between products (Chuang & Chen, 2008). Wang and Chin (2017) used Correspondence Analysis (CA) to identify key product features and construct perceptual maps to effectively display the association between samples and variables, visualising product differentiation. In summary, perceptual maps are a visualisation method for expressing multi-attribute perceptual spaces. Additionally, there are other methods for extracting user perceptual dimensions and establishing evaluation systems. For example, Dimitrios et al. (2014) in their study of automotive user demand dimensions creatively combined the Future Workshop method with Resource Grid Technology (RGT) to refine 19 dimensions related to user needs from the obtained 390 attribute structures. Li et al. (2023) studied the semantic pools and categories of automotive styling stance. In the first phase, the study used online data collection and expert questionnaires to define semantic pools. In the second phase, the study combined card sorting with two clustering methods and visualised high-dimensional data using "t-SNE" to achieve the acquisition of semantic categories. Chen et al. (2023) constructed a novel visual art quality evaluation index system for the future metaverse, consisting of 4 primary indicators and 12 secondary indicators.

In summary, we believe that perceptual research for DF practice cases consists of three main aspects: (i) Establishing a multi-attribute perceptual space, which may involve the acquisition of a range of proprietary semantics (such as text mining or manual acquisition) and semantic filtering (such as TF-IDF, C.I., expert screening, and even the use of BERT pre-trained models, among other methods). (ii) Designing perception evaluation experiments to obtain evaluation data. In addition to designing the data collection methods, this may also include effectively presenting cases to enhance the audience's experience. (iii) Selecting appropriate dimensionality reduction methods (PCA, MDS, t-SNE, etc.) to process the high-dimensional data, establishing perceptual mapping through visualisation tools, and revising and proposing a typology of practice cases for DF by combining the audience's perceptual results.

21.3 METHODOLOGY

21.3.1 Stage I: Constructing Semantic Perceptual Dimensions for DF Practice Cases

Stage I: The research in Stage 1 revolves around **RQ1**.

21.3.1.1 Data Acquisition and Processing

In the initial data collection phase, this study retrieved 102 adjectives through desktop research from relevant field-specific research papers and books. Subsequently, through brainstorming and expert consultation, a total of 149 adjectives were obtained. The data processing phase was conducted manually and primarily involved standardisation, merging, and supplementation procedures. Standardisation involved converting word forms and presenting the final outcomes using a single semantic word (Li et al., 2023; Schütte, 2005; Warell, 2008). Merging refers to eliminating repetitive adjectives and combining rare words with their synonyms (Li et al., 2023). For instance, close synonyms such as proprietary, individual, and personal were merged into personal, and so on. Supplementation entailed further expanding the word base. Through these procedures, a final list of 133 adjectives was obtained. Compared to the 149 adjectives collected in the data collection phase, these adjectives exhibited more explicit connotations and encompassed more semantic dimensions.

21.3.1.2 Subjects

The expert subjects consisted of expert practitioners from major corporations (studios, design firms, tech companies, etc.) and researchers from Tsinghua University, with a total of 20 experts, 45% male and 55% female, who formed an authoritative panel of experts in the field. The experts from companies include not only the managers, founders, and design directors of a series of innovation firms, innovation labs, and innovation platforms (aHybrid Lab, ReSpace Lab, etc.), but also the founders of product communities and clubs (Futurist Club), as well as the experts from technology companies and consulting firms (Huawei UCD, Accenture, Tang Shuo), and so on.

21.3.1.3 Materials

Guided by the proposed typology, which encompasses speculative design, design fiction, transition design, visionary design, and technological innovation, we have gathered a total of 111 case examples. Simultaneously, we have developed specific collection sources and criteria to ensure their relevance and accuracy. These cases have been uploaded to Notion and made widely accessible to all subjects, fostering their understanding of the practice cases and allowing them to select more appropriate words from the questionnaire.

21.3.1.4 Process

In the questionnaire, 133 adjectives were systematically categorised into seven groups (I–VII), each containing approximately 20 words. Subjects were requested to browse the case library on Notion and subsequently select more appropriate terms from the list of 133 words to describe the practice cases (allowing for multiple selections) (Figure 21.1).

21.3.1.5 Data Analysis

In order to optimise the semantic space, further screening of semantic words is required. The study first evaluates the importance of adjectives through IF-IDF, then divides the word frequency threshold through confidence interval (C.I.), and, finally, combines the two methods to obtain the word set and invites experts to make the final selection. The steps are as follows:

Step (i): Filtering words by TF-IDF

First, construct the TF matrix: create a matrix of 20 rows (questionnaires) × 133 columns (adjectives), where the value of each cell is the number of times the adjective appears in the corresponding questionnaire.

Second, compute TF-IDF: for each adjective *t* and questionnaire *d*, compute TF(*t*, *d*) and IDF(*t*). TF(*t*, *d*) is the ratio of the number of times word t

FIGURE 21.1 133 adjectives were systematically categorised into seven groups (I–VII) in the questionnaire.

appears in questionnaire d to the total number of times all words appear in questionnaire d (133). IDF measures the prevalence of an adjective, with the denominator being the number of questionnaires that contain the adjective, and the numerator being the total number of questionnaires (20). TF-IDF is the product of TF and IDF and is used to assess the importance of an adjective across all questionnaires:

$$\text{TF-IDF}(t) = \text{TF}(t) \times \text{IDF}(t)$$

Third, adjective ordering: all adjectives are sorted in descending order according to their TF-IDF values. By sorting, the word set **A** is obtained.

Step (ii): Filtering words by C.I.

First, the confidence level is chosen and the C.I. is calculated. Depending on the situation, it is taken as $\alpha = 0.95$ or $\alpha = 0.99$. Subsequently, the C.I. $(1-\alpha)$ is calculated for the frequency of each word:

$$\text{Confidence Interval } = \left[f_w - Z_{\alpha/2} \times \sqrt{\frac{f_w \times (1-f_w)}{n}},\ f_w + Z_{\alpha/2} \times \sqrt{\frac{f_w \times (1-f_w)}{n}} \right]$$

where f_w is the frequency of the word w, $Z_{\alpha/2}$ is the two-tailed critical value of the standard normal distribution corresponding to the chosen confidence level α, and n is the total number of words.

Second, adjective screening: to ensure that only statistically significant and frequent words are selected, this study uses the upper limit of the C.I. as the word frequency threshold, and all words below this threshold are rounded off. By filtering, the word set **B** is obtained.

Step (iii): Takes the intersection and invites expert selection

First, the intersection of word set **A** and **B** is taken to get word set **C**.

Second, as this study involves more domain knowledge, it is necessary to invite experts to manually revise the results of data analysis.

21.3.2 Stage II: Reflecting Audience Cognitive Structures through Perceptual Maps, thus Providing Insights for Typological Revision

Stage II: The research in Stage 2 revolves around **RQ2**.

21.3.2.1 Subjects

A total of 85 subjects were recruited to participate in the questionnaire. The subjects came from colleges, studios, design firms, consulting firms, and technology firms. Among them, 54 were from universities, five from studios, eight from design firms, four from consulting firms, and 14 from technology firms. Unlike the first experiment, in which the participating experts were important driving forces in the development of the field in China, the subjects in this experiment were mainly professional audiences.

21.3.2.2 Materials

Since 111 cases are too large, to reduce the burden of case evaluation, we first screened typical cases. Based on mean sorting, 15 key cases were obtained. Furthermore, we restructured the content of the cases into segments, following the "definition-background-what-why-how-so what" framework to make it easier for the subjects to read and understand (Ramos et al., 2019) (Figure 21.2).

21.3.2.3 Process

Subjects were requested to evaluate 15 typical cases based on extracted adjectives. The study used the Semantic Differential Method (SDM) (Osgood et al., 1978), which lists the semantic attributes of the product to analyse and conduct user testing.

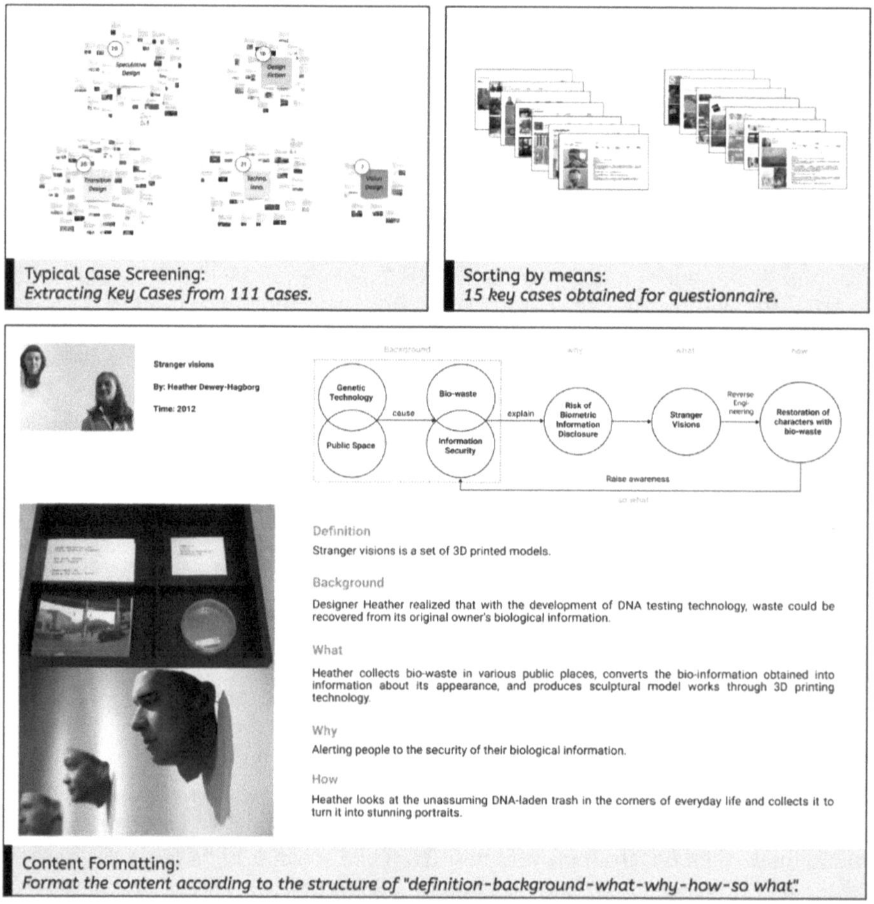

FIGURE 21.2 Following the "definition-background-what-why-how-so what" framework, we restructured the contents of 15 key cases into segments.

Users were then required to assess the cases according to these semantic attributes (Petiot & Yannou, 2004). The 5-point scale, ranging from 1 to 5, represented varying degrees of relevance, with scores of 1 indicating very irrelevant, 2 representing irrelevant, 3 denoting average relevance, 4 signifying relevant, and 5 indicating very relevant.

21.3.2.4 Data Analysis

The study proposes to visualise the relationship between 15 key cases and the acquired semantic terms through a perceptual map. The purpose is to examine the descriptive effect of the extracted semantic dimensions on the cases, to obtain the audience's cognitive structure of the subtypes, and to provide inspiration and assistance for refining the typology. The perceptual map is essentially a PCA biplot. The perceptual map is constructed as follows:

Step (i): Principal component analysis

The mean data of 15 cases (cum taking the mean of 85 subjects' ratings of the cases) were taken for principal component analysis;

$$\text{score}_{ik} = \sum_{j=1}^{p} z_{ij} \cdot e_{jk}$$

where score_{ik} is the score of case i on principal component k, z_{ij} is the standardised score of case i on variable j, e_{jk} is the factor loading of variable j on principal component k, and p is the number of variables.

Step (ii): Selection of axes

Select the two principal components with the largest eigenvalues as the x-axis and y-axis;

$$\lambda_1 \geq \lambda_2 \geq \ldots \geq \lambda_p$$

where λ_1 and λ_2 are the two largest eigenvalues.

Step (iii): Constructing perceptual maps

In this study, perceptual maps are drawn by Graphpad Prism (a scientific mapping software). In this case, the scores of 15 typical cases with 21 semantic words on two principal components are their coordinates, and a biplot can be generated. The coordinates of the cases and semantic words on the perceptual map are calculated as follows:

$$\text{Case coordinates}_i = \left(\text{score}_{i1}, \text{score}_{i2} \right)$$

$$\text{Word coordinates}_j = \left(e_{j1} \times \sqrt{\lambda_1}, e_{j2} \times \sqrt{\lambda_2} \right)$$

With the above method, all cases and semantic words can be displayed in a two-dimensional space for further exploration and judgement.

21.4 RESULT

21.4.1 The Semantic Perceptual Dimensions for DF Practice Cases

In the TF-IDF method, based on the TF-IDF values of 133 words, the top 40 words were taken to form the word set A. In the C.I. method, the 95% C.I. was [7.0807, 8.4562] and the 99% C.I. was [6.7900, 8.6869]. The study divided the word frequency thresholds according to the upper limit of the 99% C.I., and any adjectives with a frequency less than 9 were filtered out, resulting in the word set B with a total of 53 adjectives. On the basis of taking the intersection of the two word sets, we also used a manual method to further merge these adjectives (researcher from Tsinghua University).

21.4.2 Perceptual Evaluation Results: Perceptual Maps of Semantics and Cases

The study conducted a principal component analysis based on the mean data of the case perception evaluation, and two principal components were extracted. Among them, the eigenvalue (λ) of PC1 is 11.67 and the proportion of variance is 55.5%; the eigenvalue (λ) of PC is 2.582 and the proportion of variance is 12.29%. The cumulative proportion of variance of the two principal components has been as high as 67.85%, so it is persuasive and explanatory to draw a perceptual map by these two principal components (Figure 21.3).

Figure 21.4 is a perceptual map generated based on the aforementioned data, consisting of two main components, which shows the positional relationship between 21 semantics and 15 cases. In the perceptual map, the load of each semantic term is displayed by an arrow, and the scores of each case are displayed by dots.

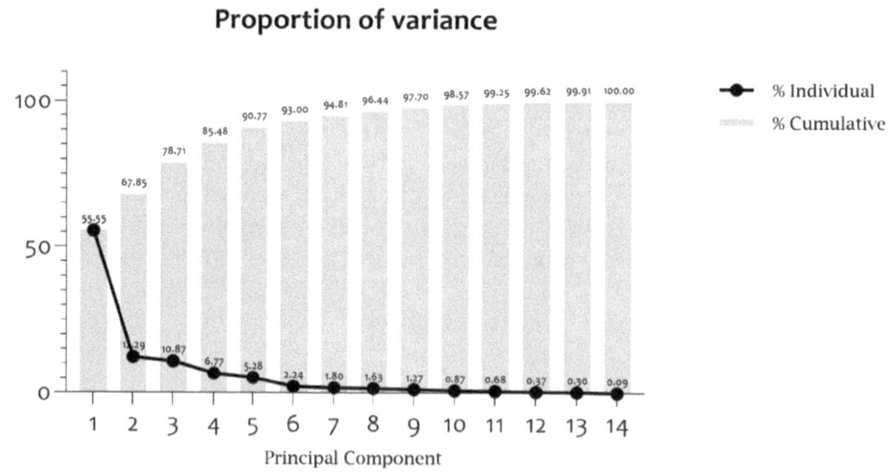

FIGURE 21.3 The results of the principal component analysis show that the cumulative proportion of variance of the two principal components has been as high as 67.85%.

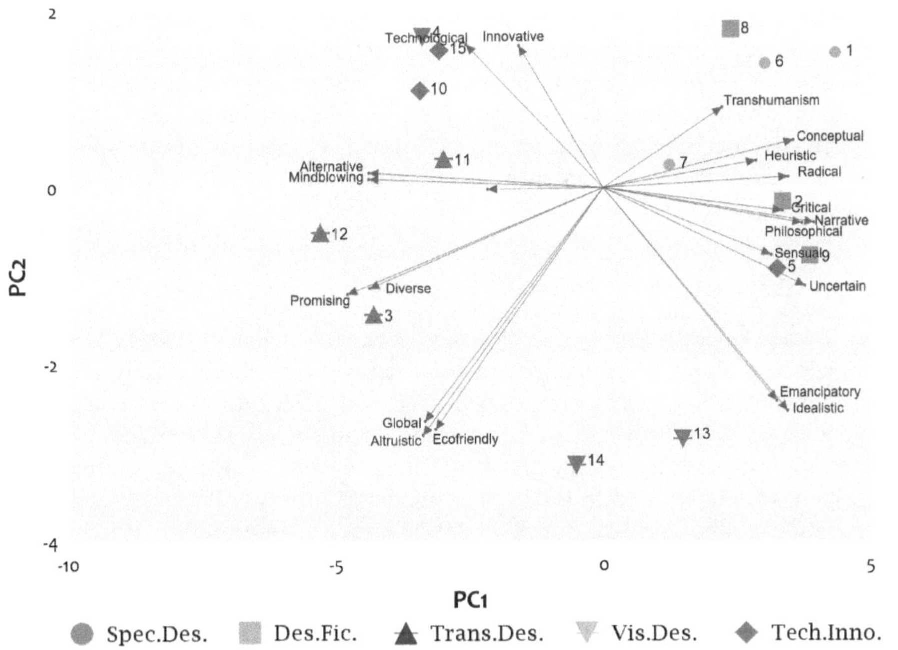

FIGURE 21.4 Perceptual map of semantics and cases (the eigenvalue of PC1 is 11.67 with a variance ratio of 55.5%, and the eigenvalue of PC2 is 2.582 with a variance ratio of 12.29%).

21.5 DISCUSSION

21.5.1 Speculative Design: Pioneering Forms and Marginal Issues

Cases related to speculative design are closest to the semantic dimensions of "transhumanism", "conceptual", and "heuristic". This not only reflects the perceptual results of the cases but also reveals, to some extent, the audience's cognition of this type. "Transhumanism" implies overcoming the assumption of human supremacy to embrace a post-human worldview (Ferrando, 2013), such as co-designing with participants who transcend humanity (Romani et al., 2022). Dunne and Raby argue that once designers step away from industrial production and the marketplace, they enter the realm of the unreal, the fictional, or what they prefer to think of as conceptual design – "design of ideas" (Rijkens-Klomp et al., 2017). If understood from the perspective of design heuristics, the semantics of heuristic seems to be less related to the creative approach and conceptual expression of speculative design and more oriented towards the traditional design field of creative stimulation, etc. This situation raised our concern. Through interviews with typical users, we found that some users understood "heuristic" as speculative design aiming to provoke debate and make the audience think about what might be possible (Malpass, 2016).

Therefore, in the audience's perception, speculative design is highly conceptualised and goes about stimulating debate through counter-conventional emergent forms. This reflects that there seems to be a need for further revision of the findings of this study. Anyway, the audience is less positively focused on the critical function of speculative design and more on the conceptualised presentation of speculative design and external forms such as pioneering forms and marginal issues (transhumanism, etc.).

21.5.2 Design Fiction: Critical Traditions and Narrative Devices

Cases related to design fiction are closest to concepts such as "critical", "sensual", "philosophical", and "narrative", with some cases being more closely related to "transhumanism". "Critical" represents a critical design thinking to challenge the ideological constraints that limit designers and society (Jakobsone, 2017). In future studies, there has always been a critical and emancipatory tradition that corresponds to the utilitarian tradition (Ahlqvist & Rhisiart, 2015). In terms of its purposeful function, critical design is not objective or interpretive. It is intended to provoke debate and make the viewer think about what might be. As such, narratives are intrinsically subjective, focusing on intersubjectivity and propositions, and the narrator will have his or her own opinions, attitudes, and worldview. It is precisely because of this that narrative stories, while not generating new or emerging technologies of visualisation per se, integrate different disciplinary perspectives through archetypal visual representations, allowing new insights to emerge from the storytelling process (Gummesson et al., 2014; Yu & Sangiorgi, 2018). The association of transhumanism with speculative design and design fiction suggests that "transhumanism" may be emerging as a topic of active discussion and interest. It also reflects the critical tradition and reflective function of speculative design and design fiction. In conclusion, speculative design and design fictions as types of critical design have numerous consistencies, which are also reflected in users' perceptions. However, it is equally important not to lose sight of the fact that these are more subtle descriptions beyond semantics and thus constitute a more precise understanding of typology.

21.5.3 Transition Design: Positive Action for Well-Being

Examples related to transition design are close to the concepts of "alternative", "promising", "diverse", and "mindblowing". These semantics suggest that transformative design, in the public's perception, is a type of design that actively seeks positive alternative solutions to address challenging problems, with the primary goal of promoting future well-being. We need to use design as a means of promoting social and environmental change, to address wicked problems, to work more consciously towards creating potential, alternative, more hopeful futures (GarduñoGarcía & Gaziulusoy, 2021), and to integrate this action into systemic change. In our definition, transition design not only emphasises more compelling future-oriented visions to inform and inspire current projects, but also calls for the development of these visions through the tools and methods of design. The audience interprets the process or outcome as "mindblowing". We argue that this "mindblowing" experience is different from the conceptual stimulation provided by critical means. Instead, it is a tangible vision and expectation for a consented future. Therefore, in the public's view, transformative design is a positive action oriented towards well-being.

21.5.4 Visionary Design: A Confusingly Intertwined Field

Visionary design is the most dispersed on the perceptual map. Cases are either close to the "technological" semantics or in the middle of the two clusters of semantic dimensions, "ecofriendly-altruistic-global" and "emancipatory-idealistic", but not close to any of the semantic dimensions. Visionary design is a new typology proposed in this paper that synthesises the reality of the practice cases as a form of strategy and design combined, as opposed to scenario planning, which is more oriented towards strategic foresight. In our view, visionary design is based on extensive or focused research on factors or key areas such as business, education, future and socio-political, to develop a set of scenarios for the future of the country, industry to form a cohesive and consensual programmatic content of mission, goals, and visions, to demonstrate the culture, values, and mission for a government department, business, institution, or cooperative (Amer et al., 2013; Muiderman, 2022; Szocik et al., 2020). However, visioning only maps out a blueprint for the future without presenting a roadmap for transformation or practice, for example. This definitional deficiency perhaps creates confusion in the perception of the audience. We believe that this lack of definition is limited by the level of development of the practice cases, and moreover, it is the main manifestation of the intertwined and unclear relationship between futures studies and design research. We will keep a close eye on related trends and further revise the typology.

21.5.5 Technological Innovation: the "New Engineering" in Disciplinary Integration

Technological innovation is closest to "technological" semantics and also to "sensual" and "uncertain". Technology has several levels of meaning; technology in technological innovation focuses on the combination and application of technology in new scenarios, involving interdisciplinary cooperation in order to form relevant collaborative strategies and design solutions (Gandy et al., 2017); however, technology in technological innovation does not stop with technological solutions, but also actively integrates technological meanings (Felt et al., 2009) and potentially technological future perspectives (Nabuurs et al., 2023). This is a way of thinking that is different from solving a specific problem from the beginning of the intention. It can be seen that technological innovation is not only oriented towards the solution of specific technical problems, but also an emerging interdisciplinary approach that involves the combination of foresight methodology, humanistic thinking, and technological

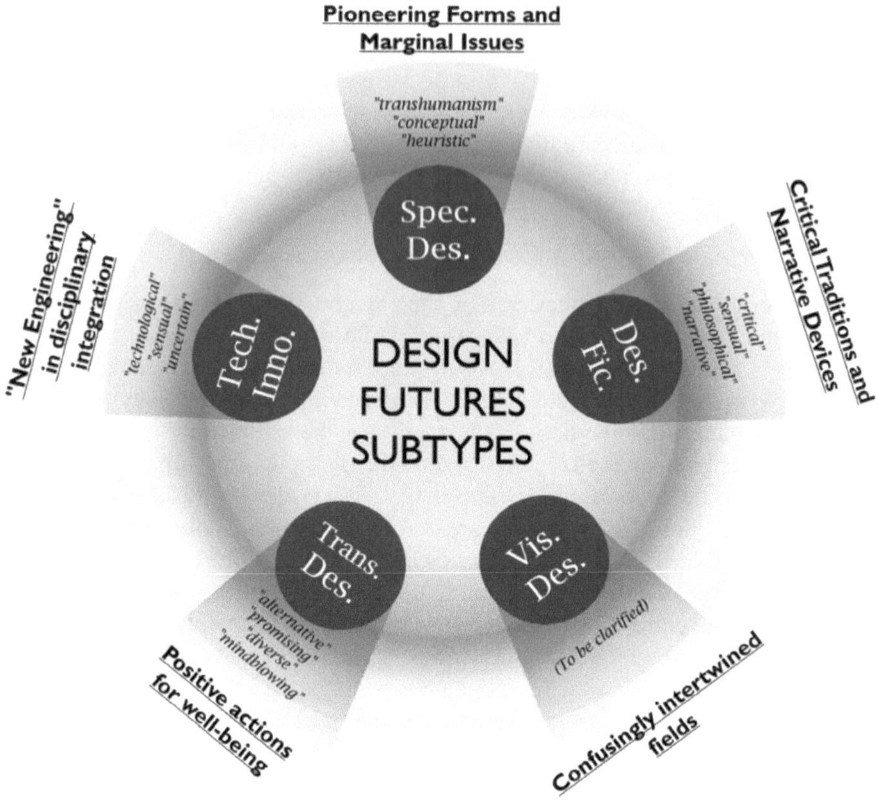

FIGURE 21.5 Type differences in the five subtypes of DF derived from the audience perception perspective, including key semantic dimensions for each subtype.

application, as well as the reshaping of processes by this combination, and ultimately ending up in concrete technological prototypes. This provides some inspiration for understanding the shape of future technological innovation (Figure 21.5).

21.6 LIMITATIONS

First, quantitative research methods and the results they present are a reflection of the audience's cognitive structure. This is extremely demanding on the subjects, who must have a certain level of knowledge of the field; otherwise, the quality of the research results will be greatly compromised. However, in contradiction to this, some important concepts or terms in the field of DF have been developed, established, and popularised by a series of representatives or pioneers from different fields and schools of thought. The specialisation of the audience is related to the extent to which they are exposed to or aware of these terms and concepts. However, to a large extent, the proliferation of terminology is limited to publications or within the academic market (which is perhaps why concepts cannot be clearly defined). As a result, the expertise of the general audience is limited, and groups of experts or pioneers may lack a mass base (limited to the fact that some concepts in the discipline remain unfamiliar to the general public). At the same time, the DF vocabulary itself is non-systematic (although we tried to collect it through a systematic approach). The validity of this study depends on the audience's understanding of the non-systematic character of DF vocabulary and the phenomenon of proliferation (the audience as the influenced), which varies from person to person rather than being purely objective.

Second, the alternative approach of this study, which emphasises the study of cases, also contains another paradox. DF is an emerging field, and in order to study cases, a number of concepts need to be constructed (a typology of cases). While it is possible to follow a rigorous line of research and sift through the cases from the proposed typology, the final typology is still being revised and refined. This is inevitably an iterative and dynamic process. Therefore, this study can only serve as a milestone. We hope to revise the typology further.

Again, although this study relies on quantitative analysis methods such as perception mapping techniques, this study, particularly the interpretation of the occurrences and their possible meanings, is also a discursive analysis. This interpretation is, to a large extent, plausible. Even if strong assertions can sometimes be made, they are not conclusive.

It is also important to note that this paper is not intended to replace literature studies (nor can it) through perceptual analyses. Literature research can provide prerequisites, depth, and inspiration for quantitative analyses and research design, whereas technological tools tend to be utilitarian and instrumental. We believe that a more systematic literature study will also help to build an ontology of knowledge in the field of "DF". We hope that the combination of audience perception and literature research will contribute to the development of the field.

21.7 CONCLUSION

DF is the integration of design and futurology (Bhatti et al., 2014; Bleeker, 2009; Candy, 2010; Dunne & Raby, 2013). The integration of design and futures initially came about when people began to envisage the expansion of futures studies in the field of practice to accommodate materials, methods, and approaches from design and art (Candy, 2010; Dunagan et al., 2011; Valtonen, 2020). However, DF combines a large number of paradigms and approaches from different fields, driven by individuals or organisations from different perspectives. For the purposes of the research, case studies were needed to further build the ontology of knowledge of DF as a discipline. A series of studies have been conducted to achieve this aim. The perspective of this paper is to collect evidence of typological differences by visualising the audience's perception of the cases and thus revise the typology as a complement to the literature studies and initial definitions. The main contributions and outlook of this paper are as follows:

i. The research in this chapter is grounded in and revised by the five subtypes (DF subtypes) of the DF. Cases, a typology that is emphasised at the beginning of this chapter and then further discussed at the end, around empirical research and discourse analysis of the initial definitional semantics. However, if we subvert these five typological divisions, does the basis for everything cease to exist? After all, there must be a more perfect way of dividing the typologies, which remains to be further developed by subsequent research.
ii. In this study, 133 semantic terms describing cases of DF practices were obtained through literature research and other methods, which were then further compressed into 21 semantic dimensions (RQ1). We hope that both the 133 semantic words and the 21 semantic dimensions will be of greater value for further research. For example, in developing the case base, we plan to adopt the Bayesian method (MCMC) to dynamically assign the extracted key semantics so as to cope with the continuous proliferation of cases, thus facilitating case retrieval.
iii. The results of this study depend on the participation of domain experts and professional audiences. In Experiment 1, 20 domain experts from China (innovation platform masterminds, community initiators, etc.) enhanced the value of this study. In Experiment 2, 85 professional audiences provided credibility to the study's findings. With the proliferation of cases and the continuous enhancement of research tools, we hope to further leverage the power of data-driven research and to drive the professionalism and research depth of DF disciplinary research through data and knowledge together.
iv. The research study conducted a perception evaluation experiment to visualise the relationship between the 15 typical cases and the acquired semantic terms by drawing a perception map from the two extracted principal components (RQ2). This allowed us to summarise and infer the characteristics presented by many genres of DF from the perspective of audience perception. Among them, speculative design is characterised by "pioneering forms and marginal issues"; design fiction is "critical traditions and narrative devices"; transformative design is "positive action towards well-being"; and visionary design is "a confusingly intertwined field". Visionary design is a "confusing intertwined field", reflecting the intertwined fusion of design and futures studies, which needs to be further revised; technological innovation is a "new form of engineering through the integration of disciplines", which provides some inspiration for us to understand the shape of future technological innovation. It provides some inspiration for us to understand the shape of future technological innovation. The results of this research will help to establish a more scientific typology of practice cases, thus contributing to solving the "fatal" problem of the development of this discipline.

ACKNOWLEDGEMENTS

We sincerely thank all the individuals who made significant contributions to this project, with a special mention to Mao Yuxi, Li Guohao, Wang Chenxi, Wang Kexing, Wang Jiayu, Su Tao, and Tian Xiaoxi. A special thanks is due to Zhang Chenfan for her support in this study. We would also like to express our sincere gratitude to all the experts and practitioners who participated in the questionnaires.

REFERENCES

Ahlqvist, T., & Rhisiart, M. (2015). Emerging pathways for critical futures research: Changing contexts and impacts of social theory. *Futures*, *71*, 91–104.

Amer, M., Daim, T. U., & Jetter, A. (2013). A review of scenario planning. *Futures*, *46*, 23–40. https://doi.org/10.1016/j.futures.2012.10.003

Angheloiu, C., Sheldrick, L., & Tennant, M. (2020). Future tense: Exploring dissonance in young people's images of the future through design futures methods. *Futures*, *117*, 102527. https://doi.org/10.1016/j.futures.2020.102527

Bhatti, Y., Kimbell, L., Ramirez, R., & Selin, C. (2014). *The Fourth Oxford Futures Forum Report*. Saıd Business School, University of Oxford, Design & Scenarios.

Bleeker, J. (2009). *A Short Essay on Design, Science, Fact and Fiction*. Los Angeles, CA: Near Future Laboratory.

Candy, S. (2010). *The Futures of Everyday Life: Politics and the Design of Experiential Scenarios* [Doctoral Dissertation, Doctoral Thesis]. University of Hawai'i, Manoa.

Candy, S., & Dunagan, J. (2017). Designing an experiential scenario: The people who vanished. *Futures*, *86*, 136–153. https://doi.org/10.1016/j.futures.2016.05.006

Chen, X., Tian, X., & Ye, H. (2023). Quality improvement path and countermeasures of visual arts for future metaverse: Based on G1-entropy value method. *Journal of Intelligent & Fuzzy Systems*, *44*(3), 4999–5019. https://doi.org/10.3233/JIFS-223351

Chuang, Y., & Chen, L.-L. (2008). How to rate 100 visual stimuli efficiently. *International Journal of Design*, *2*(1), 31–43.

Dunagan, J., Jeffery, L., Fidler, D., & Maguire, R. (2011). *The Magic of Kidstech*. Institute for the Future.

Dunne, A., & Raby, F. (2013). *Speculative Everything: Design, Fiction and Social Dreaming*. Cambridge, MA: The MIT Press.

Felt, U., Fochler, M., Müller, A., & Strassnig, M. (2009). Unruly ethics: On the difficulties of a bottom-up approach to ethics in the field of genomics. *Public Understanding of Science*, *18*(3), 354–371.

Ferrando, F. (2013). Posthumanism, transhumanism, antihumanism, metahumanism, and new materialisms: Differences and relations. *Existenz*, *8*(2), 26–32.

Gandy, M., Baker, P. M., & Zeagler, C. (2017). Imagining futures: A collaborative policy/device design for wearable computing. *Futures*, *87*, 106–121.

GarduñoGarcía, C., & Gaziulusoy, İ. (2021). Designing future experiences of the everyday: Pointers for methodical expansion of sustainability transitions research. *Futures*, *127*, 102702. https://doi.org/10.1016/j.futures.2021.102702

Gkouskos, D., Normark, C. J., & Lundgren, S. (2014). Investigating dimensions in automobile user needs. *International Journal of Design*, *8*(1), 13.

Gummesson, E., Kuusela, H., & Närvänen, E. (2014). Reinventing marketing strategy by recasting supplier/customer roles. *Journal of Service Management*, *25*(2), 228–240.

Holak, S. L. (1994). *Design and Marketing of New Products (2d edition): By Glen Urban and John Hauser. Englewood Cliffs, New Jersey: Prentice Hall, 1993. 701+ xxvii pages*. No longer published by Elsevier.

Jakobsone, L. (2017). Critical design as approach to next thinking. *The Design Journal, 20*(sup1), S4253–S4262.

Kelliher, A., & Byrne, D. (2015). Design futures in action: Documenting experiential futures for participatory audiences. *Futures, 70*, 36–47. https://doi.org/10.1016/j.futures.2014.12.004

Li, T., Li, Y., & Rampino, L. (2023). Construction of semantic pool and acquisition of semantic categories for automobile styling stance: A domain knowledge perspective. *Advanced Engineering Informatics, 56*, 101995.

Malpass, M. (2016). Critical design practice: Theoretical perspectives and methods of engagement. *The Design Journal, 19*(3), 473–489.

Mankoff, J., Rode, J. A., & Faste, H. (2013). Looking past yesterday's tomorrow: Using futures studies methods to extend the research horizon. In *Proceedings of the SIGCHI Conference on Human Factors in Computing Systems*, pp. 1629–1638. Montréal Québec, Canada. https://dl.acm.org/doi/proceedings/10.1145/1124772

Muiderman, K. (2022). Approaches to anticipatory governance in West Africa: How conceptions of the future have implications for climate action in the present. *Futures, 141*, 102982. https://doi.org/10.1016/j.futures.2022.102982

Nabuurs, J., Heltzel, A., Willems, W., & Kupper, F. (2023). Crafting the future of the artificial womb Speculative design as a tool for public engagement with emerging technologies. *Futures, 151*, 103184.

Osgood, C. E., Suci, G. J., & Tannenbaum, P. H. (1978). *The Measurement of Meaning*. University of Illinois Press. Manufactured in the United States of America. Library of Congress Catalog Card No. 56-5684. https://gwern.net/doc/psychology/1957-osgood-themeasurementofmeaning.pdf

Petiot, J.-F., & Yannou, B. (2004). Measuring consumer perceptions for a better comprehension, specification and assessment of product semantics. *International Journal of Industrial Ergonomics, 33*(6), 507–525. https://doi.org/10.1016/j.ergon.2003.12.004

Pierce, J. (2021). In tension with progression: Grasping the frictional tendencies of speculative, critical, and other alternative designs. In *Proceedings of the 2021 CHI Conference on Human Factors in Computing Systems*, pp. 1–19. https://doi.org/10.1145/3411764.3445406

Ramos, J., Sweeney, J. A., Peach, K., & Smith, L. (2019). *Our Futures: By the People, for the People*. London, UK: NESTA.

Rijkens-Klomp, N., Baerten, N., & Rossi, D. (2017). Foresight for debate: Reflections on an experience in conceptual design. *Futures, 86*, 154–165.

Romani, A., Casnati, F., & Ianniello, A. (2022). Codesign with more-than-humans: Toward a meta co-design tool for human-non-human collaborations. *European Journal of Futures Research, 10*(1), 1–9.

Schütte, S. (2005). *Engineering Emotional Values in Product Design: Kansei Engineering in Development* [PhD Thesis]. Linköping: Institutionen för konstruktions-och produktionsteknik. https://www.diva-portal.org/smash/get/diva2:20839/FULLTEXT01.pdf

Sterling, B. (2009). COVER STORYDesign fiction. *Interactions, 16*(3), 20–24. https://doi.org/10.1145/1516016.1516021

Szocik, K., Abood, S., Impey, C., Shelhamer, M., Haqq-Misra, J., Persson, E., Oviedo, L., Capova, K. A., Braddock, M., Rappaport, M. B., & Corbally, C. (2020). Visions of a Martian future. *Futures, 117*, 102514. https://doi.org/10.1016/j.futures.2020.102514

Valtonen, A. (2020). Approaching change with and in design. *She Ji: The Journal of Design, Economics, and Innovation, 6*(4), 505–529. https://doi.org/10.1016/j.sheji.2020.08.004

van der Duin, P. (n.d.). *Foresight in Organizations*. Retrieved December 31, 2023, from https://api.taylorfrancis.com/content/books/mono/download?identifierName=doi&identifierValue=10.4324/9781315728513&type=googlepdf

Wang, C.-H., & Chin, H.-T. (2017). Integrating affective features with engineering features to seek the optimal product varieties with respect to the niche segments. *Advanced Engineering Informatics*, *33*, 350–359.
Warell, A. (2008). Multi-modal visual experience of brand-specific automobile design. *The TQM Journal*, *20*(4), 356–371.
Yu, E., & Sangiorgi, D. (2018). Service design as an approach to implement the value cocreation perspective in new service development. *Journal of Service Research*, *21*(1), 40–58.

22 Speculative Proposals
Resources for Designers to Work with Alternative Futures

David Palma and Marco Neves

22.1 INTRODUCTION

A 'common' sense has been observed in design, characterized by established and shared practices closely aligned with a commercial context, although there is no unanimous term that defines it. Sangiorgi (2011) and Hill (2017) refer to it as 'traditional,' Dunne and Raby (2013) define it as 'affirmative design,' while Dorst (2019) classifies it as 'normal,' and Tharp and Tharp (2018) use the expression 'conventional design.' These terms aim to describe the activity of most designers, regardless of their area of specialization, as everyday professions. Professions that have grown accustomed to their role, surrounded by specific technologies and a production system that directed them toward expected forms and functions (Koskinen & Dorst, 2015), in a close relationship between design and industry (Fallman, 2008), where design is a process resulting in material achievements (Sanders & Stappers, 2008). In general, these authors reflect on the divergence of professional design activities from critical and reflexive thinking, assuming that design professional practices are oriented and shaped toward using methods and resources that assist in problem-solving and commercial objectives (Mitrović et al., 2021).

However, from a broader perspective, these terms can also refer to one of several generations of design approaches (Jones, 2014). Specifically, if we understand that designers face complex situations today. Thus, we observe the emergence of areas that seek to deal with these situations and bring us different understandings, such as design for social innovation (Manzini, 2015), transition design (Irwin, 2015), and system design (Jones, 2014), or that seek to contribute to a diversity of possibilities in activities where designers work nowadays, such as speculative design (Dunne & Raby, 2013).

In the specific case of speculative design, as stated by Dunne and Raby (2013), it involves the exploration of alternative futures to understand current problems and debate the type of future we desire. Similarly, it allows for disputing and critically reflecting on relevant social, political, and ethical issues (Dong et al., 2020). However, the term 'speculative' is continually correlated with other interpretations, such as critical design, fictional design (Auger, 2013), and discursive design (Tharp & Tharp, 2018). These are domains of design that primarily operate in the territory of

DOI: 10.1201/9781032719290-26

experimentation, distancing themselves from design practices oriented toward commercial problem-solving and centered on client needs (Tharp & Tharp, 2018).

Given the relatively unexplored nature of a speculative design approach within the design community, this article focuses, first, on observing the types of methods and resources associated with speculative design and subsequently on the development, experimentation, and analysis of a prototype. This prototype is intended for designers and design students, aiming to enable the understanding and construction of alternative futures. Moreover, it aims to stimulate critical thinking and devise alternatives to confront adverse problems.

22.2 SPECULATIVE DESIGN IN NEED OF RESOURCES

To deepen the understanding of the practical application of speculative design, references, articles, and projects were gathered to enable an analysis of how this approach is introduced and applied from the perspective of a group of researchers and designers. We found that speculative design not only leads to the creation of reflective and critical artifacts but also manifests as a method and research resource. Its usage stands out in participatory contexts, involving different types of communities and issues such as racial concerns and social exclusion. This involvement provides participants with a reflection on the future of culture and the potential trajectories to be pursued (Baumann et al., 2017; Bray & Harrington, 2021). Similarly, it can adopt a stance capable of combining with other research methods. For instance, Dong et al. (2020) investigate the trajectory of Chinese marriage using an approach that combines speculative design and ethnography across four stages, encompassing the analysis of the present, the past, and the projection of trends, with the subsequent purpose of conceiving future scenarios. This sequence of distinct stages aims to facilitate the construction of speculative scenarios while avoiding the ideation of excessively utopian and decontextualized scenarios (Dong et al., 2020).

Moreover, Stals et al. (2019) also illustrate how speculative design, coupled with ethnography, is employed to collectively construct hypothetical futures. In this case, researchers focused on personal and emotional relationships within the urban environment, facilitating the joint discovery of new technological devices and services, in the same way that the social, political, and ethical implications of emerging technologies and urban transformations are explored (Stals et al., 2019).

However, the use of participatory methods linked to speculative design does not always occur seamlessly. Simply establishing empathy and involvement with communities is not enough; it is imperative to strategize how and when they should participate throughout the process (Tironi, 2018). Likewise, the application of participatory methods to random and unstructured audiences may compromise outcomes and the genuine speculative approach (Dolejšová, 2018). An example of this is the work of Tironi (2018), who, alongside design students, developed a speculative design project with homeless communities in Santiago, Chile. Throughout the process, they encountered setbacks related to misunderstandings and obstacles between the communities and the students, due to a lack of planning and connection with the homeless, resulting in incompatibility and collaboration difficulties. Nevertheless, these setbacks were resolved during the process, ultimately allowing for discussions on the ethical and social implications in participatory projects. Despite the process not being linear and seamless, this study also demonstrates how prototyping a scenario with a speculative focus can be used as a resource to involve marginalized groups in creating solutions for their predicaments, underscoring the significance of critical reflection and active community participation. This case reinforces the importance of ethical and social implications in speculative projects, not exclusively focusing on technological issues, scenarios, and visions of futures inaccessible to most of the population or projects that exclude marginalized groups (Galloway & Caudwell, 2018).

Nevertheless, the introduction of resources and approaches that allow for the extrapolation of future scenarios in design education can contribute to an awareness of social transformation. By sensitizing young people to exploring different possibilities for the future, it not only prepares them to anticipate possible, preferable, and probable scenarios but also equips them to deal with various potential futures (Bell, 1998). It also increases students' awareness of the impact of their actions as individuals on society (Bell, 1998). Furthermore, it reinforces the idea that design as an activity allows work in areas beyond the commercial focus typically associated with professional practices, such as graphic or industrial design (Buchanan, 1992).

22.3 SPECULATIVE RESOURCES FOR DESIGNERS

To comprehend, develop, and explore new resources of speculative design aimed at design professionals and students, it becomes relevant to map and identify resources already developed in this context. It is possible to identify different types of resources, which can take on different formats and applications.

One notable resource is *The Prediction Generator*, developed by De La Campos and Castorena (2020), which assists in generating and validating ideas and potential speculative design projects. This resource resembles the traditional canvas model, where the designer begins with an idea and continuously validates at each filled block. The project has been primarily applied to students, aiming to facilitate speculation and idea validation for designers.

In the academic field, Dong et al. (2020), as already mentioned, combine the extrapolation of future scenarios with participatory ethnographic methods to explore the future of Chinese weddings. In this case, the incorporation of ethnographic methods into the speculative design approach generates a process made up of different phases, allowing interaction and the collection of information from different audiences. This resource aims to start by gathering information on current behaviors and habits, then trace the past in order to understand the origins of current habits, and, consequently, generate speculative scenarios based on the cultural focuses and trends identified. By mapping the traditions and cultural customs of Chinese weddings from the present to the past, they projected what the vision and reality of weddings might be in the future, combined with new technologies (Dong et al., 2020).

Design studios specializing in speculative and critical design approaches, such as Normals and Futur 2 Studio, not only explore future scenarios and provide critical reflections, but also develop their processes and, in some cases, create resources to support their work.

For example, Normals developed a set of cards (Figure 22.1) specifically designed for use in *World-building workshops*, a participatory activity that combines tools from futurists and designers. This resource helps participants from diverse backgrounds construct and reflect on future scenarios. Specifically, it assists in the design of future neighborhoods, where participants must consider different types of inhabitants, new technologies to integrate, possible fictional trends that may emerge, and the kinds of products or services that might arise in future societies or neighborhoods. Furthermore, it encourages critical reflection on environmental and sustainability issues (Normals, n.d.).

FIGURE 22.1 Example of cards designed by Normals Studio (Normals, n.d.).

In this set of cards developed by Normals, different types of cards can be randomly combined during workshops: 'Trait cards,' 'Mod cards,' and 'Superspecial cards.' **Trait cards** represent profiles of different types of people and their obsessions, such as nutritionists or gamers. **Mod cards** introduce unpredictability and can be selected multiple times during a session. These cards can trigger sudden environmental changes or impose material restrictions. **Superspecial cards** have drastic effects on the activity. For example, groups may have to change projects, switch character types, or mix the consciousness of different characters (Normals, n.d.).

This type of resource facilitates collaboration between designers and participants, helping to generate and discuss ideas, objects, and various speculative future scenarios. Furthermore, it improves debate and increases awareness of the future and the potential lifestyles that may emerge (Sanders & Stappers, 2008).

Likewise, Futur 2 Studio, an agency that combines future scenario projection with design to develop more conscientious products and services (Futur 2 Studio, n.d.), has created a unique set of trend cards and engaged in critical reflection, such as the *Happy Little Accidents – Dystopian Trend Cards* (Figure 22.2). This set of cards projects emerging dystopian future scenarios based on various contemporary issues: corporate, economic, environmental, political, social, and technological. Themes are complex, multidisciplinary, and challenging to resolve, what Buchanan (1992) refers to as *wicked problems*, where design can play a transformative role. Buchanan (1992) suggests that design thinking is a potential approach to addressing these *wicked problems*. However, speculative design approaches can also offer alternative perspectives, anticipate situations, and counteract aspects of future negativity. Similarly, speculative design enables deeper critical reflection on emerging issues (Dunne & Raby, 2013).

The introductory card (Figure 22.3) contains the following message:

> Given all the issues that are unfolding in today's world, it's difficult not to become cynical. While we generally believe in the goodness of mankind and a positive future for everyone, as designers, we need a medium to articulate critical observations and express our concerns.
>
> What you are holding is a compilation of our random, dark interpretations of significant signals across the spectrum – intended for your inspiration, entertainment, or as a conversation starter.
>
> Good night and good luck.
>
> *(Futur 2 Studio, n.d., Card 1).*

This message from Futur 2 Studio underscores the necessity for critical reflection and demonstrates how these cards engage to provoke thought, challenge dystopian projections, and facilitate discussions about uncertain future scenarios.

Similar to these examples, there are other accessible resources for constructing alternative futures that can be applied in the designer's ideation processes, such as *The Tarot Cards of Tech*. This deck of cards contains a set of provocations that aid designers in reflecting, predicting the impact that products and ideas may reveal in scenarios, and discovering new opportunities (Artifact, n.d.). Another example is *The Thing from the Future*, a game developed by Candy and Watson (2015), which can be used by anyone, either in groups or individually, to stimulate participants to describe

FIGURE 22.2 From Happy Little Accidents: Dystopian Trend Cards, packaging.

emerging objects from alternative futures. This encourages the development of participatory structures and the importance of enriching the collective vocabulary of anticipation (Candy, 2018). Additionally, there is the *Future Scan*, which facilitates the discussion and analysis of different topics regarding the future. It is a map proposing over 150 possible future scenarios based on technologies and social changes, where one can select a topic and subsequently discuss what the future might entail (Lutz, 2022).

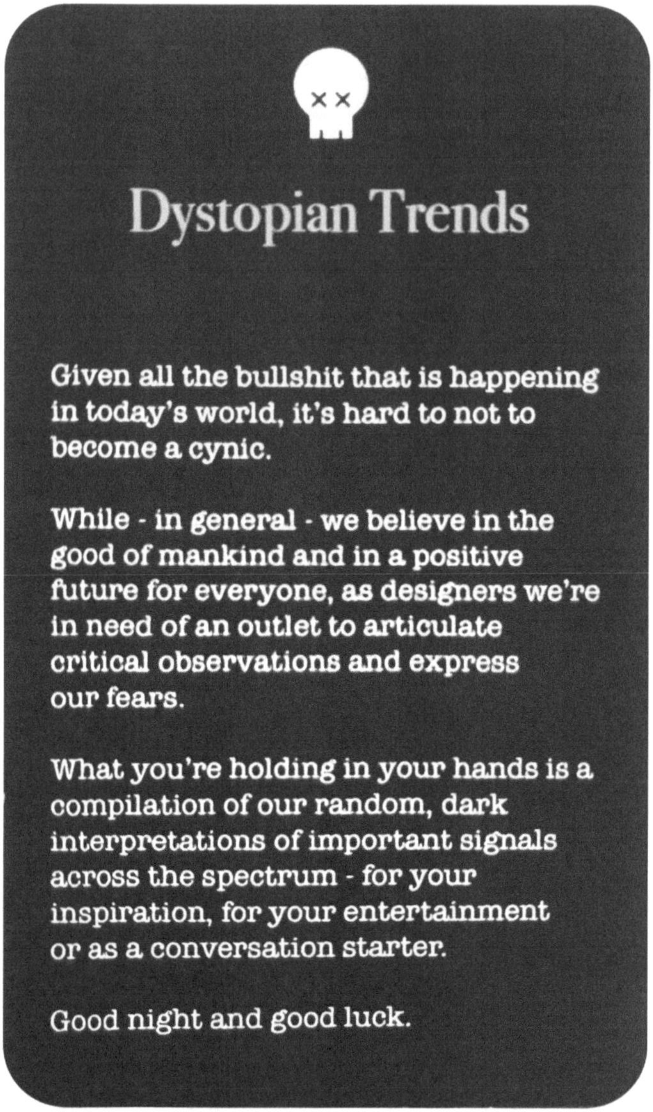

FIGURE 22.3 Introductory card from *Happy Little Accidents: Dystopian Trend Cards* (Futur 2 Studio, n.d., Card 1).

Despite the diversity of existing resources for exploring and prospecting the future, not all necessarily adopt a wholly speculative approach. For instance, the *Future Scan*, by initially presenting 150 future scenarios, while allowing reflection and criticism, tends to condition the extrapolation of ideas since it already presents possible scenarios. On the other hand, *The Tarot Cards of Tech* assist designers in questioning and thinking about how their ideas and projects will endure in the future, yet the set of cards is oriented toward technology logics and reflecting on ideas already structured beforehand. Moreover, there is a lack of evidence and scientific investigation into the effects and outcomes that these resources and games may contribute to the exploration and creation of future scenarios when applied to a real design project.

22.4 SPECULATIVE CARDS AND DICES AS DESIGN RESOURCE

As a result of the conducted review, we were driven to develop resources starting in 2023 that aim to introduce speculative design into the workflow of designers and design students. In this regard, we have introduced *Speculative Proposals*, a resource designed to encourage participants to explore hypothetical and utopian future scenarios, allowing them to detach from solutions applicable only in the present. This resource has been developed and prototyped in two versions: one in the form of cards and another in the form of dice. Both formats have been designed to serve as support in group activities that promote interaction and discussion on speculative design.

Speculative Proposals enables the creation of different future scenarios by combining four categories: What, When, Where, and Who (Figure 22.4). For instance, with the cards, participants are required to select one card from each category. Starting from the *What*, a specific problem is presented for which they need to design solutions for the indicated period on the *When* card and in the location indicated by the *Where* suit, while simultaneously considering the type of person depicted in the *Who* category. Figure 22.5 illustrates a potential combination example that might arise during the activity, prompting participants to explore, discuss, and present potential future scenarios incorporating all the variables.

In the case of the dice, the logic remains the same; however, the different options are represented by icons and with fewer variables (Figures 22.6 and 22.7).

22.5 SPECULATIVE PROPOSALS ACTIVITY

The *Speculative Proposals* prototypes were presented and tested exclusively with first-year doctoral design students and professors at the Faculty of Architecture of the University of Lisbon in 2023. The activity was developed within an educational environment. In this context, speculative, critical, or design fiction methods can assist students in distancing themselves from ideas and practices considered 'conventional' (Helgason et al., 2020), while educational leaders can promote creative confidence by encouraging openness to experimentation and allowing for different conceptualizations (Boer & Jenkins, 2022).

The activity was planned and structured to introduce the developed resources and evaluate their relevance. For this purpose, the activity was divided into three stages. The first phase involved a 15-minute introduction, explanation of interaction rules, and the purpose of the developed prototypes. Subsequently, participants were instructed to form groups of three to four members, to whom cards or dice were distributed, with a 30-minute deadline to conceive a response to the presented challenge, using either the cards or dice. It is worth noting that groups had the option to use various research and idea generation resources. In the third phase, the projected scenarios were presented and discussed, with an estimated 10-minute time slot for each group. Following this, a brief interval was proposed, followed by a second round

CARDS WHERE

_BIG CITY IN THE DESERT
_CAMPING TENT PARK
_DESERT
_FOREST
_INTANGIBLE LAND
_MARS
_MIDDLE OF THE PACIFIC OCEAN
_MOON
_REMOTE SMALL ISLAND
_RURAL VILLAGE
_SKY
_SMALL CITY NEAR THE SEA
_SMALL DESERTED TOWN
_SMALL MOUNTAIN TOWN
_SUBWAY
_TROPICAL FOREST

CARDS WHO

_BABY, 1 YEAR OLD
_BOY, 10 YEARS OLD
_BOY, 15 YEARS OLD
_COCKROACHES
_COWS
_DOLPHINS
_GENDER-FLUID, 25 YEARS OLD
_MAN, 65 YEARS OLD
_NON-BINARY, 18 YEARS OLD
_PET CAT
_PET DOG
_SEA FISH
_WOMAN, 30 YEARS OLD
_WOMAN, 36 YEARS OLD
_WOMAN, 45 YEARS OLD
_WOMAN, 80 YEARS OLD

CARDS WHAT

_ COMMUNICATION
_ DEATH
_ EDUCATION
_ EMPLOYABILITY
_ FASHION
_ FOOD
_ HEALTH
_ HOUSING
_ NATALITY
_ PLASTICS
_ PRIVACY
_ SECURITY
_ SOCIAL
_ TRANSPORTATION
_ TRASH
_ WATER

CARDS WHEN

_ 10 YEARS FROM NOW
_ 15 YEARS FROM NOW
_ 20 YEARS FROM NOW
_ 25 YEARS FROM NOW
_ 30 YEARS FROM NOW
_ 35 YEARS FROM NOW
_ 40 YEARS FROM NOW
_ 45 YEARS FROM NOW
_ 50 YEARS FROM NOW
_ 55 YEARS FROM NOW
_ 60 YEARS FROM NOW
_ 65 YEARS FROM NOW
_ 70 YEARS FROM NOW
_ 80 YEARS FROM NOW
_ 90 YEARS FROM NOW
_ 100 YEARS FROM NOW

FIGURE 22.4 Set of categories presented in the cards *What*, *When*, *Where*, and *Who*.

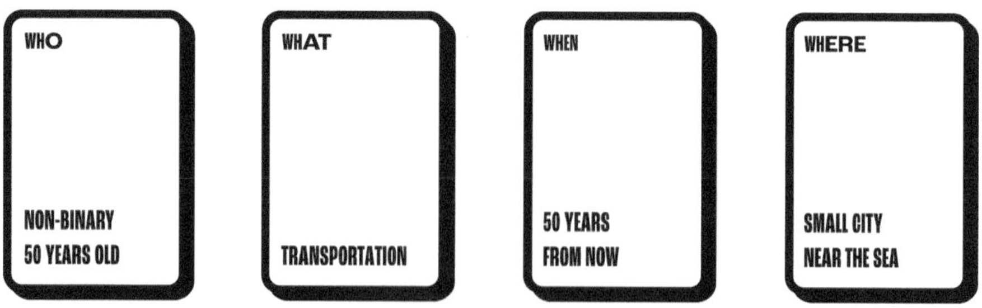

FIGURE 22.5 Example of a set of cards to be solved.

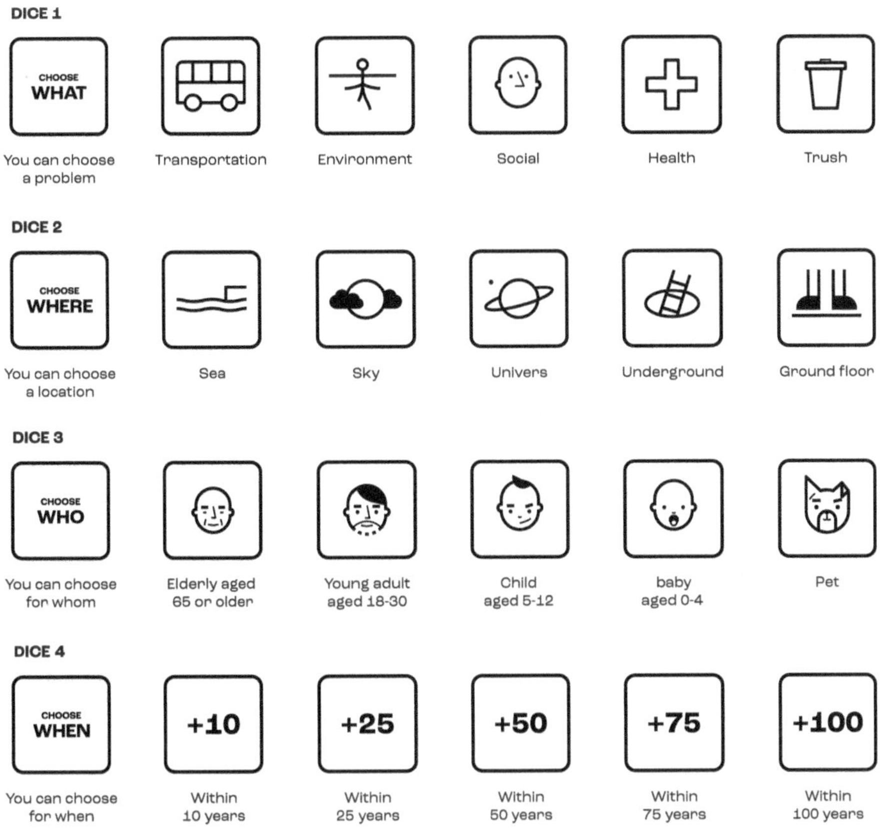

FIGURE 22.6 Different categories presented on the faces of the dice.

FIGURE 22.7 Example of a set of cards to be solved.

where participants alternated between using cards and dice. At the conclusion of the activity, a debate was held to gather feedback on the activity and the achieved results.

22.6 ACTIVITY RESULTS

The realization of this activity in an academic context with the *Speculative Proposals* prototypes enabled us to observe, map, and record the different approaches and dynamics adopted by the groups during the activity and the use of prototypes in extrapolating and constructing future scenarios (see Table 22.1). We can observe, for

TABLE 22.1
Data Collected from the Activity

	Group Members	Research Resources	Presentation
Group 1	**4 Designers:** 2 PhD Students 2 Design Professors	Internet Notebook	Narrative Presentation and Augmentation of AI-Generated Images Based on the Story
Group 2	**3 Designers** **1 Marketeer** All PhD design students	Internet AI Notebook	Narrative Presentation and enhancement of Sketches on a Whiteboard
		Set of Cards Explored	**Set of Dices Explored**
Group 1		**What** _Transportation **Who** _ Women 30 years old **When** _ 60 years from 2023 **Where** _Tropical forest	**What** _ Environment **Who** _ Old person+65 years **When** _ 100 years from now **Where** _ Sea
Group 2		**What** _ Employability **Who** _ Men 65 years old **When** _ 45 years from 2023 **Where**_ Tropical Forest	**What** _ Trash **Who** _ Kid 5-12 years old **When** _ 100 years from 2023 **Where** _ Sky

instance, how the groups adopted distinct approaches when conceptualizing their proposed future scenarios, as well as in the way they presented the results. While Group 1 resorted to artificial intelligence (AI) as a supplement to generate narratives for future scenarios and presentation of results, Group 2 focused more on idea discussion and scenario description. Additionally, this group chose to use a whiteboard to draw and outline their scenarios, thus reinforcing the explanation of the scenarios created during the activity and the presentation. However, despite different approaches, both groups managed to articulate and develop distinct scenarios, achieving the objective of the activity.

Group 1, which used AI as a complementary research resource, created a story that encompassed and responded to the following data combination they received: **What**_ Environment, **Who_** Old person +65 years, **When_** 100 years from 2023, and **Where_** Sea. However, the group, until getting to the presented result, explored different topics and scenarios, such as the possibility of living at sea for 100 years. They identified problems and questioned possible scenarios related to the environment, such as global warming, which they later integrated into a story depicting a possible alternative future scenario. Together with AI, the following story was written:

> In 2125, John is 80 years old, he lives in a floating house in Jakarta, part of a floating city, the city of PhD; he works as an underwater farmer at 80, he likes to swim and dive; the sea temperature has risen 5 degrees in 100 years, the coastal cities have moved; John has worked in multiple careers and studied multiple subjects (highlander); he has 1 kid and 1 grandchild; families are small; he has some sea pets he takes care of (sea pets?); he sells the sea production locally, to make algae burgers, medicine, drinks, etc.;

he exchanges and shares the products because everyone has access to a basic income; although he does need it, he found that he loves to work and meet his co-workers, neighbors, and friends.

The group, to complement their story about a possible alternative future scenario for 2123, generated a set of images also produced by AI to reinforce their vision of the proposed scenario.

On the other hand, Figures 22.8 and 22.9 characterize Group 2 in their way of working and presenting the result obtained with the set of cards: **What_** Employability, **Who_** Men 65 years old, **When_** 45 years from 2023, and **Where_** Tropical Forest. This allowed the group to explore and discuss, in a short period, possibilities for

FIGURE 22.8 Ideation moment.

FIGURE 22.9 Presentation of a speculative future scenario.

new hypothetical urban ecosystems that could be implemented in tropical forests. Additionally, through scenario description, they identified new professions that could emerge in the future, such as the 'Nature Artist.'

Both approaches adopted by the groups achieved the proposed objective of extrapolating ideas for alternative futures. Similarly, this type of exercise allowed for contestation and critical reflection on relevant social, political, and ethical issues.

After conducting and discussing the activity, some participants identified how conditioned they are to think of solutions for the present, highlighting the difficulty and demand involved in extrapolating ideas and hypothetical future scenarios. Additionally, both groups identified a preference for using cards instead of dice. Despite both resources having a similar purpose, cards allow for a greater diversity of combinations and variations to explore future scenarios.

Regarding the time allocated for ideation and group discussion, participants identified that 30 minutes was not sufficient. It was also suggested that, after presenting the proposed future scenario solution, there should be a scale of classification by categories, such as creativity and project advocacy, which could be assigned by an activity moderator or debated among the different participants.

22.7 CONCLUSION

The activity carried out with the *Speculative Proposals* prototypes allowed for the observation and analysis of how the introduction of these resources can stimulate the exploration and discussion of plausible futures. By challenging participants to depart from 'traditional' approaches and instead focus on critical and speculative reflection (Dunne & Raby, 2013; Dorst, 2019), the activity enabled them to develop and present hypothetical and alternative future scenarios. These scenarios were capable of challenging conventional ideas, developing alternatives centered on social and cultural issues, and distancing themselves from purely commercial solutions. However, one of the main challenges faced by most students during the activity is the difficulty in disconnecting from design solutions based on the current reality. Even though it was an exercise in speculation, there was an initial concern about ensuring the feasibility of materializing their ideas, something that was not within the scope of the proposed objectives. This became a relevant and challenging aspect during the activity: being able to move beyond this spectrum and focus not on materialization but on issues that could foster societal change, valuing and awakening the role of design as an agent capable of enhancing such societal change.

This experiment was conducted with a small group of design doctoral students, from whom various distinct insights and feedback were obtained. However, the aim is to apply the same activity to a larger number of students from different academic levels and designers, seeking to obtain a more comprehensive and comparative sample of results. By involving more participants in this prototyping phase, we hope to understand and evaluate if there are areas in design where speculative design approaches are applicable. Additionally, it will be beneficial to analyze if different participants resort to additional resources for idea generation, beyond, for example, AI. Likewise, this interaction highlighted the need to optimize the conduct of the activity, especially concerning the timing of introducing the topic and the ideation phase. We expect that *Speculative Proposals* will become a resource capable of integrating different design contexts and disciplines, allowing for the exploration and creation of space for what does not yet exist. This will drive critical thinking regarding challenges and possible future scenarios, even challenging ways of life that have not yet been considered or explored.

ACKNOWLEDGMENTS

This work is financed by national funds through FCT – Fundação para a Ciência e a Tecnologia, I.P., under the Strategic Project with the references UIDB/04008/2020 and UIDP/04008/2020.

REFERENCES

Artefact. (n.d.). *The Tarot Cards of Tech | The Power of Predicting Impact*. Artefact. Retrieved July 2, 2023, from https://www.artefactgroup.com/case-studies/the-tarot-cards-of-tech/

Auger, J. (2013). Speculative design: Crafting the speculation. *Digital Creativity*, *24*(1), 11–35.

Baumann, K., Stokes, B., Bar, F., & Caldwell, B. (2017). Infrastructures of the imagination. In *Proceedings of the 8th International Conference on Communities and Technologies* (pp. 1629–1638). Troyes, France.

Bell, W. (1998). Making people responsible: The possible, the probable, and the preferable. *American Behavioral Scientist*, *42*(3), 323–339. https://doi.org/10.1177/0002764298042003004

Boer, L., & Jenkins, T. (2022). Fostering creative confidence with SCD in interaction design education. *Interaction Design & Architecture(s)*, *51*, 270–302.

Bray, K., & Harrington, C. A. (2021). Speculative blackness: Considering afrofuturism in the creation of inclusive speculative design probes. In *Proceedings of the 2021 ACM Designing Interactive Systems Conference (DIS 2021)* (pp. 1793–1806). Virtual Event, USA. https://doi.org/10.1145/3461778.3462047

Buchanan, R. (1992). Wicked problems in design thinking. *Design Issues*, 8(2), 5–21.

Candy, S. (2018). *Designing the Future: How Experiential Futures Work.* PhD thesis, University of Hawai'i at Manoa.

Candy, S., & Watson, J. (2015). Designing an experiential scenario: The people who vanished. In A. Curry (Ed.), *The APF Foresight Manual: Volume 1 – Knowing the Past, Creating the Future* (pp. 147–160). Singapore: Asia-Pacific Foresight.

De La Campos, J. O., & Castorena, D. G. (2020). The prediction generator: A tool for creating meaningful speculations. In *Proceedings of the 22nd International Conference on Engineering and Product Design Education, E and PDE 2020*. The Design Society, Herning, Denmark (10–11 September 2020).

Dolejšová, M. (2018). Edible speculations in the parlour of food futures. In *Extended Abstracts of the 2018 ACM CHI Conference on Human Factors in Computing Systems (CHI EA '18)*. Association for Computing Machinery, Montreal, QC, Canada, April 21–26, 2018. https://doi.org/10.1145/3170427.3188406

Dong, F., Sterling, S., Schaefer, D., & Forbes, H. (2020). Building the history of the future: A tool for culture- centred design for the speculative future. In *Proceedings of the Design Society: Design Conference* (vol. 1, pp. 1883–1890), Online (DESIGN 2020). Cambridge University Press, Cavtat, Croatia. https://doi.org/10.1017/dsd.2020.63

Dorst, K. (2019). Design beyond design. *She Ji: The Journal of Design, Economics, and Innovation*, *5*(2), 117–127.

Dunne, A., & Raby, F. (2013). *Speculative Everything: Design, Fiction, and Social Dreaming*. Cambridge, MA: The MIT Press.

Fallman, D. (2008). The interaction design research triangle of design practice, design studies, and design exploration. *Design Issues*, *24*(3), 4–18.

Futur 2 Studio. (n.d.). *Futur 2 Studio*. Retrieved August 17, 2024, from https://futur2studio.com

Futur 2 Studio. (n.d.). Happy Little Accidents: Dystopian Trend Cards [Card Set]. Berlin, Germany: Futur 2 Studio.

Galloway, A., & Caudwell, C. (2018). Speculative design as research method. *Undesign*, 85–96.

Helgason, I., Smyth, M., Encinas, E., & Mitrović, I. (2020, July). Speculative and critical design in education: Practice and perspectives. In *Companion Publication of the 2020 ACM Designing Interactive Systems Conference* (pp. 385–388). Virtual Event, USA.

Hill, D. (2017). *Slowdown Papers: On the agency of speculation in the design of urban futures*. London: Bartlett School of Architecture, University College London.

Irwin, T. (2015). Transition design: A proposal for a new area of design practice, study, and research. *Design and Culture*, *7*(2), 229–246.

Jones, P. (2014). Design research methods for systemic design: Perspectives from design education and practice. In *Proceedings of the 58th Annual Meeting of the ISSS-2014 United States*.

Koskinen, I. K., & Dorst, K. (2015). Academic design. In *DS 80-11 Proceedings of the 20th International Conference on Engineering Design (ICED 15) Vol 11: Human Behaviour in Design, Design Education*; *Milan, Italy, 27–30 July 2015* (pp. 227–234). Glasgow, UK: The Design Society.

Lutz, D. (2022, November 3). *Future Thieving #2 – Speculative Design Methods & Tools*. UX Collective | UX Collective. Medium. Retrieved July 2, 2023, from https://uxdesign.cc/future-thieving-2-speculative-design-methods-tools-45b655096d95
Manzini, E. (2015). *Design, When Everybody Design: An Introduction to Design for Social Innovation*. Cambridge, MA: The MIT Press.
Mitrović, I., Auger, J., Hanna, J., & Helgason, I. (Eds.). (2021). *Beyond Speculative Design: Past – Present – Future*. Split, Croatia: SpeculativeEdu; Arts Academy, University of Split
Normals. (n.d.). *Cards of the Normal Future*. Retrieved October 22, 2024. https://normalfutu.re/co/cards-of-the-normal-future/
Sanders, E. B. N., & Stappers, P. J. (2008). Co-creation and the new landscapes of design. *Co-Design*, *4*(1), 5–18.
Sangiorgi, D. (2011). Transformative services and transformation design. *International Journal of Design*, *5*(2), 29–40.
Stals, S., Smyth, M., & Mival, O. (2019). UrbanIxD: From ethnography to speculative design fictions for the hybrid city. In *Proceedings of the Halfway to the Future Symposium 2019 (HTTF '19), November 19–20, 2019* (pp. 1–10). Association for Computing Machinery, Nottingham, United Kingdom. https://doi.org/10.1145/3363384.3363486
Tharp, B. M., & Tharp, S. (2018). Discursive Design: *Critical, Speculative, and Alternative Things*. Cambridge, MA: The MIT Press.
Tironi, M. (2018). Speculative prototyping, frictions and counter-participation: A civic intervention with homeless individuals. *Design Studies*, *59*, 117–138.

Part 5

Thematic – CONVERT

Subthemes: Production – Industry – Customization

23 Micro and Small Furniture Industry in Emerging Economies

Development of Strategies for Sustainable Design Product

Juliana Cardoso Braga, Fernando Moreira da Silva, Juliano Aparecido Pereira, and Luis Carlos Paschoarelli

23.1 INTRODUCTION

Since the emergence of the Industrial Revolution until the present day, the industrial sector has exhibited a growing and constant evolution marked by the improvement of new technologies and, consequently, by the progress of human society. On the other hand, this same evolution generated a very costly ecological liability for future generations, and even for life on the planet itself, whose alternative solutions need to be urgently studied. As a result, there seems to be a consensus that the planning phases of industrial production must involve the search for these solutions, among which sustainable design strategies stand out, whose purposes involve minimizing the impacts of this ecological liability.

Sustainable design strategies seem to be present in the planning of large companies and industrial conglomerates. A good example can be seen with Apple (2023), which demonstrates to society, including as a powerful marketing instrument, its sustainable design strategy. Unfortunately, this culture of sustainability has not yet reached MSBs, especially in countries with emerging economies whose economic participation and ecological impact in these societies are quite significant.

The objective of this reflection is to study and discuss sustainable design strategies based on general guidelines for MSBs, especially those in the industrial sectors whose economic participation and environmental impact are significant in countries with emerging economies.

DOI: 10.1201/9781032719290-28

23.2 PRINCIPLES ON SUSTAINABLE DESIGN STRATEGIES

It is understood that the insertion of sustainable design strategies in local contexts and on small production scales is very relevant for MSBs and is essential to reduce the generation of solid waste, waste of materials, pollution, and environmental damage caused by the production of small-scale products. According to Fiksel (2009), sustainable design practices represent ethical behavior and good citizenship, and mainly a value strategy for companies' businesses. This principle of generating value and economic benefits for companies is known as eco-efficiency – that is, doing more with less. Part of this perspective is to reduce the consumption of resources (energy, materials); reduce environmental impact (emissions, waste, and the use of renewable resources); and create value (provide better products, provide additional services, offer benefits to customers) (Bhamra & Lofthouse, 2007). Several corporations recognize sustainability as a strategy for the development, growth, and survival of economic activities and its ability to stimulate new business opportunities, improve competitiveness, increase customer confidence, and motivate employees, local communities, partners, etc. (Bhamra & Lofthouse, 2007, Fiksel, 2009). Organizational changes to include sustainable design strategies also favor innovation and can offer a greater capacity to attract customers since they improve the organization's image and provide greater profitability through the reduction of environmental impacts, among other issues. However, all these benefits, obtained by adopting design and sustainability principles, need to be integrated at a strategic level so that they can act effectively (Bhamra & Lofthouse, 2007).

Strategies for developing sustainable products are characterized by a systemic analysis of design that involves the search for clean production technologies and strategies with the lowest possible impact. Such strategies are capable of predicting the desired result in all aspects of product development, determining the type of use, the target market, costs, and execution possibilities, generating flexible, durable, modular or multifunctional, adaptable, or recyclable products (Barbero & Cozzo, 2009). Some authors call these strategies: Ecodesign (Bhamra & Lofthouse, 2007, Barbero & Cozzo, 2009, Wimmer et al., 2010); Strategies of Life Cycle Design (Manzini & Vezzoli, 2008); Design for Sustainability (D4S) (Crul & Diehl, 2006), or even Design for Environment (DFE) (Fiksel, 2009). In the opinion of these authors, most of the environmental impact of a product is defined during the generation of the concept, that is, in the initial stage of the project development. For Jeswiet and Hauschild (2005), the application of sustainable concepts and strategies is important since the impact of a product is determined during the design phase.

According to Leonard (2011), design is responsible for determining various aspects of a product's environmental impact, such as the materials that need to be extracted or created, the amount of energy used in production and usage, the presence of toxic substances, the product's lifespan, the ease of repair, recyclability, and the harm caused by improper disposal, such as burying or burning the product.

Luttropp and Lagerstedt (2006) and Bhamra and Lofthouse (2007) agree and point out that the project development phase is the most important moment since it is when the requirements and specifications are decided. Manzini and Vezzoli

(2008) reinforce that, therefore, it is important to include and integrate environmental requirements from the beginning of the project development process, be it a product or service. Therefore, strategies that aim for a systemic approach involving all product life stages must be considered during product design. As a result, whenever possible, designers should apply a set of ecodesign strategies involving the entire life cycle of products, that is, considering the stages of pre-production, production, distribution, use, and disposal.

23.3 STRATEGIES FOR DEVELOPING SUSTAINABLE PRODUCTS

Among the strategies for developing sustainable products, we can highlight (i) the choice of low-impact materials; (ii) reduction in material consumption (raw materials and energy); (iii) production optimization; (iv) distribution optimization; (v) optimization during use; (vi) extending product life; and (vii) optimizing the end of product life.

The sustainability guidelines presented in the next sub-items were compiled from the different strategies presented by the authors: Behrendt et al. (1997), Brezet and Hemel (1998), Tischner et al. (2002), Gisbert and Garcia (2004), Bhamra and Lofthouse (2007), Manzini and Vezzoli (2008), Barbero and Cozzo (2009), Fiksel (2009), Fuad-Luke (2009), and Vezzoli (2018).

23.3.1 The Importance of Selecting Low-Impact Materials

The selection of low-impact materials must occur mainly during the project design phase. At this stage, the designer must specify the materials considering all life cycle stages of products, processes, and services. Among its basic assumptions are the following: use renewable raw materials produced by sustainable methods (of certified origin); prefer regional raw materials; avoid using non-renewable or low-abundance raw materials; prioritize recycled materials supported by product collection and collection systems or mixed with virgin materials; use materials that can be recycled or reused; reduce toxic and hazardous materials, additives, or finishes; use more efficient machines and equipment; choose low-impact transformation technologies; prioritize materials and resources that require less resources, water, and energy; prioritize natural lighting; and use renewable energy sources.

However, it is necessary to consider that most materials cause some type of impact. Therefore, the selection of low-impact materials should not be considered in isolation; it is also necessary to consider their impacts and consequences throughout the production chain, considering the function of the products, durability, and their context of use. A material may have a greater environmental impact during the production and disposal phases, but it can extend the product's lifespan more efficiently. Products with longer useful lives reduce the need for new materials for manufacturing, transportation, and disposal, thus lessening the associated environmental impact. The environmental impact of a product depends on various factors, including the materials used, the energy consumed in production and use, the presence of toxic chemicals, the ease of repair, its recyclability, and the harm caused by disposal methods such as burying or burning (Manzini & Vezzoli, 2008).

23.3.2 The Importance of Reducing Materials and Resources

The environmental impacts of a product can be considerably reduced by reducing the consumption of materials and energy (ABNT, 2011). Consequently, this can also influence cost reduction and generate environmental gains at other stages of the product's life cycle. Fiksel (2009) considers the reduction in material and energy consumption as a dematerialization strategy, while Lovins et al. (2007) call it a resource productivity strategy. For Lovins and collaborators, this strategy offers three significant advantages: it slows the depletion of resources, reduces pollution, and generates new prospects for job growth. Therefore, the designer must rethink production systems so that fewer resources are used and the need for extraction is reduced, contributing to minimizing the use of virgin raw materials (Leonard, 2011). This proposal encompasses a wide spectrum of possibilities that depend, to a large extent, on decisions made during the design phase.

These design strategies should reduce the amounts of materials and resources used; minimize waste; design in a modular way to make good use of raw materials; avoid excessive sizing and components or parts that are not strictly functional; reduce energy consumption in production; reduce the number of components and assemblies; minimize types or diversity of materials; integrate functions; simplify assemblies; minimize packaging; reduce the thickness of a product's material (as long as its resistance is not affected); design services, not products; and assess packaging needs.

Therefore, reducing is the watchword of this strategy that aims to reduce the environmental burden of products, since according to Manzini and Vezzoli (2008), by using less material or energy, environmental impacts are considerably reduced because fewer materials are explored or produced; that is, their transformation, transportation, and need to discard them are avoided.

23.3.3 The Contribution of Optimizing Production Techniques

At this stage, the objective is that the products meet the shape, dimensions, and quality requirements in a way that is compatible with the capabilities of the production process, facilitating the manufacture of the product. This strategy presents a broad list of good design practices for optimizing production and some of them are as follows: design with the minimum number of parts; minimize the number of fixations; design with a modular approach; design easy-to-manufacture parts; use software to carry out projects, modeling and prototyping; facilitate assembly; and eliminate or simplify adjustments.

The tactics of simplifying assembly and reducing the number of parts of a product directly influence the reduction in assembly and disassembly time, which, consequently, in the latter case, would facilitate a future recycling process. In addition to strategies aimed at product development, it was considered essential to incorporate strategies aimed at reducing steps in the manufacturing process and improving the layout, system, and production techniques.

23.3.4 The Contribution of Optimizing the Distribution (Transport) System

Transport optimization is also an important strategy for developing sustainable products. As Leonard (2011) and Papanek (2021) remind us, the transport of materials and articles consumes enormous amounts of fossil fuels and contributes to pollution around the world and creates the need for many roads, railways, airports, and warehouses. The improvement of transport systems can occur with the development of projects that facilitate storage and movement. This perspective includes a wide spectrum of design possibilities that can integrate the following strategies: prioritize the acquisition of local raw materials; optimize the developing detachable, stackable, and foldable products; reduce product weight and/or packaging volume; encourage consumer assembly of the product; reduce transport urgency allowing for longer delivery times; avoid return trips without cargo; and prioritize transport systems that use renewable energy sources.

Shortening the distances required to transport raw materials or products is also important for reducing CO_2 emissions and the consumption of fossil fuels. Furthermore, some of these alternatives can reduce the volume of the object to be transported and stored, considerably reducing the need to build large warehouse areas, resulting in a reduction in the number of trips during transport and also generating a significant reduction in the consumption of fuels and emissions. According to Barbero and Cozzo (2009), the greater the number of products transported on each trip, the lower the CO_2 emissions. In addition to the above, it was considered that it is also important to add two fundamental strategies to reduce fuel consumption and emissions: perform correct vehicle maintenance and plan transport routes in advance.

23.3.5 Reduced Impacts During Use

The reduction in resource consumption during use refers mainly to products that consume energy, water, or other consumable materials during their practical use. Therefore, improving the efficiency of this type of object during the use stage can reduce its environmental impact. As a result, they can be adopted as design requirements, among others: reduce energy, water, fuel, or other resource consumption during use; prioritize renewable energy sources (solar and wind energy); reduce the consumption of materials during use; design refillable products; and design for shared use.

23.3.6 Extending the Useful Life of Products

Extending the useful life of products refers to increasing their duration under normal conditions of use. In the opinion of Fiksel (2009), increasing the longevity of a product is one of the ways to improve environmental performance because this strategy reduces the consumption of new resources through the acquisition of new products. According to Manzini and Vezzoli (2008), the lower the obsolescence (technological and aesthetical) of a product, the greater the reduction in the production of new objects to meet the same demands (needs). Durable products generally have a smaller impact because of the delay in their elimination. Thus, product design can extend the useful life of the product, combating planned obsolescence or rapid disposal by providing alternatives such as the following: increase the durability of products and/or components; design an appropriate lifespan; facilitate use and provide instructions for use and maintenance; facilitate cleaning; facilitate maintenance or repair (facilitate disassembly); creation of objects with timeless design (avoiding premature aesthetic obsolescence); develop products or systems for updating or upgrading (avoiding premature technological obsolescence); design aesthetically timeless products; develop products that users identify with (personalized); avoid fragile joints; create adaptable, reconfigurable, and multifunctional products; and increase the warranty period.

The increase in useful life brings an important reduction in the consumption of materials and the generation of waste if we consider the consumption of materials necessary for the manufacture of new products. However, it is important to consider that not all types of products must have a long life, for example, products that fulfill their function only once, or those that are subject to obsolescence due to the greater efficiency of new products as explained by the same authors: regarding the use phase of products, sometimes the extension of the useful life does not determine the reduction of the environmental impact; on the contrary, it may result in an increase in the impact [...]. In other words, for some products, whose impact is greatest during the use phase, it may happen that there is a beneficial limit to their existence. It also happens when, for the same quality of service offered, technological development provides new products with better environmental efficiency (less consumption of energy and materials, or reduction of emissions and pollutants) (Manzini & Vezzoli, 2008, p. 183). And they continue this reasoning, launching the following proposition regarding the determination of the extension of the life of objects: products with the longest useful life are those that require few resources during their use cycle.

23.3.7 Optimization of the End of Life of Materials

When approaching the issue of sustainability, its objectives, and materials life cycle, we cannot fail to highlight the importance of the 5Rs: Reduce – reduce the amount of waste produced, wasting less and consuming only what is necessary; Reuse – save materials that can be used again; Recover – find degraded materials and reuse them; Renew – reinvent materials, repair them; and Recycling – transform materials into new products or raw materials in order to reduce the amount of waste eliminated. In this article, and given the materials and products we focus on, we have chosen to address two of these 5 Rs: Reuse and Recycling. These are ways to easily extend the useful life of materials, optimizing their end and obtaining new products with added value, that is, with usable conditions. According to Papanek (2011), to achieve the reuse or recycling of materials after the end of their useful life, the design of a product must consider the importance of ease of disassembly, i.e., to be designed according to the Design to Disassemble (DTD) approach, facilitating the collection, separation, disassembly, cleaning, and maintenance processes.

Both reuse and recycling make it possible to reduce the impacts of early disposal, minimizing the consumption of new raw materials, with recycling being a necessity in many countries, given the increasing scarcity of raw materials: nothing should be wasted. Post-consumer recycling is the process in which recycled products are converted into new materials. If it is found that the resources for reusing or recycling the materials have already been exhausted or it is impossible to use them due to their toxicity, or the lack of technology, or even adequate resources, the final disposal must be environmentally correct, depositing the industrial waste in specific and duly regulated industrial landfills. If this cannot be verified, the waste must be sent for controlled incineration, preferably for energy generation. The following are some possible design interventions for product reuse and recycling.

23.3.7.1 Reuse

As mentioned above, the reuse of a product can help to optimize its end of life. Nowadays, we already know that multifunctional, adaptable, and reconfigurable products have great potential to be reused for the same or another function. There are some strategic guidelines for the design of products that can extend their end of life through their reuse, of which the following stand out: facilitate disassembly; increase the resistance of parts subject to breakage; facilitate the removal of parts and components that can be reused; design standardized parts and components, that can be reused more easily; design aiming for the reuse of auxiliary parts; design reusable packaging; design aesthetically timeless products; and designing for more than one type of use.

In addition to the presented strategies, as a result of the work already implemented, adopting the strategies recommended by Project CACO, it was considered important to add another strategy to encourage reuse, that is, to highlight the importance of existing collection and donation systems.

23.3.7.2 Recycling

Recycling is another important R, which can be made possible both by reusing materials within the production process (pre-consumer recycling) and after discarding products (post-consumer recycling). These processes tend to mainly reduce the environmental burden directly related to the consumption of virgin raw materials and energy expenditure (Papanek, 2021; Manzini & Vezzoli, 2008). However, so that recycling can also be made viable, it is essential that products are developed that facilitate this process through the adoption of fundamental strategies: design products facilitating operations for disassembly and separation; use removable joining systems; ease of access for joining elements; minimize the number of joining elements to reduce disassembly time; choose materials with efficient recycling technologies (thermoplastics instead of thermosets, glass, aluminum, etc.); minimize the diversity of materials used (monomaterial); minimize materials that are incompatible with each other; facilitate the separation of incompatible materials; facilitate cleaning, avoiding finishes that are difficult to remove; identify materials through coding; provide additional information about the date of manufacture of the material, the number of recycling already carried out and the additives used; facilitate collection and transportation after use; use materials with high combustion power in articles that must be incinerated; and avoid artifacts that produce dangerous substances during combustion. In addition to the above, it is also a fundamental strategy to encourage and indicate existing collection and recycling systems, separating the materials.

Although many authors consider recycling to be the best solution to environmental problems, we cannot fail to highlight that the vast majority of recycling processes also have an environmental impact. However, in general terms, recycled products always present environmental gains, especially if they are designed to reduce disassembly and separation time of their component materials. Papanek himself (2021) presents examples of design products in which it is almost impossible to separate materials, such as a motorcycle or car body, which mix glass, metals, plastic, paint, shellac, and fillers. The same goes for its classification, which would be almost impossible and very expensive. This leads us to products in which individual parts are made of simple materials rather than composite materials, such as bi-directional fasteners, pop-in and pop-out rivets, for example, rather than using screws, glue and other mastic agents, and many welding methods, which nullify detachable technology.

23.3.7.3 The 4Ms

Karell and Niinimäki (2020) are authors that underline that, in parallel with the 5Rs, we should start to adopt the 4Ms: Making-off, Maker, Re-Make, and Makeover. Making-off refers to the optimization of the manufacturing process; Maker is directly related to user's involvement in both the manufacturing process and the reuse; Re-Make is the second life cycle of the product; and Makeover refers to recycling.

Over time, additional Ms have sometimes been added to the traditional 4Ms. Whether to direct more industry specific thinking or to demonstrate evolution, the main 4Ms can also be designed according to the 5.0 industry objectives: HuMan (Skill, Technology, Organization, Resources): to have the skills (and the will) to do what is expected, safely, consistently and effectively; machine (Equipment): the machines must be capable to achieve safe and reliable output at the desired quality, avoiding breakdowns, defects, or unplanned stoppages, which inhibit their ability to meet the goals; method (Process, Schedule, Procedure): to adopt standard work methods in place which ensure and support consistent, safe production; and material (Information, Raw Materials, Consumables, Quality): meeting the required specifications.

23.4 DISCUSSION

The present study proposed a reflection on sustainable design strategies based on general guidelines for MSBs, with emphasis on those in the furniture sector, whose economic and environmental impacts are significant in countries with emerging economies.

Initially, the theoretical basis highlights important benefits highlighted by sustainable design strategies. On the other hand, Braungart and McDonough (2002) strongly criticize these practices that aim to make products or systems less destructive. According to the authors, these are fatally limited strategies and they defend the "cradle to cradle" approach – from cradle to cradle, in which materials and products must be developed as biological nutrients (which can be consumed by microorganisms or other animals) or even as technical nutrients (free from toxic materials and that can return to the industrial cycle from which they come unlimitedly). In this approach, materials can be reused; no waste is produced or can be completely recycled; therefore, there would be no negative impacts because of the closed cycle of materials.

However, the "cradle to cradle" approach will certainly go a long way, and, in the meantime, we need to deal with a large number of existing materials and products that cannot be returned to biological systems due to their toxicity. Thus, it is necessary to minimize the environmental impacts arising mainly from this type of product, and, therefore, according to Manzini and Vezzoli (2008), recycling is seen as an inevitable procedure in which both the producer and consumers must pay both for the product, as well as its final disposal.

Other important aspects of sustainable design strategies involve operational efficiency and the interest of the stakeholders involved, especially in emerging economies or countries that share the same economic and social structure (Aftab et al. 2022). In this sense, the use of a certain strategy aimed at reducing the environmental impacts of the product, service, or process may positively or negatively affect other phases of the life cycle, requiring the designer to make a critical assessment of the possible environmental gains. Each design solution will require that the proposed alternatives be weighed against each other with the aim of determining the best design choices. Adomako et al. (2022) indicate that, in relation to the contexts of countries and regions with limited resources, the interests of these stakeholders should be mediated by sustainable design strategies aimed at achieving innovation through ecological products. The establishment of efficient systems for material collection, separation at source, and environmentally appropriate final disposal must encompass shared responsibilities for the life cycle of products. These initiatives require collective participation involving all generators – community, manufacturers, importers, distributors, traders, and public authorities – the latter of which should stimulate others, both through tax incentives, the valorization of recycled raw materials, educational campaigns, and through stricter legislation, inspection, or even penalties. According to Brown (1991), this would alleviate the problems of waste disposal since areas for its storage are becoming increasingly scarce. And to corroborate this thought, Gutberlet (2018) points out that this is a problem with a major impact on societies in emerging economies, seen mainly in large urban centers.

Furthermore, according to the order of priorities established by the National Solid Waste Policy (PNRS), recycling is not the ideal solution for the final destination of products since it is also a polluting process to the extent that trucks are needed to collect and transport items to the transformation industries – a process that also consumes energy and causes waste emissions. Therefore, the correct final destination must be used as one of the last resorts for objects.

First of all, it is necessary to avoid disposal as much as possible, developing products that use fewer resources for production, that are durable, light, foldable, and stackable, that use materials from renewable sources, that can be reconfigurable and adaptable, that can be standardized, dismantled, upgradable, renovated, repaired, and reused, that have greater efficiency in the consumption of energy, water, and materials, and that have more compact, reusable, or biodegradable packaging. Most of the time, it is impossible to adopt all of these strategies for the same product because these characteristics will be determined according to performance and due to the context of use for which the object is intended.

Another point to be considered regarding sustainable design strategies is directly related to the circumstances and context in which it is proposed. An important example occurs in Brazil, a country with an emerging economy, in which MSBs in the furniture sector have particularities that need to be considered carefully, since in general, they have a low degree of specialization and technological level (SENAI, 2006), facing difficulties to maintain itself in a market whose economy is not completely stable. Despite this, it is a sector whose revenue tripled in the first decade of this century, as well as being responsible for the direct generation of more than 275 thousand jobs (Oliveira et al., 2012). Therefore, it presents a relevant economic and social situation for local society, requiring creative planning and actions to remain active in the system. Due to Brazil's continental characteristics, it was decided to define a specific region to consider an analysis based on sustainable design strategies. Furniture MSBs in *Triângulo Mineiro* and *Alto Paranaíba* (MG – Brazil) have significant environmental liabilities, mainly due to the irregular disposal of waste on land on the outskirts of cities in the region. This practice arises from factors such as the following: local politics that are not sensitive to environmental issues; lack of local management aimed at the appropriate disposal of this type of waste; the resignation of poor populations living near these disposal sites; the reduced or no interest in assuming their share of environmental responsibility on the part of entrepreneurs in MSBs in the region; and the weak awareness (social and environmental) of designers regarding the environmental impacts of their projects (Nunes, 2013). According to a diagnosis by SENAI (2006), 90.8% of furniture MSBs in the region discard industrial waste, 3.7% sell it, and 5.5% usually carry out recycling. This entire framework helps to justify the development of sustainability strategies, using the principles of sustainable design.

23.5 PROPOSAL FOR SUSTAINABLE STRATEGIC DESIGN: CACO PROJECT

Braga (2019) introduces the "CACO Project," a sustainable strategic design initiative aimed at reducing the social, environmental, and economic impacts of furniture MSBs. This project promotes proper waste disposal, public space preservation, and municipal cost savings and mitigates issues from irregular waste disposal. It incorporates digital technologies to enhance the efficiency and optimization of sustainable design strategies, facilitating faster circular economy processes and fostering industry-wide sustainable solutions (Rusch et al., 2022; Geissdoerfer et al., 2018). The CACO model includes six sustainable design strategies:

1. Material and equipment selection: choosing environmentally friendly materials and equipment.
2. Resource consumption minimization: reducing material and resource usage during production.

3. Production and transportation optimization: improving production and logistics efficiency.
4. Product lifespan extension: enhancing furniture durability.
5. End-of-life management: ensuring proper waste disposal and promoting recycling.
6. Extending product life: prolonging the period of use of the furniture.

The model also features 47 Sustainable Guidelines for practical application and nine support tools, such as templates and checklists. The project began with a detailed diagnosis of participating furniture companies' current practices, leading to a gradual, supported implementation with training and ongoing technical assistance.

Results include waste reduction: lower waste volumes due to minimization and recycling practices; improved waste management: correct disposal in licensed landfills; pollutant reduction: decreased emissions and improved air quality; enhanced working conditions: safer, healthier work environments; community benefits: less irregular waste disposal, positively affecting local communities; competitive advantage: differentiation through sustainable product attributes; cost reduction: lower production and disposal costs, boosting profitability.

Evaluated by ten experts, the model was praised for its tailored approach and effectiveness in minimizing waste and impacts. Challenges include ongoing stakeholder engagement and model adaptation to market changes. The CACO Project demonstrates how sustainable design strategies can transform industrial practices, benefiting the environment, society, and the market (Figure 23.1).

23.6 CONCLUSION

Only with a deep awareness of the impacts linked to the life cycle of products and with the interference of political forces aimed at global resource management, will society be able to considerably reduce its impacts. Therefore, it must act preventively, in a systemic and strategic way, even before the first stage of a product's life cycle, encompassing the principles of ecodesign, even during the development of the project. These principles must also consider the geographic, economic, social, and cultural differences of each location, implementing sustainability requirements in companies from an environmental, social, economic, and cultural point of view, respecting the specificities and characteristics of each region.

Sustainable business model innovation and sustainable design strategies are a relatively nascent field of research. The research presented in this chapter has addressed this issue, underlying concepts and identifying a research gap in the implementation of strategies to achieve a sustainable strategic design and a business model innovation process, as well as the challenges of this process, and the tools to address these challenges.

Probably, with Industry 5.0 (called the Fifth Industrial Revolution), human society will reach a new phase in which the technological level, associated with human potential, can generate not only the improvement of products, systems, and services, but also innovation in clean production and ecologically efficient alternatives.

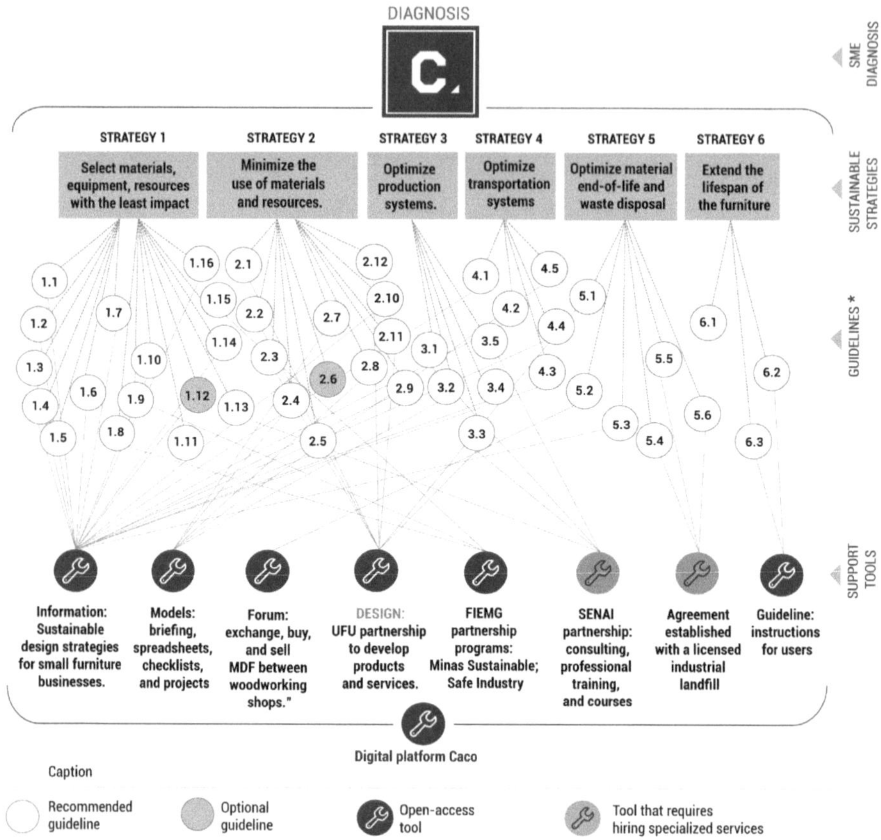

FIGURE 23.1 Caco project: sustainable strategic design model (Braga, 2019).

ACKNOWLEDGMENTS

Acknowledgments to CNPq – Brazilian National Council for Scientific and Technological Development (Grant 232345/2014–7); CIAUD – Research Centre for Architecture, Urbanism and Design, Lisbon School of Architecture, Universidade de Lisboa; and FCT – Foundation for the Science and Technology, Portugal (Grant UIDB/04008/2020 and UIDP/04008/2020).

REFERENCES

ABNT (2011). *NBR ISO 14001 Sistemas de gestão ambiental: requisitos com orientações de uso* [in Portuguese]. Rio de Janeiro: ABNT. https://www.normas.com.br/visualizar/abnt-nbr-nm/10259/nbriso14001-sistemas-de-gestao-ambiental-requisitos-com-orientacoes-para-uso

Adomako, S., Simms, C., Vazquez-Brust, D., & Nguyen, H. T. T. (2022). Stakeholder green pressure and new product performance in emerging countries: a cross-country study. *British Journal of Management*, 34(1), *299*–320. https://doi.org/10.1111/1467-8551.12595

Aftab, J., Abid, N., Cucari, N., & Savastano, M. (2022). Green human resource management and environmental performance: the role of green innovation and environmental strategy in a developing country. *Business Strategic and the Environment*, 32 (4), *1782*–1798. https://doi.org/10.1002/bse.3219

Apple (2023). *2023 Environmental progress report*. Cupertino, CA: Apple Inc. https://www.apple.com/environment/pdf/Apple_Environmental_Progress_Report_2023.pdf

Barbero, S., & Cozzo, B. (2009). *Ecodesign*. Köln: HF Ullmann.

Behrendt, S., Jasch, C., Peneda, M. C., & Weenen, H. van. (1997). *Life cycle design: a manual for small and medium sized companies*. Berlin: Springer.

Bhamra, T., & Lofthouse, V. (2007). *Design for sustainability: a practical approach*. Aldershot: Gower Publishing.

Braga, J. (2019). *CACO project: insertion of sustainable strategic design in micro and small furniture enterprises* [PhD Thesis – in Portuguese]. Lisboa: University of Lisbon.

Braungart, M., & McDonough, W. (2002). *Cradle to cradle: remaking the way we make things*. New York, NY: North Point Press.

Brezet, H., & Hemel, C. van. (1998). *Ecodesign: a promising approach to sustainable production and consumption*. Paris: UNEP.

Brown, L. R. (1991). *Building a self-sustaining society* [in Portuguese]. São Bernardo do Campo: Metodista.

Crul, M. R. M., & Diehl, J. C. (2006). *Design for sustainability: a practical approach for developing economies*. Paris: UNEP.

Fiksel, J. R. (2009). *Design for environment a guide to sustainable product development*. New York, NY: McGraw-Hill.

Fuad-Luke, A. (2009). *The Eco-design handbook: a complete sourcebook for the home and office* (3th edition). London, UK: Thames & Hudson.

Geissdoerfer, M., Vladimirova, D. & Evans, S. (2018). Sustainable business model innovation: a review. *Journal of Cleaner Production*, 198, *401*–416. https://doi.org/10.1016/j.jclepro.2018.06.240

Gisbert, P. F., & Garcia, R. L. (2004). Las estratégias de disenõ respetuoso com el medio ambiente. In T. G. Navarro & S. C. Rizo (Orgs.), *Ecodiseno: ingenieria del ciclo de vida para el desarrollo de productos sostenibles* [in Spanish]. Ed. Valencian: Universidad Politécnica de Valencia.

Gutberlet, J. (2018). Waste in the city: challenges and opportunities for urban agglomerations. In M. Ergen (Org), *Urban agglomeration*. London: IntechOpen. https://doi.org/10.5772/intechopen.72047

Jeswiet, J., & Hauschild, M. (2005). EcoDesign and future environmental impacts. *Materials & Design*, 26(7), *629*–634. https://doi.org/10.1016/j.matdes.2004.08.01

Karell & Niinimäki (2020). A mixed-method study of design practices and designers' roles in sustainable-minded clothing companies, June 2020. *Sustainability*, 12(11), *680*. https://doi.org/10.3390/su12114680

Leonard, A. (2011). *The story of stuff: the impact of overconsumption on the planet, our communities, and our health-and a vision for change*. New York, NY: Free Press.

Lovins, L. H., Lovins, A., & Hawken, P. (2008). *Natural capitalism: creating the next industrial revolution*. Boston, MA: Back Bay Books.

Luttropp, C., & Lagerstedt, J. (2006). Ecodesign and the ten golden rules: generic advice for merging environmental aspects into product development. *Journal of Cleaner Production*, 14(15–16), *1396*–1408. https://doi.org/10.1016/j.jclepro.2005.11.022

Manzini, E., & Vezzoli, C. (2008). *Development of sustainable products: the environmental requirements of industrial products* [in Portuguese]. São Paulo: EDUSP.

Nunes, V. G. A. (2013). *Design pilot project as a boundary object: a strategy to foster sustainable design policies for Brazilian MSEs* [PhD Thesis]. Milan: Polytechnic Institute of Milan.

Oliveira, P. R. S., Paes, F. A. S. V., Rezende, J. B., Alvarenga, A. de P., & Mendonça, F. M. de. (2012). *Furniture production chain: the Triângulo Mineiro furniture cluster* [in Portuguese]. Belo Horizonte: EPAMIG.

Papanek, V. (2021). *The green imperative: ecology and ethics in design and architecture.* London: Thames & Hudson.

Rusch, M., Schöggl, J.-P., & Baumgartner, R. J. (2022). Application of digital technologies for sustainable product management in a circular economy: a review. *Business Strategic and the Environment*, 32(3), *1159*–1174. https://doi.org/10.1002/bse.3099

SENAI. (2006). *Business diagnosis of the furniture industries in Uberlândia region* [in Portuguese]. Belo Horizonte: Sistema FIEMG.

Tischner, U., Schmincke, E., Rubik, F., Prosler, M., Dietz, B., Maßelter, S., & Hirschl, B. (2002). *How to do ecodesign? A guide for environmentally and economically sound design.* (G. F. E. Agency, Org.). Basel: Birkhauser.

Vezzoli, C. A. (2018). *Design for environmental sustainability – life cycle design of products.* Cham: Springer.

Wimmer, W., Züst, R., & Lee, K. M. (2010). *Ecodesign implementation: a systematic guidance on integrating environmental considerations into product development.* Dordrecht: Springer.

24 IKEA Hacking and Mass Customization

Redefining the Role of Designers through Parametric Design and Additive Manufacturing

Domenico Di Fuccia

This chapter is an extended version of the article presented in the Design Commit '24 conference proceedings.

24.1 INTRODUCTION

In recent years, we have been hearing more and more about IKEA hacking, a phenomenon that started out as a need and soon became a real movement, with a community of enthusiasts sharing designs and instructions. This phenomenon is not only of interest to the amateur world, but the topic has also been addressed by professional designers. All this suggests that IKEA hacking is not destined to disappear, but rather follows a growing interest in do-it-yourself (DIY) and customization. However, merely replicating individual hacks does not utilize the full potential of the phenomenon. The next step could be to allow the user who wants to replicate a hack to be able to participate in the creative process. Today's technologies allow for this advancement, as tools for digital fabrication are increasingly cheap, available, and accessible to the average consumer, foremost among them additive manufacturing, i.e., 3D printing. This is facilitated by the spread of the maker phenomenon, the new digital artisans, and FabLabs, laboratories that make digital fabrication technologies available to the public. An important contribution is made by Web 2.0 technologies, which facilitate the sharing of information, and by the open culture, which makes this information free and usable by anyone. IKEA hacking represents an expression of the consumer's desire to take part in the creative process again. This chapter aims to explore the designer's ability to design not only the finished product, but tools that enable the user himself to design and produce.

DOI: 10.1201/9781032719290-29

24.2 DEMOCRATIZATION OF THE DESIGN PROCESS

24.2.1 Open Design: A Collaborative Methodology

The concept of open design is not new, nor does it originate independently, but is the result of an evolutionary process starting from the Free Software movement through the open-source movement.

The Free Software movement was born in the early 1980s thanks to Richard Stallman with the GNU project, which had the intent of creating an operating system composed of Free Software, later evolving into the non-profit organization Free Software Foundation in 1985. It was thus born in response to the "deprivation of freedom" due to changing U.S. antitrust legislation in the 1970s–80s, which had led companies to make it impossible for users to modify software and adapt it to their own needs. In 1998, a pragmatic version of the Free Software movement spread through the open-source Initiative, which focuses on the practical consequences of "open" principles rather than the values related to it.

In this context, we can see that what is open is not the software itself, but the process and rights, i.e., software development and intellectual property; this has allowed this approach to extend outside the field of computing. In fact, since the early 1990s, the open approach has been spreading into various sectors, encouraged by the digitization of these sectors, the development of new digital technologies, and the democratization of the high-speed Internet. The digitization of sectors is a necessary condition for the development of an open approach, but not sufficient, as those who take part in such initiatives are also driven by ideology and opportunity (Boisseau et al., 2018).

The different open approaches are all based on the same principles, according to the Open Knowledge Foundation: "Open means that anyone can freely access, use, modify, and share for any purpose (subject, at most, to requirements that preserve provenance and openness)".

The foundation itself defines the following core principles:

- Availability and access: Data must be available in their entirety and at a reasonable cost of reproduction, preferably by downloading from the Internet. They must also be available in a convenient and modifiable form.
- Reuse and redistribution: Data must be provided in terms that allow reuse and redistribution, including mixing with other data sets. They must be machine-readable.
- Universal participation: Everyone must be able to use, reuse, and redistribute; there should be no discrimination against fields of activity or against people or groups. For example, "non-commercial" restrictions or restrictions on use for certain purposes (e.g., only in education) are not allowed.

These principles emphasize inclusivity and accessibility, enabling a wider audience to engage in design activities and modifications. They also foster innovation, as diverse perspectives contribute to the creative process. The importance of these principles is further highlighted by their application in various domains, from product design to art and education, where collaboration can lead to innovative solutions and products that resonate with users' needs.

These principles are also echoed by Creative Commons, "an international nonprofit organization that provides free licenses, tools that copyright and related rights holders can use to enable others to legally share, reuse and remix their works" (creativecommons.org). These licenses are contracts through which the copyright holder grants anyone permission to use the work while retaining authorship; he, therefore, does not assign his rights but grants enjoyment of them to third parties under certain conditions.

There are four such conditions, and it is the owner who decides which rights to reserve and which to grant:

- Attribution (BY): Credit must be given to the creator of the work;
- Share alike (SA): Adaptations must be shared under the same terms;
- No derivative works (ND): No modifications or adaptations are allowed;
- Non-commercial (NC): Only non-commercial uses are allowed.

When applied to physical objects, the open approach takes the name of open hardware; this implies sharing the original project files in order to be modified or reproduced by others; nevertheless, there remains a fundamental difference between open software and open hardware: While software can be reproduced indefinitely for free, objects, despite digital files, need to be manufactured. This distinction is critical as it reflects the additional challenges associated with open hardware, such as the need for access to materials and fabrication resources, as well as the skill sets necessary for physical production.

To date, a wide variety of products are open hardware, e.g., microcontrollers like Arduino, 3D printers like RepRap, cars like Local Motors, smartphones like OpenMoko, satellites like Ardusat, furniture like Opendesk, various objects like Thingiverse, and even prosthetics like Open Prosthetics Project.

Open products also tend to have an open design process; just as software does with bug reporting, debugging, and requests for new features and additions, those who rely on an open approach have a supportive community available to them during design. This collaborative environment fosters innovation, as participants can build upon each other's ideas and make improvements, leading to a dynamic ecosystem where design evolves continuously. This community consists of the users themselves, who then take part in the design process, gaining the ability to fix the product and possibly to be able to "hack" it by improving it or changing its original function. With open design, the open approach meets product design; however, referring back to the above definition, we can see that almost no real case fully meets it. We always find parts of a design that are open and parts that are not, deliberately or not. So, it is more correct to assess the degree to which a design is open, rather than considering it to be completely open or completely closed.

So, open design is not a one-size-fits-all reality; rather, it is composed of different nuances that are more or less and differently open. For example, crowdsourcing, that is, the collective development of a project by a multitude of people from outside the company or institution, possesses an open design process even if the results of the latter are not. At the opposite end of the spectrum, however, we find downloadable design, which refers to products that can be freely downloaded, whose design process, however, is not necessarily open.

An example of this type of approach is Opendesk, a project that consists of a globally spread network of designers and workshops that aims to produce furniture objects locally and that makes the files for making them available online for free. The user can then choose the model and material with which to make the object. Opendesk will manufacture it in the workshop nearest to him; the user can, if he has the necessary technologies, also download the design files and produce it himself. Open design can thus embrace a wide variety of areas, one possible outcome being platforms that generate the design of an object based on user input. These platforms democratize the design process, allowing users to contribute their unique ideas and requirements, which can lead to a more diverse array of products that better meet the varied needs of different user groups. An example of this approach is Nervous System, a design studio that relies on generative design, releasing online design applications that allow customers to co-create products in an effort to make design more accessible. Inspired by natural processes and patterns, these tools allow for endless design variations and customization by creating unique pieces of art, jewelry, and home goods. The users, for example, can log on to the website, choose the object they want to create from those proposed, modify it to their liking thanks to the parameters present, choose the material from a list of possibilities, and finally commission their unique piece from Nervous System, which will create it through digital fabrication technologies.

Companies decide to "open up" their design process because they are incentivized to do so. There are several reasons behind this choice, the first being adaptability, that is, the ability to adapt to subjective needs, individual users, or specific environments. Open design also helps address niche needs that would not be filled otherwise. In addition, other reasons are increased speed of development and thus decreased costs, faster adoption of technology, and increased efficiency in the design process. By leveraging open design principles, companies can tap into the creativity of their customers, leading to innovative products that are closely aligned with market demands. This shift not only enhances customer satisfaction but also fosters brand loyalty, as consumers feel more invested in products they helped create. Open design appears, therefore, to be a major shift in design projects. It is driven by changes in society itself, and for some, "Openness, in short, is more than a commercial and cultural issue. It is a matter of survival" (Thackara, 2011).

24.2.2 Communities of Makers and FabLabs

In the last few decades, the focus has been entirely on the virtual representation of the real: from 3D movies to renderings to various social networks. Now the reverse process is taking place; a set of technologies and cultural processes are making it easy to move from virtual content to a physical object; it is the consumer who becomes a producer again. These new producers are the so-called makers, a term that is difficult to give an unambiguous definition to because it reflects a multiplicity of very different practices, values, and stakeholders.

The website of the Maker Faire in Rome (the largest makers' fair outside the United States, second only to the one in San Francisco) defines makers as "technology enthusiasts, educators, thinkers, inventors, engineers, authors, artists, students, artisans including digital ones. All those, in short, who create and amaze with the power of their ideas. They are people, startups, businesses that solve problems and contribute to creating a more sustainable society".

Strictly speaking, this term refers to an American subcultural movement related to *Make* magazine. The original culture draws on the idea of DIY, but over time, it has become much more than just DIY; it has to do with a new form of independence: Makers are people who rediscover the possibility of realizing ideas on their own without merely being the end link in a manufacturing chain. This shift has given rise to a culture of innovation that encourages experimentation and exploration, where the process of making is as important as the final product. This newfound independence fosters a sense of empowerment among individuals, allowing them to take control of their creative output and challenge traditional notions of production. They represent an attitude that has historically always been present in humans, namely the taste for solving problems for themselves through inventiveness and the instinctive pleasure of DIY. All this has been made possible by the development of the web and especially file sharing; all knowledge is available online, and those who desire can easily find it and freely dispose of it. In addition, what has allowed this virtually shared knowledge to take the next step, which is to become real objects, has been the development and dissemination in desktop format of technologies that were once exclusive to large corporations. These technologies that have become democratic are computer numerical control (CNC) machines, laser cutters, but first and foremost rapid prototyping through 3D printers. These are the tools for digital fabrication, tools that enable the transformation of data into real objects.

Nevertheless, access to knowledge and production techniques alone are not enough, what distinguishes maker culture is the sharing of ideas, projects, and opinions; it is the creation of a community: "Their motto is 'do it yourself' but, more than anything, 'let's do it together'" (makerfairerome.eu). This is a key aspect; in makers, there is a willingness to recognize the value of sharing one's work, often as opposed to classic craftsmanship where we find the opposite attitude. The need to share is motivated by several reasons, to keep track of what has been done, to enhance one's reputation, and to receive improvements for one's project (Maker, 2014). This culture of sharing not only promotes personal growth but also leads to collective progress, as ideas are iterated upon and improved in collaborative environments. As such, maker communities often become incubators of innovation, where individual contributions can significantly impact the group's overall output.

The experience of sharing objects and material things is the fruit of the work done in this direction in the field of computing, as stated in the previous paragraph, with open-source software, but also with Creative Commons licenses.

> Digital files can be shared and copied endlessly at practically no cost and without the slightest loss of quality; but the most important thing is that they can be modified just as easily. We live in a culture of 'remix': everything is inspired by something that came before, and creativity is seen in the reinterpretation of pre-existing works no less than in the originals. [...] This ability to easily remix digital files is the engine that drives communities. What it offers is an invitation to participate. You don't have to invent something from scratch or have an original idea. Instead, you can participate in collaborative improvement of existing ideas and models.
>
> *(Anderson, 2013)*

The aggregation of makers also gave rise to FabLabs (Fabrication Laboratories), small-scale laboratories that offer all the tools needed to carry out digital fabrication projects. The first FabLab (fabfoundation.org) was born in Boston in 2003 from an idea of Neil Gershenfeld, a professor at the Massachusetts Institute of Technology, and his colleagues. The idea is related to a lab that can collaborate remotely and process projects digitally. After the first lab, there was a rapid spread around the world, and to date, there are more than 2,000 distributed in more than 120 countries. A FabLab is generally equipped with a set of computerized tools capable of making a wide range of objects. These include technological products generally considered to be the prerogative of mass production only. These tools usually include 3D printers, CNC milling machines, laser cutters, electronic boards, and microprocessors. These labs serve as community hubs, providing access to advanced fabrication tools that might otherwise be cost-prohibitive for individuals. But despite their high technological content, FabLabs are based on people before machines; the main purpose is to share ideas, designs, and know-how. This community-driven approach ensures that knowledge is passed down and built upon, creating a sustainable cycle of innovation.

The principles commonly shared by these labs are:

- the choice to democratize public access to digital fabrication technologies;
- the signing of the Fab Charter, the set of requirements and questions that a FabLab must answer to define itself as such;
- the sharing of a common set of production tools and processes with the aim of sharing their use practices, favoring open-source software and hardware;
- collaboration with the global network of FabLabs and the sharing of knowledge, design, processes, ideas, and prototypes.

24.2.3 Prosumer: Project, Participation, Sharing

One of the most obvious consequences for businesses since the advent of the Internet is the increased participation of users in the production process. This has been particularly noticeable since the emergence of Web 2.0; in fact, for some of the most successful Web 2.0 sites (e.g., Facebook, Instagram, Twitter), user-provided content constitutes the majority of the value of the service. This increased user participation blurs the line between consumption and production, as users consume and produce content at the same time. No longer exclusively consumers, users have become prosumers.

The concept of prosumer is not new; it was coined by futurologist Alvin Toffler in 1980 in his essay The Third Wave. At this early stage, it focuses primarily on the growing role of the consumer in self-producing or self-service activities, or in providing input for personalization. Prosumers are thought to provide only "labor power" as part of the production process.

More recently, however, the advent of Web 2.0 and significant opportunities for user participation have led to a wave of interest in the prosumer phenomenon. This issue was discussed by Don Tapscott and Anthony D. Williams in their essay Wikinomics (2006). They extend the prosumer view to situations where consumers take part in the production process in a more creative way, such as when the consumers "hack" or "remix" existing products. However, looking at recent developments, it is clear that consumers have gone further. Nowadays, in fact, when they take part in the production process, they increasingly do so with their own means of production. Starting with their own computers, smartphones, and cameras that are used to produce content for Web 2.0 platforms.

> The consumer actually co-innovates and co-produces the products they consume. In other words, customers do more than customize or personalize their wares; they can self-organize to create their own. The most advanced users, in fact, no longer wait for an invitation to turn a product into a platform for their own innovations. They just form their own prosumer communities online, where they share product-related information, collaborate on customized projects, engage in commerce, and swap tips, tools, and product hacks.
>
> *(Tapscot & Williams, 2006)*

This collaborative dynamic within prosumer communities fosters a culture of shared creativity, where users build on each other's ideas and solutions. These interactions often lead to the rapid evolution of products as prosumers identify and address gaps or inefficiencies.

This new role of consumers is even more evident when one considers their involvement in the production of physical objects. With 3D printing technologies, the role of consumers in the process of producing physical objects ceases to be limited to minor contributions and becomes central.

> Beyond the question of the omnipresent role of consumers in the production process as customers, home printing also raises the question of consumers as producers. Indeed, over the past decades, the role of consumers in production processes has changed from an almost entirely passive role to a far more active and, in some cases, dominant role.
>
> *(Rayna & Striukova, 2016)*

Indeed, the production of an object typically involves three stages: design, manufacture, and distribution. Traditionally, consumers have been involved in all three stages but only to a limited extent (e.g., color selection, assembly of prefabricated parts, transportation from the store to the home). However, recent advances in 3D printing technologies have the potential to completely overturn this model. Domestic 3D printers allow consumers to take full charge of the manufacturing and distribution stages; even if some parts are produced in local labs, the consumer remains responsible. In addition, the rise of co-creation platforms also allows consumers to take an active role in the design phase. Over the years, consumers have been increasingly involved in all stages of production; nevertheless, consumer input has never been decisive. However, digital technologies have accelerated the "prosumer trend", and online consumers now play a leading role in all stages. The spread of 3D printing will enable consumers to take a leading role "offline" as well.

24.3 TOOLS AND TECHNOLOGIES OF DIGITAL FABRICATION

24.3.1 Parametric Design

The term parametric comes from mathematics and refers to the use of certain parameters or variables that can be changed to alter the result of an equation. While today the term is used in reference to computational design systems, the concept of parametric has implications apart from the tools being used. We find precedents for these modern systems in the works of architects such as Antoni Gaudi, who used "analog" models to explore design space in his model of upside-down churches, and Luigi Moretti, who coined the term "parametric architecture", describing for the first time an architectural design process based on mathematical algorithms. These early explorations laid the groundwork for today's parametric design tools, which allow designers to manipulate variables dynamically, making design processes more flexible and adaptable.

Designers have always designed with paper, pencil, and eraser; the pencil adds "marks" while the eraser removes them. Traditional design systems are emulations of this centuries-old working method; it is easy to create an initial model, just add "parts". Modifying it, however, can be difficult: Even changing one dimension can require adjusting many other parts, and all this reworking is done manually. The more complex the model, the more work is required. Tools like this can limit exploration and thus the project itself. Parametric modeling introduces a fundamental change; "marks" are related to each other and change in a coordinated way. The idea is that the designer establishes the relationships through which the parts are connected, builds a design using those relationships, and modifies those relationships by observing and selecting from the results produced. The system takes care to keep the design consistent with the relationships and thus allows the designer to explore by avoiding manual reworking of each change. This process allows for a high degree of flexibility, encouraging more experimentation. Designers can create adaptable systems that generate multiple variations of an object with ease, pushing the boundaries of creativity.

This comes at a cost; parametric design depends on defining relationships and the designer's willingness to consider the step of defining them as an integral part of the larger design process. Defining relationships is a complex act, involving strategies, skills, and abstraction capabilities. It initially requires the designer to take a step back from direct design activity and focus on the logic that binds the project together. The advantage, however, is in the ability to mutate the design countless times without the need to rework it each time. Not only does this allow the designer greater freedom in the design process, but it also introduces the possibility of creating a theoretically infinite series of objects that are different from each other but still linked by the same relationships and consequently belonging to the same family. The parameters, in fact, are by definition variable, and this variability can be controlled by the designer himself or it can be automated and controlled by the computer: For example, a program can be required to generate any number of variations, either randomly or as a function of external factors (generative design). In practical terms, this means that a single design system can be adapted to suit different functions, environments, or user needs without having to start from scratch. This makes parametric design a powerful tool not just for efficiency but for personalization and customization in fields such as architecture, product design, and digital fabrication. However, there is also a third possibility: Some parameters can be changed by someone other than the author. "Openness and interactivity are concepts related to the notion of digital variability, but this participatory approach to digital design has only recently gained wider recognition in the new techno-cultural environment of the so-called Web 2.0" (Carpo, 2011). This allows the average user, who lacks the knowledge and skills to design an object from scratch or to use certain modeling programs, to take part in the process, to become a designer themselves, to adapt the object to their own tastes or needs. Parametric design combined with Web 2.0 capabilities and digital fabrication technologies, especially additive manufacturing, is capable of turning the traditional production chain upside down. By allowing non-experts to engage in the design process, parametric design opens up the possibility for a more inclusive and participatory design culture. As tools become more accessible, users are empowered to create objects that reflect their individual needs and preferences, effectively democratizing the design process.

24.3.2 Additive Manufacturing, Materials, and Technologies

Additive manufacturing, also known as 3D printing or rapid prototyping, has been around for decades. The first working 3D printer was created in 1984 by Charles W. Hull with the invention of stereolithography (SLA). Since 1986, 3D printing has evolved and differentiated with the introduction of new printing techniques and countless materials that have enabled the spread of this technology to many fields and industries. In particular, we owe the invention of the fused deposition modeling (FDM) method in 1989 to S. Scott Crump. But it is since 2009, with the expiration of a number of patents, that the cost of this technology has been drastically reduced, making it economically accessible to small and medium-sized businesses and encouraging its entry into the amateur market.

All 3D printers use a process that relies on overlaying layers of material to create a three-dimensional object; however, there are different technologies for creating these layers. The most common are SLA, selective laser synthesizing (SLS), and FDM.

SLA technology uses special photosensitive resins solidified through a ultraviolet (UV) source as the material. There are various declinations of this technology, but in principle, they all use a source of ultraviolet light, often a laser, that cures the resin contained in a tank on a plate according to a specific pattern. When the first layer is finished, the plate moves slightly, and a second layer is cured by the laser; the process is repeated until the object is completed. When finished, the object is removed from the liquid resin, washed with a solvent to remove excess resin, and finally placed in an ultraviolet light oven to complete polymerization.

SLS technology is conceptually similar to SLA; however, it uses a high-power laser instead of a UV source and very thin powders instead of resins. This technique synthesizes such powders by automatically aiming the laser at points in space defined by a 3D model, binding the material together to create a solid structure. The powders can be made of a wide variety of materials: polymers, metals, ceramics, and glass materials.

Finally, FDM technology is a process that, unlike the previous methods, uses thermoplastic materials that are heated to a viscous state and then deposited on a plane to form layers. The thermoplastic material is usually in the form of continuous filament in spools; the filament is unwound and pushed through a nozzle into a heated head that is moved by software to form the various layers on a plane, which is often also heated. A wide range of thermoplastic polymers can be processed with FDM technology. The most common are polyvinyl alcohol and acrylonitrile butadiene styrene, but NYLON, PET, PC, flexible materials such as thermoplastic polyurethane, and others are also popular. There are also filaments made of water-soluble materials, which can be used as supports during printing, such as polyvinyl alcohol, or made of composite materials, consisting of a fiber-reinforced polymer matrix, e.g., NYLON reinforced with carbon fiber or fiberglass.

It is the latter technology that has opened the world of 3D printing to the general public, starting with the Fab@home and RepRap projects. The former was born with the intention of producing a low-cost printer to spread the culture of 3D printing; the latter adds to the desire for dissemination, the aspiration to be self-reproducing, representing a significant step toward a future where additive manufacturing can become fully decentralized, empowering individuals to produce and even replicate the machines themselves, leading to a potential revolution in home and small-scale production. Since then, more and more companies have entered the 3D printer business, bringing the price of 3D printers down dramatically in a few years, allowing more and more enthusiasts to access this technology. As of today, the price of a non-industrial 3D printer ranges from as low as 150 up to about €1,000; a mid-range model is around €300–€400.

However, apart from domestic use, even at the industrial level, 3D printing has brought enormous possibilities. It has moved from its original use of rapid prototyping as a design aid to become a direct manufacturing technology with applications in various industries. The reasons for this lie in its advantages over traditional manufacturing processes. One of the key advantages of additive manufacturing is the ability to produce highly complex geometries that would be impossible or prohibitively expensive using traditional manufacturing methods. Moreover, by its very nature, it allows new design freedoms, direct manufacturing from 3D models, and more flexible fabrication. These advantages enable the manufacture of bespoke and on-demand products that are economically attractive compared to mass production methods (Ford et al., 2015). Furthermore, additive manufacturing offers distinct environmental benefits, such as reducing material waste. Traditional subtractive manufacturing processes cut away excess material from a larger block, resulting in significant waste, while 3D printing only uses the material necessary for each layer. This efficiency not only lowers material costs but also contributes to more sustainable production practices.

Hence, additive manufacturing reverses the economic logic of traditional manufacturing. In mass production, almost all costs go to the initial tooling of the machine, and the more complex the product, the higher the cost. With digital manufacturing, on the other hand, it is exactly the opposite; variety is free, differentiating each individual product costs no more than making identical copies; complexity is free, an extremely complex product with a lot of detail can be printed in the same way as a simple block of plastic; and flexibility is free, changing a product after production has already started means only changing the instruction program.

> It is the inverse of mass production, which favors repetition and standardization. Three-dimensional printing, on the other hand, favors individualization and customization. The big benefit brought by the digital manufacturing era is that we can now choose between the two without having to go back to very expensive craftsmanship: with automated methods, you can have both mass production and customization.
>
> *(Anderson, 2013)*

In addition to enabling mass customization, 3D printing technologies also create significant opportunities for co-creation between businesses and customers. Co-creation and mass customization are two important vectors of user innovation, which is itself a crucial source of radical innovation. In addition, these machines can make available a more sustainable way of manufacturing objects. By making the product locally, there are virtually no transportation costs. Waste is minimized because only the amount of raw material strictly necessary is used. Finally, because the product is personalized and tailored to the user, it will be more appreciated and last longer.

> Similar to the way IKEA hacks adapt existing products, desktop manufacturing will give end users the tools to make professionally crafted products tailored to their preferences, without the need for compromises aimed at satisfying a large market.
>
> *(Saakes, 2011)*

24.3.3 Distributed Manufacturing

Large-scale manufacturing, from the Industrial Revolution to the present day, has been characterized by the progressive centralization of operations; productivity advantages, cost optimization through economies of scale, and trends toward globalization have allowed this form of manufacturing to dominate the global industrial landscape. Factories have thus become increasingly efficient, but this centralized paradigm is also characterized by long supply and distribution chains, often associated with inefficient use of resources. Centralized manufacturing typically requires vast amounts of raw materials and labor to be concentrated in one location, resulting in increased energy consumption, transportation costs, and delays in production due to supply chain disruptions.

Distributed manufacturing, on the other hand, subverts this system; it is a decentralized form of production using a network of geographically dispersed production facilities connected online. In this model, production is not restricted to a single, large facility but is instead spread across smaller, localized sites that are connected digitally. This allows for more responsive, on-demand production that can be tailored to local markets. It could be argued that this is not a new concept, referring to craft production carried out before the Industrial Revolution; however, there are some key differences. A good craftsman may be consistent in what he produces in one place, but there is unlikely to be consistency across different geographic areas. An important characteristic of distributed manufacturing is geographic dispersion and the ability to produce the same product identically anywhere in the world. In addition, by producing close to the final destination, logistics costs and environmental impact are reduced.

All of this is possible thanks to digital fabrication technologies, which break down economies of scale; 3D printing or machining a product with a numerically controlled machine has the same cost per piece regardless of the number of copies made. This contrasts with traditional manufacturing, where the cost per unit decreases only as production volumes increase. Digital fabrication technologies allow companies to produce small batches, even single units, without facing prohibitive costs. An example of a company operating in this way is the aforementioned Opendesk: "an online marketplace that hosts independently designed furniture and connects its customers to local manufacturers around the world. Rather than mass producing and shipping around the world, we are building a distributed, ethical supply chain through a global network of manufacturers" (Opendesk.cc). After selecting one of the designs available on the site, users will receive home furniture produced by the workshop closest to them using digital technologies such as CNC milling and using local raw materials.

Distributed manufacturing is capable of disrupting the corporate world as we know it, but it is in relation to the maker movement that it expresses its full potential.

> Additive manufacturing technologies have particular potential for customer-centric and personalized production systems as they allow consumer to produce parts, products, and machines, as users of dedicated AM technologies. In such a model, end users of AM can use the technology to produce a variety of parts for themselves and others, which then implies a fundamental change to the global structure of manufacturing, amplifying the change from centralized to decentralized supply chains.
>
> *(Bogers et al., 2016)*

An even smaller-scale production, often of individual units, made by the consumer himself through digital fabrication tools, foremost among them 3D printing, becomes possible. Indeed, thanks to the spread of this technology, it is now possible for the user to freely download designs online from one of several sites that collect 3D models and produce them themselves.

> A major advantage of 3D printing is the separation of product design from manufacturing capabilities. Since design and manufacturing can be easily outsourced in 3D

> printing, designers can contract with firms such as Shapeways to produce, ship, and collect proceeds for goods based on their designs. Alternatively, a consumer can download a CAD software design for a replacement part online as easily – as he/she would download digital music – and then download and print the part on his/her 3D printer.
>
> *(Berman, 2012)*

This ability to separate design from manufacturing opens new opportunities for designers who no longer need access to large factories to see their products brought to life. It also allows for unprecedented flexibility, where a product can be designed in one country, modified in another, and manufactured in a third location, all within a short period.

Additive manufacturing is emblematic of the shift to distributed manufacturing because it allows production to take place only on demand, uses only the material needed, and eliminates transportation time and costs. Furthermore, it fosters innovation by allowing individuals and small businesses to enter markets previously dominated by large manufacturers. Distributed manufacturing, enabled by digital fabrication technologies, presents a sustainable, flexible, and more consumer-centric approach to production.

24.4 THE IKEA CASE

24.4.1 IKEA Hacking: Reinterpretation, Customization, Upgrade

The term hacking is usually associated with the digital world, where it acquires the meaning of taking advantage of software weaknesses; however, when we talk about IKEA hacking, we are referring to a phenomenon of taking a pre-existing object, altering it to fit one's needs, customizing it, improving it, or simply making it one's own. This trend reflects a broader cultural shift toward personalization and creative reuse, where individuals feel empowered to modify mass-produced goods to better reflect their personal tastes or needs.

Robson et al. (2019) defined the motivations that led consumers to engage in creating new products from pre-existing products. The primary motivations that lead them to the idea behind their creativity are a problem to be solved or a creative exploration. The motivations behind the implementation of the idea are cost savings, unavailability of commercial options but also enjoyment, ego (wanting to receive validation), and escape (e.g. from work, reality, life in general). Even the choice of products to modify is dictated by several motivations: low cost, accessibility, ease of modification, and emotional connection. So, IKEA hacking can be viewed not just as a means of individual expression, but as a response to the limitations inherent in mass-produced objects, which are often designed for the "average" user but do not account for unique preferences or specific use cases.

IKEA has always been a source of raw materials for projects; the company's products are cheap and available in most parts of the world, making them ideal candidates for modification. The IKEA hacking phenomenon started with individuals driven by creativity or a need not met by commercially available products, but turned into a full-fledged movement when these hacks were shared online. The rise of social media and DIY platforms, such as Instructables, has amplified the visibility of this movement, making it easier for individuals to share their projects and inspire others. This allowed other individuals to replicate them, share information, instructions, to the point of creating a real community. In fact, websites have even sprung up that collect the projects and allow hackers from all over the world to interact, such as ikeahackers.net. The standardization of IKEA products might seem like a hindrance to creativity, but it is instead a blessing to communication and sharing. Since IKEA products are designed to fit within certain modular standards, they provide a predictable foundation upon which hacks can be built. This uniformity helps make instructions easier to follow and universally applicable, regardless of the user's geographical location. Hacks range from small modifications, such as repainting a piece of furniture or changing the feet on a sofa, to substantial changes that completely transform both form and function of a product. What they have in common is the pleasure that comes from transforming an existing object into something unique and personal. Indeed, IKEA hacking reflects an increasingly popular interest in DIY and customization. This interest can also be seen as a form of resistance to the homogenization of consumer culture, where mass-produced goods dominate the market. IKEA hacking allows individuals to reclaim authority in a world of standardized products.

The Swedish company initially clashed with this phenomenon and in 2014 was even involved in a legal dispute with the website ikeahackers.net. IKEA initially sent a cease-and-desist letter demanding that the site be taken down. However, after a public backlash, IKEA reversed its decision and allowed the site to continue operating. Over time, however, the company has come to welcome the phenomenon, even incentivizing it and telling its story. In fact, in 2019, an exhibition was held at the IKEA Museum in Älmhult, Sweden, entitled precisely "IKEA Hacked: Our Products. Your Ideas", which featured the work of more than 30 international artists, architects, and designers, all of whom used IKEA products as a starting point. This official recognition of the hacking movement by IKEA marked a significant shift in how the company viewed consumer interaction with its products, moving from passive consumption to active participation. Exhibition curator Camilla Junger said, "We are proud of the fact that people from all over the world like to unleash their creativity using our products – this is the phenomenon we want to analyze in the exhibition".

24.4.2 Experiences and Experiments in Design

The IKEA hacks, then, are not just about the amateur world; some designers have also tackled the subject. Professional designers, recognizing the potential of IKEA hacking, have engaged with this trend, not just for personal projects but as a method of exploring broader design possibilities. By using IKEA products as raw materials, they can push the boundaries of design within the constraints of mass-produced components. Below are some relevant case studies.

Daniel Saakes, with Platonic Sun, a reinterpretation of the iconic Lampan lamp made by joining together 6, 12, or 32 lamps following the geometries of platonic solids. Instructions for replicating the design were then shared online at instructables.com: "For me, sharing the design proved more challenging than making the design.

I was designing not only for users, but also for makers. I wanted to consider the availability of materials and the level of expertise my makers would have, with the goal of designing for optimal reproducibility" (Saakes, 2011). This highlights a key aspect of designing the hack, thinking not only about the final object but also about how others can easily reproduce and adapt it. It requires simplifying the construction process while maintaining flexibility, encouraging broader participation from the community.

Design duo Francesca Lanzavecchia and Hunn Wai created an IKEA-inspired manual on how to hack the Swedish company's products to better serve people with dementia and their caregivers. The Hack Care book, structured like an IKEA catalog, contains not only more than 50 projects, but also essays and stories about living with the disease. "The book takes a hands-on, do-it-yourself approach to inspire and encourage caregivers to invent their own hacks suited to their homes and personal needs. The goal is to help keep their loved ones with dementia engaged, empowered and able to age as long as possible" (lanzavecchia-wai.com). This project underscores the potential of IKEA hacking beyond aesthetics or personal creativity, using it as a tool to address real-world challenges. It showcases how small design interventions on IKEA products can improve the daily lives of people with specific needs, making home environments more adaptable and user-centered.

The Home Back Home project by PKMN Architectures and Enorme Studio, presented at the Oslo Architecture Triennial 2016, addresses the growing trend of young adults returning to live with their parents, a phenomenon that is often seen as a personal setback. Through its design, the project aims to challenge the stigma associated with this situation by providing flexible, adaptable living spaces that meet the needs of returning individuals while promoting independence within a shared family environment. For the exhibition, the architects presented three case studies, all based on the concept of IKEA hacking to add a series of bespoke furniture elements through a process of customization. For example, IKEA's Bekvam stool was used to create a raised platform with storage space underneath, while other additions include a folding bed, a study area that can be hidden behind a curtain, an old table, and shelving to create a flexible sewing station and custom shelving units. For all three case studies, the architects have produced illustrated instruction manuals for easy replication. By providing clear guidance, the project encourages others to implement similar adaptations in their own homes, fostering a community of shared solutions.

The UNIKEA project by designer Kenyon Yeh aims to create personal identity out of ubiquitous uniformity. In this project, the designer assembles pieces from a standard IKEA Rast flat pack without following the typical assembly instructions. Instead, he adapts each piece to his personal taste, changing the way in which components combine and adding unique details such as colored legs. The collection consists of four structures featuring geometric shapes that can be used as bookstands and bookcases. This project shows how individuals can embrace the essence of IKEA's modular design while simultaneously personalizing it to reflect their identity, demonstrating that even standardized products can be transformed into unique expressions of creativity.

Stockholm-based creative team teenage engineering, after a collaboration with IKEA in 2020 that saw the production of a collection of portable speakers and modular

LED spotlights, released a series of 3D printable accessories to the aforementioned collection the following year. The 12 accessories consist of handles, feet, wheels, containers, and more, to be added to the collection's modules, three-dimensional models of which were made freely available online along with their assembly instructions. "I wanted the collection to be both playful and provocative. Without ifs and buts" (teenage.engineering). The introduction of 3D printable accessories further demonstrates the possibilities of integrating IKEA hacking with digital fabrication technologies. This enables users to modify and enhance their products in ways that would have been impossible or prohibitively expensive through traditional manufacturing methods. By providing these accessories, designers actively engage users in the customization process, empowering them to shape products according to their own preferences.

Designer Adam Miklosi created a collection of 3D printable accessories with the aim of improving pre-existing IKEA products. The collection, called Uppgradera, Swedish for upgrade, consists of six 3D-printed accessories that modify the original design of as many IKEA products while improving their functionality. Digital files of the designs can be purchased online and printed on their own. "My goal was to design products that solve real problems and with a little 'uppgradera', make the user experience more pleasant and efficient" (uppgradera.co). This collection was followed by two others containing five new accessories each, also based on the same concept of improving existing products. The third collection that has been designed in collaboration with the platform Enable 3D is available to download for free. These collections show how even small, well-designed changes can drastically alter the user experience of a product, improving its usability and extending its life cycle.

24.5 ÖPPNA: HOW TO HACK YOUR WORKSTATION

The project aims to fit into the framework of IKEA hacking, the alteration of the Swedish company's products in order to improve or customize them, focusing on the opportunities offered by digital fabrication, especially 3D printing, to allow the user to produce the object themselves. In addition, through parametric design, it aims to make the user part of the creative process, allowing them to alter the design according to their own needs and tastes. By engaging users in this way, the project not only encourages customization but also fosters a deeper sense of ownership over the final product, shifting the role of the designer from the sole creator to a facilitator of personal expression.

The project consists of a collection of 3D-printed accessories that allow four IKEA products (NISSAFORS trolley, LAPPVIKEN cabinet door, NÄVLINGE clamp spotlight, and MARIUS stool) to be aggregated together, providing a new function, different from that of the initial products. This collection is available for download from the Internet in an open-source mode thanks to a Creative Commons license. Some of the items in the collection were designed through parametric design and thus allow the user not only to produce the item themselves, through 3D printing, but to be able to have a voice in the design process. This aspect of user-driven modification empowers individuals without formal design training to participate in design, promoting inclusivity and creativity in the development of personalized furniture solutions.

For example, the lampshade can be customized in height, diameter, and tilt by the user simply by moving sliders in a web interface; the user can then download the 3D file and produce it himself, if he has a 3D printer, or rely on a nearby FabLab. Assembly instructions have also been created following IKEA's iconic style, which should be familiar to users. With this approach, the project ensures that users feel comfortable navigating the assembly process, further lowering the barriers to entry for engaging with DIY projects. This simplicity in the user experience encourages experimentation and gives users confidence in adapting the designs to fit their needs.

The parametric design programs were created using Grasshopper, a language and visual programming environment that runs within Rhinoceros 3D, a CAD application developed by Robert McNeel & Associates. This integration of advanced design tools demonstrates the potential for professional design techniques to be made accessible to everyday users, combining computational design with DIY culture. By allowing for easy modification, the project promotes a balance between structured design and user freedom, where both the designer's intent and the user's creativity can co-exist (Figures 24.1–24.8).

24.6 DISCUSSIONS

Öppna is an example of how, thanks to technology, it is possible to involve the user in both the creative and the production process, and how the designer's task is not always to design a finished product but can also be to design tools that enable users to design for themselves. This shift in the designer's role, from creator to enabler, reflects the broader democratization of design in which users can engage with and adapt products to meet their individual needs. By leveraging digital tools such as parametric design and 3D printing, the boundaries between designer and consumer are increasingly blurred.

In these instances, there is a need to reconsider the role of the designer and his work processes; he must in fact take into consideration that the project will be modified and produced by the user himself. Hence, it is necessary to take a step back and not think in terms of the finished product but from a co-creation perspective, giving the user the means and tools to take part in the creative and production processes. This co-creation model not only fosters innovation but also allows for a deeper connection between users and the products they create, as they become part of the design process rather than passive consumers. This level of involvement can lead to products that are more meaningful, functional, and personalized to the user's life.

It is therefore required to consider the ease with which the object can be modified and produced by the user. For example, in the design of Öppna, the aim was to make the components as simple as possible, not too complex to modify, limiting the variables in the parametric design to those that are strictly necessary. Especially with a home production perspective in mind, simple-to-print components were designed, suitable to be printed even by an entry-level machine and an inexperienced person.

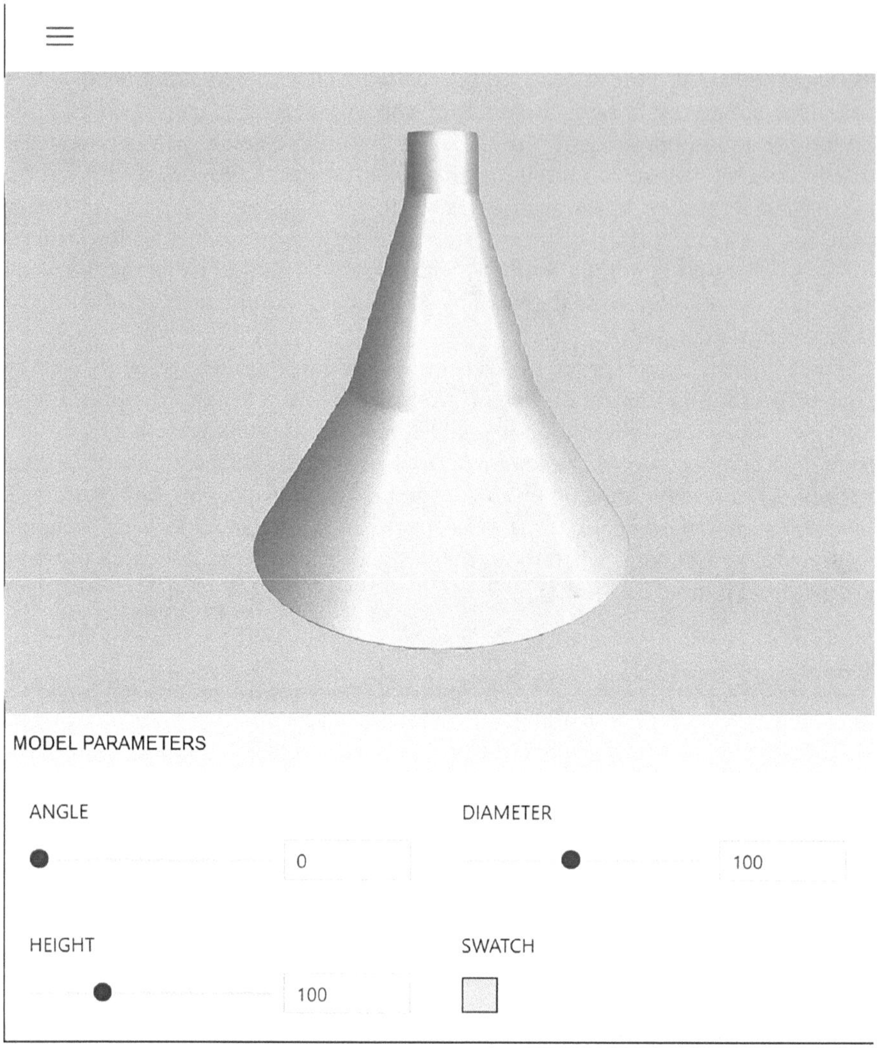

FIGURE 24.1 Graphic interface for component customization. (Image by the author.)

By focusing on accessibility in both the design and production process, the project ensures that a wide range of users can engage with the product, regardless of their technical expertise or access to high-end fabrication equipment.

Consequently, it is necessary to think in an inclusive perspective that can involve both makers and prosumers, who are the main target audience of the project, but also an average user who wants to approach the world of hacking. The selection of IKEA components was also made with inclusivity and simplicity in mind; in fact, the selected products are very inexpensive and extremely easy to modify, and the need for tools and skills, except for those already required in the assembly of the selected

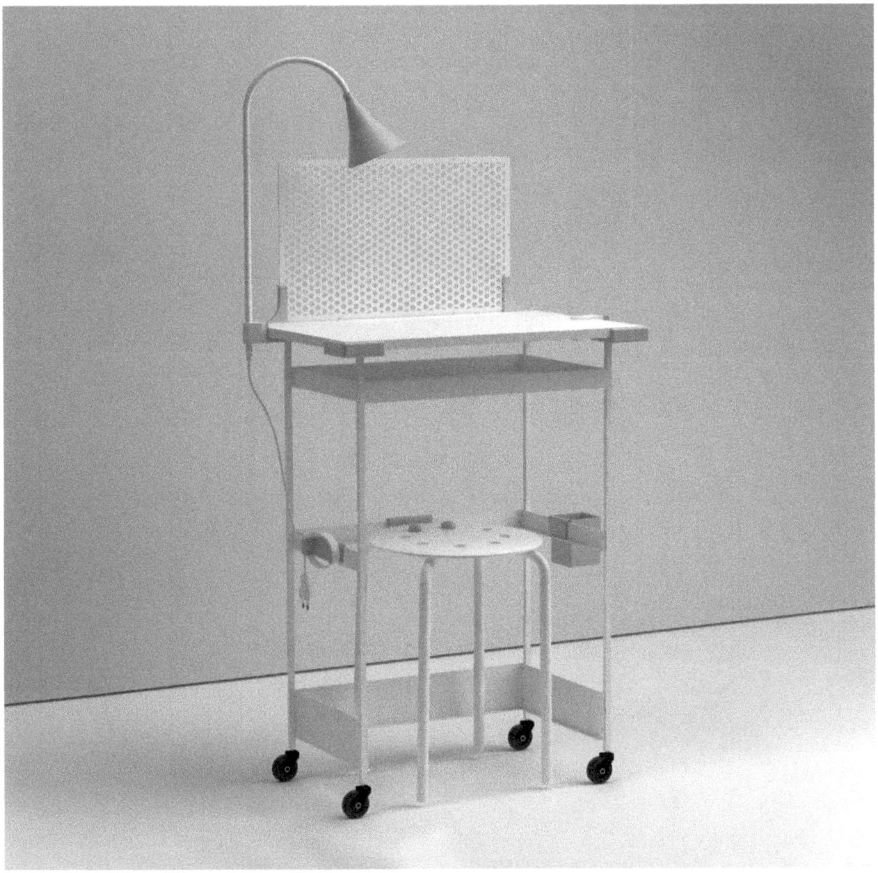

FIGURE 24.2 Render of the full configuration assembled. (Image by the author.)

IKEA products, was intentionally avoided. This approach promotes inclusivity by lowering the barrier to entry, ensuring that both experienced makers and novices alike can engage with the project. By offering a straightforward and accessible starting point, Öppna encourages creativity and customization, inviting users to explore further hacks and adaptations.

All of this is possible thanks to parametric design, which allows users to modify key elements of the project, adapting it to their own needs and tastes, without altering its essence. This, combined with the domestic diffusion of additive manufacturing, allows for mass customization; indeed, in opposition to the logic of mass manufacturing, producing different objects has the same cost as producing the same object over and over again. This shift challenges traditional manufacturing models by offering flexibility and customization without the added costs usually associated with bespoke production. It represents a significant evolution in how products are conceived, produced, and consumed, emphasizing user participation and individualization.

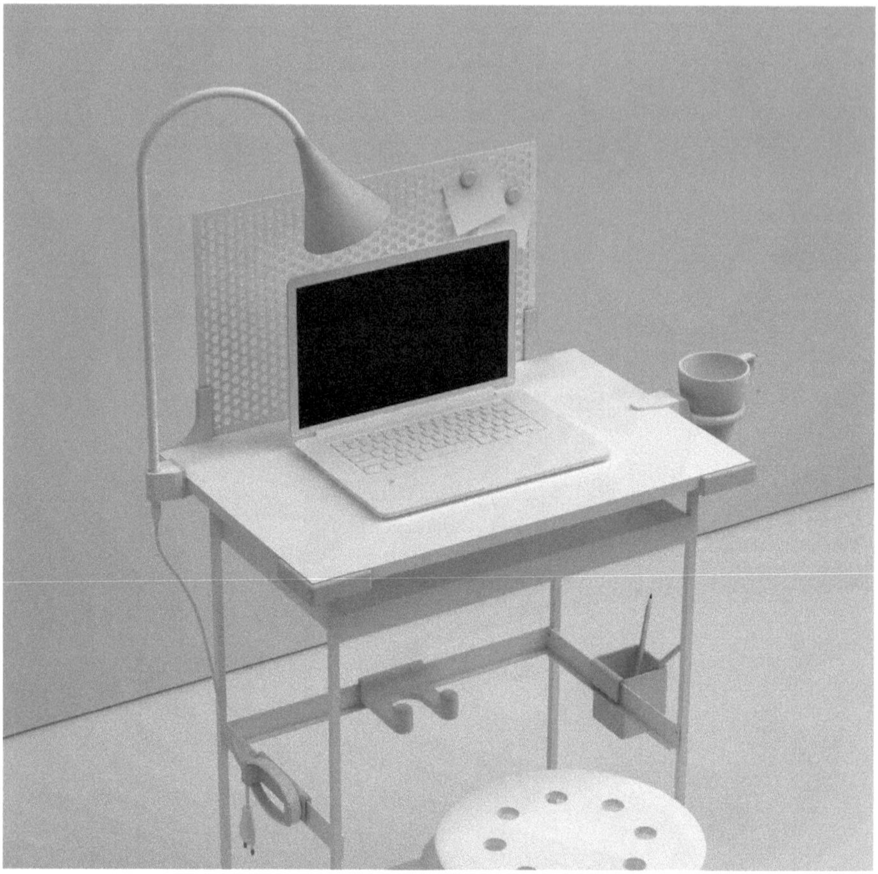

FIGURE 24.3 Render of the full configuration with accessories. (Image by the author.)

Moreover, this system represents a more sustainable way of producing objects; the product is made locally, using only the amount of material needed, thus allowing waste and transportation costs to be reduced. By doing that, the project aligns with growing concerns about sustainability in design and manufacturing. Additive manufacturing, when combined with localized production, offers a clear environmental benefit by minimizing excess material and energy usage. In addition, because the product is personalized and tailored to the user, it will be more appreciated and last longer, thus extending the product life cycle.

The project is intended not as an end in itself but stands as an example of the endless possibilities that arise from the combination of democratic design products, in this case from IKEA, and digital manufacturing technologies. It serves as a proof of concept for how design can evolve into a more collaborative approach, where users have the power to shape the objects they use in their everyday lives. Öppna lays the groundwork for a future where personalization and user involvement become central to the design and production of everyday products, redefining the relationship between designer and consumer.

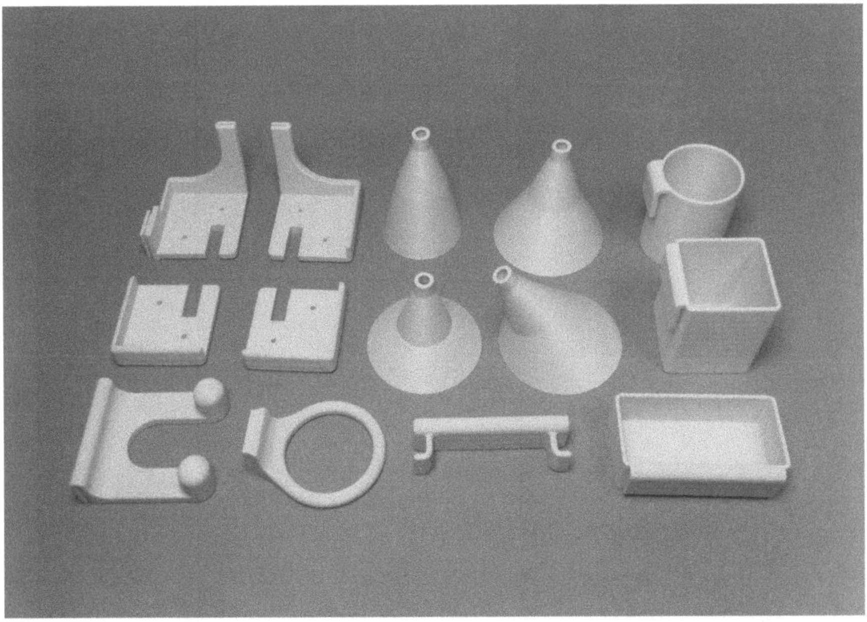

FIGURE 24.4 The collection of 3D-printed components with some of the different configurations of the parametric pieces. (Image by the author.)

FIGURE 24.5 Picture of the prototype with the human figure. (Image by the author.)

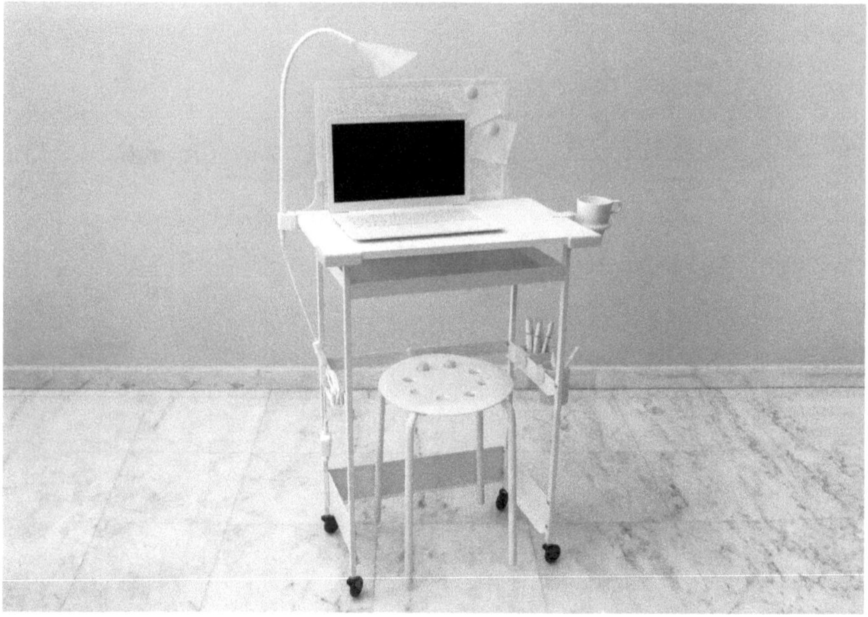

FIGURE 24.6 Picture of the prototype with accessories. (Image by the author.)

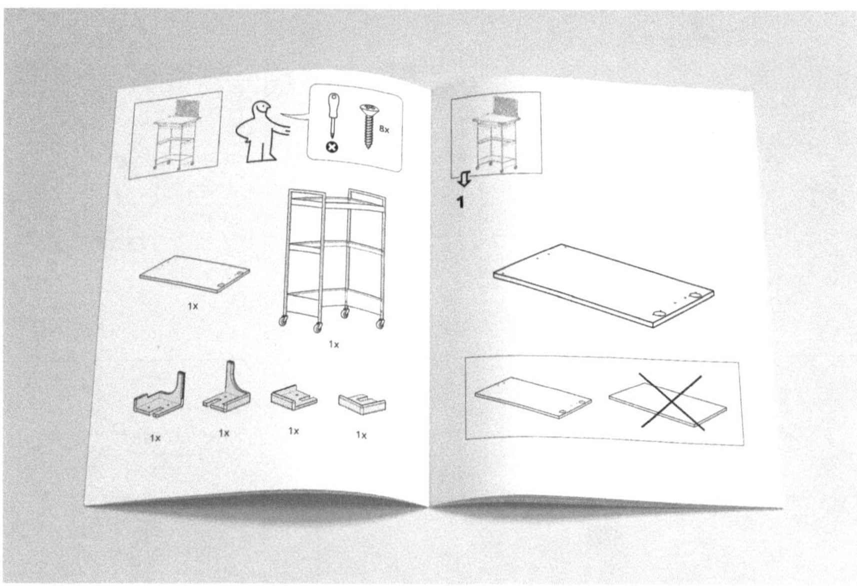

FIGURE 24.7 Some pages of assembly instructions. (Image by the author.)

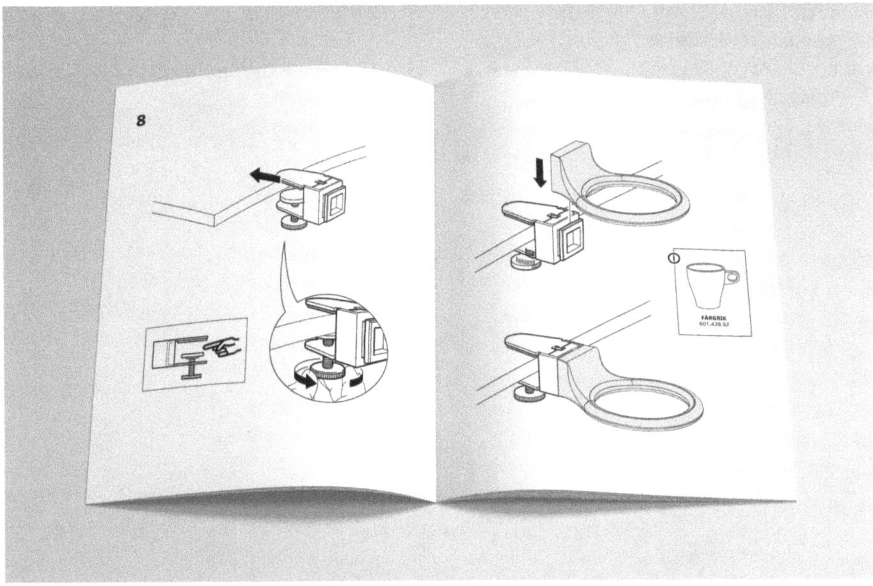

FIGURE 24.8 Some pages of assembly instructions. (Image by the author.)

REFERENCES

Anderson, C. (2012). *Makers. Il ritorno dei produttori. Per una nuova rivoluzione industriale.* Segrate: Rizzoli.

Attaran, M. (2017). The rise of 3-D printing: the advantages of additive manufacturing over traditional manufacturing. *Business Horizons*, *60*(5), 677–688. https://doi.org/10.1016/j.bushor.2017.05.011.

Berman, B. (2012). 3-D printing: the new industrial revolution. *Business Horizons*, *55(2)*, 155–162. https://doi.org/10.1016/j.bushor.2011.11.003.

Bogers, M., Hader, R., & Bilberg, A. (2016). Additive manufacturing for consumer-centric business models: implications for supply chains in consumer goods manufacturing. *Technological Forecasting and Social Change, 102,* 225–239. https://doi.org/10.1016/j.techfore.2015.07.024.

Boisseau, É., Omhover, J.-F., & Bouchard, C. (2018). Open-design: a state of the art review. *Design Science*, *4*(e3). https://doi:10.1017/dsj.2017.25.

Carpo, M. (2011). *The alphabet and the algorithm. Cambridge*: The MIT Press.

Ford, S., Mortara, L., & Minshall, T. (2016). The emergence of additive manufacturing: introduction to the special issue. *Technological Forecasting and Social Change*, *102*, 156–159. 10.1016/j.techfore.2015.09.023.

Maker, A.-Z. (2014). *Arduino, stampanti 3D, FabLab: La rivoluzione degli artigiani digitali per una tecnologia democratica.* Pergine Valsugana: Altra Economia Soc. Coop.

Rayna, T., & Striukova, L. (2016). From rapid prototyping to home fabrication: how 3D printing is changing business model innovation. *Technological Forecasting and Social Change, 102,* 214–224. https://doi.org/10.1016/j.techfore.2015.07.023.

Robson, K., Wilson, M., & Pitt, L. (2019). Creating new products from old ones: consumer motivations for innovating autonomously from firms. *Technovation*, *88*. https://doi.org/10.1016/j.technovation.2019.05.001.

Saakes, D. (2011). Platonic Sun. Instructables.com. Retrieved from https://www.instructables.com/Platonic-Sun/

Tapscot, D., & Williams, A. D. (2006). *Wikinomics, how mass collaboration changes everything*. New York: Portfolio.

Thackara, J. (2011). *Open design now*. Amsterdam: BIS publishers.

Toffler, A. (1980). *The third wave*. New York: Bantam Book.

Index

F

J

K

L

M

T

FSC
www.fsc.org
MIX
Papier | Fördert gute Waldnutzung
FSC® C083411

Kolibri 360 GmbH
Ferdinand-Jühlke-Straße 7
99095 Erfurt, Deutschland
produktsicherheit@kolibri360.de